T0212903

Lecture Notes in Artificial Intelligence   9405

Subseries of Lecture Notes in Computer Science

LNAI Series Editors

Randy Goebel
University of Alberta, Edmonton, Canada
Yuzuru Tanaka
Hokkaido University, Sapporo, Japan
Wolfgang Wahlster
DFKI and Saarland University, Saarbrücken, Germany

LNAI Founding Series Editor

Joerg Siekmann
DFKI and Saarland University, Saarbrücken, Germany

Henning Christiansen · Isidora Stojanovic
George A. Papadopoulos (Eds.)

Modeling and Using Context

9th International and Interdisciplinary Conference, CONTEXT 2015
Lanarca, Cyprus, November 2–6, 2015
Proceedings

 Springer

Editors

Henning Christiansen
Department of Computer Science
Roskilde University
Roskilde
Denmark

George A. Papadopoulos
Department of Computer Science
University of Cyprus
Aglantzia
Cyprus

Isidora Stojanovic
ENS-Pavillon Jardin
CNRS-Institut Jean-Nicod
Paris
France

ISSN 0302-9743   ISSN 1611-3349 (electronic)
Lecture Notes in Artificial Intelligence
ISBN 978-3-319-25590-3   ISBN 978-3-319-25591-0 (eBook)
DOI 10.1007/978-3-319-25591-0

Library of Congress Control Number: 2015951758

LNCS Sublibrary: SL7 – Artificial Intelligence

Springer Cham Heidelberg New York Dordrecht London

Printed on acid-free paper

Springer International Publishing AG Switzerland is part of Springer Science+Business Media
(www.springer.com)

Preface

This volume includes the papers presented at CONTEXT 2015: the 9th International and Interdisciplinary Conference on Modeling and Using Context held during November 2–6, 2015, in Larnaca, Cyprus.

Since its start in 1997, the CONTEXT conferences have become the world's prime forum for presentation and exchange of insights and cutting-edge results from the wide range of disciplines concerned with context.

Previous editions of the series were held in Rio de Janeiro, Brazil (1997), Trento, Italy (1999; LNAI 1688), Dundee, UK (2001; LNAI 2116), Palo Alto, USA (2003; LNAI 2680), Paris, France (2005; LNAI 3554), Roskilde, Denmark (2007; LNAI 4635), Karlsruhe, Germany (2011; LNAI 6967), and Annecy, France (2013; LNAI 8175). The main theme of CONTEXT 2015 was "Back to the Roots," focusing on the importance of interdisciplinary cooperations and studies of the phenomenon. Context, context modeling, and context comprehension are central topics in linguistics, philosophy, sociology, artificial intelligence, computer science, art, law, organizational sciences, cognitive science, psychology, etc., and are also essential for the effectiveness of modern, complex, and distributed software systems.

CONTEXT 2015 included a Doctorial Symposium, organized by Peter Eklund and Rebekah Wegener, and three workshops: Smart University 3.0, organized by Thomas Roth-Berghofer, Samia Oussena, and Martin Atzmüller; CATI: Context Awareness and Tactile Design for Mobile Interaction, organized by Patrick Brézillon, Alexandre Kabil, Sébastien Kubicki; and SHAPES 3.0: The Shape of Things, organized by Stefano Borgo, Oliver Kutz, and Mehul Bhatt.

We are grateful to our keynote speakers for sharing their insights and perspectives, and providing inspiration for new research in the field.

Emma Borg (University of Reading, UK):
Linguistic Meaning, Context and Assertion
Enrico Rukzio (Universität Ulm, Germany):
Mobile Interaction with Pervasive User Interfaces

We also wish to thank all the authors who submitted their research papers to the conference, thereby demonstrating the interdisciplinary strengths and common threads in context-focused research. Our special thanks go to the Program Committee members, our sponsors, and everyone who contributed at any level to the organization of CONTEXT 2015, including the staff at Easy Conferences, Cyprus, especially Kyriakos Georgiades and Marios Christou for taking care of all practical aspects.

August 2015

Henning Christiansen
Isidora Stojanovic
George A. Papadopoulos

Organization

Program Chairs

Henning Christiansen   Roskilde University, Denmark
Isidora Stojanovic   Institut Jean-Nicod (CNRS-ENS-EHESS), France

Program Committee

Varol Akman   Bilkent University, Turkey
Maria Aloni   Universiteit van Amsterdam, The Netherlands
Patrick Barlatier   Université de Savoie, France
Leonor Becerra-Bonache   Jean Monnet University, France
Luciana Benotti   Universidad Nacional de Cordoba, Argentina
Anne Bezuidenhout   University of South Carolina, USA
Patrick Blackburn   Roskilde University, Denmark
Marcos Borges   Federal University of Rio de Janeiro, Brazil
Paolo Bouquet   University of Trento, Italy
Torben Braüner   Roskilde University, Denmark
Patrick Brezillon   University Pierre and Marie Curie (UPMC), France
Sylvie Calabretto   LIRIS-INSA, France
Keith Cheverst   Lancaster University, UK
Henning Christiansen   Roskilde University, Denmark
Antonio Corradi   University of Bologna, Italy
Richard Dapoigny   LISTIC/Polytech'Savoie, France
Klaus David   Kassel University, Germany
Anind Dey   Carnegie Mellon University, USA
Bruce Edmonds   Manchester Metropolitan University Business School, UK
Jérôme Euzenat   Inria and University of Grenoble, France
Xiaoliang Fan   Lanzhou University, China
Christian Freksa   University of Bremen, Germany
Chiara Ghidini   FBK-irst, Italy
Yonathan Ginzburg   Université Paris-Diderot (Paris 7), France
Avelino Gonzalez   University of Central Florida, USA
Nathaniel Hansen   University of Reading, UK
Ruth Kempson   King's College London, UK
Anders Kofod-Petersen   Alexandra Instituttet, Denmark
David Leake   Indiana University, USA
Karen Lewis   Columbia University, USA
José Manuel Molina   Universidad Carlos III de Madrid, Spain

Ivelina Nikolova	LMD, IICT, Bulgarian Academy of Sciences, Bulgaria
Nearchos Paspallis	UCLan Cyprus, Cyprus
Angel Pinillos	ASU, USA
Joëlle Proust	Institut Jean-Nicod (CNRS-ENS-EHESS), France
Thomas R. Roth-Berghofer	University of West London, UK
Flavia Maria Santoro	NP2Tec/UNIRIO, Brazil
Hedda R. Schmidtke	University of Oregon, USA
Julie Sedivy	University of Calgary, Canada
Isidora Stojanovic	Institut Jean-Nicod (CNRS-ENS-EHESS), France
Matthew Stone	Rutgers, USA
Roy Turner	University of Maine, USA
Carla Umbach	Zentrum für Allgemeine Sprachwissenschaft (ZAS), Germany
Kristof Van Laerhoven	University of Freiburg, Germany
Ivan José Varzinczak	COPPE-PESC, Brazil
Renata Wassermann	University of São Paulo, Brazil
Rebekah Wegener	RWTH Aachen University, Germany
Arkady Zaslavsky	CSIRO, Australia

Additional Reviewers

Carlos Areces	Universidad Nacional de Córdoba, Argentina
Thomas Barkowsky	Universität Bremen, Germany
Nik Nailah Binti Abdullah	Monash University, Malaysia
Stefano Bortoli	Okkam SRL, Italy
Loris Bozzato	Fondazione Bruno Kessler, Italy
Chiara Di Francescomarino	FBK-IRST, Italy
Zoe Falomir	Universität Bremen, Germany
Helmar Gust	Universität Osnabrück, Germany
Katherine Metcalf	Indiana University, USA
Stella Tagnin	Universidade de São Paulo, Brazil
Jasper van de Ven	Universität Bremen, Germany

General Chair

Henning Christiansen	Roskilde University, Denmark

Local Arrangements Chair

George A. Papadopoulos	University of Cyprus, Cyprus

Workshop Chair

Samia Oussena	University of West London, UK

Chairs of the Community of Context

Patrick Blackburn	Roskilde University, Denmark
Patrick Brezillon	Pierre and Marie Curie University (UPMC), France
Henning Christiansen	Roskilde University, Denmark
Richard Dapoigny	Université de Savoie, France
Thomas R. Roth-Berghofer	University of West London, UK
Hedda R. Schmidtke	University of Oregon, USA

Sponsors

Austrian Airlines
Cyprus Tourist Organization
University of Cyprus

Abstracts of Keynote Speeches

Linguistic Meaning, Context, and Assertion

Emma Borg

University of Reading, UK

I explore the impact of contextual features on linguistic communication. In particular, I am interested in the difference between asserting and implying, and the question of the role the context of utterance plays in determining either kind of content. I begin by introducing two distinct approaches to the difference between asserting and implying: one that locates the difference in the language, as it were, and one that locates it in the mind. However, I argue that both these approaches face problems. I argue that the distinction is better captured in sociolinguistic terms, whereby the difference between asserting and merely implying emerges from the distinct kinds of social role these speech acts play and the kinds of culpability or responsibility a community assigns to a speaker for a conveyed content. I conclude by sketching how this socially orientated view of assertion relates to certain philosophical analyses of assertion and how the view might impact on our understanding of more applied issues in philosophy of language, such as informed consent and libel or slander.

Mobile Interaction with Pervasive User Interfaces

Enrico Rukzio

Universität Ulm, Germany

Mobile human–computer interaction has been a very active research area in the last decade, and yet there are still many unsolved questions regarding the limited input and output capabilities of mobile devices, the interaction with objects in the environment, and the consideration of the usage context. Many of these issues could be addressed through mobile interactions with displays in the user's vicinity and with additionally available wearable computers. The talk discusses current issues, analyzes corresponding research activities, and focuses on recently published systems for mobile interaction with pervasive user interfaces, such as AMPD-D (CHI 2014), HoverPad (UIST 2014), or Belt (CHI 2015).

Contents

Semantics and Philosophy

Logic and Meaning

Context and Cognition

Experimental Methods in Linguistics

Short Papers

Knowledge Representation
and Reasoning

Situation Awareness Meets Ontologies: A Context Spaces Case Study

Andrey Boytsov[1(✉)], Arkady Zaslavsky[2], Elif Eryilmaz[1], and Sahin Albayrak[1]

[1] Distributed Artificial Intelligence Laboratory (DAI-Labor),
Technical University of Berlin (TU-Berlin), 10587 Berlin, Germany
{andrey.boytsov,elif.eryilmaz,sahin.albayrak}@dai-labor.de
[2] Digital Productivity Flagship,
Commonwealth Scientific and Industrial Research Organization (CSIRO),
Melbourne, Australia
arkady.zaslavsky@csiro.au

Abstract. Efficiency and appeal of pervasive computing systems strongly depends on how well and robustly they represent and reason about context and situations. Populating situation search space and inferring situations from context which, in turn, is computed from fusing sensor data and observations remains a major research challenge. This paper proposes to use ontologies as representation of domain knowledge to generate situation search space and then match context with already defined situations. To illustrate the feasibility, a context spaces approach is used to represent, generate and reason about situations as abstractions in a multidimensional space. The proposed approach is evaluated and discussed.

1 Introduction

Efficiency and appeal of pervasive computing systems strongly depends on how well and robustly they represent and reason about context and situations. Populating situation search space and inferring situations from context which, in turn, is computed from fusing sensor data and observations remains a major research challenge. In this paper we propose a novel approach to build fast and intuitive approach for context awareness, situation awareness and sensor fusion in pervasive computing system.

Bazire and Brézillon [4] analyzed a set of 150 definitions of context for different subject areas. The authors noted that there is no consensus about some aspects of context and there is no single definition. However, common understanding is that *"context is the set of circumstances that frames an event or an object"* [4] and *"context acts like a set of constraints that influence the behavior of a system (a user or a computer) embedded in a given task"* [4]. It complies with the most popular definition of context in pervasive computing, which was proposed by Dey and Abowd [11]. They defined context as *"any information that can be used to characterize situation of an entity"* [11]. It should be noted that this definition was one of the definitions analyzed by Bazire and Brézillon [4]. The definition is very general and yet it provides sufficient criteria to identify context-related information among sensor data and user input. Location, identity, activity and time are the main aspects of context. It should be also specifically noted that the word *situation* in the

© Springer International Publishing Switzerland 2015
H. Christiansen et al. (Eds.): CONTEXT 2015, LNAI 9405, pp. 3–17, 2015.
DOI: 10.1007/978-3-319-25591-0_1

definition by Dey and Abowd means *circumstances* or *conditions*, and does not imply rigorous definition of situation in pervasive computing, which will be presented later in this section. Pervasive system is context aware *"if it uses context to provide relevant information and/or services to the user, where relevancy depends on the user's task"* [11]. Context awareness is a key feature of any pervasive computing system.

Context awareness can be enhanced by appropriate situation awareness. Throughout the paper we are going to use the definition of situation proposed by Ye et al. [19]. From pervasive computing perspective Ye et al. [19] define a situation as *"external semantic interpretation* of sensor data". Like a definition of context, this is a very general definition, yet it provides criteria for technical implementation of situation recognition. *Semantic interpretation* means that "situation assigns meaning to sensor data" and *external* implies "from the perspective of applications" [19]. Therefore, situation recognition can be viewed as one of the final stages of context awareness – extracting the essential meaning of context. For example, context processing results like "user is sitting" or "presence in the living room" can be viewed as situations.

The algorithm of *semantic interpretation* of sensor data takes sensor values as input and provides the interpretations as output. In the output, each possible situation is assigned a value, which can be Boolean value (whether a situation is occurring), or probability of a situation occurrence, or fuzzy confidence level. Sometimes reasoning algorithm can be formulated as a single mathematical expression, and this expression will be referred to as *situation formula* in this paper.

Ontologies are domain-independent way of storing semantic information [14]. There are multiple ontologies for sensor networks [3, 8, 9] and for pervasive computing context [2, 18, 19]. Context ontologies often include situation awareness aspects.

In this paper we propose a novel approach and a set of algorithms that generate situation formulas. Those formulas enable fast and lightweight situation awareness in pervasive computing system. As an input the proposed approach uses pervasive computing system description in terms of context and sensor ontologies. As a part of situation generation approach we propose a method that bridges the gap between context and sensors and introduces an integrated ontology of a pervasive computing system.

The rest of the paper is structured as follows. Section 2 contains related work overview. Section 2 also identifies the scope of work to obtain situation formulas out of context and sensor ontologies. Section 3 proposes the main principles of situation formulas generation. Section 4 further elaborates on those principles and proposes the detailed situation formula generator algorithm. Section 5 introduces context spaces approach, which is essential for demonstrating generated situation formulas. Section 6 finalizes generation of situation formulas out of integrated ontologies. Section 7 provides discussion and evaluation of the proposed approach. Section 8 proposes future work directions and concludes the paper.

2 Related Work

Ontology in artificial intelligence can be defined as *"explicit specifications of a conceptualization"* [14]. Ontologies provide a domain-independent way for knowledge representation, sharing and reasoning [14]. Multiple surveys [2, 18, 19] provide overviews of ontology-based context reasoning approaches.

SOUPA [7] (Standard Ontology for Ubiquitous and Pervasive Applications) is one of the earliest examples of ontologies in pervasive computing. SOUPA was based on CoBrA-ONT [6] (ontology for Context Broker Architecture). SOUPA already had the vocabulary to describe *situational conditions* like "in a meeting" or "out of town". Situational conditions were actually situations by the definition of Ye et al. [19].

Dapoigny and Barlatier [10] used ontological modelling to overcome the lack of expensiveness in logic-based theories. They formalized the context by using the Calculus of Inductive Constructions in the lower layer and an ontological layer in the upper layer.

Wang et al. [17] developed CONON (CONtext ONtology) ontology. CONON used the concept of context entities, which is a general term for concepts like location, person or activity. Situations were defined in terms of rules, like in formula (1).

$$(\text{?u locatedIn Bedroom}) \ \& \ (\text{Bedroom lightLevel LOW}) \ \& \\ (\text{Bedroom drapeStatus CLOSED}) \ \rightarrow \ (\text{?u situation SLEEPING}) \tag{1}$$

Gu et al. [15] introduced SOCAM system (Service Oriented Context Aware Middleware). Along with other features SOCAM contained context ontology, which was based on CONON. The concept of situation in SOCAM is similar to CONON.

Anagnostopoulos et al. [1] proposed situation awareness technique based on combination of ontologies and fuzzy logic. Situations were represented as concepts, and the relation "person is involved in a situation" (that effectively means "situation is occurring for this person") was reasoned about using fuzzy logic (see formula (2)).

$$\bigwedge_{i=1}^{N} context\left(x_i, user\right) \rightarrow IsInvolvedIn\left(Situation, user\right), N > 1 \tag{2}$$

Formula (2) is quite similar to formula (1) – situations are reasoned about in terms of rules. Anagnostopoulos et al. [1] also introduced the relation "disjoint", which means that situations cannot occur at the same time (like "Meeting" and "Jogging").

Ejigu et al. [13] proposed ontology-based generic context management model (GCoMM). GCoMM mainly relied on rules and queries for inference and decision making. Although the concept of situation was not explicitly introduced, some rules could be viewed as situation inference rules. For example, consider formula (3) [13].

$$[\text{rule1: ?user1 nsp:locatedIn ?roomN}) \ \& \ (\text{?user2 nsp:locatedIn ?roomN}) \\ - > (\text{?user1 nsp:coLocatedWith ?user2})] \tag{3}$$

According to the related work, there are two available concepts of situations:

1. Potential situation is an entity. The relation connects context entity to potential situation, if that situation is occurring. This approach is applicable if a situation refers to a single context entity like *"User* is sleeping" or "Presence in the *Living Room*". Formulas (1) and (2) provide an example. In formula (1) the relation is called

situation [17], while in formula (2) the relation is called *IsInvolvedIn* [1]. However, the meaning of both relations is similar.

2. Potential situation is a relation that connects context entities. The relation exists if the situation is occurring. This approach is applicable if a situation refers to two context entities at once. For example, *"user:Alice* coLocatedWith *user:Bob"* or *"user:Alice* locatedIn *location:LivingRoom"* (formulas (2) and (3)). Relationship *locatedIn* and *nsp:coLocatedWith* are situations according to the definition, i.e. "external semantic interpretation of sensor data" [19].

Note that context ontology sometimes combines both approaches, and the chosen approach for any particular situation is determined by whether the situation is parametrized by one or two context entities. For example, formula (1) combines situation *Sleeping*, defined as an entity, and situation *locatedIn*, defined as a relation. Therefore, situation formula generator should be able to work with both approaches at once.

The common aspect of both approaches is that situation reasoning is performed in terms of rules. This common aspect is an important factor that allows situation formula generation using both situation representation approaches.

Usually context ontologies do not contain information about the sensors. The facts like "Bedroom lightLevel LOW" (see formula (1)) are assumed to be already inferred from the sensed information. Context ontology does not contain information about reasoning in case there are multiple sensors that measure light level or in case there are none. Therefore, context ontology should be augmented with information about sensors in order to generate situation formulas, which take sensor values as input.

There are some research efforts to construct ontology-based sensor description and data modelling [8, 9]. The challenge is in the generic representation of sensor data and characteristics to have a common level of sensor representation. The main approach how to capture the critical characteristics of a sensor accurately is the metadata annotation for sensor characteristics [3]. Although there are many efforts on defining sensor meta-information, the description of observations measured by sensors has not been addressed much. The W3C Incubator Group released Semantic Sensor Network XG Final Report, which defines SSN ontology [21] allowing describing sensors, including their characteristics. SSN ontology focuses on providing a domain independent ontology which is generic enough to adapt to different use-cases and compatible with the OGC standards [22] at the sensor and observation levels.

To summarize the related work, situations in most context ontologies are represented in terms of parametrized rules. Therefore, situation generator should accept parametrized rules as input. Moreover, context ontologies usually do not contain sensor information. Therefore, situation formula generator should accept both context and sensor ontology as input. Next section takes into account the requirements and proposes general approach to situation formula generation.

3   Fusion of Context and Sensor Ontologies

Previous section identified the input format for situation generator. The input includes sensor and context ontologies, which includes situations as parametrized rules.

Context and sensor ontologies contain much common knowledge that does not depend on exact pervasive computing system. For example, CONON [17] contains information about context entities (like concepts of location, person or activity) and their subclasses. In this article we use CONON ontology as an example, but proposed principles and algorithms can be applied to other context ontologies.

Situation in formula (1) is applicable to any smart home. The only requirement is that parts of the situation like "light level in bedroom is low" and "user is in bedroom" can be inferred from sensor data. For any particular pervasive computing system only the instances of classes should be added: what kind of rooms are there in the smart home (instances of *location*), who lives there (instances of *person* class), etc.

In turn, SSN sensor ontology [21] mainly consists of general information about sensors, the values they measure and measurement capabilities of the sensors. The only thing left to add for particular pervasive computing system (like smart home) is the exact sensors we have and where they are.

The approach to situation formula generation can be summarized in Fig. 1.

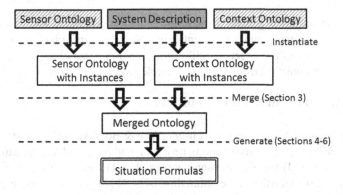

Fig. 1. High-level system description for situation formulas generator. Pervasive system description is an input. Context and sensor ontologies are static input information, which does not depend on the system under analysis. Situation formulas are the output of the approach.

Context and sensor ontologies can be integrated as follows. Each sensor can be viewed as either measuring user characteristic (e.g. mobile phone measuring user's location and acceleration) or place characteristic (e.g. light level or motion sensor in the room). Location and person are context entities in CONON context ontology. Therefore, we need a relation that specifies that some instance of sensor belongs to some context entity. We define relation *belongsTo*, which connects hardware platform with context entity (Fig. 2). As will be shown in Sects. 4–6, this relation is enough to integrate context and sensor ontology for situation generation purpose.

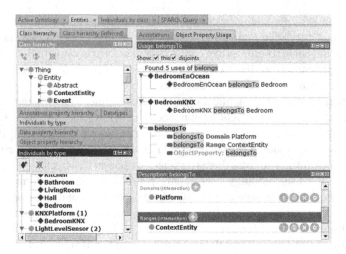

Fig. 2. Proposed connection between SSN ontology [21] and context ontology. The example was implemented in PROTÉGÉ [23] tool and shows the definition of new *belongsTo* relation.

This section proposed general approach to situation generation and described the steps to integrate context and sensor ontology. Next section further elaborates that approach and provides the exact algorithm for situation generation, i.e. (in terms of Fig. 1) details for transition from integrated ontology to exact situation formulas.

4   Generation of Situation Formulas

As an example context, we assume that users Alice & Bob live in the smart home, and the smart home has locations living room, bedroom, bathroom and hall. Situations from expressions (1) and (3) are used as running examples. The example was implemented in PROTÉGÉ [23] tool using the combination of SSN [21] ontology and elements of CONON ontology, recreated using the description in the paper [17].

The first step involves recognizing the dependencies between situations. For example, in formula (1) the situation "User is sleeping" is defined as conjunction of conditions "User located in bedroom", "bedroom light level low" and "bedroom drape status closed". The condition "User located in bedroom" (for each particular user) is a situation on its own, and its inference differs a lot depending on what kind of indoor localization systems are used. Therefore, situation "User is sleeping" depends on situation "User located in bedroom" Those dependencies can be summarized in the following directed graph (Fig. 3).

Dependency graph can be obtained by relatively straightforward approach - connection is drawn whenever the situation rule mentions another situation. We assume that the dependency graph is acyclic.

Situation formulas should be generated in the order defined by dependency graph: each situation is processed only after all its dependencies are processed. The assumption that the graph is acyclic ensures that such order of processing should exist. Therefore, for the running example situation "User located in" should be generated first, and then situations "User is sleeping" and "User is co-located with another user" can be generated in any order. For the example we take only formulas (1) and (3), we assume that situation "User located in" is already generated. A formula for a single situation can be generated using the algorithm below (see the pseudocode).

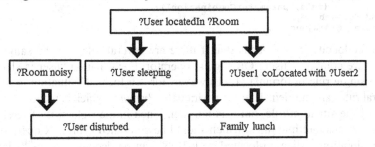

Fig. 3. Dependencies between situations. Apart from situations from formulas (1) and (3), it contains several more situations. This figure is designed for illustration purpose, and does not account for reasoning uncertainty that inevitably occurs in real-life scenarios.

Main situation generator algorithm – pseudocode

```
generateSituations(rulesList)
    results = new List<ContextSpacesSituation>();
    forEachCombination
            parameterCombination:rulesList.getParameters()
        rulesListSituation = rulesListToSituation
            (rulesList, parameterCombination)
        results.add(rulesListSituation);
        addToSituationCache(rulesListSituation);
    end forEachCombination
    return results;
```

To summarize the pseudocode, the first task for situation generator is unrolling the parameters. For situation "User is sleeping" there is only one parameter – the person. Therefore, there will be two situations generated "Alice is sleeping" and "Bob is sleeping". For the second example there will be formally four situations generated – "Alice co-located with Alice", "Alice co-located with Bob", "Bob co-located with Alice", and "Bob co-located with Bob". Detection of symmetric situations ("Alice co-located with Bob" vs "Bob co-located with Alice") and always true situations ("Alice co-located with Alice") is a part of future work.

The pseudocode above manages situation dependencies using the cache. Situation "Bob is in bedroom" is generated before "Bob is sleeping" (see Fig. 3) and put into the cache (*addToSituationCache*). Later, when generating "Bob is sleeping", the situation "Bob is in bedroom" is retrieved from cache and used as a condition.

The function *rulesListToSituation* will be called for each of the situations mentioned above. Note that there might be several rules for the same situation, corresponding to multiple ways to infer the same situation. As will be shown further, sometimes the presence of multiple rules is implicit.

Generator algorithm for instantiated situations – pseudocode

```
rulesListToSituation (rulesList, parameterCombination)
    disjuncts = new List<Situation>();
    foreach rule : allSituationRules
        disjuncts.add(ruleToSituationFormula
            (rule, parameterCombination));
    end foreach
    return OR(disjuncts);
```

The pseudocode above means that if there are several rules for the same situation, they are combined using OR logical operator. For situation "user is sleeping" there is a single rule in formula (1). However, for "co-located" situation in formula (2) several rules are implied. There is a parameter *?roomN*, which is mentioned in the left side of the situation rule, but it is not mentioned in the right side. Therefore, it implicitly creates multiple rules, which should be combined using OR operator. For example, situation "Alice co-located with Bob" can be decomposed as "Alice co-located with Bob in the kitchen" OR "Alice is co-located with Bob in the living room" OR "Alice co-located with Bob in the hall" etc. At least one of those subsituations should be true in order for "Alice co-located with Bob" to be true. Therefore, they should be combined using logical OR.

Next pseudocode introduces situation generation from a single rule.

Transformation algorithm for a single rule – pseudocode

```
ruleToSituationFormula(rule, parameters)
    conjuncts = new List<Situation>();
    foreach condition : rule.getConditions()
        if (situationInCache(conditon, parameters))
            conjuncts.add(situationFromCache
                (conditon, parameters));
        else
            situationForCondition = conditionToSituation
                (conditon, parameters)
            conjuncts.add(situationForCondition);
            addToSituationCache(situationForCondition);
        end if
    end foreach
    return AND(conjuncts);
```

The pseudocode above combines all the conditions using AND operator. It follows the rules quite straightforwardly. However, the generated situations for each condition are cached and then retrieved from the cache if necessary. Note that for situations like "User is sleeping" there are several parameter-independent parts. In this example those are conditions "Bedroom light level low" and "Bedroom drape status on". Once calculated for a single user, there is no need to recalculate them for another user. So, the situation "Bedroom light level low", generated as a part of "Alice is sleeping" situation will be stored in the cache and later reused as a part of "Bob is sleeping" situation.

Some conditions are represented as situations. Those conditions are already stored in the situation cache as a part of situation dependency management (see earlier in this section). Therefore, those situations will be retrieved from the cache and reused.

The function *conditionToSituation* needs to process the concepts like "light level low". It also needs to handle the case like, for example, several light level sensors in the bedroom. Before introducing the pseudocode for *conditionToSituation*, we need to introduce context spaces approach.

5   Context Spaces Approach

There are following requirements to situation reasoning approach:

- The approach should support fuzzy concepts. This requirement follows from the presence of rule conditions like "light level low" (formula(1)).
- The approach should support logical operators like AND and OR. It follows from the pseudocode for *ruleListToSituation* and *ruleToSituationFormula*.
- The approach should allow combining the data from multiple sensors. It follows from considerations discussed in Sect. 3.

For this purpose we propose to use context spaces approach [16]. In order to fully incorporate fuzzy concepts, we are going to use an extension of context spaces approach proposed by Delir et al. [12]. Context spaces approach represents all possible contexts as a multidimensional space, which is referred to as *context space*.

A domain of values of interest (i.e. an axis in *context space*) is referred to as a *context attribute*. For example, air temperature, light level, air humidity can be relevant context attributes in a smart home. Non-numeric values, like on/off switch position or open/closed drape status, can also be context attributes. In our approach every sensor will have its own corresponding context attribute, and sensor fusion will be done on situation awareness level.

The situations are represented using *situation spaces*, which can be viewed roughly as subspaces of multidimensional context space. The output of situation reasoning is confidence level, which falls within [0;1] range. A situation confidence value in context spaces approach is viewed as a combination of contributions of multiple context attributes. Confidence level can be determined using formula (4).

$$conf_S(X) = \sum_{i=1}^{N} w_i * contr_{S,i}(x_i) \qquad (4)$$

The term $conf_S(X)$ represents confidence level for a situation S. The context X contains the context attribute values x_i, and each of those context attributes are assigned the importance weights w_i (all the weights sum up to 1). The contribution function of i-th context attribute into situation S is defined as $contr_{S,i}(x_i)$. Figure 4 presents plausible contribution functions for "User is sleeping" situation.

Note that rule-based situations are represented not as single context space situations, but as logical combinations of context spaces situations. For that kind of reasoning the context spaces approach contains the concept of situation algebra. Formula (5) provides the definition of basic operators of situation algebra. The definitions are compliant with Zadeh operators [20].

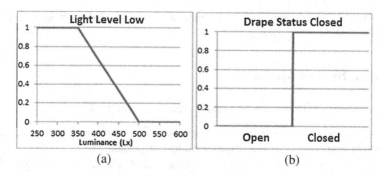

Fig. 4. Plausible fuzzy concepts for "User is Sleeping" situation. (a) Confidence level for "Light Level Low" condition depending on light level, (b) Confidence level for "Drape Status Closed" condition depending on drape status.

$$AND:(A \& B)(X) = min(A(X), B(X))$$
$$OR:(A \mid B)(X) = max(A(X), B(X)) \tag{5}$$
$$NOT:(\neg A)(X) = 1 - A(X)$$

As a result, context spaces approach complies with the requirements that are summarized in the beginning of this section. Next section finalizes situation generation algorithm and finished extraction of situation formulas from context and sensor ontologies. The extracted formulas are compliant with fuzzy context spaces approach [12].

6   Transformation of Rule Condition

Sections 3 and 4 provided the general approach to situation formula generator, and there was only one component missing: transformation of rule condition into the situation formula for runtime reasoning. That method was referred to as *conditionToSituation* in the pseudocode. This section finalizes *conditionToSituation* method and concludes the situation generator algorithm.

SPARQL [24] is a query language for ontologies. All the sensors that belong to certain context entity can be retrieved by the following SPARQL query.

Sensor list query - SPARQL

```
SELECT ?sensor WHERE {
  ?sensor :onPlatform ?platform .
  ?platform :belongsTo <ContextEntity> .
  ?sensor rdf:type ?sensorClass .
  ?sensorClass rdfs:subClassOf* ?widerSensorClass .
  ?widerSensorClass owl:onProperty :observes .
  ?widerSensorClass owl:hasValue <MeasuredProperty>
}
```

The SPARQL query effectively asks "Which sensors measure < *MeasuredProperty* > for a certain < *ContextEntity* >?". The query has two parameters. For example, for the condition "Bedroom lightLevel LOW" the parameter *MeasuredProperty* is substituted by *Luminance*. We assume that there is a map, which connects the name of relation in context ontology (like *lightLevel*) to the name of characteristic, measured by sensor (like *Luminance*). That map can be implicitly encoded into ontology itself by introducing some additional connection between context and sensor ontology.

ContextEntity parameter is replaced by the exact context entity of interest. For example, for condition like "Bedroom lightLevel LOW" the context entity of interest is *Bedroom*, the instance of location class.

The results of the SPARQL query are used by the following pseudocode.

Transformation algorithm for a single condition - pseudocode

```
conditionToSituation(condition, parameterCombination)
   result = new ContextSpacesSituation();
   instantiatedCondition =
         condition.instantiate(parameterCombination);
   sensorList = SPARQLquery(
         instantiatedCondition.getContextEntity(),
         instantiatedCondition.getProperty());
   if (sensorList.size()==0)
     fail 'Property not measured`
   end if
   weight = 1/sensorList.size();
   foreach attribute : sensorList.attributes()
      result.addContextAttribute(attribute, weight,
         sensorList.getFuzzyFunction(attribute));
   end foreach
   return result
```

So, the pseudocode above generates a subsituation for each condition. At first, the method *SPARQLquery* requests the list of sensors that measure certain characteristic of a certain context entity. Sensor readings correspond to certain context attributes in context space. For example, for situation "User is Sleeping" it can retrieve 2 sensors that measure light level in the room. Let's say, those are *BedroomLightSensor1* and *BedroomLightSensor2*. We assume that sensor names correspond to the names of context attributes, i.e. axes in multidimensional context space. Each of those sensors is taken with equal weight, and each context attribute is given the same contribution function. In that case it will be "Light level low" contribution function from Fig. 4(a). To summarize, condition *"Bedroom light level LOW"* will be transformed to the following situation (formula 6). Other conditions will be processed in the similar manner according to the same algorithm.

$$Conf\,(BedroomLightLevelLow) =$$

$$=0.5 * contr\,(BedroomSensor1) + 0.5 * contr\,(BedroomSensor2)$$

$$contr\,(x) = \begin{bmatrix} 1, & x < 350Lx \\ \frac{500-x}{150}, & x \in [350, 500]Lx \\ 0, & x > 500Lx \end{bmatrix} \quad (6)$$

This concludes the situation formula generator algorithm. Situation formula can be generated from ontologies using a set of methods described in Sects. 3–6. Next section will provide evaluation and discussion of the proposed approach.

7 Evaluation and Discussion

This article proposes a novel approach to situation awareness in pervasive computing systems. The proposed approach contains two main steps. The first step, integrating the ontologies, connects context aspects to the methods of measuring the context features. The second step, transforms the representation of situations from the format of ontology into the format of context space. The algorithms were designed to maintain the equivalence of situation representations, i.e. the same sensor readings should result in the same reasoning outcomes, whether the situations are represented as ontology or as context spaces. Context space representation provides additional synergy and introduces the following benefits.

Identifying Essential Information. The ontology aims to capture all available information about pervasive computing system. This information gives a lot of insights for many aspects of pervasive system, including situation awareness. However, for situation awareness at the runtime only small part of that information is relevant. The proposed approach preprocesses the ontology information, singles out the information necessary for situation reasoning and leaves the rest out of reasoning process.

Efficient Lightweight Reasoning. The complexity of situation reasoning is summarized in Table 1. Complexity estimations are based on formula (4), where each contribution function is a piecewise linear function like in Fig. 4. So, the total complexity is of order $O(\sum_{i=1}^{M} N_i)$, where N_i is the number of context attributes involved in i-th situation and M is the number of situations in the context space. It should be noted that M can be larger than the number of situations in ontological representation. It can happen due to the unfolding of parameters ("User is sleeping" corresponds to "Alice is sleeping" and "Bob is sleeping") and due to the fact that each ontological rule is represented as a logical formula over situations in the context space (like "Bob is sleeping" is represented as AND-operation over 3 small situations – "Bob is located in the bedroom", "Draping status - closed" and "Light level in Bedroom is low").

Sensor Fusion. Situation rules in context ontologies do not specify what to do if there are multiple ways to measure the context feature, which influences situation reasoning (e.g. what if there are multiple light sensors for situation "user is sleeping"). Generation of context space situations allows fusing the values of multiple sensors. Smart sensor fusion is one of the directions of future work.

Reasoning Under Uncertainty. Proposed situation reasoning approach is robust to uncertainty, both in terms of measurement uncertainty (which is partially covered by previous paragraph) and uncertainty in terms of concepts (like "Light level Low").

Focusing on Most Relevant Situations. With growing amount of entities in the context (like users, locations, etc.), the number of possible situations can grow polynomially. Context spaces allow explicitly selecting situations of interest and, therefore, saving computational power by reasoning only about relevant situations.

Table 1. Complexity of situation reasoning – single situation

Operations	Order	Explanation
Additions	O(N)	The sum in formula (4) contains N summands. Each calculation of contribution function contains one more addition at most.
Multiplications	O(N)	Formula (4) contains N multiplications – one in every summand. Contribution function contains at most one more multiplication.
Division	None	The formulas for each linear fragment of contribution function can be precompiled in the form $y = k*x + b$. It eliminates the need for any divisions.
Comparisons	O(N)	There are N contribution functions, each calculation requires at most 4 comparisons – interval of piecewise function needs to be determined (see Fig. 4).
Time	O(N)	Complexity for each relevant operation grows linearly with the number of involved contribution functions or subsituations.

Generation of context spaces situations also introduces some challenges:

Preprocessing. Preprocessing can be done only once for any particular pervasive system. However, situations should be generated again if context model or sensors changed significantly (e.g. new sensor, new user, or user moving out).

Reasoning Explanation. Situation reasoning rules provide very intuitive representation of situations. Moreover, if the situation is triggered, it is relatively easy to identify and present to the user the factors that lead the pervasive system to assume that the situation is occurring. Context spaces provide intuitive representation of the situations. However, generation of explanations (like "Why did the system decide that it is a family lunch?") is slightly more complicated comparing to ontologies.

Next section identifies the direction of future work and concludes the paper.

8   Conclusion and Future Work

In this paper we proposed and developed fast, lightweight and robust approach to context awareness, situation awareness and sensor fusion in pervasive computing systems. As an input, the information about different aspects of pervasive computing systems was represented by two separate ontologies – context ontology and sensor ontology. We proposed and proved a method to integrate those ontologies, and achieve unified representation of complete pervasive computing system. On top of the unified representation we developed a method to extract all the necessary information for runtime situation awareness and generate lightweight and intuitively clear situation formulas, which allows handling sensor fusion, reasoning under uncertainty, and working with fuzzy concepts.

We identified the following main directions of the future work:

- **Generation of Additional Situations**. From the rule (1) we generated formulas for situations like "Alice is sleeping" and "Bob is sleeping". It is also possible generate situation "Someone is sleeping", i.e. leave the parameters undefined. That situation might require different set of sensors (for example, it can be inferred using motion sensors without any involvement from indoor localization system).
- **Smart Sensor Fusion**. At the moment readings from several sensors are combined using equal weights. However, sometimes other approaches are preferred. For example, movement is detected if at least one motion sensor is triggered. Or if there is severe disagreement between sensors, pervasive system might consider discarding some sensor values as unreliable.
- **Using Derived Measurement**. Some context characteristics can be obtained even if they are not directly measured. For example, absolute acceleration might be derived using relative acceleration and orientation sensors, which both reside on user's mobile phone. Those derived values can be introduced by extending SSN ontology with artificial virtual platforms that will correspond to derived values.
- **Handling Disjoint Situations.** Situations like "User is standing" and "User is sitting" should not co-occur together. Some context ontologies [1] allow declaring situations as disjoint. In order to generate situations as disjoint, situation verification [5] can be extended and used it in conjunction with situation formula generation.

References

1. Anagnostopoulos, C., Ntarladimas, Y., Hadjiefthymiades, S.: Situational computing: an innovative architecture with imprecise reasoning. J. Syst. Softw. **80**(12), 1993–2014 (2007)
2. Bettini, C., Brdiczka, O., Henricksen, K., Indulska, J., Nicklas, D., Ranganathan, A., Riboni, D.: A survey of context modelling and reasoning techniques. Pervasive Mob. Comput. **6**(2), 161–180 (2010)
3. Barnaghi, P., Meissner, S., Presser, M., Moessner, K.: Sense and sens' ability: semantic data modelling for sensor networks (2009)
4. Bazire, M., Brézillon, P.: Understanding context before using it. In: Leake, D.B., Kokinov, B., Dey, A.K., Turner, R. (eds.) CONTEXT 2005. LNCS (LNAI), vol. 3554, pp. 29–40. Springer, Heidelberg (2005). doi:10.1007/11508373_3
5. Boytsov, A., Zaslavsky, A.: Formal verification of context and situation models in pervasive computing. Pervasive Mob. Comput. **9**(1), 98–117 (2013)
6. Chen, H., Finin, T., Joshi, A.: An ontology for context-aware pervasive computing environments. Spec. Issue Ontol. Distrib. Syst. Knowl. Eng. Rev. **18**, 197–207 (2003)
7. Chen, H., Finin, T., Joshi, A.: The SOUPA ontology for pervasive computing. In: Tamma, V., Cranefield, S., Finin, T. (eds.) Ontologies for Agents: Theory and Experiences. Whitestein Series in Software Agent Technologies, pp. 233–258. Springer, Switzerland (2005)
8. Compton, M., Henson, C.A., Lefort, L., et al.: A survey of the semantic specification of sensors. In: CEUR Workshop Proceedings, October 2009

9. Calbimonte, J.P., Yan, Z., Jeung, H., et al.: Deriving semantic sensor metadata from raw measurements. In: 5th International Workshop on Semantic Sensor Networks, in conjunction with the 11th International Semantic Web Conference (ISWC), November 2012
10. Dapoigny, R., Barlatier, P.: Formalizing context for domain ontologies in Coq. In: Brézillon, P., Gonzalez, A.J. (eds.) Context in Computing, pp. 437–454. Springer, New York (2014)
11. Dey, A.K., Abowd, G.D.: Towards a better understanding of context and context-awareness. In: CHI 2000 Workshop on the What, Who, Where, When, and How of Context-Awareness, pp. 304–307 (2000)
12. Delir Haghighi, P., Krishnaswamy, S., Zaslavsky, A., Gaber, M.M.: Reasoning about context in uncertain pervasive computing environments. In: Tröster, G., Lombriser, C., Kortuem, G., Havinga, P., Roggen, D. (eds.) EuroSSC 2008. LNCS, vol. 5279, pp. 112–125. Springer, Heidelberg (2008)
13. Ejigu, D., Scuturici, M., Brunie, L.: An ontology-based approach to context modeling and reasoning in pervasive computing. In: Proceedings of the Fifth IEEE International Conference on Pervasive Computing and Communications Workshops, IEEE Computer Society, pp. 14–19 (2007)
14. Gruber, T.R.: A translation approach to portable ontology specifications. Knowl. Acquis. 5(2), 199–220 (1993)
15. Gu, T., Pung, H.K., Zhang, D.Q.: A service-oriented middleware for building context-aware services. J. Netw. Comput. Appl. 28(1), 1–18 (2005)
16. Padovitz, A., Loke, S.W., Zaslavsky, A., Burg, B.: Verification of uncertain context based on a theory of context spaces. Int. J. Pervasive Comput. Commun. 3(1), 30–56 (2007)
17. Wang, X.H., Zhang, D.Q., Gu, T., Pung, H.K.: Ontology based context modeling and reasoning using OWL. In: Proceedings of the Second IEEE Annual Conference on Pervasive Computing and Communications Workshops, pp. 18–22 (2004)
18. Ye, J., Coyle, L., Dobson, S., Nixon, P.: Ontology-based models in pervasive computing systems. Knowl. Eng. Rev. 22(4), 315–347 (2007)
19. Ye, J., Dobson, S., McKeever, S.: Situation identification techniques in pervasive computing: a review. Pervasive Mob. Comput. 8(1), 36–66 (2012)
20. Zadeh, L.A.: Fuzzy sets. Inf. Control 8, 338–353 (1965)
21. w3.org: Semantic sensor network XG final report: w3c incubator group report. http://www.w3.org/2005/Incubator/ssn/XGR-ssn-20110628/, June 2011. Accessed 14 Aug 2015
22. OGC observations and measurements: http://www.opengeospatial.org/standards/om. Accessed 14 Aug 2015
23. Protégé ontology editor: http://protege.stanford.edu/. Accessed 14 Aug 2015
24. SPARQL 1.1 overview: http://www.w3.org/TR/sparql11-overview/. Accessed 14 Aug 2015

Modeling Expert Knowledge and Reasoning in Context

Patrick Brézillon[✉]

University Pierre and Marie Curie (UPMC), Paris, France
Patrick.Brezillon@lip6.fr

Abstract. Based on the results of studies on cancer diagnosis and battle simulation, we discuss the role of context in mental representation and reasoning that operators hold during task realization. Mental representations are considered under the form of expert maps, which are semi-structured expressions of the mental representations, and reasoning is represented in a context-based model that provides a uniform representation of pieces of knowledge, reasoning and contexts that make possible a task realization oriented approach for systems. This work is based on two applications, one in medicine, and the other in battle simulation. Our conclusion is that the Contextual Graphs formalism allows the modeling of all operators' practices in a structured way, while expert maps are unique and require a process of contextualization-decontextualization and recontextualization for representing operational knowledge and experience of operators for various needs.

1 Introduction

As part of the digitalization of the battlespace, the Command & Control systems (C2 systems) are already widespread in the army. These information and communication systems provide an operator with a view on an operational situation, which typically shows a map with symbols representing units. The system gathers information from various sources and allows users to interact and give orders directly. C2 systems are also used or being adopted in non-military areas, for example in civil safety or by large private operators. In the TACTIC project, three complementary sources of information commonly are used together: spatial coordinates of objects (the field map), temporal coordinates (the chronology) and socio-technical coordinates (ODB). It seems appropriate to consider them as a cognitive tridimensional referential in which the events take place in a specific context.

Operator-Simulator interaction goes through an interface (the place of cognitive interaction with the operator) and the screen (the place of physical interaction for the visualization of the simulation). During a simulation, the operator has to face three interrelated challenges: (1) Collecting relevant data and information from several sources; (2) Translating data and information into knowledge to produce contextual understanding of the events and behaviors of interest in relation to particular goals, capabilities, and policies of the decision makers; and (3) Using that knowledge for making relevant decision in the working context. As a consequence, operators must deal with an interpretation of the domain (the simulation resulting of the operator-simulator interaction) intertwined with an interpretation of the interface functioning for translating actions on the simulation into commands to the interface.

H. Christiansen et al. (eds.): CONTEXT 2015, LNAI 9405, pp. 18–31, 2015.
DOI: 10.1007/978-3-319-25591-0_2

By focusing on the process that leads to an action (including the decision-making part), rather than the result of the action execution only, it is possible to make explicit the context in which the operator works effectively. Generally, context is used to eliminate or reduce ambiguity, detect inconsistencies, explain observations, and constrain processing [4].

A context is associated with the operator's focus of attention (a task realization, a new event) during the simulation. It contains two types of knowledge, namely, the contextual knowledge that is more or less related to the current focus in a flat way and the external knowledge that has nothing to do with the focus at the time at which the operator considers the focus. The proceduralized context is a structured subset of contextual knowledge that is explicitly used to address the focus of attention. The focus of attention evolving, its context evolves too: there are exchanges between contextual knowledge and external knowledge and, thus, the frontier between them is porous.

Brézillon [5] introduces the notion of contextual element for representing context information coming from heterogeneous sources in a uniform way. A contextual element is an "element of the nature" for which it is necessary to know its value in the current focus (i.e. its instantiation). Contextual elements come from different highly heterogeneous sources like the operator, the task, the situation and the local environment where available resources are. The distinction between a contextual element and its values is important for the reuse of experience because a difference between a past context and the working context can be a difference of either contextual elements (e.g. a contextual element only exists in one context) or instances (e.g. the same contextual element has different instantiations in the two contexts).

Hereafter, the paper is organized in the following way. The next section discusses the type of knowledge that is used in task realization and the way to represent it. The following section presents the modeling of reasoning in task realization, especially when operational knowledge intervenes. The section after positions this work with related works and emphasizes the role of context.

2   Knowledge Representation in Task Realization

2.1   Mental Representation

Experts rely on a highly compiled experience because they generally act under temporal pressure and are very concerned by the consequences of their decision. Experience reuse is never direct because the context of any decision-making is unique, and thus any experience must be adapted in a process of contextualization-decontextualization-recontextualization [4] to be efficient in another context. Fan et al. [11] used this process in finding a scientific workflow (SWF) in virtual screening: A researcher extracts from a repository a SWF close to his working context (phase of contextualization), extracts the SWF model (phase of decontextualization), and, finally, looks for instantiating the SWF model in his working context (phase of recontextualization). Thus, the experience acquired by the researcher during this process (and thus experience management) relies on context management. As a consequence, it is more important to model task realization than a task model. This context-based modeling makes operators' behavior explicit.

An operator receives a lot of information on events occurring in his environment, but only a small number of events enter operator's focus of attention. Most of events concern what the operator is doing and thus are processed automatically (often unconsciously). A good example of the importance of the focus of attention on the selection of events judged relevant by the operator is given on the video of cognitive blindness[1] with the gorilla. Events in the focus of attention correspond to either information proactively searched by the operator (e.g. information about the mission of a specific unit) or unpredicted events (e.g. an action of the enemy) not explained in his mental representation of the focus, the situation, the resources available. The goal is to integrate the corresponding information in the mental representation of the task realization. Other events are put in the periphery of operator's attention because they are not directly related to the focus of attention.

2.2   Expert Maps

A mental representation depends on expert's experience with the realization of tasks attached to his role. Experience contains knowledge accumulated by the expert during his practical use of the domain knowledge along a number of realizations of the same task in different contexts. Thus, experience relies more on operational knowledge than domain knowledge and, thus, experience is highly contextual. We hypothesize that the mental representation corresponds to contextual knowledge, which is related to the operator (the expert), the task at hand, the situation of the work, and the local environment in which resources are available. By expressing contextual knowledge with contextual elements (and thus mental representation too), it is possible to externalize the mental representation with a classical knowledge-management tool like Freemind[2] as a cognitive map, that is, a semi-structured expression of the mental representation.

Operators may start from a domain map to develop their expert map. The domain map is a representation of the whole context (i.e. the contextual and external knowledge). A domain map can be obtained by different ways, from a state-of-the-art on the use of domain knowledge, or by developing a glossary to fix the terminology among experts [1]. All elements of the domain map belong to operators' knowledge, but operators use only the part that is operational for task realization.

The expert map corresponds to the selection of the operational part of the domain knowledge effectively used by operators during the realization of their tasks. As a result, the expert map is a tree representation of the elements considered by operators. In terms of context, the expert map is a representation of the contextual knowledge (the part of the context) that operators relate more or less directly to their focus (i.e. task realization).

Figure 1 presents the general shape of an expert map expressing the organization of contextual elements in the mental representation of an operator of the experiment of the TACTIC project. This expert map is strongly inspired by the domain map proposed initially as a bootstrap. (Some operators introduced links between different leaves of the map).

---

[1] http://en.wikipedia.org/wiki/Inattentional_blindnesshttp://en.wikipedia.org/wiki/Inattentional_blindness

[2] http://freemind.sourceforge.net/wiki/index.php/Main_Page

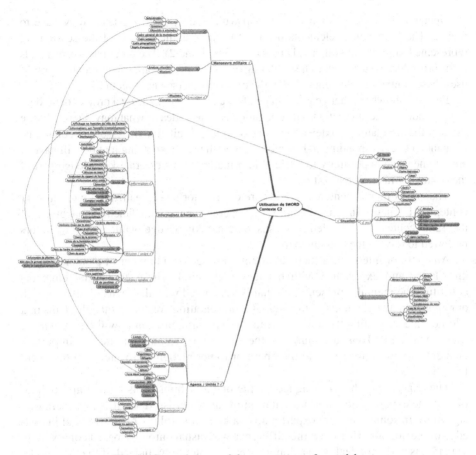

Fig. 1. General shape of the expert map of a modeler

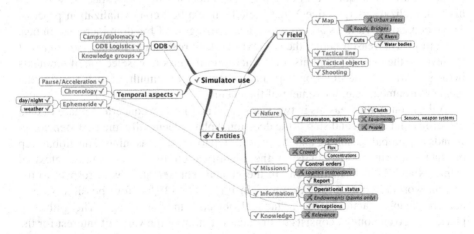

Fig. 2. General shape of the expert map of a project manager

Operators established their expert map with respect to all the tasks they have to realize. The (contextual) elements not used in the specific context of the session are represented in grey and with a red cross (see Figs. 1 and 2). The other part corresponds to the proceduralized context. The modeling of operators' reasoning as contextual graphs uses these contextual elements kept in the expert maps (see next section).

Figure 2 shows another general shape of an expert map made by a project developer (i.e. a vendor of the simulator). The operator has a clear idea of what may interest future users of the simulator, and developed an expert map limited to the essential contextual elements (i.e. corresponding to his interaction with users), knowing that he will have at any moment the opportunity to retrieve information in his external-knowledge part of the context of his task realization.

The expert map is centered on the core of operator's task realization with only 41 contextual elements. In term of context, the operator makes an efficient management of the ratio contextual knowledge versus external knowledge while other operators preferred to rely on the domain map.

An operator interprets a focus of attention based on a number of elements in his knowledge and experience with his task realization. These contextual elements constitute the mental representation that is associated with the focus of attention. In our approach, the elements of the mental representation are contextual elements that "do not intervene directly in the operator's task but constrain how the task will be performed" [8]. Thus a failure in the mental representation (some important elements are missing) can lead the operator to not recognize (or not see) the event for what it is.

The expert map, being designed by the operator himself, is a good approximation of the expert's mental representation of his expertise field. It is the operator's signature. In some sense, an expert map is a kind of ontology of operational knowledge of the domain. However, the difference with usual ontology-based context (e.g. see [17]) is to do not deal with a domain-specific ontology. Indeed, if the task model (or procedure) is unique for all operators, each operator develops specific ways for his task realizations (practices) that include an explicit contextualization process. These contextual variants appear as soon as the degree of freedom increases in task realization. The reason is that the context of a task realization includes elements at the level of the operator (e.g. his preferences), of the task (e.g. selection of moments in the scenario), of the situation (e.g. many events occurs simultaneously), and of the local environment (e.g. movement of the enemy).

A decision-maker reasons in two steps [7]. The first step concerns a phase of data gathering, and the second step is the decisional phase. Generally, the first step corresponds to a global reasoning and the second step to a local reasoning. The global step of data gathering corresponds to the identification of the relevant contextual elements and their "instantiation" in the mental representation with respect to the current working context. For example, in Fig. 2 "Field" has four possible values, namely "map", "tactical issue", "tactical objects" and "shooting". The gathering phase in a given context consists of the identification of the value of interest for the focus at hand (i.e. its instantiation). If the operator focuses on information retrieval about a given unit, "tactical objects" will be the instantiation of "Field". Note that "tactical object" is itself a contextual element with different values (including "unit"). Thus, the expert map is assimilated to a search space where the operator looks for instantiations of relevant contextual elements in real time.

2.3 Example: Modeling Operator-Simulator Interaction

Interaction is a *phenomenon* between a user and a computer that is controlled by the user interface running on the computer. Designing interaction rather than interfaces implies that interfaces are the means, not the end [3]. This supposes to combine and understand the context of use with a particular attention to the details of the interaction. Different users work differently and a given user applies different interaction patterns according to the context of use [13]. As a result, no single interaction technique works identically in all contexts, and the best solution is to provide a range of interaction techniques and let users decide which one must be used according to the working context, although users may have context-aware support for choosing an interaction technique.

Figure 3 shows the two main changes in order to simplify operator's task realization with a simulator. The first one concerns a clear distinction of the interface with operator-simulator interaction. The consequence is the separation of "domain_actions" and "interface_actions" and a simple mechanism of translation between domain-actions and interface-actions by shifting the main problem of translation at the level of the exchange of interfaces. (This part will be described in another paper). This would facilitate the change of computer (say, from a PC to a touchpad). The second change is to compare the expert map of the operator with the "expert map" of the simulator for making them compatible, even across a translation in terms of "interface_actions", for transmitting commands to the simulator and, conversely, for presenting results to the operator.

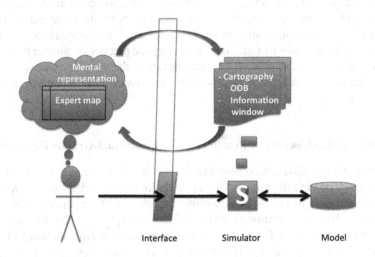

Fig. 3. A model of user-simulator interaction

Thus, domain_actions will be more easily associated with interface_actions by taking into account the (shared) context of interaction, resulting in greater flexibility of the interface, not only with respect to the operator, but also with respect of task realization. This leads us to propose a "task realization centered approach" for designing the interface as a shared space (or shared context) between the operator and the simulator.

3   Reasoning Modeling in Task Realization

3.1   Introduction

The effective application of a procedure supposes to account for the working context in which the task must be realized. This leads operators to establish practices that are tailored to specific contexts. A practice is the way in which operators contextualize a procedure for taking into account their preferences, the particularities of the task to realize, the situation where the task is realized and the local environment where resources are available. The essence is to understand and model how work actually goes done (i.e. the practice or task realization), not what is supposed to happen (i.e. the procedure or a task model). This means to identify which contextual elements are important, and what their values are for the current focus of attention.

Contextual elements structure experiences (practices) differently, on the one hand, from the knowledge bases of expert systems represented in a flat way because context is not represented explicitly, and, on the other hand, from knowledge organization in an ontology where links between concepts depend on the domain (is-a, kind-of, etc.) while elements in our context model concern the operator, the task, the situation and the local environment.

A practice is developed jointly with the building of a proceduralized context, i.e. a context-specific model [6]. Thus, there are simultaneously the development of the practice (by looking for instantiations of contextual elements orienting the choice of the path to follow) and the realization of the task. Finding a good practice consists of the progressive assembling of the components that are relevant in the working context during the development of the practice. A "best practice" thus has a meaning only in a specific working context.

3.2   Representing Operators' Practices in the Contextual-Graphs Formalism

A Contextual Graph (CxG) is a context-based formalism for representing all the different practices developed in different working contexts for a task realization [5]. Operator's experience is represented by an organization of practices that are structured by contextual elements. Thus, a contextual graph is similar to an experience base focusing on the realization of a given task. This supposes, first, to use a formalism allowing a uniform representation of knowledge, reasoning and context, and, second, to have support systems with powerful functions for processing such a representation.

Formally, contextual graphs are acyclic and series-parallel due to the time-directed representation that garanties algorithm termination. Each contextual graph has exactly one root and one end node because the decision-making process starts in a state of affairs (i.e. a working context) and ends in another state of affairs and the branches express only different contextually dependent ways to achieve this goal. Each path in a contextual graph corresponds to a practice effectively developed in a working context leading to a specific solution.

Rogova [15] describes the Contextual-Graphs formalism as incorporating action and context nodes (variables and relationships) as well as paths through them. Although a contextual graph is not free of weaknesses, e.g. there is a problem with the lack of direct time representation. However, the formalism offers certain advantages over other approaches since it allows a representation of knowledge and reasoning in a way that is directly comprehensible by users. Thus, information in contextual graphs is useful and usable for users.

3.3   Modeling Reasoning

A practice represents the result of the application of reasoning with its choices (instantiation of contextual elements) and actions executed. Generally graph traversal in reasoning is lead between a "depth-first" strategy and a "breadth-first" strategy. The "depth-first" strategy goes to the finest possible granularity on a line of reasoning in order to anticipate the course of events. It assumes that we know what to do and how to get there quickly. This strategy allows studying the technical feasibility of an approach as well as the needs in terms of resources, and, in a second step, gradually expands this approach. Figure 2 gives an example of an expert map of this kind.

Conversely, the breadth-first strategy is applied when it is necessary to consider all possible situations first. The breadth-first strategy is observed in expert maps of operators evolving at a strategic decisional level that maintain important contextual elements, even if not directly necessary in the realization of their tasks. Figure 1 gives an example of expert map of this kind. For example, "environment" was considered as a part of "situation", even if environment is the main source of contextual elements on the battlefield map. Indeed, the operator considers environment in expert maps through what the operator needs to extract of it for realizing his task, that is, operational knowledge. For example, it is only when the focus is on a zone of interest that operator performs a global reasoning, switching from a global reasoning (e.g. finding a zone of interest) to a local reasoning (e.g. exploring the zone of interest for detecting relevant features). The operator thus is interested by an external event more through its effect on his task rather than by event origin.

The CxG formalism allows the representation of a system at the tactical and operational levels. The contextual graph represents at the tactical level the different practices used to realize a task in various contexts, while the development of a practice in a specific context is made at the operational level.

3.4   An Example of Modeling "Manage a Unit"

Figure 4 gives the contextual graph of the mission "Giving an order of recognition". It has three actions and three activities. The activity "unit manager" is found in two locations (pink ovals 62 and 67). The reason is that an operator initially chose a shielded but in seeking to define the scope of recognition around enemy, the operator realized that the enemy were in a city that was not screened the most appropriate unit for what he wanted to do (blue contextual element 65).

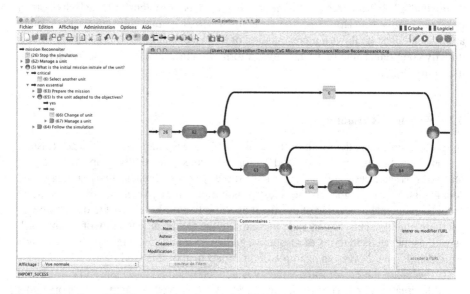

Fig. 4. Modeling of "Giving an order of recognition" in the Contextual-Graphs formalism (Color figure online).

Figure 5 represents the modeling of the activity "Manage a unit" in Fig. 4 as a contextual graph accompanied by its legend. "Manage a unit" is rather a sequential activity beginning by the choice of an area where to select a unit (first contextual element), followed by the manner to choose the unit (second contextual element), continuing by checking if the selected unit may realize the required recognition mission, and finishing by positioning the unit in the center of the window of the field map. Brézillon [5] shows that such a sequential structure of a contextual graph corresponds to a pragmatic approach, while a parallel structure is closer of a procedure (like self-diagnose of a piece of equipment in its user manual).

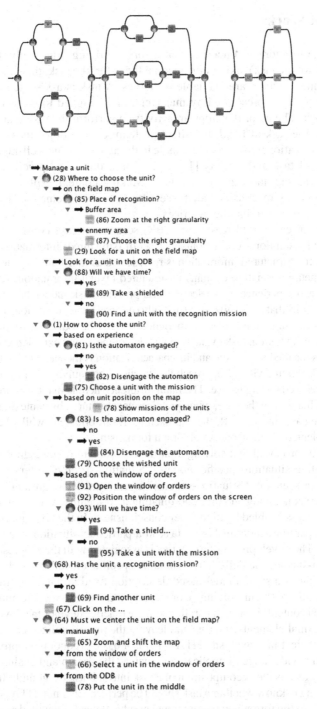

➡ Manage a unit
▼ ⬤ (28) Where to choose the unit?
  ▼ ➡ on the field map
    ▼ ⬤ (85) Place of recognition?
      ▼ ➡ Buffer area
        ▦ (86) Zoom at the right granularity
      ▼ ➡ ennemy area
        ▦ (87) Choose the right granularity
      ▦ (29) Look for a unit on the field map
  ▼ ➡ Look for a unit in the ODB
    ▼ ⬤ (88) Will we have time?
      ▼ ➡ yes
        ▦ (89) Take a shielded
      ▼ ➡ no
        ▦ (90) Find a unit with the recognition mission
▼ ⬤ (1) How to choose the unit?
  ▼ ➡ based on experience
    ▼ ⬤ (81) Isthe automaton engaged?
      ➡ no
      ▼ ➡ yes
        ▦ (82) Disengage the automaton
    ▦ (75) Choose a unit with the mission
  ▼ ➡ based on unit position on the map
    ▦ (78) Show missions of the units
    ▼ ⬤ (83) Is the automaton engaged?
      ➡ no
      ▼ ➡ yes
        ▦ (84) Disengage the automaton
    ▦ (79) Choose the wished unit
  ▼ ➡ based on the window of orders
    ▦ (91) Open the window of orders
    ▦ (92) Position the window of orders on the screen
    ▼ ⬤ (93) Will we have time?
      ▼ ➡ yes
        ▦ (94) Take a shield...
      ▼ ➡ no
        ▦ (95) Take a unit with the mission
▼ ⬤ (68) Has the unit a recognition mission?
  ➡ yes
  ▼ ➡ no
    ▦ (69) Find another unit
  ▦ (67) Click on the ...
▼ ⬤ (64) Must we center the unit on the field map?
  ▼ ➡ manually
    ▦ (65) Zoom and shift the map
  ▼ ➡ from the window of orders
    ▦ (66) Select a unit in the window of orders
  ▼ ➡ from the ODB
    ▦ (78) Put the unit in the middle

Fig. 5. The contextual graph of the activity « Manage a unit »

4 Related Works

The challenge to address concerns what actors are doing effectively, that is, their activity (and not the task) taking into account the actor, the task, the situation and the local environment where are available resources. A task is associated with a set of objectives, which are prescribed by managers, and assigned to the actor that must realize the task. The actor develops an activity to perform the task that includes his mobilization, the task at hand, the situation features, the objectives and the technical and organizational resources available to the actor. It is the well-known problem of separation of task and activity [10, 12], of procedures and practices [5], and of logic of functioning and logic of use [14]. Making context explicit as contextual elements allows us to consider all these heterogeneous elements of context in a uniform way and thus to handle practices.

In a classical case-based reasoning (CBR) scenario, a case consists of a problem description and a solution. A case contains a set of (structured) information entities, and optional artifacts. Structured information is represented as (attribute – value pairs), while the optional meta-information contains unstructured textual information. Atzmueller [2] uses stored cases (experiences) for selecting an appropriate task and method, reusing stored task-configurations that are similar to a (partially) defined characterization. The process of capturing and reusing complex task-experiences is lead in four main steps: experience retrieval, task instantiation, task evaluation and deployment, and experience maintenance. Thus, a case is recalled as a whole and its characterization is then adapted to the context at hand. In the Contextual-Graphs approach the practice (the equivalent of the case) is identified and developed during its use. The main difference here is that cases are represented in a relatively flat way in the base, while practices are structured by contextual elements in the experience base. In the CBR, the approach is "result-oriented" while the approach is "reasoning-oriented" in the Contextual-Graph formalism.

Clancey [9] proposed that solving a particular problem (e.g. diagnosing a patient) involves creating situation-specific models. "Situation-specific" refers to a particular case, setting, or scenario. "Situation-specific" is not "situated cognition" that refers to how people are conceiving and thus coordinating their identity, values, and activities in an ongoing process enabled by high order consciousness. In the CxG approach, context concerns an operator accomplishing a task in a particular situation in a specific local environment. The development of a practice is associated with the progressive building of a "context-specific model". The "situation-specific model" is embedded in the problem solving as a static model-based description fixed initially and filled progressively during the problem solving. Conversely, the context-specific model (i.e. the proceduralized context) is built in parallel with the practice development with the movement of contextual elements entering and leaving the proceduralized context.

The hierarchical task analysis (HTA) [16] is a basic ergonomic approach used for thirty years in a wide range of applications, such as the design and evaluation of interfaces. The key idea is the decomposition of a task into subtasks at granularities finer and the rules relating to know whether a task should be performed or not. HTA is very similar to what is done in the formalism of contextual graphs. Indeed, a detailed example in [16] —the passage of a customer at the checkout of a supermarket—was the subject of a translation of the HTA in a contextual graph (one activity being actually a finer granularity of sub-task) with rules translated in terms of contextual elements.

5   Conclusion

Taking into account the end-user in the design loop is not sufficient. The design loop must also integrate end-user variability. Our work on expert maps—as expressions of mental representations—shows that each expert map is unique (the operator's signature) and corresponds to a specific view on operational knowledge used in a task realization. Thus, the expert map can help to make a real contextualization of the interface to meet the needs of each operator (as illustrated for operator-simulator interaction in the TACTIC project). Expert maps can be extended by introducing domain_actions as instantiations that the operator has the habit of applying to each object in the domain. This would provide a description of the domain_actions linked to the operational knowledge on operator-side (in his expert map) and help a more or less automatic translation (at least for ranking them along operator's preference) of domain_actions into interface_actions.

Operators' reasoning, as shown in the contextual graph presented in Fig. 4, relies on operators' operational knowledge for contextualizing the task "Give a recognition mission". Operators contextualize at the operational level the procedures coming from the tactical level (for example, the various means used to monitor the progress of the mission). If different operators have specific views on a task realization (a result of our study with the large spectrum of expert maps), a contextual graph, which is the accumulation of practices developed by all operators realizing the task, may be used as a tool for sharing experiences among operators performing the same task, explanation generation purposes and a training tool for future operators. This has already been done in other areas where contextual graphs are used (e.g. see [1]).

The change from "operator communicating with the simulation" to "operator communicating with the simulator about the simulation" allows giving the interface the role of a flexible communication medium equivalent to a shared context through which the operator (with domain_actions in his reasoning) and the simulator (with domain_actions in his model of the battlefield) communicate with a simple translation in interface_actions (essentially, mouse clicks actually). In the TACTIC project, three complementary sources of information are commonly used together: spatial coordinates of objects (the field map), temporal coordinates (the chronology) and socio-technical coordinates (ODB). It seems appropriate to consider them as a cognitive tridimensional referential in which events take place. Thus, clicking on a unit, all information related to this unit would be extracted automatically for presentation according to the operator's request. In the cognitive referential, a unit would be associated with a knowledge network with close information, such as the life bar, and other more distant information such as belonging to an automaton. Such a knowledge network should allow different views adapted to the desired level of aggregation of the context.

An important lesson is that design and development of an interface would be made along a "task-realization oriented" rather than "user-oriented" approach, thanks making context explicit. This finding is interesting because it would be easier to design a task-oriented interface than to develop a generic user-oriented or multiple user-oriented interfaces for the presentation of information.

A next step would be to model the relationships between domain_actions and interface_actions. A path to explore is to develop an "interface map" for representing operator's mental representation of the interface in a similar way to the expert map that represents operator's mental representation of the domain.

An expert map corresponds to the operational knowledge used by an actor in a task realization. There are as many expert maps as actors, but all expert maps have a large nonempty intersection. This intersection corresponds to the context of the task realization that is shared by actors. There are two lessons to retain. First, the common part of expert maps could be assimilated to a generic expert map for training new actors and may be the basis for developing a support system. Second, private parts of expert maps (i.e. the not shared parts) correspond to actors' personal experience. However, the differences of organization of expert maps have to be studied first.

Depth-first and breadth-first strategies are another challenging aspect for modeling reasoning. Expert maps based on a breadth-first strategy are interesting for detecting weak signal because the expert keeps an "open mind" when analyzing a situation. His focus is not limited to the task realization in an isolated way, but replaced in the context of the task realization. Such actors reason at a strategic level and do not consider details of lower levels of the task realization. Conversely, expert maps based on a depth-first strategy correspond to experts that are able to decide rapidly at the tactical level which reasoning must be held (units managed by the simulator are at the operational level).

Acknowledgments. This work is supported by grants from the TACTIC project funded by the ASTRID program of Délégation Générale aux Armées. We especially thank partners at MASA Group and Lab-STICC for rich discussion during our collaborative work.

References

1. Attieh, E., Capron, F., Brézillon, P.: Context-based modeling of an anatomo-cyto-pathology Department Workflow for Quality Control. In: Blackburn, P., Dapoigny, R., Brézillon, P. (eds.) CONTEXT 2013. LNCS, vol. 8175, pp. 235–247. Springer, Heidelberg (2013)
2. Atzmueller, M.: Experience management with task-configurations and task-patterns for descriptive data mining, KESE (2007). http://ceur-ws.org/Vol-282/02-AtzmuellerM-KESE-Paper-CRC.pdf
3. Beaudouin-Lafon, M.: Designing interaction, not interfaces. In: AVI 2004 Proceedings of the Working Conference on Advanced Visual Interfaces, pp. 15–22 (2004)
4. Brézillon, P.: Focusing on context in human-centered computing. IEEE Intell. Syst. **18**(3), 62–66 (2003)
5. Brézillon, P.: Task-realization models in contextual graphs. In: Kokinov, B., Leake, D.B., Turner, R., Dey, A.K. (eds.) CONTEXT 2005. LNCS (LNAI), vol. 3554, pp. 55–68. Springer, Heidelberg (2005)
6. Brézillon, P.: Context-based development of experience bases. In: Blackburn, P., Dapoigny, R., Brézillon, P. (eds.) CONTEXT 2013. LNCS, vol. 8175, pp. 87–100. Springer, Heidelberg (2013)

7. Brézillon, P., Pasquier, L., Pomerol, J.-Ch.: Reasoning with contextual graphs. Eur. J. Oper. Res. (EJOR) **136**(2), 290–298 (2002)
8. Brézillon, P., Pomerol, J.-Ch.: Contextual knowledge sharing and cooperation in intelligent assistant systems. Le Travail Humain **62**(3), 223–246 (1999)
9. Clancey, W.J.: Model Construction operators. Artif. Intell. J. **53**, 1–115 (2002)
10. Clancey, W.J.: Simulating activities: relating motives, deliberation, and attentive coordination. Cogn. Syst. Res. **3**(3), 471–499 (2002)
11. Fan, X., Zhang, R., Li, L., Brézillon, P.: Contextualizing workflow in cooperative design. In: Proceedings of the 2011 15th International Conference on Computer Supported Cooperative Work in Design (CSCWD-2011), Lausanne, Switzerland, pp. 17–22 (2011)
12. Leplat, J., Hoc, J.-M.: Tâche et activité dans l'analyse psychologique des situations. Cahiers de Psychologie Cognitive **3**, 49–63 (1983)
13. Mackay, W.E.: Which interaction technique works when? Floating palettes, marking menus and toolglasses support different task strategies. In: Proceedings of Conference on Advanced Visual Interfaces, AVI 2002, pp. 203– 208, ACM Press (2002)
14. Richard, J.-F.: Logique du fonctionnement et logique de l'utilisation. Rapport de Recherche INRIA N° 202 (1983)
15. Rogova, G.L.: Context-awareness in crisis management. In: 5th IEEE Workshop on Situation Management (SIMA-2009) (In Conjunction with MILCOM-2009) Paper ID # 900748 (2009)
16. Shepherd, A.: Hierarchical task analysis and training decisions. Innov. Educ. Train. Int. **22**(2), 162–176 (1985)
17. Wang, X.H., Gu, T., Zhang, D.Q., Pung, H.K.: Ontology based context modeling and reasoning using OWL. In: Proceedings of the Workshops of the Second IEEE Annual Conference on Pervasive Computing and Communications, pp. 18–22 (2004)

Towards a Conceptualization of Sociomaterial Entanglement

Roberta Ferrario$^{(\boxtimes)}$ and Daniele Porello

Institute of Cognitive Sciences and Technologies, CNR, Trento, Italy
{daniele.porello,roberta.ferrario}@loa.istc.cnr.it

Abstract. In knowledge representation, socio-technical systems can be modeled as multiagent systems in which the local knowledge of each individual agent can be seen as a context. In this paper we propose formal ontologies as a means to describe the assumptions driving the construction of contexts as local theories and to enable interoperability among them. In particular, we present two alternative conceptualizations of the notion of sociomateriality (and entanglement), which is central in the recent debates on socio-technical systems in the social sciences, namely critical and agential realism.

We thus start by providing a model of entanglement according to the critical realist view, representing it as a property of objects that are essentially dependent on different modules of an already given ontology. We refine then our treatment by proposing a taxonomy of sociomaterial entanglements that distinguishes between ontological and epistemological entanglement. In the final section, we discuss the second perspective, which is more challenging form the point of view of knowledge representation, and we show that the very distinction of information into modules can be at least in principle built out of the assumption of an entangled reality.

1 Introduction

Starting from the 60's, scholars began to use the concept of socio-technical system [11] to describe and analyze workplaces and wider infrastructures in which the technological component played a pivotal role in the production system, but it could also be turned into a threat to the very same system and to its participants, due to its interlacement with the human dimension of work. Roughly speaking, such approaches and those taking inspiration from them see a socio-technical system as a hybrid system, constituted by interacting components that are heterogeneous and which, in order to be analyzed, require different theoretical instruments, those of the social and natural sciences.

With the continuous advances in science and technologies, some devices inhabiting socio-technical systems have been endowed with functionalities that are so complex and effective that their behavior can be more and more rightly

---

Supported by the VisCoSo project grant, financed by the Autonomous Province of Trento through the "Team 2011" funding programme.

H. Christiansen et al. (Eds.): CONTEXT 2015, LNAI 9405, pp. 32–46, 2015.
DOI: 10.1007/978-3-319-25591-0_3

dubbed as "intelligent" and "autonomous". In fact, already in the 1990's, scholars in distributed artificial intelligence started to use the multiagent paradigm [8,17] to model socio-technical systems and nowadays intelligent agents play an important role in many work environments.

In the artificial intelligence community, and especially scholars embracing the multiagent approach, in order to represent (among other things) the knowledge and perspective of agents [16] have been using contexts, intended in rough terms as local theories that the agents entertain with respect to the domain of interest. The contextual approach has shown to be very useful in representing both the "local" perspective of each agent and if and how such perspectives can be integrated through the use of "lifting" [23] or compatibility [16] rules.

Ideally, each local context representing the perspective of an agent can have its own language and inference rules, and each language bears an ontological commitment with respect to how the terms of the vocabulary should be interpreted to express the agent's perspective on the domain of interest and which is the intended meaning of the terms that are used in the local language.

The idea of using formal ontologies to constrain the interpretation of the local languages of artificial agents is not new [13]; in this paper we would like to show how they can be used also to express different ways in which the agents conceptualize the socio-technical systems they inhabit, starting from a specific notion that has recently been much debated, especially in sociology and organization science, that of sociomateriality and the connected notion of entanglement.

For many years, the studies on socio-technical systems have been viewing such components as interacting but sharply separated, resulting in approaches focused on the degrees of adoption of new material technologies by the operators of socio-technical systems, or on the reshaping of the organizational structures induced by the introduction of technology. Such view is in fact the one that has been widely adopted by the knowledge representation community for modeling socio-technical systems.

Nonetheless, more recently, in social and organizational studies a different reading of socio-technical systems has been spreading, more focused on work practices, which are seen as sociomaterial, where the two dimensions of sociality and materiality are so tightly intertwined that are not really separable. And it is in these studies that the concept of entanglement has been employed for the description of sociomateriality:

> Such an alternative view asserts that materiality is integral to organizing, positing that the social and the material are *constitutively entangled* in everyday life. A position of constitutive entanglement does not privilege either humans or technology (in one-way interactions), nor does it link them through a form of mutual reciprocation (in two-way interactions). Instead, the social and the material are considered to be inextricably related—there is no social that is not also material, and no material that is not also social [24, 1437].

These studies have succeeded in explaining, through theoretical and empirical analyses conducted in real case studies, how entanglement is enacted in organizational life. A scholar who, taking inspiration from a notion elaborated in quantum physics, developed a new conception of entanglement to be applied to social analysis is Karen Barad:

> [...] the primary ontological unit is not independent objects with inherent boundaries and properties but rather phenomena. In my agential realist elaboration, phenomena do not merely mark the epistemological inseparability of observer and observed, or the results of measurements; rather, *phenomena are the ontological inseparability/entanglement of intra-acting "agencies."* That is, phenomena are ontologically primitive relations—relations without preexisting relata [1, 139].

In social and organization studies this idea is being applied in manyfold ways, to talk about the inseparability of social and material dimension, of the scholar (observer) and the object of his/her observation, of the actors, their agencies and the networks they belong to, etc.

However, besides Barad's *agential realism*, another philosophical position has been proposed as a theoretical foundation for the studies on sociomateriality, namely that of *critical realism* [19]. Even though both approaches hold sociomateriality as a proper lens to look at organizations, they differ in attributing a metaphysical primacy to the sociomaterial over the social and the material (agential realism) or vice versa (critical realism).

Leonardi states very clearly such difference:

> The main crux of the difference in theoretical foundation offered by agential realism and critical realism is that the former treats the "sociomaterial" as something that pre-exists people's perceptions of it while the latter argues that the "social" and the "material" are independent entities that become "sociomaterial" as they are put into relationship with one another through human action [19, 69].

When we come to knowledge representation and multiagent modeling, we can say that formal modeling approaches to socio-technical systems have mostly neglected the usefulness of this notion and this is quite surprising, as knowledge representation is mainly concerned with the way in which a rational agent may conceive and represent a certain domain of discourse. It is exactly from this perspective that sociomateriality is an important analytical category, either considered as the explanation of how socio-technical systems are metaphysically constituted, or of the "imbrication" [19] of social and material resulting from agents' action within the system.

Concerning this point, we would like to stress the fact that our final aim is to build an ontological model of socio-technical systems acknowledging the importance of sociomateriality and able to represent both positions as the result of alternative ontological choices. The rationale is that, if one wants to represent how artificial agents interpret the features of socio-technical systems by local theories as contexts and then wants to compare and put them into communication or, in other words, wants these theories to be interoperable, it is necessary to make explicit the assumptions behind each one of these local theories. Formal ontologies are a powerful tool for this endeavor and, in the case of socio-technical systems, a good starting point is to make explicit which are the assumptions behind the notions of sociomateriality that the agents use. Therefore, in our framework the ontological analysis precedes and guides the modelization of the local knowledge of agents with the use of contexts by making explicit which are the most important ontological choices with respect to the domain of interest. One of such choices for socio-technical systems is how to conceptualize the relation between sociality and materiality.

In this paper we present the proposal to formalize one of the possible interpretations of the term "entanglement", namely that of the critical realism and we leave the formalization of the alternative ontological choice, that of agential realism, for future work.

In formal ontologies, a methodology has been devised to deal with very complex models (as is the case for models of socio-technical systems), which goes under the name of "modularization". This consists in isolating meaningful fragments of an ontology, which can be used as stand-alone sub-parts of the ontology, where reasoning is facilitated by the fact of being applied only to the categories of the module and not to all the categories of the ontology. Modules are thus good candidates to represent different aspects or realms characterizing a socio-technical system, like the physical, the mental and the social. But what happens when properties belonging to different realms of information, for instance from the physical and from the normative social realm are ascribed to the same entities? We call "entanglement" the dependence of an entity from entities belonging to different realms (as the social and the material).

The first step of our proposal for capturing entanglement in knowledge representation under the critical realist view is by representing the different realms as different modules in an ontology and entanglement as the interdependency of entities belonging to different modules of an ontology. In other terms, we conceptualize entanglement in ontologies as the need, in order to characterize certain entities belonging to a certain module of an ontology, of categories belonging to different modules. Thus, entanglement becomes a feature of an object (or a concept) that exhibits an essential interdependence between different realms, such as the social and the physical, or the physical and the mental.

Furthermore, we will introduce a first classification of types of entanglement, distinguishing:

- *ontological entanglement:* an object that is ontologically dependent on objects in different modules (e.g. physical and social, physical and mental);
- *epistemological entanglement:* an object or concept such that every possible definition of the object/concept requires concepts/objects belonging to different realms.

What we will try to provide with this work, rather than a sharp classification of phenomena under one or the other type of entanglement, is a framework enabling modelers to represent various types of entanglement by choosing the type that they deem more appropriate to represent the phenomena they want to model.

The remainder of this paper is organized as follows. Section 2 introduces some fundamental features of a foundational ontology. We place our treatment within DOLCE [21] and discuss how to classify the elements of the ontology into modules (i.e. physical, social, mental module). Section 3 presents a formalization of entanglement under the critical realist view and the further distinction between ontological and epistemological entanglement and, in Sect. 4 we sketch some preliminary ideas for formalizing the agential realist view and we foresee the need to this aim to "turn upside down" the way in which we model in knowledge representation.

2   Ontological Analysis: DOLCE

We present some basic features of DOLCE-CORE, the ground ontology, to show that they allow for keeping track of the rich structure of information in a sociotechnical system. For an introduction to DOLCE-CORE, we refer to [4], here we simply point at the relevant features. The ground ontology is designed to be general and domain independent. This is motivated by the need of a common language to talk about very general properties that are ascribable to entities belonging to different domains.

The ontology partitions the objects of discourse, labelled *particulars* PT into the following six basic categories: *objects* O, *events* E, individual qualities Q, *regions* R, *concepts* C, and *arbitrary sums* AS. The six categories are to be considered rigid, i.e. a particular cannot change category through time. For example, an object cannot become an event. In particular, we shall focus on the following categories.

Objects represent particulars that are mainly located in space, e.g. a screwdriver. On the other hand, *events* have properties that are mainly related to time, e.g., the boarding of flight 717. The relation that links objects and events is the *participation* relation: "an object x participates in an event y at time t" $\mathsf{PC}(x, y, t)$.

An *individual quality* is simply an entity that we can perceive and measure, which inheres to a particular (e.g. the weight of a hammer, the temperature inside waiting room 3...). The relationship between the individual quality and its (unique) bearer is the *inherence*: $\mathsf{I}(x, y)$ "the individual quality x inheres to the entity y". The category Q is partitioned into several *quality kinds* Q_i, for example, color, weight, temperature, the number of which may depend on the domain of application. Each quality kind Q_i is associated to (one or more) *quality spaces* $S_{i,j}$ that provide a measure for the given quality. We say that individual qualities are *located* at a certain point of a space S at time t: $\mathsf{L}(x, y, t)$: "x is the location of quality y at time t".

Spaces allow for evaluating relationships between objects from the point of view of a given quality. For example, "the temperature inside room 3 (q) is higher than the temperature inside room 4 (q')" is represented in the ontology by assuming spaces of values with order relations and by saying that the location of the individual property q is lower than the location of q'. Spaces may be more structured objects and they may be specified along several dimensions[1].

The axioms that define the relationships between individual qualities, locations, and spaces state for example that every individual quality must be located in some of its associated spaces and that the location in a particular space must be unique, cf. [4]. The category of *regions* R includes subcategories for spatial locations and a single region for time, denoted T: $T(x)$ means "x is a time location" (e.g. October 10, 2012, 12:31 PM). The relation $\mathsf{PRE}(x,t)$, where t is a time location, allows to specify that "x is present at time t".

Arbitrary sums AS allow for talking about mereological sums of particulars. We shall apply sums directly when we will approach ontological entanglement.

The category of *concepts* is used in particular to model social objects. Concepts are reified properties that allow for viewing them as entities and to specify their attributes. In particular, concepts are used when the intensional aspects of a property are salient for the modeling purposes. The relationship between a concept and the object that instantiates it is called *classification* $\mathsf{CF}(x,y,t)$ "x is classified by concept y at time t".

We represent the DOLCE taxonomy as a tree (cf. Fig. 1, where we listed only the categories that are relevant for the current argument).

In the next paragraphs, we informally describe three modules: the physical, the mental, and the social module. We present them briefly, in order to use them to exemplify our definitions of entanglement.

2.1 Physical Module

In order to simplify the presentation, we assume that the physical module includes physical objects, such as rocks, chairs, planets. Moreover, we include physical qualities (PQ), i.e. measurable qualities, such as weight, length, or temperature. A category of interest for the present discussion that we list among the subcategories of physical object is that of technical artifact, which includes tools, like for instance screwdrivers. Technical artifacts are of course non-reducible to mere physical objects, in the sense that they have some specific properties that characterize them for what they are that are evidently not physical. For instance, they have a specific function that has been attributed to them by intentional agents [3]. We shall come back to the discussion of artifacts as they provide an interesting general example of entanglement [5].

Other kinds of artifacts can also be viewed as exemplifications of entangled objects, like pieces of arts. Let's take a classical example in philosophy, that of a statue constituted by a lump of clay and its shape: while the clay certainly belongs to the physical realm, the shape could probably be seen as the result of the intentions of the sculptor. Modeling pieces of arts should thus rely both on the physical and on the mental module.

---

[1] Quality spaces are related to the famous treatment of concepts in [15].

A further hint we would suggest is that, under *constructivist* perspectives, also physical qualities could be seen as entangled, given the dependance of their measurement on the attuning of the outcomes of apparatuses and the conventional ascription of values to such outcomes [7]. Under such perspective, modeling measurable qualities involve the physical and the social module. This is not the way in which physical qualities are currently characterized in DOLCE, our claim here is just that entanglement would be required in a constructivist ontology of qualities.

2.2 Mental Module

The mental module includes particulars that are in general ascribable to and dependent on specific individual agents. For instance, beliefs, desires, intentions belong to the mental module of DOLCE. In [12], they are all collected under the category of *computed* objects, to render the idea that they are indirect, depending on other mental objects, and that they are distinct from percepts, which depend on something "external", for instance the physical world, if we take a realist stance.

If we follow this line of thought, we can see that percepts belong to the mental realm, but also depend on the physical one, so in order to model them also the physical module is required.

2.3 Social Module

One predicate that is particularly important for modeling socio-technical systems is the classification predicate: $CF(x, y, t)$, "x is classified as y at time t". By using CF, we can define a special type of social object, namely the notion of *role*, e.g. student, the president of the US. Roles are supposed to be contextual properties, which are characterised by *anti-rigidity* (AR) and *foundational dependence* (FD): roles are concepts that classify entities at a certain point in time, but not necessarily classify them in each moment or each possible world in which they are present (AR) and that require a level of definitional dependence on another property (FD). In this sense, roles are social objects as they are grounded in a sort of Searlian *counts as*.

For instance, someone who is a student at a certain point, not necessarily will be a student all throughout his/her life and there are possible worlds in which he/she is not a student; in order for someone to be classified as an employee, we need someone else who is classified as an employer.

Given these characteristics, roles are essential to model organizations, as they allow to talk about properties that an individual acquires by virtue of the fact that she/he is member of an organization or has some rights/duties connected with the role he/she is playing in that very moment. Moreover, roles may also classify aggregates of individuals: in this case, we can model specific types of roles (e.g. ORG and GRP) in order to model groups and organizations (cf. [6,25]). In DOLCE it is also possible to treat norms and plans, cf. [2].

2.4   Classifying Information into Modules

We have briefly described a number of modules that compose DOLCE. We want now to classify information according to the module it belongs to. We define the predicates PM, MM, SM, which stand for physical, mental, and social module respectively. We thus proceed to classify the elements of the ontology according to their module. For instance, by using the three modules we have specified in the previous section, we can group the particulars of the foundational ontology as follows. For the sake of example, we just propose the following very simple grouping.

a1 $PM(x) \leftrightarrow PO(x) \vee PQ(x)$

a2 $MM(x) \leftrightarrow MO(x)$

a3 $SM(x) \leftrightarrow SO(x) \vee SC(x)$

By the subsumption relation, we can infer for instance that every particular that is below the category of social object is within the social module. For instance, assume that *customer* is a role, that is, it belongs to RL, RL(*customer*). Since RL is included in SC, then also SC(*customer*). Thus, by axiom 3, customer is in the social module.[2] Therefore, the particulars in the ontology can be simply classified according to their module.[3]

3   A Conceptualization of Entanglement

We present now the formal treatment of entanglement from the perspective of critical realism, that is, by assuming that it is possible to provide a prior separation of the social and the material reality. The separation of social and material reality is reflected by the distinction of pieces of information into modules.

For simplifying our presentation, we restrict our formal description of entanglement to objects. We coin here the property of being entangled, or manifesting entanglement, we label it $ent(x)$, and we specify its different types.

a4 $ent(x) \rightarrow O(x)$

In principle, other types of entities, such as concepts, properties, or events, may exhibit entanglement. We leave this for future work.

---

[2] Note that, although we predicate on concepts, e.g. RL(*customer*), all the definitions are in first order logic. As usual in DOLCE, concepts are reified in order to describe them.

[3] In case we want to extend the classification into modules to *propositional* information concerning a specific domain, we need a little more caution. For instance, if we want to say that Mary is a customer at time t, we use the classification relation CF(Mary, customer, t). Which is then the module of the proposition CF(Mary, customer, t)? One way to cope with that is to extend the definition of modules to the predicates in DOLCE and assume that it is the predicate that determines the module.

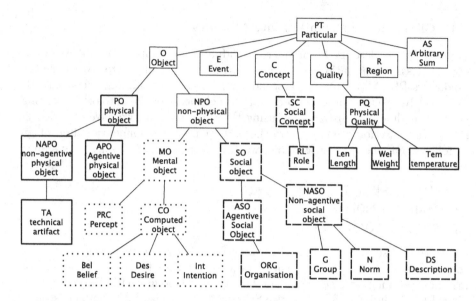

Fig. 1. An excerpt of the DOLCE Ontology. The thick lines indicate the physical module, the dotted lines indicate the mental module, and the dashed lines indicate the social module.

3.1 Ontological Entanglement

A level of ontological entanglement seems to be supported by [1] in terms of "mutual constitution" of entangled entities. We approach the model of this mutual constitution by means of the notion of *grounding* [9,14,20]. Grounding is a relation between particulars that is assumed to be a strict partial order (i.e. transitive and asymmetric). Here, we assume that grounding is defined on objects or on aggregates of objects. In [20], the notion of grounding is related to the possibility of distinguishing different ontological *levels* that arrange the particulars of an ontology into hierarchies.

We say that an object y is grounded (existentially depends on) x and write $x \prec y$ to express that the existence of x grounds the existence of y. We view the concept of grounding as specifying the essential constituents of an object.

We are here assuming that an object may be grounded in the mereological sum of objects, denoted by $x + y$.[4] For instance, a statue existentially depends on the mereological sum of the lump of clay and the shape that is given to it. The example of the statue guides our definition of ontological entanglement. We view a statue as showing the essential dependence on objects coming from different realms. That is, at any time, a statue cannot exist without its amount

---

[4] In [20], grounding is defined only for one-level objects, that is, objects that are on the same level of their parts. By assuming that a merelological sum such as $x + y$ may ground another object, we implicitly depart from [20] (in particular, we are rejecting axiom 14).

of matter and its given shape. Since amounts of matter and shapes belong to different modules, the statue suggests in its nature the entanglement of two objects coming from different realms. By generalizing this example, we define the *ontological entanglement* as follows:

d1 $ent_O(x) \leftrightarrow \exists y, z(y + z \prec x \land M(y) \land M'(z))$

We assume that M and M' label distinct modules.[5] In words, definition (d1) means that an object x manifests entanglement if it is grounded on two objects that belong to different modules.

For instance, one view of technical artifacts is that they are dependent on their physical substratum as well as on the intentional selection of the physical object with the purpose of attributing it a specific property, e.g. a rock becomes an artifact, a paperweight, by selecting it and attributing such function to it [3]. Thus, our approach formalizes the view of the dual nature of artifacts [18]. That is, a paperweight p may be viewed as grounded on a physical object, the rock r, and on a mental object, the designer's goal g: $r + g \prec p$. Therefore, by axiom (d1), we can infer that a paperweight manifests entanglement.

On a closer inspection, the function of an artifact is not a mere mental object, that is, its function has to be sharable in principle by a community of users who acknowledge the physical object as a tool for a given purpose. By elaborating on this point, we can view artifacts as exhibiting entanglement between physical, mental, and social objects, i.e. they are also grounded on socially recognizable functions.

A further important example of ontological entanglement comes from social ontology. By endorsing a multiplicative view, we can see an organization as a distinguished object with respect to the aggregate of individuals that are members of the organization at a given time [25]. Therefore, if we assume that organizations exist as objects in our ontology, we may ask what are their constituents. In [25], we viewed organizations as grounded on aggregates of individuals as well as on normative constraints that bind and specify the organization. In this case, organizations exhibit entanglement of a physically dependent object, the aggregate of physical persons (in the category of AS), and the norms that are in place in the organization.

3.2   Epistemological Entanglement

Epistemological entanglement concerns the possible ways in which an agent knows an object of the ontology. We model the epistemological relationship between an agent and an object by means of the relation of definition between a concept and a description. For the formal details, we refer to [22]. The main difference with [22] is that we apply descriptions to objects and not just to concepts. A description of an object is the way in which the agents may understand the object, may talk about it, may refer to it.

---

[5] We use M and M' just as a shortcut to distinct modules.

a5 $DS(x) \rightarrow NASO(x)$

a6 $US(x, y) \rightarrow DS(y)$

a7 $DF(x, y) \rightarrow (C(x) \vee O(x)) \wedge DS(y)$

a8 $DF(x, y) \wedge DF(x, z) \rightarrow y = z$

Descriptions are non-agentive social objects, (a5). By making them social objects, we assume that their meaning is shared among agents. Thus, descriptions are not private mental entities, they are accessible by communities of agents. Axiom (a6) requires some explanation. Descriptions may only use concepts in [22]. In general, a description may *use* heterogenous particulars, such as other objects, events, properties, etc. For instance, I may define the number 2 by means of the description "the successor of 1" that uses another object, i.e. the number 1, and a property, "being the successor of". We are not claiming that a description "contains" those particulars, we simply exploit the generality of the relation *use* in order to avoid a demanding characterization of descriptions. For instance, the description of a physical object may appeal to physical properties or to other physical objects. That does not mean that an abstract object such as the description contains physical objects. Moreover, a description of a social object such as a contract may use the individuals that are bounded by the contract, the roles that the individuals may play, the actions that each agent is bounded to perform, and so on. That is why in axiom (a6) we do not restrict the type of entities that may be used in a description. This is due to the quite abstract nature of *use*.[6] By (a7), we say that a concept or an object may be *defined by* a description. For the sake of simplicity, we assume that there is only one description of an object, (a8).[7] By assuming a single description of objects and concepts, we are implicitly assuming that descriptions completely specify their objects, that is, we are not including partial descriptions.

We are ready now to present the definition of epistemological entanglement.

d2 $\mathsf{ent}_E(x) \wedge DF(x, y) \rightarrow \exists v, w \ US(v, d) \wedge US(w, d) \wedge \mathsf{M}(v) \wedge \mathsf{M}'(w))$

In words, an object manifests epistemological entanglement iff the description that defines it uses concepts (or objects, qualities, etc.) belonging to different modules. An example of epistemological entanglement is given by the objectification of a role. Assume that in our ontology roles, such as student, are distinct from the persons who are classified as students, that is roles are instantiated as particulars in the ontology. A description of a student then must refer to its physical substratum as well as to the normative features involved in the social concept of student.

It is possible to show that the ontological entanglement entails the epistemological one. Intuitively, if an object essentially depends on objects in different

---

[6] One way to cope with this problem is to view descriptions as only composed by *concepts* and to assume that DOLCE contains concepts for any particular, i.e. concepts of individuals, physical objects, properties, n-ary relations, qualities, etc.

[7] We view this assumption as a mere simplification move. In general, one may think about many descriptions of the same objects and discuss whether they are equivalent.

realms, then its definition should use elements coming from different modules. We can formalize this intuition by means of the following axioms.

a9 $M(x) \wedge DF(x, d) \rightarrow \exists y \, US(y, d) \wedge M(y)$

a10 $x_1 + \cdots + x_m \prec x \wedge DF(x, d) \wedge DF(x_1, d_1) \wedge \cdots \wedge DF(x_m, d_m) \wedge US(y_1, d_1) \wedge$
$US(y_m, d_m) \rightarrow US(y_1, d) \wedge \cdots \wedge US(y_m, d)$

Axiom (a9) states that if an object belongs to a certain realm, then its description must use at least one concept from that realm. For instance, a description of a physical object may refer to its weight, its colors, etc. This does not mean that the description must use only concepts from that realm. For instance, a description of my pen may be "the pen my father gave me for my birthday", that includes a physical property, i.e. being a pen, as well as social concepts, i.e. presents and birthday.

Axiom (a10) states that if an object x is grounded on other objects $x_1, \ldots,$ x_m each defined by its respective description, then a description of x must include elements from those descriptions. For instance, the description of a statue shall use concepts referring to its matter and to its shape. Axiom (a10) makes sense if we, as we do, assume complete descriptions. In case of partial descriptions, for instance, one may partially define a statue by mentioning its shape. Assuming the hypothesis that there is in our ontology a description of x, by (a9) and (a10), we can infer that:

t1 $ent_O(x) \wedge \exists d \, DF(x, d) \rightarrow ent_E(x))$

The assumption concerning the existence of a definition of x is redundant if we assume that every object in the ontology must be defined by a description.[8] That (t1) follows can be easily shown. Assume that d defines x. Then, since x is ontologically entangled, by definition (d1), x is grounded on at least two objects y and z belonging to different modules. By axiom (a10), the description of x must use elements from the descriptions of y and z. By axiom (a9), such elements belong to different modules, so by definition (d2) we conclude. Therefore, if we assume that every object of the ontology is definable, ontological entanglement entails epistemological entanglement. Note that (t1) also depends on the fact that we are excluding partial descriptions. In case our (unique) description of an entangled object is partial – for instance, it refers only to a specific realm – we would have ontologically entangled objects that do not exhibit epistemological entanglement.

Viceversa, we ask whether epistemological entanglement entails ontological entanglement. Formally this is not the case, as we have no means to infer that an object is grounded on other objects. We believe that this is conceptually correct and it may account for the following case. Suppose we do not want to make a distinction in our ontology between actions and events, that is, in

---

[8] Ontologies are here motivated by knowledge representation, thus it is reasonable to assume that every object has a description. However, it is up to the modeler whether to make this choice.

our ontology, actions are of the same type as events. Even if it is so, actions can be conceptually separated from events because they have to be described as intentional. This is in a nutshell Davidson's point [10]. Hence, although we do not view actions as separated with respect to events – and accordingly we assume that actions do not exhibit ontological entanglement between a physical and a mental realm – any possible description of actions must include elements coming from the intentional, mental, or social module. Thus, actions do exhibit epistemological entanglement.

4 Conclusion: Towards Agential Realism

We have seen that by presupposing that information can be divided into modules, it is possible to provide an ontological model of entanglement as a property of objects: entangled objects are those that manifest properties belonging to different modules. We conclude by providing a few elements for approaching an agential realist position on the distinction between social and material. By rephrasing agential realism within our framework, there is no separation of objects into modules that is prior to the act of a knowing subject. Therefore, an ontological separation into modules appears illegitimate from the perspective of agential realism. On a closer inspection, however, there is indeed a way to express agential realism by means of DOLCE. One of the main motivations of DOLCE is to provide a clear representation of how a *cognitive* agent looks at the reality. Therefore, DOLCE can be viewed as the result of the categorization made by a knowing agent of a certain domain of interest. From this point of view, then, we can justify the separation of information into distinct modules of DOLCE as the result of an "agential cut" [1] that a knowing agent has performed on a domain of interest. The challenge of representing agential realism from the point of view of the ontological modeling is to capture and model how a knowing agent performs the categorization of reality into modules. That is, the challenge is to model the very notion of the agential cut.

The general idea of an ontological model of agential realism is based on an inverse process with respect to the one we have depicted in Sect. 3: instead of starting from an already given separation into modules and then defining entanglement, we need to start from a general notion of possibly entangled object and retrieve the modules as the agent's categorization choices.

We can use our previous definition of ontological entanglement (cf. Definition (d1)) to give at least an idea of a possible modellization. The intuition leading our view is that information is going to be separated into modules, whenever an agent needs to express properties of an object that she/he is viewing as incompatible. If there is no object that may in principle share incompatible properties, there is no need for separating pieces of information into distinct modules. Thus, suppose that an agent is assuming that an object a may have two incompatible properties $P(a)$ and $Q(a)$. The agent now can decide whether the incompatibility lies at an ontological or at an epistemological level, that is, whether P and Q represent incompatible *constitutions* of a (ontological entanglement) or they are prone to represent incompatible *descriptions* of *a* (epistemological entanglement). By using definitions (d1) or (d2), the agent is led to separate the properties P and Q by associating them to distinct modules. In the situation we have described, modules are not given, they are *constructed* starting from the agent's view of incompatibility of pieces of information. Building on that, we can also consider as an important challenge for multiagent systems the problem of coordinating different conceptualizations given by different agential cuts.

This simplified example of how to define modules as the outcome of an agential cut of reality poses a fundamental challenge in knowledge representation and ontology modeling and it is the aim of our future work. By taking seriously the motto "there is no social that is not also material, and no material that is not also social" [24, 1437], entanglement is no longer a property of some entity of the ontology, rather it becomes the designing principle for approaching the general construction of an ontology. According to this view, the task of a foundational ontology is not to separate the domain of discourse into rigid categories, relations, and modules that group agent-independent entities, it rather has to model the relationship between agents who focus on aspects of reality and the entities that are postulated or produced by the agents' activities. Accordingly, categories and modules become features of entities that may be intentionally selected, rather than rigid categorizations of the domain of discourse. In order to approach the development of such a view, that is a subject aware ontology, we need to embrace a constructivist view of entities and model their production. This is of course a long term plan, however we believe that it is worth pursuing.

References

1. Barad, K.: Meeting the Universe Halfway. Qunatum Physics and the Entanglement of Matter and Meaning. Duke University Press, Durham (2007)
2. Boella, G., Lesmo, L., Damiano, R.: On the ontological status of plans and norms. Artif. Intell. Law **12**(4), 317–357 (2004)
3. Borgo, S., Franssen, M., Garbacz, P., Kitamura, Y., Mizoguchi, R., Vermaas, P.E.: Technical artifacts: an integrated perspective. Appl. Ontol. **9**(3–4), 217–235 (2014)
4. Borgo, S., Masolo, C.: Foundational choices in DOLCE. In: Staab, S., Studer, R. (eds.) Handbook on Ontologies, 2nd edn. Springer, Heidelberg (2009)
5. Borgo, S., Porello, D., Troquard, N.: Logical operators for ontological modeling. In: Proceedings of the Eighth International Conference on Formal Ontology in Information Systems, FOIS 2014, Rio de Janeiro, Brazil, 22–25 September 2014, pp. 23–36 (2014)
6. Bottazzi, E., Ferrario, R.: Preliminaries to a DOLCE ontology of organizations. Int. J. Bus. Process Integr. Manage. **4**(4), 225–238 (2009)
7. Bottazzi, E., Ferrario, R., Masolo, C.: The mysterious appearance of objects. In: Donnelly, M., Guizzardi, G. (eds.) Seventh International Conference on Formal Ontology in Information Systems, FOIS 2012, pp. 59–72 (2012)
8. Clancey, W.J., Sachs, P., Sierhuis, M., Hoof, R.V.: Brahms: simulating practice for work systems design. Int. J. Hum.-Comput. Stud. **49**(6), 831–865 (1998)
9. Correia, F., Schnieder, B.: Metaphysical Grounding: Understanding the Structure of Reality. Cambridge University Press, Cambridge (2012)

10. Davidson, D.: Essays on Actions and Events. Oxford University Press, Oxford (1980)
11. Emery, F.E., Trist, E.: Socio-technical systems. In: Churchman, C.W., Verhust, M. (eds.) Management Sciences: Models and Techniques, vol. 2, pp. 83–97. Pergamon, Oxford (1960)
12. Ferrario, R., Oltramari, A.: Towards a computational ontology of mind. In: Varzi, A.C., Vieu, L. (eds.) Proceedigs of the International Conference on Formal Ontology in Information Systems, FOIS 2004, pp. 287–297. IOS Press (2004)
13. Ferrario, R., Prévot, L.: Formal ontologies for communicating agents. Appl. Ontol. **2**(3), 209–216 (2007)
14. Fine, K.: Guide to ground. In: Correia, F., Schnieder, B. (eds.) Metaphysical Grounding: Understanding the Structure of Reality, pp. 37–80. Cambridge University Press, Cambridge (2012)
15. Gärdenfors, P.: Conceptual spaces - the geometry of thought. MIT Press, Cambridge (2000)
16. Ghidini, C., Giunchiglia, F.: What is local models semantics? (Chapter 2). In: Bouquet, P., Serafini, L., Thomason, R.H. (eds.) Perspectives on Context, pp. 19–42. CSLI Publications, Stanford (2008)
17. Joslyn, C., Rocha, L.M.: Towards semiotic agent-based models of socio-technical organizations. In: Sarjoughian, H.S., et al. (eds.) Proceedings of AI and Simulation 2000, pp. 70–79 (2000)
18. Kroes, P.: Engineering and the dual nature of technical artefacts. Camb. J. Econ. **34**(1), 51–62 (2010)
19. Leonardi, P.M.: Theoretical foundations for the study of sociomateriality. Inf. Organ. **23**, 59–76 (2013)
20. Masolo, C.: Understanding ontological levels. In: Proceedings of the 12th International Conference on the Principles of Knowledge Representation and Reasoning (KR 2010) (2010)
21. Masolo, C., Borgo, S., Gangemi, A., Guarino, N., Oltramari, A.: Wonderweb deliverable d18. Technical report, CNR (2003)
22. Masolo, C., Vieu, L., Bottazzi, E., Catenacci, C., Ferrario, R., Gangemi, A., Guarino, N.: Social roles and their descriptions. In: Proceedings of the 6th International Conference on the Principles of Knowledge Representation and Reasoning (KR 2004), pp. 267–277 (2004)
23. McCarthy, J.: Notes on formalizing context. In: Proceedings of the 13th International Joint Conference on Artifical Intelligence, IJCAI 1993, vol. 1, pp. 555–560. Morgan Kaufmann Publishers Inc., San Francisco (1993)
24. Orlikowski, W.J.: Sociomaterial practices: exploring technology at work. Organ. Stud. **28**, 1435–1448 (2007)
25. Porello, D., Bottazzi, E., Ferrario, R.: The ontology of group agency. In: Proceedings of the Eighth International Conference on Formal Ontology in Information Systems, FOIS 2014, Rio de Janeiro, Brazil, 22–25 September 2014, pp. 183–196 (2014)

When owl:sameAs isn't the Same Redux: Towards a Theory of Identity, Context, and Inference on the Semantic Web

Harry Halpin[1]([⊠]), Patrick J. Hayes[2], and Henry S. Thompson[3]

[1] W3C/MIT, 32 Vassar Street, Room 32-G515, Cambridge, MA 02139, USA
hhalpin@w3.org
http://www.ibiblio.org/hhalpin
[2] Institute for Human and Machine Cognition,
40 South Alcaniz Street, Pensacola, FL 32502, USA
phayes@ihmc.us
http://www.ihmc.us/groups/phayes/
[3] School of Informatics, University of Edinburgh,
10 Crichton Street, Edinburgh EH8 9AB, UK
ht@inf.ed.ac.uk

Abstract. The Web changes knowledge representation in a number of surprising ways, and decentralized knowledge representation systems such as the Semantic Web will require a theory of identity for the Web beyond current use of *sameAs* links between various data-sets, including the fact that entities can not be linked across semantic roles. We empirically analyze the behavior of identity and inference on the Semantic Web currently in order to analyze the size of the problem. Lastly, some of the problem due to identity statements operating over different domains of discourse, and propose a modest extension to RDF (RDFC, or RDF with Contexts) that can formally distinguish different contexts in RDF, including contexts for identity statements.

Keywords: Semantic Web · Identity · Context · Pollarding

1 Introduction

With the beginning of the deployment of the Semantic Web, the problem of identity in knowledge representation – sometimes assumed trivially solved – has returned with a vengeance. In traditional logic names are arbitrary strings, but by definition, one criterion that distinguishes a Web logic from traditional logic is that in Web logic names are URIs (Uniform Resource Identifiers, such as http://www.example.org/) (henceforth abbreviated as *ex:*). On the 'actually existing' deployed Semantic Web known as Linked Data, an URI not only can refer to things, but can be accessed, so that accessing *ex:Paris* in a Web browser results in receiving a bundle of statements in a Web logic such as RDF (Resource Description Format, where statements are composed of 'triples' of

© Springer International Publishing Switzerland 2015
H. Christiansen et al. (Eds.): CONTEXT 2015, LNAI 9405, pp. 47–60, 2015.
DOI: 10.1007/978-3-319-25591-0_4

subject-property-object names). When users of Semantic Web find another URI about their item of interest, they connect their two different URIs with an *identity link*. This link in practice is given by a *owl:sameAs* property, henceforth just called *sameAs* [11].

Yet the results of inference using *sameAs* are often surprising, either 'smushing' up distinct individuals or failing to 'smush' individuals due to their semantic role. One solution would be to forget the semantics of the Semantic Web entirely, treating *sameAs* as just an English language mnemonic. Those that do not remember history are doomed to repeat it; for it was precisely the forgoing of logical semantics that led to semantic networks having their infamous crisis over divergence in meaning in *IS-A* links [3]. One can only imagine that such problems would only increase given the purported global scale of the Semantic Web. We point out that a new formal semantics is needed for identity statements if Semantic Web inference is going to succeed in deployment. While there has been deployment of Linked Data, use of inference 'in the wild' across data-sets is nearly non-existent. In Sect. 2 we explicate the concept of identity in formal semantics, and show that equality across different semantic roles can be implemented formally using a technique known as 'pollarding.' We empirically analyze identity links on the Semantic Web, with a focus on *sameAs*, in Sect. 3. An experimental investigation of whether real identity links and their concomitant inferences are accurate or not is presented in Sect. 4. After analyzing the results, it appears that much of the variance between identity is due to domain-specific contextual uses of identity. Finally, the use of RDF with contexts is proposed to allow RDF statements to semantically define contextual identity.

2   Identity in Logic

2.1   Identity in Terms of Logical Equality

Identity is logically defined as equality, $A = A$. The basic relation of identity holds only between a thing and itself. So, it is important not to think of an equality sentence such as $A = B$ as saying that there are two things, A and B, which are equal. This is a category error of believing that two separate *names* A and B in the syntax can be confused with the *things* in the universe of discourse of the semantics. Properly, $A = B$ says that there is one thing which has two names, 'A' and 'B' respectively. To say A equals B, $A = B$, is to say that A is identical with B, so that if $A = B$ is true then the set $\{A, B\}$ is identical to the set $\{A\}$ in the semantics with $A = B = [\text{some individual in the semantics}]$. This means that all notions such as 'approximately equal' or 'equal for some purposes' or 'equal with respect to certain facets,' etc., are not the same as being logical *equal*, and semantic identity is defined in terms of logical equality. All of these other kinds of relations require two entities to be at least logically distinguishable in order to be not equal for all purposes, such as being not equal with regards facets and the like. All such 'nearly equal' notions are relations between two things, and so are fundamentally different from logical equality.

A hallmark of logical equality is that anything said about a thing using one of its names should be just as true when said using the other name. Formal semantics usually operates entirely in terms of reference to a model, and mapping names within that model. For Frege and generations of philosophers after him, while two names may be about the same thing and yet these two names have different properties [6]. For Frege, there are different senses of 'Morning Star' and the 'Evening Star,' despite the fact they refer to the same star [6]. This kind of concept of sense as separate from reference applies to natural language – including its myriad attendant social, epistemological, and ontological issues – rather than the more barebones world of logic, where formal semantics is defined in terms of mathematically defined reference to a model rather than more encompassing and informal notion of sense. This is not to say that some kind of cognitive meaning or sense does not exist, but this kind of analysis is not currently bound formally to logical semantics.

To restate, anything logical that can be said about a thing using one of its names should be just as true when said using the other name. If some sentence with a syntax $...A...A...A...$ containing A is true, and $A = B$ is true, then the same sentence replacing 'A' with 'B' syntactically one or more times must also be true, i.e. $...B...A...B...$ should be true. This is the basic inference rule, substitution, associated with equality. Notice that this rule makes no mention of how the name 'A' is used in the sentence, of its semantic role. For true equality, substitutivity should apply to names *regardless of their semantic role* (i.e. what it is that A denotes, be it a function, a class, a proposition, an individual), since the intuitive argument for the validity of the rule is semantic: the meaning of a sentence is determined by the things referred to by the names in the sentence rather than the names themselves. Since logical sentences are *de re* (about the thing), it follows that the choice of which name to use to denote a thing must be irrelevant to the truth of the sentence.

2.2 Punning

In terms of logical identity, there are essentially two kinds of semantics so far proposed for as logics for the Web: logics such as OWL DL (Description Logic)[11] and OWL2 [10] that only allow substitivity in the same semantic role and logics such as RDF [9] and ISO Common Logic [4] that allow substitivity regardless of semantic role. One can think of logics as either partitioning the universe of discourse into disjoint sets of things or having *all* things in a single universe of discourse. Languages that partition the universe of discourse must have separate equality operators for each kind of thing in the universe of discourse, so OWL has *owl:sameAs* for individuals, *owl:equivalentClass* for classes, and *owl:equivalentProperty* for properties, although the later are not heavily used in practice [11]. It is incorrect to use *sameAs* across a class and an individual in OWL DL, and if one does so, one immediately falls into OWL Full. This semantic partitioning is reflected then in the partitioning of names in the language itself into lexical roles. So a name in OWL DL had always to refer to either properties, classes, or individuals, but identity can never be made between different semantic roles.

This inability to have the same name used across different semantic roles was corrected in OWL2 by the introduction of 'punning.' [10] This refers to a technique for reducing the number of lexical categories used in a formalism by allowing a single name to be used in a variety of semantic roles. For example, allowing a property like *ex:Latitude* to serve as *both* the subject and property of a statement, an important use-case for meta-modelling. The language may allow punning between class and property names only if the syntax unambiguously assigns a class or property role to every occurrence of a name. This ensures that any language which uses either of these techniques can - in principle - be replaced by an equivalent language which does not use it by replacing each occurrence of a punned name by an alternative name of a recognizable type which is reserved for use in that particular semantic role. We will call such a language *segregated*. Conventional textbook accounts of logic usually define segregated languages. For example, a conventional syntax for FOL distinguishes $1 + 2\omega$ categories of names, with the roles being individual, function of arity n and relation of arity n, which map respectively to elements of the universe, n-ary functions over the universe and n-ary relations over the universe. Only the first kind of name can be bound by a quantifier, since FOL allows quantification only over elements of the semantic universe, which is required to not contain relations and functions.

Punning works by allowing the name categories to intersect (or simply to be identified) while retaining the conventional interpretation mappings. This amounts to the use of multiple interpretation mappings between names and the semantics. Each name is given several denotations in an interpretation, and the immediate syntactic context is used to 'select' which one of these to use in the semantic truth recursion. Thus equality statements across all different kinds of lexical roles can be made, and the name will be given an interpretation that fits into the necessary semantic role. If a name is used in multiple semantic roles, then the name will denote a different thing in each semantic role.

2.3 Pollarding

'Pollarding' refers to a different, but closely related, technique which is used in various forms within the semantics of RDF and RDFS [9], OWL Full [11], and ISO Common Logic [4]. Pollarding retains the single denotation mapping of a conventional interpretation, but treats all names as denoting individuals, and the other semantic constructions are related to individuals – not names! – by extension mappings, which are treated as part of the interpretation. Just as with punning, the immediate syntactic context of a name is used to determine whether it is intended to be interpreted as the immediate denotation (that is, an individual) or one of the 'extensions' associated with that individual. Therefore any name that is syntactically declared equal to another can be substituted across all sentences while maintaining the truth value of the sentences, and unlike punning, in any interpretation both names denote the same individual, albeit the same individual may have multiple extension mappings.

The two schemes are illustrated in Fig. 1. Figure 2 illustrates how the two schemes reduce to a classical model theory when the language is, in fact, segregated. The case of punning is trivial: the extra mappings are simply ignored; and given any classical interpretation, one can create a punning interpretation simply by adding the extra mappings in some arbitrary way. Pollarding is a bit more complicated. The classical mappings for non-individual names are created by composing the interpretation and extension mappings. In the other direction, given a classical interpretation, one has to select an individual in the universe to be the 'representative' of each non-individual semantic entity, map the name to it, and assign the individual to be its appropriate extension. One way to do this is to use a Herbrand-style construction by putting the name itself into the universe, use the classical interpretation mapping as the extension function, and treat the name as denoting itself.

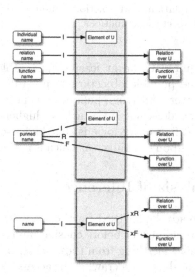

Fig. 1. For an individual: lexical and semantic segregation (top), punning (middle), and pollarding (bottom).

As this shows, the pollarding technique does impose one requirement on the classical interpretation, viz. that its universe is large enough. If the cardinality of a classical universe is smaller than the number of non-individual (class, relation, property or function) semantic entities required to interpret the language, then that universe cannot be used as the basis of a pollarded interpretation of the same language. An example illustrating this is $\forall x.y.(x = y)$, $P(a)$, and $\neg Q(a)$. This set of sentences is satisfiable, but cannot be satisfied in a pollarded interpretation since the two relations P and Q must be distinct, but there is only one entity in the universe, which is therefore too small to provide the distinct relational extension mappings required. This is only a concern in cases where the universe can be consistently made small. Languages such as RDF and Common Logic,

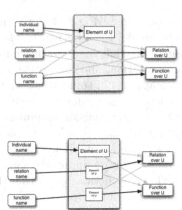

Fig. 2. For an individual, relation, and function: punning (top) and pollarding (bottom). Grey arrows are interpretation functions.

whose universes are required to be at least denumerable, will always have a sufficient cardinality. One possible objection to pollarding is that it seems to make the logic 'higher-order.' Although pollarding is not used in textbook first-order logic [1], pollarding does not make a logic higher-order. The difference between higher-order and first-order logic is that higher-order logics all impose comprehension principles upon the universe, which is not necessary in pollarding.

3   Empirical Analysis of Inference

However, *sameAs* is already being used 'in the wild' in Linked Data, and an inspection of its behavior is in order before considering improving its semantics. First, can one actually infer anything from these *sameAs* links on Linked Data? To test this, we used as our data-set a crawl of the accessible Linked Data Web, so that our results would be an accurate 'snapshot' of the use of *sameAs* on the Web. This resulted in 10,850,606 *sameAs* statements being found (forming 7,229,140 equivalence classes), with the average number of URIs in an equivalence class having 2.00 *sameAs* statements. No blank nodes and 9 literals were found in the set of equivalence classes (the literals from the treatment of a URI as a string in RDF), and the total number of distinct URIs was 14,461,722. The largest equivalence class explicitly declared was of size 22, the particular case being an equivalence class for Semantic Web researcher Denny Vrandenic. The number of *sameAs* statements substantially outweigh *owl:equivalentProperty* (1,451 occurrences) and *owl:equivalentClass* statements (106,305) on the Semantic Web. Do any of these statements violate the segregation of individuals, classes, and properties found in OWL? Our sample showed that only a small number (3,636) of them did.

While it appears small, the misuse of *sameAs* will impact inference. Then we ran *sameAs* reasoning that purposefully ignored segregation between classes,

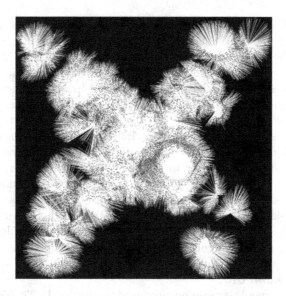

Fig. 3. Visualization of top 40 *sameAs* equivalence classes.

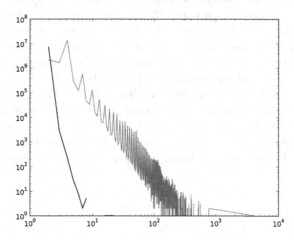

Fig. 4. Size of Equivalence class vs. frequency: No inference (blue) and inferred classes (green) on a log-log axis (Color figure online).

relations, and instances over the set of statements containing at least one *sameAs* statement. These produced, excluding symmetric statements (which would trivially create a 'double' of every statement), 60,045,705 additional inferred *sameAs* statements with 79,276,264 distinct URIs. The 40 largest equivalence classes of (or closures over) *sameAs* connections (via their connection to a single 'pivot' URI) are visualized in Fig. 3. This shows the range of sizes varies dramatically, with the largest closure containing 4,177 URIs, seemingly consisting of all sorts of biomedical data and *UniProt*, a massive protein knowledge-base. As the number of RDF-enabled biomedial databases surely does not measure in the thousands, one wonders if at some point the *sameAs* chain got out of control and distinct drugs are being declared logically equal. Upon closer inspection, one finds that in this massive equivalence class very different things - which break OWL-style segregation semantics - have been connected by *sameAs*, such as 'class of drugs' and an 'individual instance of that drug applied' in open Linked Drug Data. *UniProt* is not the only one; the *bio2rdf* Semantic Web triple-store produces a *sameAs* equivalence class of size 1,367. Even the ubiquitous DBPedia, a Semantic Web-export of Wikipedia, produces a closure of size 367 around a French verb conjugation, mixing up the class of French verbs in WordNet and the class of actions those verbs are about in Cyc. Interestingly enough, there are also occurrences of 'semantic' spam, with organizations such as the New York Times whole-scale copying *Freebase*-produced RDF to their own URIs (probably an innocent move), and the possibly less innocent copying of RDF data from diverse sources by rdfabout.com. 'Semantic' spam is simply the use of *sameAs* links and copying of RDF data to other data-sources in order to increase the visibility of one's own URI.

In total, the *sameAs* reasoning produced a total of 19,230,559 equivalence classes with an average content of 4.16 URIs. This is an approximate double in size over the average size of explicitly declared *sameAs* equivalence classes. However, this interpretation assumes the size of equivalence classes is normally distributed, which it is not, as illustrated in Fig. 4 (log-log scale). The inferred *sameAs* distribution is clearly skewed towards smaller equivalence classes, but the mass of the distribution continues out towards the hundreds. Again, view Fig. 3 for their visualization. Due to this, after inference the size of the *sameAs* statements that violated the segregation semantics increased to 893,370, a small but statistically significant part of the Semantic Web. Upon closer inspection, the nature of this increase seems to be due to having some *sameAs* statements across segregated semantic roles being caught in a few of the larger equivalence classes.

4   Experiment

Another question would be, how many of these inferred *sameAs* are actually correct and obey the semantics of identity? A set of 250 *sameAs* triples (an identity link between two URIs) were chosen at random (with the chances of being chosen at random had been scaled down by the logarithm of the frequency of their domain name, in order to prevent a few major 'hubs' from dominating the entire data-set) from the previously described sample of the Semantic Web that violated the segregation semantics of *sameAs* (i.e. had an individual being *sameAs* with a class, for example). The closures of these 250 triples were generated, and from each of these about half (132) had at least two inferred identity links via transitivity. After removing 2 triples for training purposes, the remaining 130 triples were used for our experiment. Five judges (each Semantic Web experts and computer scientists) were chosen. A standard sample of properties and object (URI and literals) were retrieved from each URI and displayed to the user in a table. The judge was given a three-way forced choice, categorizing the statement as *Same*: if it was clearly intended to identify the same single thing, *Related*: if it was descriptions of fundamentally different things which none-the-less have some important properties in common, or *Unrelated/Can't tell*: if neither of the above, or one can't tell. We merged the 'unrelated' and 'can't tell' cases based on earlier work [8], which suggested that trying to make a distinction between the two cases was unreliable insofar as it seemed the difference was based mainly of personality rather than domain expertise, with some rushing to judge something different while others wanting more information.

There was also a checkbox next to each property, as well as one for each URI itself, and the judges were instructed to check the properties that were useful in forming their judgment. Overall, the judges had only slight agreement, reaching a Fleiss's κ of 0.06, which shows that the scheme is 'slightly' reliable. Looking at the distribution of judgments suggests more consistency than that, however. Figure 5 shows that more than half the time a 'consistent' judgment (unanimous, or only one disagreement) was made, and that consistent judgments of *same* and *different* are the most common. Using voting, we can determine that 52 (22 %) are the same, 36 (28 %) are related, 22 (17 %) are different or can't tell, and 20 are ties (15 %). Inspecting the useful properties triples judged as related, on an average a human judge chose 5.08 (S.D. 2.23) properties as important. This was primarily because some judges choose almost all properties to be relevant, while others would choose only a few.

A histogram showing the spread of the three categories is given in Fig. 6. Note that there is often disagreement in any of the three classes, but that there is also usually a clear majority, so an approach based on the 'wisdom of crowds' seems to handle identity judgments in a more consistent way than individual experts. According to majority-wins voting (with 4 ties removed), there were 68 identical (53 %), 34 related (27 %), and 24 (19 %) unrelated or unable to tell. Inspecting the useful properties triples judged as related, on an average a human judge chose 5.08 (S.D. 2.23) properties as important. This was primarily because some judges choose almost all properties to be relevant, while others would choose only a few.

5   Discussion

Surprisingly enough, the results with inference showed that the task itself with inference is significantly *hard* for individual experts. In fact, in comparison with the 'moderate' κ values from previous studies done by others [8], inference seems to make experts less reliable. Simply, inference forces individual experts to confront edge-cases produced by the inference chains where issues such as those caused by the semantics of *sameAs* cause real problems. However, groups of experts, despite the low aggregate reliability, could on individual cases discriminate via voting, and so could determine cases of mistaken identity via an 'identity leak' caused by a misplaced *sameAs*. In fact, compared to randomly selected

n	j1	j2	j3	j4	j5
23	**2**	**3**	**3**	**3**	**3**
17	**3**	**3**	**3**	**3**	**3**
14	**1**	**1**	**1**	**1**	**1**
11	1	2	2	2	3
9	2	2	2	3	3
8	1	2	2	3	3
7	1	1	2	2	3
6	**2**	**2**	**2**	**2**	**2**
5	**1**	**2**	**2**	**2**	**2**
5	1	1	2	3	3
4	2	2	3	3	3
4	1	2	3	3	3
4	**1**	**1**	**1**	**1**	**2**
3	1	3	3	3	3
3	1	1	2	2	2
2	1	1	1	2	2
2	**1**	**1**	**1**	**1**	**3**
1	1	1	3	3	3
1	1	1	1	3	3
1	1	1	1	2	3

Fig. 5. Distribution of Human Ratings per Judgment: 3 = same, 2 = related, 1 = not related or can't tell; 'Consistent' judgments are bold

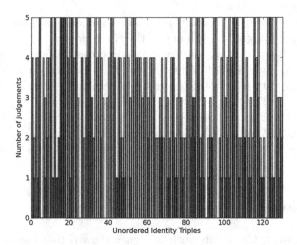

Fig. 6. Histogram of Human Ratings per Judgment: Same (green), Related(white), Not the same and can't tell (red) (Color figure online)

triples without inference [8], the inferred triples are less likely to be correct (40 % compared to 53 % after inference), slightly more likely to judge them as merely related triples (21 % compared to 28 % after inference), and a decrease in the amount of triples thought to be incorrect (29 % compared to 17 %). This decrease in the amount of incorrect is likely due to the triples being tied between two categories, which was 20 %. When this is taken into account as likely "can't tell" it appears that the number of incorrect triples may go up as high as 37 %. So in general, it seems that 'identity leaks' of misused *sameAs* statements cause the reliability of some identity statements to lower somewhat, but still remain broadly comparable with the kinds of human judgments not having inference produced.

What is also interesting is precisely when an approach based on the 'wisdom of crowds' can help disambiguate difficult results even when individual judges have difficulty. For example, a voting method could be deployed to solve *sameAs* ambiguities. Not all results are difficult: for geographical entities and individual people such as 'Lower Austria' and 'Johnny Bristol' agreement that they did not represent the same entity in the experimental data was generally high. Also, the results of incorrect inferences could also be easy to determine, such as an equivalence between a computer science researcher Tom Heath and a 'word-paving machine' or Le Flore County and the Marburg Virus in the experiment.

However, where there was more likely to be confusion was very closely related things such as the relationship between the concept of 'scoring an ace' in Cyc and the verb 'acing' in WordNet that denotes that concept, or the relationship between the Maranthi language and international code for that language. What should be done about this weaker notion of 'being related' that *sameAs* is erroneously used for? Indeed, one would be tempted to say that such a practice is simply wrong, and a virtually inference-free alternative such as *rdfs:seeAlso* or a *SKOS* (a Semantic Web vocabulary for thesauri) construct should be used for this weaker notion of "being related" where no formal inferences should be defined.

More troublesome is cases where it seemed some degree of context marked things as logically equal, such as the mereological relationship between the town of New Canaan and the South School therein. Then there were topics that were difficult due to background knowledge where classes and individuals were denoted to be the same, such as whether Francisco Franco was the ruler of Spain, which is obviously true in some temporal context but not true today. This broke OWL *sameAs* semantics but seems to be contextually true in some contexts. A closer look at the experimental data shows that what is happening is stronger than just 'being related,' as two names have *some* relationship that means they should be considered logically identical *within a given context*. So, given some social context, what appears to be two distinct entities in another context may indeed 'be the same thing' and so share all the same properties. The ruler of Spain is indeed the same individual as Franco at a given time, but in other contexts may range over other individuals or be a single other individual. In some geographical contexts, a country like Spain may be identical with some spatial territory, but in another context one may be referring to a larger cultural or administrative nexus. The key to cutting this Gordian knot is proper treatment of contexualized relatedness in terms of semantics, which requires serious domain-specific semantic treatment. The hard trick is to determine how to define domain-specific contexts in terms of formal semantics.

6   RDF Contexts and Identity

One way out of this is to add contexts to RDF in terms of formal semantics. The vision of RDF as a context logic is still being developed and numerous proposals have been put forward [2], but nonetheless a straightforward way to add 'web contexts' to RDF has been proposed. A *web context* represents a social agreement concerning the meaning of a vocabulary of URIs, which is called the reserved vocabulary of the context. Asserting a graph in a context means that one is a commitment to use the reserved vocabulary in a way that conforms to the agreement. This agreement may be explicit or implicit. RDF can be considered to be a sort of topmost context, called *rdf:*, which defines the RDF namespace as defined by the 2004 RDF specification documents [9]. This is a default, so all existing RDF graphs are understood to be asserted in it unless otherwise explicitly denoted. The other extreme is to assert a graph in itself should be considered as a context and so not share semantics with any other graphs. This effectively declares all its non-reserved URIs as reserved, and hence separates them in meaning from the same URIs used outside the graph, forbidding the usage of *sameAs* between data-sets.

There is a large middle-ground, where reserved URIs can be added that impose different semantic conditions on previous contexts or propose new semantic conditions on new reserved vocabulary terms. While a detailed working out of this is forthcoming, what is needed is a way to assert a graph in a context, a way to specify the reserved vocabulary of a context, a way to describe the semantic conditions imposed on the reserved vocabulary by the context, and a way to assert that one context inherits another. As explained in more detail elsewhere,[1] a new term called *rdf:inherits* could both indicate a graph is explicitly in a given context and import other contexts to the existing context, much like *owl:imports* is purported to do. Then new semantic conditions and lists of terms can be defined either formally or informally. This allows the semantics of RDF to be extended on a per-context basis without re-chartering an entire new Working Group.

This provides an intriguing possible solution to defining context and contextual identity in formal semantics, including not just domain-specific identity but also the more general-purpose technique of pollarding. As there is no *sameAs* in RDF, current usages of *owl:sameAs* are denoted in the OWL context with its 'punning' semantics. If we want a more universalist *sameAs* with 'pollarding' semantics, rather than wait for the RDF Working Group to re-convene, we could simply create a context that includes RDF's original semantics but then creates the new *rdf2:sameAs* semantics with 'pollarding' semantics. On the other end of the scope, we can also imagine many different new domain specific usages of

---

[1] http://www.slideshare.net/PatHayes/rdf-with-contexts

identity being created, such as *sameAs* within the context of geographical mapping, a certain historical era, or some other application that needs to be able to link data without making grand universalist claims. Contexts provide a degree of useful referential opacity, as an *sameAs* asserted in one context might cease to be true in a subcontext when more refined meanings are in use, such as in comparing chemical elements vs. chemical isotopes.

7   Related Literature

The issue of identity has a long history in philosophy and ontology [7], and more recently in knowledge representation and databases. To narrow the focus to the Semantic Web, this work directly follows Halpin et al. [8]. However, there are important differences. The analysis of Halpin et al. presents some informal alternatives to the use of *sameAs*, but the only alternative tested was a more fine-grained identity scheme with no logical semantics, which could not be replicated even by experts using directly declared *sameAs* statements [8]. In contrast, we present a preliminary articulation of a logical alternative to *owl:sameAs*, focusing our studies on the reliability of identity in Linked Data on inferred triples. Our empirical analysis in general confirms some of the results of the more detailed empirical analysis of Ding et al. [5], and variance between our results is explained by the fact that while Ding et al. used the Billion Triple Challenge, our analysis was performed over a 'live' copy of the Linked Data Cloud. There have been various attempts to put a formalization of context into the semantic web such as C-OWL [2]. A similar proposal to RDFC was put forward by Jie Bao et al.[2]

8   Conclusions

In order for a logic to be universal on the Web, it should be able to describe anything in any possible way, and still enable inference. By giving the users more freedom in modeling rather than less, one simultaneously encourages good modeling constraints, so that the language provides the necessary machinery to distinguish the world as it exists to the ontological engineer while not requiring particular metaphysical distinctions to be committed to *a priori*. An 'individual' in a metaphysical sense is one notion, whose merits can be debated; but 'individual' in the logical sense is quite another. The latter means simply 'a member of the universe of discourse' or 'within the scope of quantification'. Traditional 'good practices' typically get these two distinct notions confused, and use syntactic constraints arising from the latter to model the former. As *sameAs* as currently being used as a generic equality statement is in violation of the segregated lexical roles of OWL, it would make sense to create a different identity link that is based on a pollarding-based semantics. It's also possible if someone is using *sameAs* to mean 'related to' (and our experiments show people do this sometimes), then they are actually in error. Yet then our experiment

---

[2] http://www.slideshare.net/baojie_iowa/2010-0624-rdfcontext.

also showed for a significant minority they reasonably meant to declare equality across semantic roles and to make an identity claim that would obviously be true, but only within a delimited context. This is not all bad news: Cleavages in the use of identity would then provide strong hints for the creation of new formal contexts. The combination of consistent and unrestricted logical identity and richer use of domain-specific and formally defined identity constructs could simultaneously put the links back into Linked Data and takes advantage of the semantics in the Semantic Web.

References

1. Andrews, P.: An Introduction to Mathematical Logic and Type Theory: To Truth Through Proof. Kluwer Academic Publishers, Dordrecht (2002)
2. Bouquet, P., Giunchiglia, F., van Harmelen, F., Serafini, L., Stuckenschmidt, H.: C-OWL: contextualizing ontologies. In: Fensel, D., Sycara, K., Mylopoulos, J. (eds.) ISWC 2003. LNCS, vol. 2870, pp. 164–179. Springer, Heidelberg (2003)
3. Brachman, R.: What IS-A is and isn't: an analysis of taxonomic links in semantic networks. IEEE Comput. **16**(10), 30–36 (1983)
4. Delugach, H.: ISO Common Logic. Standard, ISO (2007). http://cl.tamu.edu/. Accessed 8 Mar 2008
5. Ding, L., Shinavier, J., Shangguan, Z., McGuinness, D.L.: SameAs networks and beyond: analyzing deployment status and implications of owl:sameAs in linked data. In: Patel-Schneider, P.F., Pan, Y., Hitzler, P., Mika, P., Zhang, L., Pan, J.Z., Horrocks, I., Glimm, B. (eds.) ISWC 2010, Part I. LNCS, vol. 6496, pp. 145–160. Springer, Heidelberg (2010)
6. Frege, G.: Sense and reference. Philos. Rev. **57**, 209–230 (1948)
7. Guarino, N., Welty, C.: Ontological analysis of taxonomic relationships. Data Knowl. Eng. **39**, 51–74 (2000)
8. Halpin, H., Hayes, P.J., McCusker, J.P., McGuinness, D.L., Thompson, H.S.: When owl:sameAs isn't the same: an analysis of identity in linked data. In: Patel-Schneider, P.F., Pan, Y., Hitzler, P., Mika, P., Zhang, L., Pan, J.Z., Horrocks, I., Glimm, B. (eds.) ISWC 2010, Part I. LNCS, vol. 6496, pp. 305–320. Springer, Heidelberg (2010)
9. Hayes, P.: RDF Semantics. Recommendation, W3C (2004). http://www.w3.org/TR/rdf-mt/. Accessed 21 Sept 2008
10. Motik, B., Patel-Schneider, P.F., Cuenca Grau, B.: OWL 2 Web Ontology Language: Direct Semantics (2009)
11. Patel-Schneider, P.F., Hayes, P., Horrocks, I.: OWL Web Ontology Language: Semantics and Abstract Syntax (2004)

Context-Aware Systems

An Automation Component for Cross-Platform, Context-Aware Applications Development

Achilleas P. Achilleos[✉], Marita Thoma, Georgia M. Kapitsaki,
Christos Mettouris, and George A. Papadopoulos

Department of Computer Science, University of Cyprus,
1 University Avenue, Nicosia, Cyprus
{achilleas,mthoma03,gkapi,mettour,george}@cs.ucy.ac.cy
http://www.cs.ucy.ac.cy

Abstract. Context-aware computing faces many challenges mainly due
to the increasing number and heterogeneity of context sources, since
the Internet of Things introduces billions of devices. The development
of context-aware applications is thus becoming a complex and cumbersome process, which is also augmented by the availability of different
mobile platforms. This requires a modular approach that aims to auto-
mate the development of these applications, by enabling developers to
easily add context-aware functionality. In this paper, an automation
component is presented that allows novice developers to select context
plug-ins (e.g., Geolocation, Facebook profile, battery level) and generate
a sample application that includes these context-aware functions. This
application serves as a basis for the development of more complex cross-
platform, context-aware applications. The code generation support of the
automation component is demonstrated through a case study. Finally, a
basic evaluation is performed to showcase the benefits, issues and identify
potential future work.

Keywords: Context-aware applications · Separation of concerns ·
Mobile computing · Web development · Code generation

1   Introduction

The widespread use of mobile platforms has led to new business models and
has urged organizations to provide applications for users "on the move". These
mobile users are best served if their needs of mobility are fulfilled in all settings
without overwhelming users with redundant information. Such a personalized
technology view has been adopted by various researchers [1] and is currently
intensified by the widespread presence of sensors in mobile devices, such as GPS
receivers, accelerometers and compasses, as well as the vast volume of data available on the Internet in the form of Web Services, information from social networks and sensors and actuators typically connected to micro-controllers. This
necessity to follow user needs to utilize efficiently software applications and services using different mobile devices at various places forms part of the general

© Springer International Publishing Switzerland 2015
H. Christiansen et al. (Eds.): CONTEXT 2015, LNAI 9405, pp. 63–76, 2015.
DOI: 10.1007/978-3-319-25591-0_5

term of context-awareness. Context-awareness defines a broad concept that is generally used to describe the process of acquiring and managing different pieces of context information to intelligently adapt the application behaviour. We adopt the definition from Dey and Abowd [2], referring to context as *"any information that can be used to characterize the situation of an entity, in which the entity can be a person, a place, or a physical or computational object that is considered relevant to the interaction between the entity and the application"*.

Mobile computing combined with context-awareness encompass various aspects spanning from sensing information at the hardware and network information level, to context-based recommendations at the application level, such as travel or music recommendations for mobile environments [3]. Context-acquisition from device components (e.g., device motion, battery level), from social networks (e.g., LinkedIn, Facebook), network information (e.g., based on Cell ID, or Wi-Fi), etc. can support and facilitate the mobile user in a variety of tasks. In fact, this information act as enablers of context-awareness, empowering applications to be adapted to end-user preferences and circumstances.

Although mobile devices offer clear-cut benefits to the user and context-awareness is a desirable feature, the diversity of platforms makes development of platform specific applications an uneconomical choice, since it requires manpower and additional resources from the provider's side to develop for each platform. The problem of platforms diversity needs to be addressed, since is important by developers to provide cross-platform applications, because even the same users own and use different devices.

Currently, a useful alternative to native applications is offered through the HTML5 standard, which allows accessing many device resources that were unavailable to web technologies in the past. This vision of pure web-based development offers many benefits to software engineers, where the key one refers to cross-platform development: develop once, deploy anywhere. Also recent developer surveys demonstrate a tendency to move towards pure HTML5 solutions[1], as mobile browsers implement the features offered by the HTML5 standard. Gartner also defines[2]: *"HTML5 is in the top 10 technologies and capabilities for 2015-16 and despite many challenges, HTML5 will be an essential technology for organizations delivering applications across multiple platforms"*.

An HTML5, context middleware was developed [4], which provides a modular approach that promotes the concept of separation of concerns and thus enables developers to reuse context plug-ins for the development of context-aware applications. This work develops a new component that assists developers with limited experience in developing context-aware applications, by automating a large part of the process. It allows selecting from an (extensible) pool of context plug-ins and generates the code for a sample context-aware application that can serve as the basis for further development. The key target is to reduce the complexity involved with developing context-aware functionality, simply by

---

[1] http://www.sencha.com/blog/the-state-of-html5-development-in-the-enterprise/ - Published: February 12, 2014.

[2] http://www.gartner.com/newsroom/id/2669915 - Published: February 24, 2014.

allowing developers to easily select and include context plug-ins that already provide this functionality. A further benefit is found in the production of compact code, having clearer separation of concerns (focus on application logic instead of context-aware functions).

The rest of the paper is structured as follows. Section 2 introduces related work, outlining the difference of our approach in comparison to existing frameworks. Section 3 provides an overview of the context-middleware. The following section introduces the developed automation component and the development pathways. Section 5 demonstrates the use of the automation component in two scenarios of a case study and presents the results of the evaluation performed. The final section outlines the conclusions and identifies potential future work.

2   Related Work

The provision of access to context sources to facilitate the development of mobile applications, has been the focus in research work. MUSIC (Self-Adapting Applications for Mobile Users in Ubiquitous Computing Environments) defines a context-management and adaptation middleware for Java [5]. MUSIC runs on top of the OSGi (Open Service Gateway initiative) framework and includes a number of context plug-ins available as OSGi components. Similarly to our solution, MUSIC includes the notion of context sensor and reasoner plug-ins, which receive and process context data from other plug-ins via the middleware in order to produce higher level data. The evolution of MUSIC, following the same rationale, but tailored to the Android platform is found in RSCM (Really-Simple Context Middleware) [6]. The RSCM architecture differentiates between context producing components (the plug-ins) and context consuming components (the applications). RSCM is used in the Professor2Student application that offers dynamic collaboration between supervisors and students [7].

A similar solution, which also focuses on web applications accessing sensor information on the Android platform, has been proposed in Ambient Dynamix [8]. Sensor access is handled through native plug-ins that can be installed on-demand on the users Android device, and are organized and managed by the Dynamix Service. Each plug-in provides its own set of context events and (optionally) an API for controlling its functionality. To support communication with web applications, Dynamix exposes two REST APIs using a customized web server embedded within the Dynamix Service. This way a web-based application is able to interact with the associated Android plug-in to trigger context sensing and/or remote device interactions.

Dynamix has characteristics that are closer to a "hybrid" approach (it does not follow a pure cross-platform approach), but it rather ties web application development to the native context plug-ins offered by the underlying Android platform. The use of hybrid technologies is very popular in mobile application development in an attempt to satisfy the cross-platform requirement. A variety of such technologies exists, such as PhoneGap, Apache Cordova (i.e., the open-source engine/version that runs PhoneGap) and AppBuilder. Hybrid technologies offer usually a set of uniform JavaScript libraries that can be invoked,

Table 1. Comparing HTML5 properties with hybrid and native frameworks

Feature	H5CM approach	Hybrid Technologies approach	Platform-specific frameworks approach
Native code use	x	✓ [PhoneGap]	✓ [RSCM, Dynamix]
Technologies used	Web Technologies	Web along with platform-specific, e.g., Android, Firefox OS, iOS [PhoneGap]	Web [Dynamix] and Android SDK [RSCM, Dynamix]
Required development knowledge	Web Technologies	Web technologies along with basic framework/platform specific understanding [PhoneGap]	Web technologies and platform specific [Dynamix)]or Platform specific [RSCM]
Pre-installation requirements	x	✓ [PhoneGap]	✓ [RSCM, Dynamix]
Access Device-specific features	Limited (Browser restricted)	Full (based on API availability per platform) [PhoneGap]	Full (based on plug-in availability) [RSCM, Dynamix]
Security and Privacy support	Browser based	Guidelines only [PhoneGap]	Via context firewall [Dynamix]

wrapping device-specific native backing code through provided JavaScript libraries. This process provides access to native device functions through JavaScript, such as the device camera or its accelerometer.

With the introduction of H5CM (HTML5 Context Middleware) [4], the vision is a pure web-based approach. Table 1 captures the key points of variability between our approach and other approaches. All cases have both advantages and disadvantages in terms of application variability and execution speed. The main criticisms for hybrid development is the learning curve, since developers need to learn how to use the native libraries for each platform, but most importantly that mobile devices are not able to smoothly run a hybrid application [9]. On the other hand, native applications offer benefits in terms of performance and API coverage, but lack in terms of instant worldwide deployment, manual installation or upgrades and flexibility to combine data from different resources [10].

Differentiating from both native and hybrid approaches, H5CM offers a pure HTML5 approach not bound to any platform or development environment employing solely web technologies [11]. Although access to device-specific capabilities is provided based on the browser support, vendors are continuously extending their support based on the popularity and evolution of HTML5. Moreover, in respect to pre-installation requirements our approach does not have any prerequisites. Finally, a feature that is missing from all approaches, is an extended security and privacy support, since this feature is handled either by the browser (as in the case of H5CM) or by the underlying platform (in hybrid and platform specific frameworks).

3   H5CM Overview

3.1   Architectural Elements

As aforementioned the main requirements fulfilled with the creation of H5CM were to provide a framework that is modular, reusable, extensible, and that can be utilised for web applications on any mobile platform. H5CM has a hierarchical structure: at the lower level there are context-sensor plug-ins that allow acquiring and distributing low-level context data. These may refer to the location of the user, the orientation of the user's device and the results from the invocation of Web APIs. At the second level of the hierarchy there are context-reasoner plug-ins that accept low-level context from one or more sensor plug-ins and apply the appropriate reasoning, in order to create high-level context information. The application is at the top of the hierarchy and is able to communicate with the sensor and/or reasoner plug-ins to acquire context information that enables the adaptation of the application's logic. More details on the architecture of H5CM can be found in our earlier work [4].

3.2   The Context Repository

The H5CM functionality is empowered by an extensible and reusable *Context Repository*. The basic point of differentiation between sensor and reasoner plug-ins, is that the former provides access to "basic" context data in the form that these can be collected directly from context sources (e.g., device accelerometer, user geographical coordinates), whereas the latter gives access to sophisticated context information that are derived from the "basic" data. These plug-ins define an extensible and reusable repository. On the one hand, they can be reused by developers, since they are generic and can be invoked from any context-aware, web application. On the other hand, the set of plug-ins can be extended by technical users that need additional functionality, as more and more features of HTML5 are continuously being supported in mobile browsers. H5CM is currently offering different reusable plug-ins, some of which are described in Table 2. The current version of H5CM and the plug-ins are available on Google Code[3].

The context plug-ins repository has been also enriched with the addition of the SensoMan plug-in. This plug-in enables access to context data retrieved from sensors connected to Arduino micro-controller boards. The SensoMan plug-in is of particular importance, since it provides access to the context data coming mainly from the environment (see Fig. 1), which was not fully covered by other plug-ins that principally retrieve data from the mobile device (internal) and the

---

[3] https://code.google.com/p/h5cm/.

Table 2. The context repository: example sensor and reasoner plug-ins

Plug-in	Plug-in type	Functionality	Access type
SensoMan	Sensor	Enables connecting and retrieving data from sensors connected to microcontroller boards through the REST API provided by the SensoMan system [16].	External: Micro-Controllers
BatteryLevel	Sensor	Retrieves and monitors the battery level (e.g., 74 %).	Internal: Mobile Device
Geolocation	Sensor	Allows detecting and continues monitoring the position of the user. Returns the location in the form of GPS coordinates (i.e., latitude and longitude).	Internal: Mobile Device GPS Receiver
DeviceOrientation	Sensor	Monitors the physical orientation of the device (e.g., the user tilts or rotates it).	Internal: Mobile Device
RestfulService	Sensor	Allows connecting to RESTful services. It requires as a parameter the URL of the service including any parameters the service may require for its functionality.	External: Internet
FacebookConnect & LinkedInConnect	Sensor	Enables user authentication with Facebook/Linkedin and requests the user to grant access permissions for retrieving data (given in comma separated string values).	External: Social Networks
FacebookInformation & LinkedinInformation	Sensor	Provides the means to acquire Facebook/LinkedIn profile data of the user (i.e., public user data) and retrieving additional data provided that the user is authenticated (requires FacebookConnect/Linked-InConnect).	External: Social Networks
FacebookPosts	Sensor	Provides the way to obtain wall posts of the user, provided that the user is authenticated (requires FacebookConnect).	External: Social Networks
ActivityRecognizer	Reasoner	Recognizes the activity of the user (Jogging, Walking, Sitting, Upstairs or Downstairs) based on the results of a decision tree classifier.	Internal: Mobile Device
BatteryAnalyzer	Reasoner	Retrieves data from the existing BatteryLevel sensor and returns TRUE if the application is able to handle computational intensive tasks based on a developer specified cutoff value (e.g., ¿60 %) passed as a parameter to the reasoner or FALSE otherwise.	Internal: Mobile Device
FacebookRestaurants	Reasoner	Retrieves data from two sources: Google Places restaurants at a specific area and Facebook posts about user-visited restaurants. It computes the intersection of the two sets in order to give user preferred restaurants (requires FacebookConnect and FacebookPosts).	External: Social Networks

user (social networks). The SensoMan plug-in enables access to a diversity of sensors and delivers context-aware functions that further support the development of context-aware Web applications. The list of sensors currently supported by the SensoMan system is presented in [16].

Furthermore, reasoner plug-ins provide the capability to perform aggregation, analysis and reasoning on "basic" context data to derive higher level information that can be useful in taking proactive actions at the application level. Hence, a simple reasoner of H5CM can include the use of information on whether the device is charging and its acceleration to decide whether the user is walking, driving a car or sitting in a room. Such mechanism can be combined with machine learning techniques of clustering or classification for drawing useful conclusions on the user, such as activity recognition addressed in previous works [12]. Such techniques can be supported by H5CM through plug-ins that include advanced processing of context information.

In that respect we have implemented the ActivityRecognizer reasoner plug-in that performs activity recognition based on training performed over raw accelerometer data obtained from the raw dataset[4] presented in [13]. The dataset includes information collected from 29 users while performing various daily activities, such as Jogging, Walking, Sitting, Upstairs or Downstairs. We have used the dataset that includes these specific activities to train a C4.5 Decision tree classifier with a confidence factor of 0.25 [14]. For this step of the process we have employed the Weka machine learning software and its decision tree implementation indicated as J48 [15]. Subsequently, a pruned version of the tree that was created from the classification process was transferred to JavaScript code and was used in the formation of a new reasoner plug-in. The current version of the ActivityRecognizer is available on the Google code website of H5CM.

4   H5CM Automation Component

The main contribution of this work is the definition and implementation of the H5CM Automation Component (HAC), which refines and extends the middleware architecture. In specific, it automates the development process, so as to support mainly developers that are not experts in the implementation of context-aware applications. In this way a developer can easily kick-start the implementation of context-aware, web-based applications using the concepts and the reusable elements provided by the middleware. This section presents the HAC component that was developed through refinement of the H5CM architecture. The refined architecture presented in Fig. 1 enables storing all developed plug-ins in an XML-based repository, which is queried by the HAC component to identify the complete list of plug-ins that the developer can use to automatically generate the new (sample) context-aware application.

In specific, the HAC component gives the ability to the developer to select which sensor and reasoner plug-ins need to be included and used in the context-aware application to be developed. Apart from selecting the plug-ins, the HAC

---

[4] http://www.cis.fordham.edu/wisdm/dataset.php.

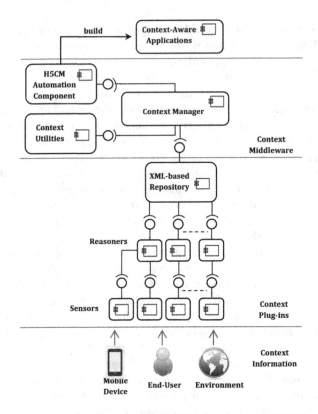

Fig. 1. H5CM Automation Component (HAC): the extended H5CM middleware.

component also queries these plug-ins so as to identify their context properties (e.g., name, birthday from FacebookInformation plug-in). This enables the developer to choose the properties needed by the application, in order to generate the code that composes the new context-aware application. A novice developer of context-aware applications, is thus able to easily include context-aware functionality in the application, simply by selecting the necessary plug-ins and properties. This allows the developer to focus on the implementation of the application logic and UIs, rather than on retrieving and analysing context information, e.g., GPS coordinates, Facebook profile, which is indeed a complex task.

The HAC component is implemented using the library JQuery Steps[5] In specific, JQuery Steps is a User Interface (UI) library that enables the developer to easily create wizard-like interfaces. The library basically groups content into sections for a more structured and orderly page view, while providing the capability for validation of steps and the information provided, so as to ensure smooth progress and quality of the generated code produced by the HAC component.

---

[5] GitHub wiki Online: http://www.jquery-steps.com/.

When a developer navigates to the first page, all available plug-ins and their properties, including an informative description in terms of the functionality, are queried, generated and presented in a visual form. The description of the plug-ins and their properties is defined in an XML file named "all-plugins.xml" (see description of Facebook plug-in in Listing 1.1), which forms the XML representation of the context repository. The HAC component was implemented using this modular and extensible approach, so as to adhere to the architecture of the H5CM. This allows any expert developer to implement a new plug-in and add it directly to the repository, simply by describing it using XML. In fact, the HAC component can query and visualise any newly implemented plug-in, and include its functionality as part of the generated application, simply by describing the new plug-in similarly to the example Facebook plug-in shown in Listing 1.1.

Furthermore, the HAC component provides the capability to the developer to select from two development pathways. The first development pathway, showcased in the form of a UML Activity diagram in Fig. 2 (A), allows developers to enter the application name, select all context plug-ins and their properties, and generate directly the new context-aware application. The generated application code can be downloaded and the application is executed and demo-ed before quitting the application. Using the generated code the developer can further modify the generated application.

Listing 1.1. XML Description of Facebook plug-in and properties.

```
1  <plugin>
2          <name>FacebookConnect</name>
3          <sources>
4              <source>modules/sensors/FacebookConnect.js</source>
5          </sources>
6          <description>This plug-in is used to allow the user to access
                 through a Facebook account </description>
7          <properties>
8              <property>
9                  <name>name</name>
10                 <description>This property retrieves the name of the
                        Facebook user</description>
11             </property>
12             <property>
13                 <name>email</name>
14                 <description>This property retrieves the email of the
                        Facebook user</description>
15             </property>
16             <property>
17                 <name>gender</name>
18                 <description>This property retrieves the gender of the
                        Facebook user</description>
19             </property>
20             <property>
21                 <name>username</name>
22                 <description>This property retrieves the username of the
                        Facebook user</description>
23             </property>
24             <property>
25                 <name>birthday</name>
26                 <description>This property retrieves the information about
                        the birthday of the Facebook user</description>
27             </property>
28         </properties>
29  </plugin>
```

The second development pathway enables the developer to enter the name of the application and then select specific plug-ins. Afterwards, the developer is able to choose context properties available in specific context plug-ins (e.g., FacebookConnect, LinkedInConnect), which are required as context information in the context-aware application to be developed. For instance, as presented in

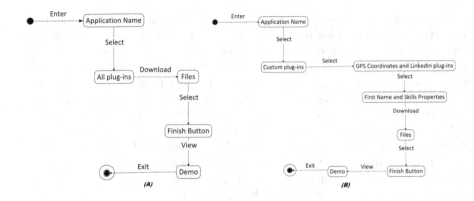

Fig. 2. UML activity diagram: the development pathways.

Fig. 2 (B), the developer has selected the Geolocation, LinkedInConnect and LinkedInInformation plug-ins. In the following step the developer chooses the first name and skills context properties associated with the LinkedIn related plug-ins. Finally, the developer is able to download the generated code and view a demo of the application prior to exiting. The developer can use and further modify the generated application based on the desired logic and UI for the final context-aware application.

5   Case Study and Evaluation of the HAC Component

This section describes the use of the HAC component for the generation of two example context-aware applications and the evaluation of the component. The evaluation is contacted by nine MSc students of the course EPL603: Advanced Software Engineering, offered at the Dept. of Computer Science, University of Cyprus. In the first scenario, all context plug-ins offered by the middleware are selected by the developer and the new context-aware application is generated. It is important to note that the SensoMan sensor plug-in and the ActivityRecognizer reasoner plug-in are still to be integrated with the HAC component, since their development was only recently completed [16]. The automatically generated application for this scenario where all the plug-ins are selected and executed using the desktop version of the Firefox Browser.

In the second scenario, the development pathway enables developers to select specific plug-ins and their associated properties. As presented in Fig. 3, during the third step the HAC component allows selecting the desired context-plug-ins. In the next step, the HAC component detects that the FacebookConnect plug-in has explicit context properties. This allows the developer to choose the necessary ones for supporting the desired functionality for the context-aware application. At the "Finish" step the summary of the process is presented to the developer, and as soon as the developer downloads the code files of the generated application, the wizard is completed. The final snapshot in Fig. 3, showcases the application running on the Android Firefox Mobile Browser.

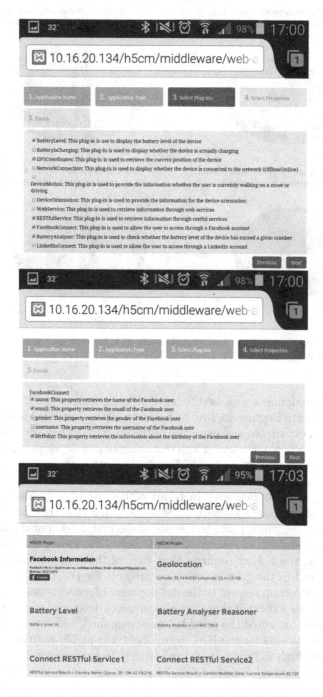

Fig. 3. Scenario 2: selecting context plug-ins and context properties.

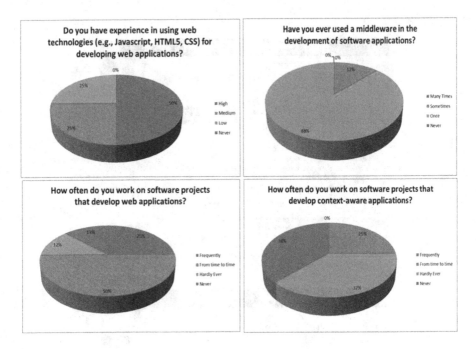

Fig. 4. Knowledge, experience and expertise on web technologies and context-aware applications development.

The HAC component was used by nine MSc students of the EPL603 course: Advanced Software Engineering. At first the evaluation gathered information on the knowledge, experience and expertise of the students in regards to prior usage of middleware technologies, use of web technologies and development of context-aware applications. As illustrated in Fig. 4, students have at 75 %, medium to high experience in the use of web technologies (e.g., Javascript, HTML5, CSS) for developing web applications. In terms of the use of a middleware for the development of software applications, 88 % indicated that they have only used such software once, which refers to the use of the first version of the middleware during the first evaluation session [4], while only 12 % have used a middleware sometimes. The third chart confirms that 75 % of the students have experience and have worked in the past on software projects for the development of web applications. Finally, in respect to the development of context-aware applications 3 out of 4 students indicated that they almost never developed such applications.

The students answered the provided questionnaire[6], which revealed that the opinion of the students for the HAC component was highly positive. In specific, the scores were high in terms of all the assessed factors: *Learnability, Memorability, Effectiveness and Functionality*. Moreover, reviewing the answers to the

---

[6] https://docs.google.com/forms/d/1DJ7FNarlpAq0FI6bqcAOsOb703NgNbRctTsSF UxhBgU/viewform, HAC Component Questionnaire.

Table 3. Evaluation results

Learnability, Memorability	Effectiveness	Functionality
4.78/5	4.63/5	4.63/5

questions dedicated to the HAC component it is evident that the students considered it as a strong point/extension for the middleware. Finally, the automation tool raised the interest of students during the evaluation session, since they actively tried it on different mobile devices to identify and assess if it was working for mobile-specific plug-ins (e.g., device motion, device orientation) and on different platforms. The results of the tests were encouraging.

The last two factors scored lower since the students indicated that they would appreciate access to additional sensors, which provided context data from the environment. The SensoMan plug-in can assist to overcome this limitation and as part of future work we will be adding this plug-in in the HAC component, so as to offer additional context-aware functions and increase effectiveness of the generated applications. Moreover, a complementary evaluation by expert developers can definitely provide additional insights in terms of functionality coverage and effectiveness in building context-aware applications. The evaluation was intentionally performed with novice developers (in terms of developing context-aware applications), since the component's key aim is to support developers that have limited experience with such kind of applications (Table 3).

6   Conclusions

Context awareness is a promising research field that refers to the acquisition, processing and reasoning on context information, so as to adapt the application mainly to the requirements of the user. Due to the diversity of context sources (i.e., mobile device, user, environment) the development of context-aware applications becomes inherently complex. The developed H5CM context middleware provides the ability to reduce complexity through the concept of separation of concerns. In specific, the middleware enables reusability of context plug-ins and favours extensibility for supporting the development of context-aware applications. Furthermore, the combination of HTML5 and context-awareness in the developed middleware addresses the issue of cross-platform development.

The work in this paper goes a step further and aims at further simplifying the development of context-aware applications. In specific, it extends the H5CM middleware by defining the HTML5 Automation Component. The HAC component is also implemented following a modular and extensible approach, which enables novice developers to define and automatically generate web-based, context-aware mobile applications. Future work aims to integrate the SensoMan context sensor and ActivityRecogniser context reasoner plug-ins with the HAC component, as there plug-ins are only recently developed [16].

References

1. Rodden, K., Hutchinson, H., Fu, X.: Measuring the user experience on a large scale: user-centered metrics for web applications. In: Proceedings SIGCHI Conference on Human Factors in Computing Systems, pp. 2395–2398 (2010)
2. Dey, A.K.D., Abowd, G.D.: Towards a better understanding of context and context awareness. In: Proceedings of Workshop: What, Who, Where, When, and How of Context Awareness, ACM Conference Human Factors in Computer Systems (2000)
3. Wang, X., Rosenblum, D., Wang, Y.: Context-aware mobile music recommendation for daily activities. In: Proceedings of ACM International Conference on Multimedia, pp. 99–108 (2012)
4. Achilleos, A., Kapitsaki, G.M.: Enabling cross-platform mobile application development: a context-aware middleware. In: Proceedings of the 15th International Conference on Web Information System Engineering (WISE 2014), pp. 304–318 (2014)
5. Floch, J., Fr, C., Fricke, F., Geihs, K., Wagner, M., Lorenzo, J., et al.: Playing USIC - building contextaware and selfadaptive mobile applications. Softw. Pract. Exp. **43**(3), 359–388 (2013)
6. Paspallis, N., Papadopoulos, G.A.: A pluggable middleware architecture for developing context-aware mobile applications. Pers. Ubiquit. Comput. **18**(5), 1099–1116 (2014)
7. Ioannides, F., Kapitsaki, G.M., Paspallis, N.: Professor2Student - connecting supervisors and students. In: 10th International Conference on Mobile Web Information Systems, pp. 288–291 (2013)
8. Carlson, D., Schrader. A.: Dynamix: An open plug-and-play context framework for android. In: 3rd International Conference on the Internet of Things, pp. 151–158 (2012)
9. Gai, D.: Hybrid VS Native Mobile Apps. http://www.gajotres.net/hybrid-vs-native-apps/. Accessed 26 Sept. 2014
10. Mikkonen, T., Taivalsaari, A.: Reports of the web's death are greatly exaggerated. IEEE Comput. **44**(5), 30–36 (2011a)
11. Mikkonen, T., Taivalsaari, A.: Apps vs. open web: the battle of the decade. In: Proceedings of 2nd Annual Workshop Software Engineering for Mobile Application Development, pp. 22–26 (2011b)
12. Abdullah, M.F.A., Negara, A.F.P., Sayeed, M.S., Choi, D.J., Muthu, K.S.: Classification algorithms in human activity recognition using smartphones. World Acad. Sci. Eng. Technol. **68**, 422–430 (2012)
13. Kwapisz, J.R., Weiss, G.M., Moore, S.A.: Activity recognition using cell phone accelerometers. SIGKDD Explor. Newsl. **12**(2), 74–82 (2011)
14. Quinlan, R.: C4.5: Programs for Machine Learning. Morgan Kaufmann Publishers, San Mateo (1993)
15. Hall, M., Frank, E., Holmes, G., Pfahringer, B., Reutemann, P., Witten, J.H.: The WEKA data mining software: an update. SIGKDD Exp. **11**(1), 10–18 (2009)
16. Paphitou, A.C., Constantinou, S., Kapitsaki, G.M.: SensoMan: remote management of context sensors. In: 5th International Conference on Web Intelligence, Mining and Semantics (WIMS 2015) (2015)

Understanding Context with ContextViewer – Tool for Visualization and Initial Preprocessing of Mobile Sensors Data

Szymon Bobek$^{(\boxtimes)}$, Sebastian Dziadzio, Paweł Jaciów,
Mateusz Ślażyński, and Grzegorz J. Nalepa

AGH University of Science and Technology,
Al. A. Mickiewicza 30, 30-059 Krakow, Poland
{szymon.bobek,gjn}@agh.edu.pl

Abstract. Mobile context-aware systems are becoming more and more popular due to the rapid evolution of personal mobile devices. The variety of sensors that are available on such devices allow building intelligent applications that adapt automatically to user preferences and needs. Together with a growth of such self-adaptable systems, number of tools for collecting, visualising and modelling context were developed. However, there is still a need for tools and methods that will support building mobile context-aware systems at the very early stage of development. Such solutions should provide mechanisms for collecting, visualising and initial preprocessing of data to allow better understanding of processes, patterns and dynamics of mobile contextual data. In this paper we propose ContextViewer– a toolkit that aims at providing such mechanisms. It is a part of a methodology for building context-aware systems that besides ContextViewer includes also modelling methods and runtime environment for executing models.

1 Introduction

According to Dey [1] context can be defined as any information that can be used to characterise a situation of an entity. Such a general definition allows to treat different types of information as context, but also makes it difficult to model and process by context-aware systems (CAS). Over the last decade several frameworks supporting this task were proposed [2]. They provide mechanism for building context-aware applications which include collecting, visualising, modelling and processing contextual data. However, most of the available solutions were crafted for the needs of stationary environments. This assumes that the contextual information is dominated by the user activities within the system, like websites visited, applications ran[1]. Although such approach seems sufficient

---

This work was funded by the National Science Centre, Poland as a part of the KnowMe project (registration number 2014/13/N/ST6/01786).

[1] Exceptions here are systems that use eye tracking for user profile discovery, or EEG signal analysis for brain-computer interactions. This types of systems however, are beyond the scope of this paper.

H. Christiansen et al. (Eds.): CONTEXT 2015, LNAI 9405, pp. 77–90, 2015.
DOI: 10.1007/978-3-319-25591-0_6

for most of the desktop context-aware systems, it cannot be considered valid in mobile systems which use dynamic and time-dependant data.

Such systems are equipped with a variety of sensors like accelerometers, gyroscopes, mobile network providers, GPS, etc., which deliver huge amounts of information in a streaming manner. This information changes very fast in the short time perspective which is mainly caused by the nature of the streaming data itself, but also in the long time perspective, which reflects changes in the user habits and routines (phenomena known in the literature as *concept drift* [3]). It is very difficult to efficiently model system that process such data without prior knowledge about patterns within the data and correlations between context providers.

Machine learning methods provide mechanisms that allow for automated discovery of knowledge from large amounts of data. However, applying machine learning algorithm to context-aware system is not always straightforward. Choosing appropriate algorithm, its parameters, and input attributes (*feature set*) is non trivial task, which requires a lot of experiments and preliminary data analysis. Therefore, the comprehensive approach is needed that will provide tools and methods supporting the designer of mobile CAS in understanding the context at the very early stage of development.

As a response to this need, we proposed CONTEXTVIEWER – a web application that allows for visualising and initial preprocessing of contextual information from mobile devices. It is integrated with AWARE[2] – one of the most popular framework for logging, sharing and reusing mobile contextual data. It allows for basic semantization of raw sensor data and provides bindings with machine learning software like WEKA[3] and ProM[4] to allow direct and immediate analysis of context. CONTEXTVIEWER is a part of a larger toolkit which consists of CONTEXTSIMULATOR [5] and HEARTDROID[6]. These tools provide respectively an lightweight simulation framework, and rule-based runtime for prototyping, testing and deployment of mobile context-aware applications.

The rest of the paper is organised as follows. Comparison of available frameworks for context management and motivation for our work was presented in Sect. 2. In Sect. 3 a detailed description of CONTEXTVIEWER tool was provided. The remaining parts of the toolkit for building mobile context-aware systems that CONTEXTVIEWER can be combined with, were briefly discussed in Sect. 4. In Sect. 5 an use-case scenario was presented. Summary and future work plans were enclosed in Sect. 6.

---

[2] See http://awareframework.com.

[3] See https://weka.waikato.ac.nz/.

[4] See http://www.processmining.org/prom/start.

[5] See http://glados.kis.agh.edu.pl/doku.php?id=pub:software:contextsimulator:start for details.

[6] See https://bitbucket.org/sbobek/heartdroid for details.

2 Overview of Context-Management Frameworks and Motivation

Context management frameworks can be divided into four groups:

1. frameworks that offer collecting, storing and distributing context within the system,
2. frameworks that provide visualisation of selected contextual information,
3. frameworks for modelling and processing context,
4. hybrid approaches that combine features of former three.

Systems that allow for collecting and storing contextual information are one of the most popular tools for building context-aware applications. They provide basic programistic interface for accessing context in a real time, as well as mechanisms for storing historical data. Examples of such frameworks are SDCF [4], AWARE, JCAF [5], SCOUT [6], and ContextDroid [7]. These frameworks do not provide any mechanisms for visualisation nor preprocessing of data. They serve as a middleware between the physical sensors or native API and the context-aware application. They usually store the data locally to preserve the privacy issues, but some of them like AWARE provides mechanisms for synchronization with external databases. This allows to transfer complex data analysis from mobile devices to the more powerful machines, integrate the knowledge from many sources and exchange this knowledge between the mobile devices in the distributed manner.

The other type of context-management frameworks are systems which focus on the visualisation of selected contextual data. In most cases these systems are commercial applications like: Ubidots[7] or Valarm[8]. However there are also open source solutions like Nimbits[9] and Freeboard[10]. The visualisation frameworks give more insight into the raw contextual data by plotting them on diagrams, or augmenting them on the map. They do not provide any tools for automated analysis of data, nor any runtime. Exception here is the Ubidots and Nimbits framework, which allows for including simple rule-based triggers that are fired when a context defined by some preconditions is met. These frameworks are mostly designed as cloud-based solutions which communicate with many mobile devices and provide web interface for visualising data.

Frameworks that support modelling context-aware systems and provide runtime environments for executing contextual models form the third class of context management systems. One of the most popular runtime for stationary context-aware application is ContextToolkit [8]. It supports rule-based modelling language and execution runtime. Other examples of rule-based frameworks crafted especially for the needs of mobile environments are Context Engine [9],

---

[7] See http://ubidots.com/.
[8] See https://play.google.com/store/apps/details?id=net.valarm.android.pro.
[9] See http://nimbits.com/.
[10] See https://freeboard.io/.

SWAN [10] and HEARTDROID [11]. The last one is used by us as a part of the runtime environment.

Different approach for modelling and executing context models was presented by Brezillon et.al. [12]. They proposed a structure called Contextual Graph. It is a directed acyclic graph that represents the actions to undertake according to the context. The action nodes represent actions to undertake to achieve a goal while the event nodes become as explained above, contextual nodes describing the possible contextual issues of a given event.

The last group of context-management frameworks are hybrid solutions like CoBrA [13] for building smart meeting rooms, GAIA [14] for active spaces or SOCAM [15] and mobileGaia [16] – middlewares architecture for building pervasive systems. They combine features of frameworks for collecting storing and distributing context with features of frameworks for modelling and processing context. They do not provide tools for visualisation and data analysis. What is more, they are usually built for very narrow, specific solutions, and can not be considered as frameworks for more general purpose than initially assumed. The exception is MUSIC [17] framework – an open platform for development of self-adaptive mobile applications, which includes a methodology, tools and middleware.

A comprehensive comparison of all the frameworks and tools discussed in this section was presented in Table 1. It is worth noting that the solution proposed by us (bolded rows in the table) covers all the features that we distinguished important for the context management toolkit.

Although there exist numerous of such solutions, they all approach the issue of building context-aware systems in a standard manner, which begins with a modelling phase. However, mobile context-aware system are not yet well established field and there is still a lot of research ongoing in the area of context modelling, automated extraction of knowledge from multiple sources, uncertainty handling, etc. Therefore, in our work we aim at providing tools that will support the initial phase of the design of mobile context-aware systems which precedes the modelling phase. During this initial phase, the designer (or researcher) should be able to understand the contextual data that later will be modelled. This involves visualisation of this data, but also semantization of raw sensor streams and integration with most common machine learning software like WEKA of ProM for deep analysis. To achieve this task we designed and implemented CONTEXTVIEWER and integrated it with AWARE – one of the most popular framework for building mobile context-aware systems. CONTEXTVIEWER is part of software bundle for building context-aware systems that includes CONTEXTSIMULATOR and HEARTDROIDsystems for simulation, testing and deploying context-aware solutions on mobile platforms. In the next section the architecture and capabilities of CONTEXTVIEWER are described.

3   Understanding Context with ContextViewer

CONTEXTVIEWER is a tool for making Aware framework data easy to browse and process. It consists of four main modules implementing corresponding functionalities:

Table 1. Comparison of available frameworks for context management

Framework	Context acquisition	Storage	Context distribution	Context visualisation	Context semantization	Context analysis	Modelling	Simulation	Runtime	Designed for mobile
AWARE	**yes**	**yes**	**yes**	**no**	no	**no**	**no**	**no**	**no**	**yes**
SDCF	yes	yes	no	no	no	no	no	no	no	yes
JCAF	yes	yes	no	no	no	no	no	no	no	yes
SCOUT	yes	yes	yes	no	no	no	no	no	yes	yes
ContextDroid	no	no	no	no	no	no	yes	no	yes	yes
ContextToolkit	no	no	no	no	no	no	yes	no	yes	no
Gimbal	no	no	no	partially	no	no	yes	no	yes	yes
Contextual Graphs	no	no	no	model only	no	no	yes	yes	yes	no
SWAN	no	runtime history	no	no	no	no	yes	no	yes	yes
Context Engine	no	no	no	no	no	no	yes	no	yes	yes
HeaRTDroid	**no**	**runtime history**	**no**	**no**	via rules	**no**	**yes**	**no**	**yes**	**yes**
Ubidots	not direct	yes	no	yes	rules	basic	basic rules	no	basic	yes
Valarm	not direct	yes	no	yes	no	yes	no	no	no	yes
Nimbits	not direct	yes	no	yes	rules	basic	basic rules	no	basic	yes
Freeboard	not direct	yes	no	yes	rules	basic	no	no	no	yes
ContextSimulator	**no**	**no**	**no**	**no**	no	**no**	**no**	**yes**	**no**	**yes**
ContextViewer	**no**	**no**	**no**	**yes**	yes	**yes**	**no**	**no**	**no**	**yes**
CoBrA	yes	no	yes	model only	ontology	no	yes	no	yes	no
GAIA	yes	no	yes	model only	ontology	no	yes	no	yes	no
MobileGaia	yes	no	yes	model only	ontology	no	yes	no	yes	yes
SOCAM	yes	no	yes	model only	ontology	no	yes	no	yes	yes
MUSIC	yes	yes	yes	model only	ontology	no	yes	yes	yes	yes

1. *Contextual data visualization module* – that aims at transforming raw Aware data to an intuitive graphical form. With it, instead of browsing database tables, user can inspect the data visually using only few intuitive controls.
2. *Basic semantisation module* – that is responsible for basic semantization of contextual data by translating raw sensor information into human readible concepts.
3. *Machine learning module*– which enables pattern discovery in contextual data. It includes WEKA data mining tool integration component and process mining component based on ProM tool.

The following section describes in details the architecture and implementation of these modules.

3.1   Architecture of ContextViewer

CONTEXTVIEWER is a web application developed using Google Web Toolkit (GWT). GWT is an open source project that makes creating JavaScript front-end applications in Java possible.

Front-end of CONTEXTVIEWER, referred later as a client-side application, is received over a network by users where it runs as JavaScript in their web browsers. At some point, the client-side needs to interact with a backend server. For example when loading or processing data. The so-called server-side of CON-TEXTVIEWER is based on Java Servlets. Client communicates with the server via HTTP, sending requests and receiving updates. The communication is done using remote procedure calls (RPC). The concept of CONTEXTVIEWER's architecture is presented in Fig. 1. There is also an external contextual data provider, which is the Aware framework.

The client-side code implements everything that is responsible for displaying context data in an end user's web browser. Simple mechanisms for browsing the data are available for the user. Handling his or her input allows for changing display dynamically to match new conditions. The main goal of this part of CONTEXTVIEWER is to transform sensors' data from tables with numbers or codes to its intuitive graphical representation. For example, numerical latitude and longitude are represented as a marker on a map using Google Maps API. For each mobile sensor type used there is a class for storing its context data. It is possible to easily add components for additional sensor types. There are methods for invoking interaction with the server, sending requests and receiving and handling results or answers. It is worth noting that to change appearance of CONTEXTVIEWER's client one mostly needs only to edit included Cascade Style Sheet (CSS) file or to replace graphics to be displayed.

The server-side code implements servlets for handling a database, exporting data to ARFF and running a process mining tool. The main goal of this part of CONTEXTVIEWER is to process the context data. Mobile sensors context data is stored in a MySQL database. Servlet loads data from a user-specified interval and sends it back to the client. Attribute-Relation File Format (ARFF) is a file type for use with Weka machine learning software. It is created by

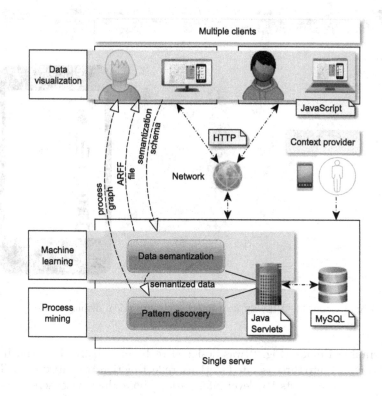

Fig. 1. The concept of CONTEXTVIEWER's architecture

invoking a Python script on the server-side. Servlet allows the user to download generated file for further data mining in Weka at his or her workstation. Process mining is performed with ProM Tools. Servlet sends result to the client where it is displayed.

There are some requirements for CONTEXTVIEWER to run. On the client-side, the end user's browser must have JavaScript enabled in order for the application to work and display correctly. To run the server-side code, CONTEXTVIEWER must be deployed to a web engine. Currently, it is working with Apache Tomcat. Besides that, the server must have Java JRE 7, Python 3 and ProM 6 installed.

3.2 Visualization of Contextual Data

Visual inspection is an effective means to understanding context data. It can give useful insights concerning the data format, structure, and meaning. Since CONTEXTVIEWER was initially designed as a visualization tool, it provides the user with several tools for data visualization. The user interface offers two interactive display modes: graphical and textual. The former is aimed to give a quick overview of collected data in an accessible form, while the latter offers a more detailed outline.

Fig. 2. CONTEXTVIEWER user interface in graphical display mode

The graphical mode (Fig. 2) was designed to be maximally clean and intuitive. It is divided into three isolated sections, reflecting tiers of context data. The rightmost section presents low level information about the device: screen state, battery level, and network information. The middle section provides higher-level information about the device (running applications, notifications) and its surroundings (weather, activity recognition). The last section is a map marking the device geographical location. Users can easily interact with the application - change the device and date, use a slider to quickly run through the data, point to any icon to get more information, or play with a fully functional GoogleMaps plugin (including StreetView).

While in the text mode, CONTEXTVIEWER lets users specify a shorter time interval with a slider and then simply displays the relevant data in tables. There are eight subsections: location, activity, weather, screen, applications, battery, wifi, and network. Although the scope of presented data is similar, the text mode is far more verbose. Take battery data for example: in graphical mode, only the battery level and adaptor status (plugged/unplugged) is presented, while the text mode includes precise measurements of voltage and temperature, as well as information about battery health, status, charges and discharges, all accurate to milliseconds. Location data are additionally plotted on a map, but now in form of splines reflecting device movement within selected time period, as opposed to a single point in graphical mode.

3.3   Basic Semantization of Contextual Data

Although data visualization is certainly a useful feature, the true power of CON-TEXTVIEWER lies in its semantization capabilities. Users can control and tweak the process by modifying the configuration file. It is a simple JSON file defining the mapping between unprocessed data from AWARE database and higher-level abstractions. Take GPS data for example. Raw values of latitude and longitude are not especially meaningful in modelling device environment or its owner's behaviour. The configuration file allows to define and label interesting areas such as workplace, home, city centre, as shown in Listing 1.1. The data is then translated accordingly.

Listing 1.1. An excerpt of the configuration file

```
{
  "label": "Work",
  "vertices": [
    { "latitude":50.066459, "longitude":19.926258 },
    { "latitude":50.061400, "longitude":19.924912 },
    { "latitude":50.067130, "longitude":19.897875 },
    { "latitude":50.071567, "longitude":19.900858 }
  ]
}
```

Time can be another example - Unix timestamp doesn't carry much information about context, but with CONTEXTVIEWER we can get extract information about time of the day, day of the week, season etc. The config file is fully customizable, well-documented and easily extendable. The output is a CSV, ARFF, or XES file, which can be easily loaded into WEKA, R, Python or ProM to perform data mining or process extraction.

3.4   Process Mining and Pattern Discovery in Contextual Data

One of CONTEXTVIEWER's modules enables process mining in contextual data. Firstly, data from an end user-specified interval is loaded. Next, on the server-side of CONTEXTVIEWER, an external tool's algorithm detects some kind of business process based on given data log. Finally, the graph visualization that resulted from the analysis is returned to the user via RPC and is displayed in the client-side application. Figure 3 shows sample output graph in CONTEXTVIEWER.

The external process mining tool invoked by CONTEXTVIEWER's server-side servlet is ProM Tools [18]. ProM is a framework supports a huge variety of process mining methods in the form of plug-ins. It is free of charge, implemented in Java and cross-platform.

Currently used version of the tool is ProM 6. A servlet invokes its methods in a batch mode using a prepared script. ProM loads the data from a XES file, generates a graph visualization of discovered process and saves it into a PNG file. Before that, contextual data is converted from MySQL tables to XES (XML-based standard for event logs, [19]) file accepted by ProM.

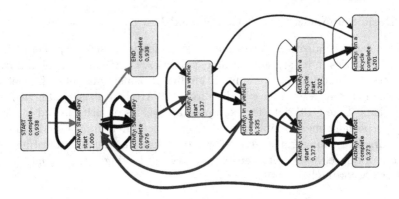

Fig. 3. Sample process discovered with ProM, visualising user transportation routines

In the following section the runtime environment was presented which allows to test and evaluate models discovered with WEKA and ProM.

4   Context Simulation and Processing

Results provided by the CONTEXTVIEWER are used to perform data analysis in WEKA or ProM, but can not be directly executed on the device. In order to allow this, system designer has to implement a related functions on the mobile-platform's side. This task could be too laborious, particularly in the phase of prototyping. Therefore, to make the whole process as seamless as possible, tools for simulating and processing data according to the declarative model provided by system designer has to be provided. This section describes CONTEXTSIMU-LATOR and HEARTDROIDtools, created specifically to fulfil these requirement.

4.1   CONTEXTSIMULATOR

The simplest way to simulate device's sensors is to reuse the historical data that was already acquired and used earlier to perform the semantization on the server's side. CONTEXTSIMULATOR is a tool implemented in Java, that fetches the required parts of the CONTEXTVIEWER's database and performs the simulation based on them. It allows system designer to repeat the series of sensors' readings in the controlled manner, first by choosing the speed of the simulation and secondly by dynamic allocation of the sensors, e.g. some of the can be turned off in order to check the robustness of the system. CONTEXTSIMULATOR was designed to extends the capabilities of the AWARE framework and therefore it is easily integrable with the existing CONTEXTVIEWER's projects.

4.2   HEARTDROID

In order to semantically process data on the device, the toolkit was enriched with the HEARTDROIDinference engine — a rule-based environment designed to meet requirements of the modern mobile platforms The most distinctive features of the engine are usage of the XTT2 knowledge representation method, which renders the knowledge as a graph of the interconnected decision tables and straightforward integration with the Android platform. The XTT2 representation is formally founded in the ALSF(FD) logic [20]. HEARTDROIDwas successfully deployed in mobile context-aware application supporting the on-line threat monitoring system for urban environments [21].

The input of the HEARTDROIDconsists of the XTT2 textual representation written in HMR— a declarative modelling language based on the Prolog syntax. The model itself has to include all the types and attributes used in the conditional and decisive parts of the rules, which are specified in the latter section of the same file as parts of the formally defined schemes; all these elements together specify clean and unambiguous semantics of the computations performed by HEART-DROID. The basic example of the HMR rule, which has semantics described by the natural language sentence: "if day attribute has value sat or sun, then set the value of the today to weekend value", is presented below:

```
xrule Today/1: [day in [sat,sun]] ==> [today set weekend].
```

More advanced modelling features of the HMR language allow handling of uncertain data (for example to represent sensors' unreliability) by usage of the certainty factors associated with the elements of the model [22] and aggregating historical data to perform statistical tests on them in the conditional part of the rules. There exist also two types of statements connecting model with the application's logic: (1) callbacks, which are used to set value of the attribute as a result of the code written in Java, (2) actions, which execute specified Java code, after the associated rule's decisional part is performed.

The following section describes briefly how the entire bundle of CONTEXTVIEWER, CONTEXTSIMULATOR and HEARTDROIDcan be used for the understanding of context and basic prototyping task.

5   Use Case Scenario

The use case scenario presented in this section was prepared using data from one month, collected with AWARE framework with LG Nexus 5 mobile phone. For the sake of simplicity and space limitation we will describe only one simple model extracted from this data with CONTEXTVIEWER. The objective of the model was to minimize energy consumption of the mobile network provider. There are four different levels of connectivity quality available on mobile devices (in descending order with respect to quality): LTE, 4G, 3G and Edge. This connectivity levels corresponds to the energy consumption levels, with LTE being most energy consuming and Edge less energy consuming.

The initial analysis of data in CONTEXTVIEWER panel revealed that there are three main location that the user is present during a day: Home, Work or in-between, called Commuting. It was also noted that the day of the user is relatively regular, therefore it is enough to partition it into: Morning, Day,

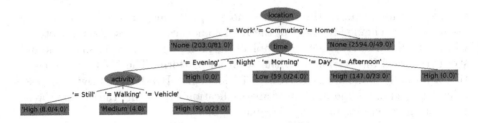

Fig. 4. Decision tree for prediction of mobile network usage with respect to user location, time and activity.

(?) location	(?) time	(?) activity	(->) network_usage		(?) network_usage	(->) mobile_status
ε {Work,Home}	ε {any}	ε {any}	:= None		= None	:= Edge
ε {Commuting}	ε {Day,Afternoon,Night}	ε {any}	:= High		= Low	:= 3G
ε {Commuting}	ε {Morning}	ε {any}	:= Low		= Medium	:= 4G
ε {Commuting}	ε {Evening}	ε {Still,Vehicle}	:= High		= High	:= LTE
ε {Commuting}	ε {Evening}	ε {Walking}	:= Medium		Table Id: tab_3 - MobileStatus	

Table Id: tab_2 - NetworkUsage

Fig. 5. XTT2 model for energy saving with respect to predicted mobile network usage. The model is based on the decision tree presented in Fig. 4

Afternoon, Evening and Night periods. More analysis of the network usage data allow to distinguish the thresholds for the usage for the: None, Low, Medium and High values. Also the activity of the user was defined based on the speed provided by the GPS sensor. All of this information was serialised into the JSON format used by the systematization module, in a form presented in the Listing 1.1 and exported to ARFF format. The decision tree generated with a WEKA is shown in Fig. 4.

The decision tree can be later transformed into the XTT2 tables as presented in Fig. 5 and evaluated with HEARTDROIDand CONTEXTSIMULATOR tools.

6   Summary and Future Work

In this paper we presented a toolkit for visualising and understanding context, which supports also rapid prototyping tasks. It is a bundle of tools integrated with AWARE – a framework designed for researched and developers for collecting, logging, and reusing context related information from mobile devices[11]. It provides also an integration with WEKA and ProM machine learning software. We focused on the phase that precedes the design and prototyping stages in a software development cycle. Our approach aim at providing tools that will allow the designer to understand the domain which alter will be modelled with appropriate methods. Although this approach could seem not scalable when we assume that the manual analysis has to be performed for every user separately, it should not be considered as a modelling phase in the development cycle. The analysis

---

[11] It is worth noting, that our work was has been acknowledged ad referenced on the AWARE framework website.

performed with CONTEXTVIEWER and Weka or ProM reveals information on the higher level of abstraction like: which attributes are good/bad for particular use-cases, what discretization ranges are good/bad for semantization purposes, what places (how many) users usually visit, etc. This phase should give the designer better insight into the data and allow to experiment and rapidly proto-type fragments of software that later will be deployed in a real-life application, possibly with an usage of automated methods.

For the future work we plan to develop tools that will automatically gener-ate XTT2 rule models used by HEARTDROIDfrom the CONTEXTVIEWER sys-tematization configuration file. This will allow to reuse the information already provided by the designer/research into the CONTEXTVIEWER and speed up the prototyping process.

References

1. Dey, A.K.: Understanding and using context. Pers. Ubiquit. Comput. **5**(1), 4–7 (2001)
2. Nalepa, G.J., Bobek, S.: Rule-based solution for context-aware reasoning on mobile devices. Comput. Sci. Inf. Syst. **11**(1), 171–193 (2014)
3. Gama, J.: Knowledge Discovery from Data Streams, 1st edn. Chapman & Hall/CRC, Boca Raton (2010)
4. Atzmueller, M., Hilgenberg, K.: Towards capturing social interactions with sdcf: an extensible framework for mobile sensing and ubiquitous data collection. In: Proceedings of 4th International Workshop on Modeling Social Media. ACM Press (2013)
5. Bardram, J.E.: The java context awareness framework (JCAF) - a service infrastructure and programming framework for context-aware applications. In: Schmidt, A., Want, R., Gellersen, H.-W. (eds.) PERVASIVE 2005. LNCS, vol. 3468, pp. 98–115. Springer, Heidelberg (2005)
6. Woensel, W.V., Casteleyn, S., Troyer, O.D.: A Framework for Decentralized, Context-Aware Mobile Applications Using Semantic Web Technology (2009)
7. van Wissen, B., Palmer, N., Kemp, R., Kielmann, T., Bal, H.: ContextDroid: an expression-based context framework for android. In: Proceedings of PhoneSense 2010, November 2010
8. Dey, A.K.: Providing architectural support for building context-aware applications. Ph.D. thesis, Atlanta, GA, USA (2000). AAI9994400
9. Kramer, D., Kocurova, A., Oussena, S., Clark, T., Komisarczuk, P.: An extensi-ble, self contained, layered approach to context acquisition. In: Proceedings of the Third International Workshop on Middleware for Pervasive Mobile and Embedded Computing. M-MPAC 2011, pp. 6:1–6:7. ACM, New York (2011)
10. Palmer, N., Kemp, R., Kielmann, T., Bal, H.: Swan-song: a flexible context expres-sion language for smartphones. In: Proceedings of the Third International Work-shop on Sensing Applications on Mobile Phones. PhoneSense 2012, pp. 12:1–12:5. ACM, New York (2012)
11. Bobek, S., Ślażyński, M., Nalepa, G.J.: Capturing dynamics of mobile context-aware systems with rules and statistical analysis of historical data. In: Rutkowski, L., Korytkowski, M., Scherer, R., Tadeusiewicz, R., Zadeh, L.A., Zurada, J.M. (eds.) ICAISC 2015. LNCS, vol. 9120, pp. 578–590. Springer, Heidelberg (2015)

12. Brezillon, P., Pasquier, L., Pomerol, J.C.: Reasoning with contextual graphs. Eur. J. Oper. Res. **136**(2), 290–298 (2002)
13. Chen, H., Finin, T.W., Joshi, A.: Semantic web in the context broker architecture. In: PerCom, pp. 277–286. IEEE Computer Society (2004)
14. Ranganathan, A., McGrath, R.E., Campbell, R.H., Mickunas, M.D.: Use of ontologies in a pervasive computing environment. Knowl. Eng. Rev. **18**(3), 209–220 (2003)
15. Gu, T., Pung, H.K., Zhang, D.Q., Wang, X.H.: A middleware for building context-aware mobile services. In: Proceedings of IEEE Vehicular Technology Conference (VTC) (2004)
16. Chetan, S., Al-Muhtadi, J., Campbell, R., Mickunas, M.D.: Mobile gaia: a middleware for ad-hoc pervasive computing. In: Consumer Communications and Networking Conference, 2005. CCNC. 2005 Second IEEE, pp. 223–228. IEEE, January 2005
17. Floch, J., Fra, C., Fricke, R., Geihs, K., Wagner, M., Lorenzo, J., Soladana, E., Mehlhase, S., Paspallis, N., Rahnama, H., Ruiz, P., Scholz, U.: Playing music - building context-aware and self-adaptive mobile applications. Softw. Pract. Exp. **43**(3), 359–388 (2013)
18. van der Aalst, W.M., van Dongen, B.F., Günther, C.W., Mans, R., De Medeiros, A.A., Rozinat, A., Rubin, V., Song, M., Verbeek, H., Weijters, A.: Prom 4.0: comprehensive support for real process analysis. In: Yakovlev, A., Kleijn, J. (eds.) ICATPN 2007. LNCS, vol. 4546, pp. 484–494. Springer, Heidelberg (2007)
19. Günther, C.W., Verbeek, H.: Xes-standard definition (2014)
20. Ligęza, A., Nalepa, G.J.: A study of methodological issues in design and development of rule-based systems: proposal of a new approach. Wiley Interdisciplinary Reviews: Data Mining and Knowledge Discovery **1**(2), 117–137 (2011)
21. Bobek, S., Nalepa, G.J., Ligęza, A., Adrian, W.T., Kaczor, K.: Mobile context-based framework for threat monitoring in urban environment with social threat monitor. Multimedia Tools and Appl. (2014)
22. Bobek, S., Nalepa, G.J.: Incomplete and uncertain data handling in context-aware rule-based systems with modified certainty factors algebra. In: Bikakis, A., Fodor, P., Roman, D. (eds.) RuleML 2014. LNCS, vol. 8620, pp. 157–167. Springer, Heidelberg (2014)

The Self-Adaptive Context Learning Pattern: Overview and Proposal

Jérémy Boes[(✉)], Julien Nigon, Nicolas Verstaevel,
Marie-Pierre Gleizes, and Frédéric Migeon

SMAC Team, IRIT, 118 rte de Narbonne, Toulouse, France
{boes,jnigon,verstaev,gleizes,migeon}@irit.fr
http://irit.fr/SMAC/

Abstract. Over the years, our research group has designed and developed many self-adaptive multi-agent systems to tackle real-world complex problems, such as robot control and heat engine optimization. A recurrent key feature of these systems is the ability to learn how to handle the context they are plunged in, in other words to map the current state of their perceptions to actions and effects. This paper presents the pattern enabling the dynamic and interactive learning of the mapping between context and actions by our multi-agent systems.

Keywords: Self-organisation · Context · Learning · Adaptation · Multi-agent system · Cooperation · Machine learning

1 Introduction

Real-world problems offer challenging properties, such as non-linear dynamics, distributed information, noisy data, and unpredictable behaviours. These properties are often quoted as key features of complexity [10]. Dealing with the complexity of the real world is an active research field. Complexity implies that models are insufficient. Systems designed for real-world have to be able to learn and self-adapt, they cannot rely on predefined models.

Thanks to their natural distribution and flexibility, self-organizing Multi-Agent Systems (MAS) are one of the most promising approaches [14]. A good way to design MASs for this type of problems is to decompose the problem following its organisation [13]. For many applications, such as bio-process control [19], engine optimization [3], learning in robotics [17], and user satisfaction in ambient systems [9], this decomposition leads to a crucial sub-problem: mapping the current state of the context with actions and their effects. Context is a word used in many domains and each domain comes with its own definition. There is several proposals for a definition of what context is [2]. In the field of problem solving, Brezillon [5] defines the context as "what constrains a step of a problem solving without intervening in it explicitly". This paper stems from the field of multi-agent systems. Agents are autonomous entities with a local perception of their environment. In this paper the *context* refers in this paper to all information

H. Christiansen et al. (Eds.): CONTEXT 2015, LNAI 9405, pp. 91–104, 2015.
DOI: 10.1007/978-3-319-25591-0_7

which is external to the activity of an agent and affects its activity. It describes the environment as the agent sees it [9].

In general a system is coupled to its environment by a cycle of observation-action. For instance for a self-adaptive system plunged in a dynamic environment, the observations become the inputs of the system and the actions its outputs. The goal of the system is to find an adequate action for the current state of inputs coming from the environment. This current state of inputs is the context. This is a mapping problem, where the current context must be mapped to an action.

When we solved the context mapping sub-problem in different applications, we have highlighted a recurrent pattern. This pattern is an abstraction of the core of several adaptive systems applied to various real-world problems and sharing the context mapping sub-problem. This recurrent pattern is named *Self-Adaptive Context-Learning Pattern* (SACL) in this paper. It is composed of an Adaptation Mechanism coupled with an Exploitation Mechanism. The Adaptation Mechanism feeds the Exploitation Mechanism with information about possible actions in the current context, while the Exploitation Mechanism is in charge of finding the most adequate action. Both mechanisms can be implemented using different approaches. In this paper, we propose an implementation of the Adaptation Mechanism based on adaptive multi-agent systems. This implementation is composed of a set of adaptive agents. Each agent represents the *effects* of a given *action* in a given *context*, the three of them being learned at runtime, in interaction with both the Exploitation Mechanism and the environment.

First, Sect. 2 details motivations and the structure of the Self-Adaptive Context-Learning (SACL) pattern. Section 3 proposes an implementation of the Adaptation Mechanism based on the AMAS approach. Section 4 shows how the Self-Adaptive Context-Learning pattern and AMAS4CL are used in two of the aforementioned applications. Finally, Sect. 5 explores an interesting relation with schema learning, before Sect. 6 concludes with some perspectives.

2   Self-Adaptive Context-Learning Pattern

The *Self-Adaptive Context-Learning Pattern* is composed of an *Exploitation Mechanism* and an *Adaptation Mechanism* both in interaction with the environment. The SACL environment is every entity which is outside SACL but affects its behaviour. So a distinction has to be done between SACL environment and the local environment of an agent (see Sect. 3). These entities can be data coming from sensors, messages from other software, effectors or systems to control. The data perceived by the pattern are called percepts. The Exploitation Mechanism is the acting entity. It has to decide and apply the action which has to be performed by a controlled system in the current context. Its decision is based on its own knowledge, including constraints from the application domain, and additional information provided by the Adaptation Mechanism. The function of the Adaptation Mechanism is to build, maintain, and provide reliable and useful knowledge about the current context and possible actions.

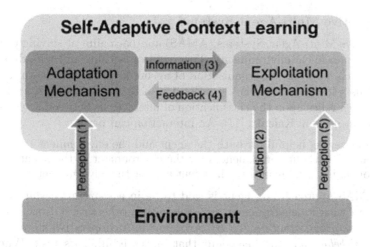

Fig. 1. The Self-Adaptive Context Learning Pattern

To gather this knowledge the Adaptation Mechanism has to correlate the activity of the Exploitation Mechanism to the observation of the environment. In order to acquire and to keep this knowledge up-to-date with the dynamics of the Exploitation Mechanism and the environment, the Adaptation Mechanism has to learn and self-adapt.

The Adaptation Mechanism and the Exploitation Mechanism are coupled entities and form a control system (Fig. 1). Adaptation Mechanism acquires (arrow 1) and provides information (arrow 3) about actions (arrow 2) made by the Exploitation Mechanism relatively to the context, while the Exploitation Mechanism relies on the information given by the Adaptation Mechanism to make these actions. Moreover, the Exploitation Mechanism sends feedback (arrow 4) to the Adaptation Mechanism to indicate whether or not the previous received information was useful. This usefulness is computed from its perceptions (arrow 5). The feedback includes two information: the last action done by the Exploitation Mechanism, and a binary qualitative evaluation of the usefulness of the previous received information.

3   AMAS4CL

The law of requisite variety states that the complexity of a control system must be greater or equal to the complexity of the controlled system [1]. This means that to control a complex system, one has to build an even more complex system. Hence, to deal with real world complexity, the SACL pattern has to be implemented with approaches able to generate and handle complex behaviours. Our proposal for an implementation of the Adaptation Mechanism is based on such an approach, called Adaptive Multi-Agent Systems (AMAS). It relies on the cooperative self-organisation of agents, and is called AMAS for Context Learning (AMAS4CL).

3.1 Adaptive Multi-Agent Systems

The Adaptive Multi-Agent Systems (AMAS) approach aims at solving problems in dynamic non-linear environments. This requires learning and self-adaptation. The approach uses the self-organisation of agents as a mean to learn and self-adapt. This self-organisation is a process driven by cooperation principles [16]. The approach relies upon the classification of interactions between a system and its environment from Kalenka [11]. An interaction can be:

- Cooperative, i.e. helpful for both the agent and the environment ;
- Neutral, i.e. without consequences for the environment or the agent ;
- Antinomic, i.e. detrimental to the agent and/or the environment.

In the AMAS approach, an agent is said to be in a *cooperative state* if all its interactions with its environment are cooperative, otherwise it is said to be in a *Non-Cooperative Situation (NCS)*. An agent in a cooperative state executes its *nominal behaviour*, the behaviour that makes it fulfill its task. When the agent detects a NCS, a specific behaviour is triggered instead of the nominal behaviour. An agent is designed to be aware of several types of NCSs:

- Incomprehension if the agent is unable to extract any information from its perceived signals ;
- Ambiguity if the agent can interpret its perceived signals in several ways;
- Incompetence if the decision process of the agent cannot be applied to its current internal representations and perceptions;
- Improductivity if the decision process of the agent leads to no action;
- Conflict if the agent estimates that its action is antinomic with its environment (including other agents);
- Uselessness if the agent estimates that its action has no effect on its environment (including other agents).

It has been shown that if all the agents of a MAS are in a cooperative state, then the MAS is functionally adequate, i.e. it is properly fulfilling its task [6]. Hence, the challenge is to design agents able to eliminate NCSs, in order to reach a cooperative state, and to always try to stay in this state despite changes in their environment. To solve the NCSs the agent changes the way it interacts with its own environment. These changes can be of three types (or any combination of them):

- Tuning: the agent adjusts internal parameters ;
- Reorganisation: the agent changes the way it interacts with its neighborhood, i.e. it stops interacting with a given neighbor, or it starts interacting with a new neighbor, or it updates the importance given to its existing neighbors.
- Openness: the agent creates one or several other agents, or deletes itself.

These behaviours are applied locally by the agents, but they have an impact on activity of the whole system. Hence, they are self-organisation mechanisms.

Applying this approach to the problem of context learning leads to a specific type of agents, called Context Agents. They are created at runtime, and self-adapt on-the fly. These agents are the core of our proposition for an implementation of the Adaptation Mechanism of the SACL pattern, described in the next section.

3.2   AMAS4CL Agents

AMAS4CL has two types of agents: Context Agents and a Head Agent.

The Head Agent is a unique agent created at the beginning allowing interactions between the Exploitation Mechanism and the inner agents in AMAS4CL. It is basically a doorman allowing messages to come in and come out. Its behaviour, described in the Sect. 3.3, enables the interoperability of the Adaptive Mechanism with any kind of Exploitation Mechanism. Although this agent performs no control over the system, its observatory position enables it to detect and repair some NCSs (see Sect. 3.4).

AMAS4CL starts with no seed, no predefined Context Agent. They are all created at runtime (see Sect. 3.4). A Context Agent has a tripartite structure: $<context, action, appreciation>$. The first part, the *context*, is a set of intervals called *validity ranges*. There is one validity range for each percept of the Context Agent.

Definition 1. *A validity range r_p, associated to the percept p, is valid if $v_p \in r_p$; where v_p is the current value of p.*

By definition, a Context Agent is valid if all its validity ranges are *valid*. When a Context Agent is valid, it sends an action proposition to the Head Agent. The *action* is a modification of the environment. For instance, it can be the incrementation or decrementation of a variable. It can also be a high level action, such as "go forward" for a robot. It is domain dependent.

The *appreciation* is an estimation of the effect of an action at the creation of the Context Agent. It may be for example the expected effects of the action on the environment, the estimated gain following a given measure, a relevance score, or a combination of these, etc. A Context Agent has to adjust this *appreciation*. For this, it possesses an evaluation function determining after a proposal if the *appreciation* is *wrong* or *inexact*. *Wrong* means that the difference between the estimated *appreciation* and the observed effect is unacceptable (from the designer point of view) whereas *inexact* means that the error is tolerable. Because the *appreciation* depends on the application domain, the estimation function is given by the designer.

A Context Agent can be seen as a tile delimited by its validity ranges. When a Context Agent sends a proposition, it basically says: "here is what I think of this action in the current context". The Exploitation Mechanism receives several propositions and uses them to decide what is the best action to be applied. This is detailed in Sect. 3.3.

3.3   Nominal Behaviour

The *nominal behaviour* is the behaviour enabling the agent to perform its tasks in a cooperative situation [4]. According to the AMAS approach, we can model the nominal behaviour of an agent through a Perception-Decision-Action cycle.

Nominal Behaviour of Context Agents:

- **Perception:**
    - 1: Receives from the environment a set of variable values (percepts).
    - 2: Receives the feedback from the Exploitation Mechanism. This feedback is composed of a binary qualitative information about the information provided by AMAS4CL at the previous decision cycle ($ENOUGH/NOT\_ENOUGH$) and the last action performed by the Exploitation Mechanism.
- **Decision and Action:**
    - 1: The Context Agent checks whether it is *valid* or not by comparing the current set of variable values to its *context*.
    - 2: If the Context Agent is valid, it sends an *action* proposition associated with the *appreciation* of this *action* to the Head Agent.

Nominal Behaviour of Head Agent:

- **Perception:**
    - 1: Receives the feedback from the Exploitation Mechanism.
    - 2: Receives action propositions from Context Agents.
- **Decision and Action:**
    - 1: Gathers all the action propositions and sends them to the Exploitation Mechanism.
    - 2: Forward the feedback from the Exploitation Mechanism only to the Context Agents that had proposed the action contained in the feedback.

There are several cases where these behaviours fail. These cases are NCS. To solve them, the agents have to self-adapt, i.e. to modify their behaviour. This is detailed in the next section.

3.4 Non-Cooperative Situations

NCS are described in two parts: the detection of the NCS, and its resolution. These mechanisms of detection-resolution are completing the behaviour of Context Agents.

NCS 1: Conflict of a Context Agent (wrong Appreciation)

- **Detection:** Thanks to the feedback from the Exploitation Mechanism, a Context Agent knows when its action is being applied. When its action is being applied, the Context Agent observes the effects of the action, to check if its *appreciation* is correct. If the agent evaluates that it has given wrong information to the the Exploitation Mechanism, it is a conflict NCS. The interaction between the Context Agent and the Exploitation Mechanism is flawed and prevents one of them to fulfill its task.
- **Resolution:** The Context Agent estimates that it should not have been valid. To solve this NCS, the Context Agent reduces its validity ranges to avoid to make a proposal in a context where it is unable to give a good *appreciation*.

NCS 2: Conflict of a Context Agent (inexact Appreciation)

- **Detection:** This NCS is similar to the conflict NCS of wrong *appreciation*. If a Context Agent considers that its *appreciation* is inexact, it is also conflict NCS. The inexactitude of the appreciation prevents an optimal activity of the Exploitation Mechanism.
- **Resolution:** This NCS is less harmful, the Context Agent estimates it was still right to make a proposal, it only needs to give a more accurate *appreciation*. Thus, the agent does not change its validity ranges, but adjusts its *appreciation*, using information from the feedback of the Exploitation Mechanism and from the observation of the environment.

NCS 3: Uselessness of a Context Agent

- **Detection:** Sometimes, after successive adjustments of their ranges, Context Agents may have one or more range greatly reduced. If this range becomes smaller than a user-defined critical size (for instance: zero, or very close to zero), the Context Agent considers itself as useless, since it has no chance of being valid again.
- **Resolution:** The agent self-destroys.

NCS 4: Improductivity of the Head Agent

- **Detection:** If the Head Agent receives a feedback containing an action that was not proposed at the previous step, the decision leads the agent to forward it to no one. It is an NCS of improductivity. It occurs when no proposal was received by the Head Agent (for instance, at the start, when there is no Context Agent in the system), or when none of the proposed actions were applied by the Exploitation Mechanism. This NCS means the Exploitation Mechanism had to explore by executing a new *action*. This exploration is domain dependant.
- **Resolution:** If the Head Agent had received proposals with the action contained in the feedback earlier in its lifetime, it requests the Context Agents that had sent it to expand their validity ranges toward the current context. If nobody is able to do that (Context Agents may reject this request if the adjustment is too big), the Head Agent creates a new Context Agent with the new action received in the feedback, and initializes it with the current context and a first appreciation that depends on the domain.

4   Applications

This section gives examples on how the context-learning pattern is used to solve real-world problems. We focus on two different applications: ALEX, a multi-agent system that learns from demonstrations to control robotic devices and ESCHER, a multi-agent system for multi-criteria optimization.

4.1 ALEX

Service robotic deals with the design of robotic devices whose objectives are to provide adequate services to their users. User needs are multiple, dynamic and sometimes contradictory. Providing a natural way to automatically adapt the behaviour of robotic devices to user needs is a challenging task. The complexity comes with the lack of way to evaluate user satisfaction without evaluating a particular objective. A good way to handle this challenge is to use Learning from Demonstrations, a paradigm to dynamically learn new behaviours from demonstrations performed by a human tutor. With this approach, each action performed by a user on a device is seen as a feedback. Through the natural process of demonstration, the user not only shows that the current device behaviour is not satisfying him, but also provides the adequate action to perform. Adaptive Learner by EXperiments (ALEX) [17] is a multi-agent system designed to face this challenge.

System Overview. Our approach considers each robotic device as an autonomous agent in interaction with other components, humans and the environment through sensors [18]. An instance of ALEX is associated with each robotic device composing the system (basically one by effector). For example in the case of a two wheeled rover, each wheel has its own controlling system. ALEX receives a set of signals coming from sensors and the activity of the human tutor performing the demonstration. The tutor can perform at runtime a demonstration by providing to ALEX the adequate action to perform. The problem is then to map the current state of sensors to the activity of the user in order to be proactive the next time a similar situation occurs and the tutor provided no action. ALEX is built on the SACL pattern and AMAS4CL (Fig. 2).

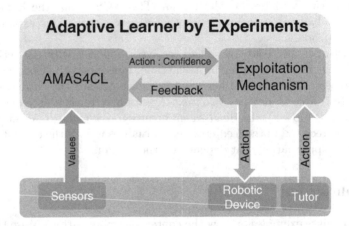

Fig. 2. ALEX is based on the Self-Adaptive Context-Learning Pattern using AMAS4CL

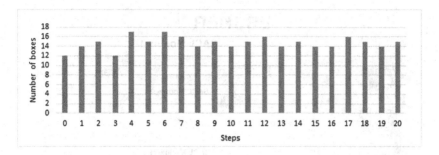

Fig. 3. Number of collected boxes each 5 min. The step 0 corresponds to the reference score.

AMAS4CL must provide to the Exploitation Mechanism the adequate action to satisfy the tutor. However, tutor satisfaction is not directly observable and it is difficult to correlate the effect of an action on tutor satisfaction without evaluating an a priori known objective. The *appreciation* part of Context Agents is a dynamically built confidence value. This confidence value is increased each time a Context Agent is activated and proposes the same action as the tutor's, and is decreased otherwise. Each Context Agent dynamically manages its confidence value thanks to the feedback from the Exploitation Mechanism. Then, at each decision cycle, the action proposed by the activated Context Agent with the utmost confidence is sent to the Exploitation Mechanism. More information on the implementation of this confidence value can be found on previous works [9].

Results. ALEX has been tested both on simulation and on a real Arduino based robot [18]. The experiment is a collecting task involving a two wheeled rover inside an arena. A user performs a five minutes demonstration of a collecting task allowing the rover to learn the desired behaviour. The number of collected artifact is computed during the demonstration and this score is compared to the score performed by the rover. Experiments have shown that ALEX increases system capacity to perform this task: the rover often collects more artifacts than the user (see Fig. 3) [18].

4.2   ESCHER

ESCHER, for Emergent Self-organized Control for Heat Engine calibRation, is a multi-agent system able to learn in real-time how to control another system. It was used to find the best control for heat engines in order to calibrate them [3]. The main problem is to find which action will increase the satisfaction of a set of user-defined criteria, regarding the current state of the controlled system. In other words, the problem includes the mapping of the controlled system state-space with actions and their effects, hence the use of the SACL pattern.

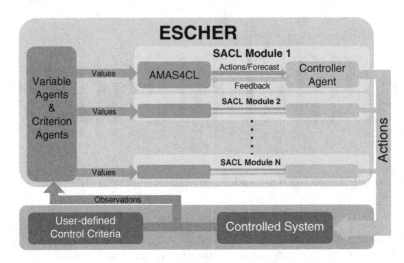

Fig. 4. ESCHER is based on the Self-Adaptive Context-Learning Pattern using AMAS4CL

System Overview. The environment of ESCHER is composed of the Controlled System along with the user's criteria for the control, such as setpoints, thresholds, and optimization needs. ESCHER performs closed-loop control, while simultaneously learning from its interactions with the environment. This learning is handled by several distributed SACL modules, implemented with AMAS4CL.

Figure 4 shows an overview of ESCHER. There is one SACL module per input of the Controlled System. The Exploitation Mechanism of each module is an agent called Controller Agent, in charge of applying the most adequate action on its associated input of the Controlled System. Along with SACL modules, ESCHER includes an internal representation of its environment in the form of Variable Agents and Criteria Agents. There is one Variable Agent per input and output of the Controlled System. A Criterion Agent represents an optimization need, a threshold to meet, or a setpoint to reach. A Criterion Agent expresses its satisfaction in the form of a *critical level:* the lower the better, zero means the criterion represented by the agent is fully satisfied.

ESCHER works with continuous systems and has an explicit representation of criteria. This has led to the following characteristics for the Context Agents of AMAS4CL instances:

- The *context* part of a Context Agent is a set of adaptive range trackers, a specific tool to learn value ranges with simple binary feedback;
- The *action* part is a modification of an input ($+\delta$ or $-\delta$);
- The *appreciation* part is a set of forecasts about the variation of the critical levels.

Controller Agents receive action proposals with their expected effects on the criteria satisfaction, and select the most adequate action to apply, meanwhile sending appropriate feedback to the Context Agents. When no proposal is received,

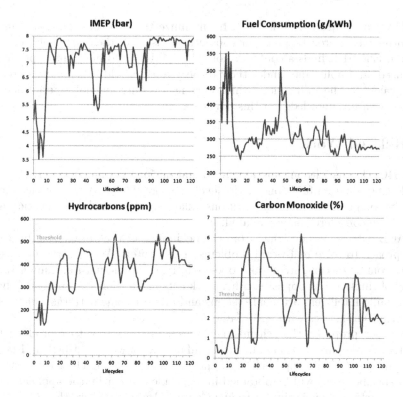

Fig. 5. An engine optimization performed by ESCHER

a Controller Agent applies a new action, chosen randomly. When no received proposal includes an adequate action, it does the same, but excluding actions known for their bad effects. There are several additional NCSs to the four of AMAS4CL. In particular, Context Agents are able to adjust their proposed *action*.

Results. ESCHER has been experimented on both artificial and real cases. Here are only presented the results of an experiment on a real 125 cc monocylinder fuel engine. More detailed results can be found in [3] and will be published in future papers. The goal of this experiment is to perform a classical optimization of the engine: maximizing the torque, minimizing the fuel consumption, and meeting pollution thresholds. ESCHER controls three parameters: the injected mass of fuel, the ignition advance, and the start of injection. Thus, there are three instances of SACL/AMAS4CL.

Figure 5 shows the evolution of the outputs of the engine over time: the Indicated Mean Effective Pressure (IMEP, an image of the torque), the fuel consumption, and the two considered pollutants: hydrocarbons and carbon monoxide. At the beginning, ESCHER has no Context Agent. Controller Agents make mistakes: during the first dozen of lifecycles, the IMEP drops, and the fuel consumption rises. After a while, ESCHER has acquired enough knowledge and is able to maximize the IMEP and to minimize the fuel consumption. During the optimization process, pollutants sometimes rise over their threshold. But, in the end, ESCHER finds a configuration where IMEP and fuel consumption are optimized, and pollutants under their respective threshold. The obtained result is equivalent to the calibration, which currently is only manually performed by experts with full knowledge of this particular engine.

5   Related Work

AMAS4CL learns without any global evaluation function about it activity, but only local evaluation functions at the level of agent. Thus, every approach using objective or global evaluation functions (like neural networks or genetic algorithms) are considered as not related.

SACL pattern was designed from the beginning as a self-organized, bottom-up approach to solve real-world problems. The Context Agents structure, associating context/action/prediction, does not try to mimic nature, human brain or way of thinking. At opposite, other work gets inspiration from human behaviour to build intelligent artificial systems and propose some patterns which have interesting similarities with AMAS4CL.

Drescher proposed the schema mechanism [8] as a way to reproduce the cognitive development of human during infancy as it was described by Piaget. The main part of this approach is the notion of schemas. Each schema asserts that a specific result will be obtained in a specific context by the application of a specific action. This is similar to our Context Agent structure (Fig. 6).

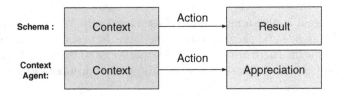

Fig. 6. Schema and Context Agent structures

Schema mechanism is able to learn about its environment by three means: induction (creating new schemas to express regularity in the environment), abstraction (creating new actions as a combination of primitive actions) and conceptual invention(creating new elements to represent unknown state of the world). As AMAS4CL, schema mechanism does not use any objective function and can learn with very few specifications about the environment. Some of the limitations of the schema mechanism were improved by later works, in particular (but not restricted to) the intense need for resource by Chaput [7] and Perotto [15], or the difficulty to handle a continuous environment [12].

Even if AMAS4CL and constructivist learning approach share some properties, their learning strategies are strongly different. It is interesting that two approaches, one starting from a representation of the child's cognitive development, and the based on cooperative interactions between agents, stemming from system theory, converge towards a similar structure for information-carrying entities (schema and Context Agent).

6 Conclusion

This paper presents the Self Adaptive Context Learning Pattern designed to handle real-world non-linear problems. This pattern relies on two parts: the Exploitation Mechanism which acts on the environment, and the Adaptation Mechanism which provides information to the Exploitation Mechanism. The Adaptation Mechanism acquires and adjusts these information by the observation of the environment.

AMAS4CL is a multi-agent system which enables this adaptation, through the use of cooperative agents. By minimizing the need of assumptions on the studied system, the approach is generic and was used in a wide variety of complex real-world problems, like complex systems control [3], robotics [18], or ambient systems [9].

The ability to handle real-world complexity could be achieved by a better understanding of the context, and by efficient mapping between current context and relevant information for the Exploitation Mechanism. But, in dynamic environments, a static mapping is somewhat limited. The SACL pattern and its implementation brings an adaptive approach to handle this problem.

Several works are currently underway using AMAS4CL, in the fields of networks control, generation of complex system models and human user behaviour understanding. These works are opportunities to improve AMAS4CL. Future work includes formalisation of AMAS4CL and comparison with other approaches, such as schema learning.

References

1. Ross Ashby, W.: An Introduction to Cybernetics. Chapman & Hall, London (1956)
2. Bazire, M., Brézillon, P.: Understanding context before using it. In: Dey, A.K., Leake, D.B., Kokinov, B., Turner, R. (eds.) CONTEXT 2005. LNCS (LNAI), vol. 3554, pp. 29–40. Springer, Heidelberg (2005)
3. Boes, J., Migeon, F., Glize, P., Salvy, E.: Model-free optimization of an engine control unit thanks to self-adaptive multi-agent systems. In: ERTS2, Toulouse, SIA/3AF/SEE, pp. 350–359 (2014)
4. Bonjean, N., Mefteh, W., Gleizes, M.-P., Maurel, C., Migeon, F.: Adelfe 2.0. In: Cossentino, M., Hilaire, V., Molesini, A., Seidita, V. (eds.) Handbook on Agent-Oriented Design Processes, pp. 19–63. Springer, Heidelberg (2014)
5. Brézillon, P.: Context in problem solving: a survey. Knowl. Eng. Rev. 14(01), 47–80 (1999)
6. Capera, D., George, J.-P., Gleizes, M.-P., Glize, P.: The amas theory for complex problem solving based on self-organizing cooperative agents. In: Proceedings of the Twelfth IEEE International Workshops on Enabling Technologies: Infrastructure for Collaborative Enterprises, WET ICE 2003, pp. 383–388 (2003)

7. Chaput, H.H., Kuipers, B., Miikkulainen, R.: Constructivist learning: a neural implementation of the schema mechanism. In: Proceedings of the Workshop on Self-Organizing Maps (WSOM 2003) (2003)
8. Drescher, G.L.: Made-Up Minds: A Constructivist Approach to Artificial Intelligence. MIT Press, Cambridge (1991)
9. Guivarch, V., Camps, V., Péninou, A.: AMADEUS: an adaptive multi-agent system to learn a user's recurring actions in ambient systems. Adv. Distrib. Comput. Artif. Intell. J., Special Issue 1(3), 1–10 (2012)
10. Heylighen, F., Bates, J., Maack, M.N.: Encyclopedia of Library and Information Sciences. Taylor & Francis, London (2008)
11. Kalenka, S.: Modelling social interaction attitudes in multi-agent systems. Ph.D. thesis, Citeseer (2001)
12. Mazac, S., Armetta, F., Hassas, S.: On bootstrapping sensori-motor patterns for a constructivist learning system in continuous environments. In: Alife 14: Fourteenth International Conference on the Synthesis and Simulation of Living Systems (2014)
13. Noel, V., Zambonelli, F.: Engineering emergence in multi-agent systems: following the problem organisation. In: 2014 International Conference on High Performance Computing & Simulation (HPCS), pp. 444–451. IEEE (2014)
14. Panait, L., Luke, S.: Cooperative multi-agent learning: the state of the art. Auton. Agents Multi-Agent Syst. 11(3), 387–434 (2005)
15. Perotto, F.S., Vicari, R., Alvares, L.O.: An autonomous intelligent agent architecture based on constructivist AI. In: Bramer, M., Devedzic, V. (eds.) Artificial Intelligence Applications and Innovations. IFIP, vol. 154, pp. 103–115. Springer, New York (2004)
16. Di Marzo Serugendo, G., Gleizes, M.-P., Karageorgos, A.: Self-organising systems. In: Di Marzo Serugendo, G., Gleizes, M.-P., Karageorgos, A. (eds.) Self-organising Software, pp. 7–32. Springer, Heidelberg (2011)
17. Verstaevel, N., Régis, C., Gleizes, M.-P., Robert, F.: Principles and experimentations of self-organizing embedded agents allowing learning from demonstration in ambient robotic. Procedia Comput. Sci. 52, 194–201 (2015). The 6th International Conference on Ambient Systems, Networks and Technologies (ANT 2015)
18. Verstaevel, N., Régis, C., Guivarch, V., Gleizes, M.-P., Robert, F.: Extreme sensitive robotic a context-aware ubiquitous learning. In: ICAART, INSTICC, vol. 1, pp. 242–248 (2015)
19. Videau, S., Bernon, C., Glize, P., Uribelarrea, J.-L.: Controlling bioprocesses using cooperative self-organizing agents. In: Demazeau, Y., Pěchouček, M., Corchado, J.M., Bajo Pérez, J. (eds.) Advances on Practical Applications of Agents and Multiagent Systems. AISC, vol. 88, pp. 141–150. Springer, Heidelberg (2011)

Method of iBeacon Optimal Distribution for Indoor Localization

Jan Budina, Ondrej Klapka$^{(\boxtimes)}$, Tomas Kozel, and Martin Zmitko

Faculty of Informatics and Management, Department of Informatics
and Quantitative Methods, University of Hradec Kralove,
Hradec Kralove, The Czech Republic
{Jan.Budina,Ondrej.Klapka,Tomas.Kozel,
Martin.Zmitko}@uhk.cz

Abstract. Currently one of the most widely used systems for GPS locating is very suitable for outdoor environment, but in buildings is practically not applicable. For indoor locating, the device based on Bluetooth can be used. It can measure and determine the approximate location. This article deals with the issue of building a network of transmitters on the basis of which the position can be determined and presents a new method for the assessment of a layout of individual transmitters in the building.

Keywords: iBeacon · Localization · Mobile context · Bluetooth 4.0 low energy

1 Introduction

Fast wireless data networks, smart mobile phones as versatile and multifunctional devices, along with the decreasing price of these components are now increasingly talking about Internet of things. The Internet of things, i.e. interconnection of commonly used devices should greatly facilitate our lives and create smart environments. Smart environments can then automatically respond to user's initiatives and help him/her solve everyday problems.

One of the characteristics of smart environments is that equipment working in it is able to interact and "intelligently" respond to the user's initiatives or automatically based on the current context to predict the user's desires. A simple example of interaction in such an environment can be e.g. automatic dampening of the lighting in the room after a user begins to watch a film from the DVD player. The smart environment enables to provide a wide range of services based on the current context (state) without asking them.

The following basic contexts are usually distinguished: [1]

- Location
- Identity
- Time
- Activity currently carried out, state

The following text will focus on the issue of localization inside buildings in order to provide contextually relevant services.

H. Christiansen et al. (Eds.): CONTEXT 2015, LNAI 9405, pp. 105–117, 2015.
DOI: 10.1007/978-3-319-25591-0_8

2 The Issue of Localization in Buildings

One of the most common ways of the localization is Global Positioning System (GPS) technology, which has basically become a standard for localization in the open field [2]. Nevertheless, the localization using GPS module device does not have in some cases sufficient accuracy and it is difficult to use it inside buildings [1, 2]. For localization inside buildings, it is therefore advisable to choose a different technology.

Bluetooth Smart 4.0 technology, which has very low energy demands on its device in comparison with previous types [3], has become an interesting medium for localization inside buildings. So as to be able to perform the localization using this technology some infrastructure should be created, similar to that which is used at GPS, where the localization infrastructure is formed by satellites. Apple's standard called iBeacon, which was established in 2013 [4], was chosen to create such a localization infrastructure.

iBeacon is a device - an electronic beacon, enabling to transmit certain data on Bluetooth 4.0 LE (low energy) protocol. These data may be information on the current ongoing discounts in the shop on a certain type of goods, data with the position etc. The device consists of a small transmitter powered with a 1.5 V-hour battery that can supply the device according to required performance for several years. The battery life varies from manufacturer to manufacturer [5, 6].

For the localization a device which can process Bluetooth 4.0 LE signal is needed. Theoretically, only Bluetooth 4.0 support is required, but practically the support in the actual operating system device is necessary. As the author is Apple, support in iOS version 7 and above is exemplary. The most widely used Google Android system supports this technology without restriction from version 5.0 and the third most frequently represented Windows Phone integration is not even in Version 8.1. iBeacon transmitters can be used for both possibilities of targeted ads and the possibilities for localization providing proper positioning of each transmitter.

Each iBeacon has its own unique identifier. With the ability to measure the signal strength from the iBeacon or more iBeacons, the position of the device which receives the signal/signals can be determined. This positioning system requires a functioning digital map of the area in which the localization will be carried out and also the well known position of individual transmitters. Nevertheless, determining the position based on the signal strength from of each iBeacon may be hiding certain issues in the form of difficulty in determining the correct position of each iBeacon in covering space.

3 Related Work

The accessible literature describes many ways of localization based on the method of collecting and processing fingerprints of signals. Two basic groups can be identified [7]:

- A calculation based on the location of the transmitters of signals.
  - Triangulation method is used to determine the positioning.

- Comparing the measured values of received signal/signals with stored reference fingerprints in the database. It can be further divided into [9]:
  - The nearest neighbor
  - K-nearest neighbors
  - Probabilistic method
  - Method of Neural Networks
  - Support vector machine
  - Smallest M-vertex polygon

Due to the diversity of different algorithms and particularly with regard to their accuracy and resistance to accidental impacts, various solutions, how transmitters should be distributed/placed, can be found in the literature [8, 13, 14].

Some solutions are based on observations of radio signal spread used when measuring the signal strength at different locations [10]. Therefore, in principle, it is an identification of locations of transmitters based on power, noise level and signal quality. To ensure reliable results, the measurement must be repeated several times.

Another approach is to use discrete mathematical methods. It is a distribution of the covered area with a cell grid. Particular transmitters are then placed in the middle of the cells. It is very important to select an appropriate accuracy of the grid [13–18].

Some models use both of these approaches. To optimize the deployment, a signal loss (path loss) or the signal strength (power) can be used. Model based on signal loss can be built on different properties which are known to spread the signal. It may be, for example, the average loss of signal on the path, the maximum loss or convex combination [19].

The model can be used both for calculating the coverage of the area of the transmitter, but also for determining the maximum distance between two transmitters. Convex combination of average and maximum loss is given by [19]:

$$F_1(a_1,\ldots,a_N) = \frac{1}{M}\sum_{i=1}^{M}\min_{j} g(a_j, r_i) \tag{1}$$

$$F_2(a_1,\ldots,a_N) = \max_{i}\min_{j} g(a_j, r_i) \tag{2}$$

where F_1 is a function of the average signal loss and F_2 is a function for maximum signal loss. Other variables are explained in Table 1. The functions of convex combination is then the following:

$$F_3 = \psi F_1 + (1 - \psi)F_2 \tag{3}$$

for which the following applies:

$$\min_{j} g(a_j, r_i) \le g_{\max} \ \forall i = 1, \cdots, M \tag{4}$$

Relationship of convex combination of F_1 and F_2 is then:

Table 1. Description of the various variables in the formulas.

Variable	Significance
a_j	Transmitter
r_i	Receiver
$d(a_j, r_i)$	Distance between the transmitter (a_j) and the receiver (r_i)
$g(a_j, r_i)$	Loss path from the transmitter (a_j) to the receiver (r_i)
g_{max}	Maximum tolerated loss path
P_t	Transmit power
P_r	Received power
R_{th}	Received threshold
μ	Limiting parameter for selecting the maximum allowable loss
$\psi \in [0,1]$	Convex combination of coefficient to define the sum of a maximum function

$$F_3(a_1, \ldots, a_N) = \psi \left(\frac{1}{M} \sum_{i=1}^{M} \min_j g(a_j, r_i) \right) + (1 - \psi) \left(\max_i \min_j g(a_j, r_i) \right) \quad (5)$$

The referred relationship describes that the signal loss is assessed against the maximum allowable g_{max} loss. This limitation ensures that the quality of the received signal will always be above a certain threshold. Determining thresholds can be done by the following relationship.

$$g_{max} = P_t - R_{th} \quad (6)$$

To determine the number (N) of required transmitters, the following algorithm is used. The initial setting $N = 1$:

```
1. Attempt to solve the problem (1) - (2)
2. Is there any solution?
        a. Then N is a number of transmitters needed
        b. Else N = N + 1
3. Repeat step 1
```

Just right the number of transmitters can be a significant parameter for the future draft of the optimum deployment of iBeacons [21], because for localization purposes, at least three transmitters should be seen at each site [19, 20]. Particular tools for deployment of transmitters for localization purposes in the interior are presented below.

3.1   Ekahau - Optimum AP Deployment for Wifi Signal Emission

Ekahau company focuses on providing a system for determining the position via received Wi-Fi signal called Site Survey. This system seems to be appropriate for a number of similarities between Wi-Fi and Bluetooth technologies. Based on the map of covered area and after defining the required parameters the Site Survey system is able to propose appropriate deployment of Wi-Fi Access Points. Defined parameters are as follows: [21].

- Minimum signal strength of the most powerful AP (Wi-Fi Access Point)
- The minimum signal to noise ratio (SNR)
- Minimum data transfer rate
- Minimum number of APs in range
- Channel Overlap (the maximum number of audible APs on the same channel)
- Minimum Ping Round Trip Time
- Packet loss (% of packet losses calculated from the last 10 packets). The system uses visualization of the estimated signal emission in the map to determine the optimal deployment of AP in the area. With this tool, it is possible to observe a number of parameters such as the number of clients per AP, access to each AP, capacity visualization, the overall coverage, data rate or the expected locations with noise. An interesting visualization is a number of audible AP at the site.

3.2   Creation of HeatMap

To search the appropriate position of transmitters, it is possible to use relatively accurate but a computationally demanding method called HeatMap formation that is formed as follows. Firstly, a 2D model building outlining the individual rooms is created while it is not necessary to record any parameters on the barriers affecting the spread of the radio signals. Then we record measured values of signals at various points of the building in the HeatMap. The values are recorded in the prepared model.

Obviously, it is not possible to perform measurements at each point of the building, but the more measurement is taken, the more accurate model is obtained. Based on these measurements, the coverage for each point in the building is calculated then. The basic algorithms work under the assumption of a linear signal loss. If we make measurements before the obstacle/barrier, the algorithm records these very different values of the received signal and is also able to process the HeatMap.

The advantage of this method is high accuracy, which is affected only by the amount of measurements performed. The main disadvantage is the difficulty of conducting a number of measurements while search for the optimal deployment was necessary at every change of deployment of transmitters to create a new HeatMap.

The advantage of this method is high accuracy, which is affected only by the amount of measurements performed. The main disadvantage is the difficulty of conducting a number of measurements while searching for the optimal deployment, it was necessary to create a new HeatMap at every change of deployment of transmitters.

4   Optimum Location of iBeacon

Indoor localization can be performed in a four-wall windowless room where it would work reliably, however real and unstructured. The environment in the form of various types of walls and other obstacles can result in complications that make it difficult to find an appropriate placement of iBeacon to get the most accurate outcome. To cover the largest possible amount of practical situations, model situations, in which measurements are created, will be developed. The measurement results will serve as a basis for further research in the use of iBeacon.

In each model situation the following data will be recorded:

- room layout (plan),
- masonry materials (concrete, glazing, etc.),
- density of built-up area (computer lab, auditorium, etc.),
- arrangement of objects in the room,
- potential sources of interference.

In the specific case, for example, it is a rectangular room where the tables and computers are placed.

Model situations will serve as a data source for subsequent assessment and resulting determination of the optimal placement for a particular situation. Several test scenarios that differ both in deployment of iBeacon equipment and in their settings will be ready for each model situation. The aim is to monitor how the signal power and other parameters of the source transmitter are changing. The main assumption, which must be met to determine the appropriate deployment of iBeacons, is visibility of three or more devices with sufficient signal power from most of the possible positions in the model situation.

4.1   Transmitted Data by iBeacons

Localization of the mobile device can be performed on the basis of broadcast data by an iBeacon transmitter. The data transmitted by iBeacon are described in detail in the following text.

Bluetooth operates in the ISM 2.4 GHz frequency band (similar to WiFi). According to [11] the communication itself consists of two parts of promotion (advertising) and connections (connecting). The promotion is one of the ways to make devices well-known. Subsequently iBeacon sends so called Transmit Packet at interval of 20 ms to 10 s. The actual packet has a length of 47 bytes. Its structure is shown in Fig. 1.

The data is further divided into:

- iBeacon prefix (9 B)
- Proximity UUID (16 B)
- Major (2 B)
- Minor (2 B)
- TX power (1 B).

TX power value that indicates the strength of the transmitted signals is the most important for localization. On the basis of the intensity of the transmitted signal (TX power) and the received signal intensity (RSSI), approximate location can be determined [22].

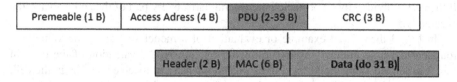

Fig. 1. Packet structure transmitted by iBeacon.

5   The Method of Optimum Allocation of iBeacon Position

The task of the proposed method is to cover as much space by the least number of iBeacon (transmitters). To obtain the optimum solution, at first the model will be created, which will then be tested in a real environment.

The proposed model is based on the division of the area, which is intended to be covered with a signal, to square shaped cells of a predetermined size. The size of cells in the model corresponds to the accuracy that is required for positioning of proposed solutions. It is recommended to choose a network of cells so that the interface of each cell coincides with obstacles in the space, e.g. the wall of a building or other obstacles, significantly suppressing electromagnetic radiation.

If the obstacle is located within a cell, similarly as it is shown in Fig. 2, the cell must be split at the point through which an obstacle passes.

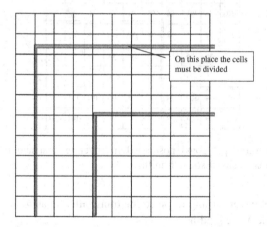

Fig. 2. Covering space (Curved corridor), divided into cells.

In the second stage, the model is equipped with transmitters whose location is chosen with respect to adequate free space around individual transmitters. Based on the predicted signal damping with increasing distance from the transmitter and on the basis of the radiation characteristics of the transmitter, the individual cells are rated on a scale expressing the estimated signal strength in the middle of the cell. The largest value is determined as the evaluation of the cell which is in reach of more than one transmitter.

If the cell is divided by an obstacle into several parts, every part of the cell is evaluated separately.

In Fig. 3 there is an example of evaluation of a model situation. The values are rounded to whole numbers by the following relationships. Evaluation of the cells in Fig. 3 is defined as the RSSI level of estimated intensity of received signals in the cell, according to the formula expressing the signal fading with increasing distance [12]:

$$C_{RSSI} = -(10 \cdot n) \log_{10}(d) + A, \tag{7}$$

where n is a signal propagation constant, d is a distance from the transmitter and A is a set value of the transmitted signal power (Tx Power). Based on experimental testing, the value $n = 3, 5$ was determined. The value of Tx Power was set to $A = -74$. Signal attenuation caused by the barrier must be determined experimentally regarding the character of the obstacles. The calculation may be further refined with respect to the transmitter radiation characteristics.

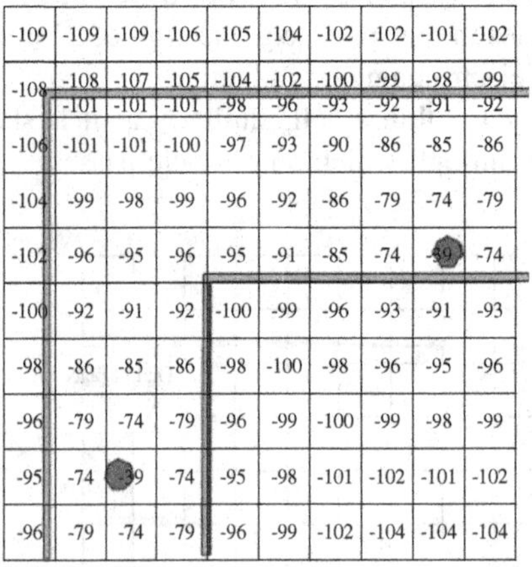

-109	-109	-109	-106	-105	-104	-102	-102	-101	-102
-108	-108	-107	-105	-104	-102	-100	-99	-98	-99
	-101	-101	-101	-98	-96	-93	-92	-91	-92
-106	-101	-101	-100	-97	-93	-90	-86	-85	-86
-104	-99	-98	-99	-96	-92	-86	-79	-74	-79
-102	-96	-95	-96	-95	-91	-85	-74	-69	-74
-100	-92	-91	-92	-100	-99	-96	-93	-91	-93
-98	-86	-85	-86	-98	-100	-98	-96	-95	-96
-96	-79	-74	-79	-96	-99	-100	-99	-98	-99
-95	-74	-69	-74	-95	-98	-101	-102	-101	-102
-96	-79	-74	-79	-96	-99	-102	-104	-104	-104

Fig. 3. Transmitters (gray polygons) positioned in covering area and the cell assessment according to the estimated signal strength in the cell.

In such a model we are trying to place the transmitter so as to achieve the greatest value of the variable S, defined as:

$$S = \sum_{i=1}^{m} \sum_{j=1}^{n} \begin{cases} c_{i,j} \in C \\ \overline{x_{i,j}} \in \overline{X} \end{cases}, \tag{8}$$

where c is evaluation of the undivided cell, C is a set of retained cells, $\overline{x}$ is a weighted average of the evaluation of all parts of the undivided cells, $\overline{X}$ is a set of all cells divided and m and n is a number of cells in the model in horizontal and vertical direction. Calculation of the weighted average $\overline{x}$ is given by:

$$\bar{x} = \frac{\sum\limits_{i=1}^{n} x_i \cdot \frac{s_i}{s_c}}{s_c}, \tag{9}$$

where n is a number of parts of the cell, s_c is an area of the whole cell, s_i is an area of the i-th part of the cell, and x_i is evaluation of the i-th part of the cell. By finding the maximum of S value, the optimum area coverage by transmitters will be obtained. K coefficient is then calculated from the total sum. The coefficient expresses the level of coverage area, as defined in the following relation:

$$K = \frac{S}{m \cdot n}, \tag{10}$$

where n is a number of cells in the horizontal direction and m is a number of cells in the vertical direction.

6   Validation of the Model

The model of transmitters deployment has been tested on a set of model situations (Fig. 4) using test scenarios. All measurements were carried out by 6 Nexus telephone with Android 1.5. Estimote (from the manufacturer) and the Locate Beacon applications were used for measurements. When calculating the values in the model, obstacles were taken into account. The expected value in the cell model and the actual measured value at a particular location were always compared for each model situation.

The measured and calculated values of a signal for the situation in Fig. 4a re depicted in Table 2 (values in the table are in the shape of *measured value/calculated data*).

From the result of measurement, it is evident that the signal does not reach from the first person to the third iBeacon over the corner of the room. Another interesting fact is that the signal attenuation between the first person and the first iBeacon is higher than −100 dB, but between the first person and the second iBeacon is −98 dB, while the difference in distance is at least five times greater. On the other hand, for the fourth person and the third iBeacon, the attenuation corresponds to the theoretical assumptions. In the second case (Fig. 4b) there has been a selection of the situations. The chosen situation corresponds to the model detecting the presence of a person in the room and has brought a lot of knowledge about the signal propagation through the walls of the building. The aim was to determine whether a person is inside the room or not. Theoretically iBeacon should be able to cover the entire room without any trouble, and at the same time not to pass through the wall. Even through around 30 cm thick walls, the signal attenuation spreads out into the hall. Measured and calculated values are recorded in Table 3.

The results have shown that differences in signal attenuation are minimum and little reveal whether the user is inside or outside the room. Fault limit was −100 dB but attenuation of −98 dB is very close to this limit.

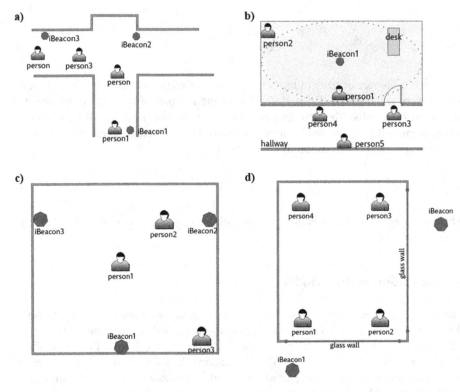

Fig. 4. (a), (b), (c), (d). Model situations in which the measurement was performed.

Table 2. Measured and calculated values for the model situation in Fig. 4a

	iBeacon1	iBeacon2	iBeacon3
Person1	−103 dB/−85 dB	−98 dB/−98 dB	—/−106 dB
Person2	−90 dB/−91 dB	−80 dB/−85 dB	−97 dB/−95 dB
Person3	−95 dB/−98 dB	−83 dB/−91 dB	−95 dB/−91 dB
Person4	−103 dB/−104 dB	−77 dB/−95 dB	−81 dB/−82 dB

Table 3. Measured and calculated values for the model situation in Fig. 4b.

	iBeacon1
Person1	−98 dB/−90 dB
Person2	−98 dB/−100 dB
Person3	−92 dB/−93 dB
Person4	−101 dB/−105 dB
Person5	−100 dB/−104 dB

In Fig. 4c there has been a selection of a room where two walls were formed by glass. The glass is metallised, which significantly influences the signal propagation.

Table 4. Measured values for model situation in Fig. 4c

	iBeacon1	iBeacon2
Person1	−100 dB/−99 dB	—/−103 dB
Person2	−92 dB/−93 dB	—/−100 dB
Person3	−100 dB/−104 dB	—/−99 dB
Person4	−96 dB/−101 dB	—/−100 dB

This situation was chosen because these materials are often used for glazing for buildings. Results are presented in the following table: (Table 4)

In the situation of Fig. 4c it was also interesting to observe how the signal is periodically lost from iBeacon2, although it was placed as iBeacon1 1.5 meters from the glass walls (the correct function of iBeacon2 has been several times verified).

The last measurement (Fig. 4d) was carried out at an ideal situation - a rectangular room without any obstacles that would affect signal propagation. The first person should theoretically have the same signal from all iBeacons, however, the actual measurements proved significant differences as shown in Table 5. Paradoxically the best result was at the third person, placed in the corner, where intensity of signals from particular transmitters were measured as expected.

Table 5. Measured values for model situation in Fig. 4d

	iBeacon1	iBeacon2	iBeacon3
Person1	−90 dB/−95 dB	−96 dB/−96 dB	−83 dB/−83 dB
Person2	−80 dB/−91 dB	−94 dB/−91 dB	−82 dB/−91 dB
Person3	−80 dB/−88 dB	−92 dB/−95 dB	−90 dB/−101 dB

The average deviations from the actual and measured are as follows:

- Situation in Fig. 4a – 4, 9 dB
- Situation in Fig. 4b – 3, 8 dB
- Situation in Fig. 4c – 2, 7 dB
- Situation in Fig. 4d – 5 dB

In calculating the average deviations, the condition, when the measurement could not detect a signal from the transmitter, were not included in the calculation. The measurement also proved an expected fact that if there are a larger number of obstacles, which significantly affect the propagation of electromagnetic signals, the model unlike the reality loses accuracy caused by signal reflections.

The proposed model could very well take into account attenuation caused by obstacles. However, due to a very difficult estimate how the signal reflects from obstacles (there are no data on the exact composition of the material of obstacles and their properties) the errors caused by signal reflections cannot be taken into account in the model.

7 Conclusion

The article has introduced localization issues in the interior using Bluetooth technology, particularly the issue of deployment of transmitters of Bluetooth signal on the basis of which reception the localization is implemented.

For designing optimal distribution of transmitters in the interior, the existing solutions have been investigated and a new method which is able to assess the proposed deployment of transmitters for the specific situation has been proposed. The method was further tested in a real environment, where the fact that the influence of various reflections and signal distortion caused by walls and equipment in the building may also occur in measuring significant deviations from the theoretical values has been proved.

Acknowledgement. This work was supported by the SPEV project, financed from the Faculty of Informatics and Management, University of Hradec Kralove.

References

1. Perera, C., Zaslavsky, A., Christen, P., Georgakopoulos, D.: Context aware computing for the internet of things: a survey. IEEE Commun. Surv. Tutorials 16(1), 414–454. http://ieeexplore.ieee.org/xpl/articleDetails.jsp?arnumber=6512846
2. Dahlgren, E.: Evaluation of indoor positioning based on bluetooth smart technology. Chalmers University of Technology, Department of Computer Science and Engineering, Göteborg, Sweden (2014). http://publications.lib.chalmers.se/records/fulltext/199826/199826.pdf
3. The Low Energy Technology Behind Bluetooth Smart. Bluetooth SIG, Inc. http://www.bluetooth.com/Pages/low-energy-tech-info.aspx
4. What is iBeacon? A guide to iBeacons. IBeaconInsider. http://www.ibeacon.com/what-is-ibeacon-a-guide-to-beacons/
5. Estimote: How to optimize battery performance of Estimote Beacons. https://community.estimote.com/hc/en-us/articles/202552866-How-to-optimize-battery-performance-of-Estimote-Beacons-
6. Estimote: What is the technical specification of Estimote Beacons. https://community.estimote.com/hc/en-us/articles/203159703-What-is-the-technical-specification-of-Estimote-Beacons-
7. Alhmiedat, T., Samara, G., Salem, A.O.A.: An indoor fingerprinting localization approach for ZigBee wireless sensor networks. Comput. Sci. Depart. Zarqa Univ. (2013). http://arxiv.org/pdf/1308.1809
8. Battiti, R., Brunato, M., Delai, A.: Optimal wireless access point placement for location-dependent services. Department of Information and Communication Technology, University of Trento (2003). http://eprints.biblio.unitn.it/489/1/DIT-03-052-withCover.pdf
9. Wong, L.W.C.: Indoor Localization Methods. Ambient Intelligence Laboratory, Interactive & Digital Media Institute, National University of Singapore. http://www.ewh.ieee.org/reg/10/conferences/Workshop201103/IndoorLocalizationMethods.pdf
10. Bahl, P., Padmanabhan, V. N.: RADAR: An In-Building RF-based User Location and Tracking System. http://research.microsoft.com/en-us/people/padmanab/infocom2000.pdf

11. Warski, A.: How do iBeacons Work (2014). http://www.warski.org/blog/2014/01/how-ibeacons-work/
12. Dong, Q., Dargie, W.: Evaluation of the reliability of RSSI for indoor localization. Faculty of Computer Science, Technical University of Dresden, Germany. http://citeseerx.ist.psu.edu/viewdoc/download?doi=10.1.1.278.3929&rep=rep1&type=pdf
13. Kamenetsky, M., Unbehaun, M.: Coverage planning for outdoor wireless LAN systems. In: IEEE International Zurich Seminar on Broadband Communications, Sweden (2002)
14. Tang, K.-S., Man, K.-F., Kwong, S.: Wireless communication network design in IC factory. IEEE Trans. Ind. Electron. **48**, 452–458 (2001)
15. Unbehaun, M., Kamenetsky, M.: On the deployment of picocellular wireless infrastructure. IEEE Wirel. Commun. Mag. **10**, 70–80 (2003)
16. Park B.-S., Yook J.-G., Park H.-K.: The determination of base station placement and transmit power in an inhomogeneous traffic distribution for radio network planning. In: IEEE 56th Vehicular Technology Conference (2002)
17. Anderson, H.R., McGeehan, J.P.: Optimizing microcell base station locations using simulated annealing techniques. In: Proceedings of the 44th Vehicular Technology, pp. 858–862 (1994)
18. Sherali, H.D., Pendyala, C.M., Rappaport, T.S.: Optimal location of transmitters for micro-cellular radio communication system design. IEEE J. Sel. Areas Commun. **14**, 662–673 (1996)
19. Kouhbor, S., Ugon, J., Kruger, A., Rubinov, A.: Optimal Placement of Access Point in WLAN Based on a New Algorithm. School of Information Technology and Mathematical Sciences, University of Ballarat (2005). http://www.researchgate.net/profile/Alexander_Kruger/publication/4167639_Optimal_placement_of_access_point_in_WLAN_based_on_a_new_algorithm/links/02bfe50e6a095393a7000000.pdf
20. Kobayashi, M., Haruyama, S., Kohno, R., Nakagawa, M.: Optimal access point placement in simultaneous broadcast system using OFDM for indoor wireless LAN. In: 11th IEEE International Symposium on Personal Indoor and Mobile Radio Communications. PIMRC 2000. Proceedings *(Cat. No.00TH8525)* IEEE. (DIT-03-052), pp. 200–204 [cit. 2015-06-04]. (2000). http://ieeexplore.ieee.org/lpdocs/epic03/wrapper.htm?arnumber=881418. doi:10.1109/PIMRC.2000.881418. ISBN 0-7803-6463-5
21. Ekahau Site Survey User Guide (2015). http://docs.ekahau.com/index.php/Ekahau_Site_Survey_User_Guide
22. Rogers, P.: H ow iBeacons Work for Indoor Location Based Services - Technical Guide and Recommendations. MobileEnterpriseStrategies.com (2014). http://mobileenterprisestrategies.blogspot.cz/2014/05/how-ibeacons-work-for-indoor-location.html

Designing Context Models for CARS Incorporating Partially Observable Context

Christos Mettouris[✉] and George A. Papadopoulos

Department of Computer Science, University of Cyprus,
1 University Avenue, Nicosia, Cyprus
{mettour,george}@cs.ucy.ac.cy

Abstract. Context modelling and context reasoning are widely used topics in Context-Aware Recommender Systems research. Based on our research, the development of context models in context-aware recommender systems is problematic in that many domain specific and application specific context models are developed with limited or no reuse and sharing capabilities. Furthermore, context-aware recommender systems that follow the representational view of context, design and model the fully observable context that is known at recommendation time but do not consider partially observable context with unknown values at recommendation time, that can nevertheless enhance the recommendation outcome. In this paper we deal with the above two issues by proposing a CARS design system that enables developers: (i) to easily and effectively design context models by defining, sharing and reusing context parameters and (ii) to utilize partially observable context at recommendation time by using an interactional approach that incorporates user feedback by applying a utility based algorithm on context models.

Keywords: Context modelling · Context model design · Context-Aware recommender systems · Interactive algorithm · Partially observable context

1 Introduction

Traditional Recommender Systems (RS) use limited or none contextual information to produce recommendations, as opposed to Context-Aware Recommender Systems (CARS) that focus in using contextual information to enhance recommendations. Context was first used within recommendation algorithms and methods by Adomavicius by proposing three approaches: the Pre-filtering approach, the Post-filtering approach and the Multi-dimensional Contextual Modelling approach [1, 2].

A contextual design issue related to CARS is the development of domain specific and application specific context models that only represent information on the particular application domain (e.g. recommendation of movies) or information regarding a particular application. The main problem with domain and application specific models is overspecialization, as well as limited or no reuse.

On another dimension, whether the contextual information being used by a context-aware recommender system is known during the recommendation characterizes the context as fully observable, partially observable, or unobservable [4, 5]. Fully

© Springer International Publishing Switzerland 2015
H. Christiansen et al. (Eds.): CONTEXT 2015, LNAI 9405, pp. 118–131, 2015.
DOI: 10.1007/978-3-319-25591-0_9

observable context refers to all contextual parameters, their structure and values being known during recommendation, partially observable refers to only part of this information being known, while unobservable context refers to the case where the context is unknown at recommendation time. Following a representational view of context in CARS design [6, 7], the context is defined through a predefined set of observable context parameters of static structure which does not change significantly over time, as opposed to the interactional view of context where the context is often unknown as it may not necessarily be an observable feature of an interaction. From the above it is evident that a representational view of context can be achieved only if the context is fully observable, i.e. its parameters and their values are known at recommendation time (can be observed, detected and retrieved) and thus the designer can include such information in context models during design. On the other hand, if not all context information is fully observable, an interactional view of context can be considered, where specific information may or may not be relevant to some activity and contextual parameters may be defined in a dynamic and occasioned manner rather than a static, predefined one [6]. We argue that the most common practice in designing CARS is to follow the representational view of context, during which CARS developers attempt to utilize the fully observable context within their recommenders by using contextual parameters known at recommendation time, which may result in omitting unknown contextual information that nevertheless may be important.

In this paper we build upon prior work [8] to address the aforementioned contextual design problems. A Context-Aware Recommender System design system is proposed which developers can use to: (i) design and build context models at the application layer for their recommender applications that incorporate both fully observable and partially observable context, (ii) apply recommendation algorithms that utilize these context models by facilitating an interactional approach for partially observable context that incorporates user feedback. Regarding designing and building context models (i), the system guides developers towards an easy, efficient and effective selection and usage of context parameters, allowing at the same time for sharing and reuse of context models among recommender applications, regardless of the domain they belong to. On applying recommendation algorithms on context models (ii), the system incorporates a method that facilitates both fully observable and partially observable context. The system supports the developer in specifying system actions using pseudo code to handle the fully observable context, while it relies on user feedback to calculate the utility of a set of possible context values for the partially observable context and recommends the most suitable context value to be used by the system. In this work we will refer to context information that is not fully observable at recommendation time (i.e. its values are not known and cannot be retrieved at the time of recommendation, although some information on the structure of the context may exists) as partially observable context. We will also refer to partially observable context with unknown values as "unknown context", as opposed to fully observable context with values known at recommendation time. The CARS design system was developed as an online web-based system using PHP, MySQL and web technologies[1].

---

[1] Online: http://www.cs.ucy.ac.cy/~mettour/phd/CARSContextModellingSystemV3/

Section 2 provides related work. Section 3 describes the context modelling design process and system actions on models. Section 4 describes the methodology used in known as well as unknown context settings. In Sect. 5 we talk about our experimentation procedure and the paper closes with conclusions and future work in Sect. 6.

2 Related Work

We have reviewed a number of context-aware recommender systems in the bibliography that use contextual and conceptual models [3, 8]. Although contextual models in the bibliography exist that attempt to facilitate a more generic and cross-domain design [9–12], the majority represent information that either concern particular application domains (e.g. tourism, movies, museums), or more abstracted domains (e.g. products in general, web services, e-learning, etc.).

Moreover, related works from the fields of Machine Learning, Data Mining and Information Retrieval incorporate contextual information within their modelling methods to enhance prediction accuracy [13–15]. While these context modelling methods unquestionably enhance recommendation accuracy, their mathematical oriented models require a good understanding of the aforementioned fields by the developer, as well as strong development skills to be implemented.

To the best of our knowledge a context design system that could facilitate the development of reusable and generic contextual models for CARS has not yet been proposed in the literature. The aim of such a system would be to facilitate context model design, model reuse among developers and applications, and model extension/update based on the needs of the developer. Such a tool would simplify the process of contextual modelling in CARS and enable context uniformity, share and reuse.

Furthermore, most of the examined systems in related work follow the representational view of context in which the fully observable context is designed in a particular structure (e.g. using semantic models, ontologies, etc.). In these cases, context parameters known at recommendation time are being modelled. We argue that this practice may not consider important contextual information with unknown values during recommendation that may be nevertheless important for the recommendation outcome. In this paper we implement an interactional approach for including partially observable context in the recommendation algorithms by incorporating user feedback.

3 Context Model Design

All context models in the CARS design system follow the design presented in the figure on the main page of the system (See Footnote 1). From top-to-bottom, the level of abstraction decreases. The "user_item_context_rating" entity represents a single complete recommendation process [1]. For each recommendation run, a recommender examines the utility of each item for a particular user in a certain context. The context variable entity specifies a contextual parameter and its value in a context model: name:value, e.g. Temperature:high. The weight property denotes the importance of a context variable. The static property refers to whether the context variable is static (cannot change dynamically, e.g. user's date of birth) or dynamic (can change, e.g. weather). A context variable may be part of the known or the unknown context. Since the context in CARS is multidimensional [1, 2], the system is able to handle the many dimensions of the context in a context model in an easy and scalable manner: we define that each unique context variable name with all its values represents a unique context dimension. A context model in the system is defined as one or more context variables, each context variable being assigned under one or more of the four context categories (Fig. 1): itemContext, userContext, systemContext and otherContext.

Context model for the Default Movie Recommender

- Edit the model
- Display the Context Dimensions of the model
- Display the Algorithms attached to this model
- Recommend similar models
- Back

Context Categories

There are **48** context variables in the model (Alphabetically Ordered per column)

(#) Num of Context Variable Appearances in Application Context Models

ITEM CONTEXT (Weight: 5)	USER CONTEXT (Weight: 5)	SYSTEM CONTEXT (Weight: 3)	OTHER CONTEXT (Weight: 7)
imdb ratings: 0-4 Static: TRUE (1)	company: alone Static: TRUE (5)	network: adequate Static: FALSE (6)	time: weekday Static: FALSE (8)
imdb ratings: 5-7 Static: TRUE (2)	company: girlfriend/boyfriend Static: TRUE (7)	network: excellent Static: FALSE (6)	time: weekend Static: FALSE (9)
imdb ratings: 8-10 Static: TRUE (2)	company: with friend(s) Static: TRUE (7)	network: poor Static: FALSE (5)	

Fig. 1. Application context model for the "Default Movie Recommender"

Three context models are defined in the system:

The *Generic Model* includes all context variables currently defined in the system.

The *Application Context* is a context model for a particular RS, e.g. a movies RS. This is a model built by a CARS developer to model the context for a RS. An application context model is built using a system interface via which all available context variables in the system are presented on screen and the CARS developer clicks on those she wants to select for her context model, or adds new context variables. It is very likely that context variables predefined by other developers are suitable for the new application context as well, which not only saves time to the developer, but more importantly it enables the developer to use context parameters that she may not have thought of using, or may think is unable to use in case the particular context is unknown at recommendation time. In the CARS design system, unknown context can be defined in the form of context variables in the same manner as known context. Note that, since each context variable has a name and a specific value (e.g. Temperature: high), the developer must select all needed variables with a particular name for the model to be accurate and complete, e.g. all of the following: Temperature:high, Temperature:medium, Temperature:low. In this manner, the context dimension Temperature is also defined in the application context model.

The *Context Instance* context model is a "screenshot" of the context during an interaction between the user and the item involved in the recommendation process. For example, for a movie recommender, a context instance is the set of context variables that constitute the context during the event of a particular user (user = "Jerry") watching a particular movie (item = "Kill Bill") at a particular time (we assume the first time that Jerry watches Kill Bill). Such a context instance may have title: "Jerry_-KillBill_C1" (C1 results from "Context1") and can be consisted of a number of context variables and dimensions, e.g.: time_of_day, day_of_week, movie_IMDB_ratings, companion_of_user (whom did the user watch the movie with), etc. In a similar way, the context instance around the second time Jerry watches Kill Bill will have a title "Jerry_KillBill_C2" and will again be consisted of a number of context variables and dimensions.

Any context model in the system is accessible and can be shared, edited, extended and reused by others. The general idea behind this concept is that, since all RS of a specific type/domain (e.g. online movie recommenders) function in similar context settings, having one context model for each recommender built by each developer may be inefficient; we propose the definition of one complete context model for all recommenders.

Supported System Actions
Building Models: The system offers to developers pre-existing application context models of similar applications to use as a baseline for their own models, instead of building a model of their own.

Validating Models: The system supports the validation of an application context model through one or more context instance models. An application context model should be able to support any context instance related to the particular recommender; otherwise, the application context model is incomplete. For example, an application context model for a movie recommender should be able to support any movie recommender related context instance, such as "Jerry_KillBill_C1". Application context model validation is a useful tool for the CARS developer to validate her application context model against a number of context instances and thus ensure that her application context is able to properly model the context (i.e. model any context instance of the recommender). A context instance can be created preferably by a CARS user who will reflect her experiences regarding the activity of watching movies in the context instance model. If this is not feasible, the context instance model can be created by another developer. The system provides information whether a context instance model was validated against the application context or not and justifications.

Context Dimensions View: The system supports a context dimensions view for each context model in the system.

Recommendation of Models: The system is able to recommend the top N (currently N = 5) most similar context models to a particular application context model. The comparison is based on the percentage of common context variables and context dimensions. This informs the developer about similar context models as well as the level of similarity. The recommendations are provided both explicitly and implicitly.

Model Comparison: Application context model comparison is supported. Though an easy to use user interface the system depicts common context variables, as well as those that belong only to one of the two application contexts. Colours are used on screen for easier comprehension of the information. The system also provides statistics regarding the percentage of participation of each application context model to the other. If many common variables are noted, the system proposes a merge of the models. This is especially interesting in cases where application context models concern RS of different domains, e.g. a movie RS and a book RS: the systems, although of different domains, use similar contextual information.

4 Assigning Algorithms to Context Models

In order to handle the known context the system supports the developer in specifying system actions using pseudo code. For unknown context however, the system relies on user feedback to calculate the utility of a set of possible context values (context variable candidates) and then recommends the context value with the highest utility to the CARS developer in order to use it as the best context setting in subsequent system recommendations to the user. This context value is considered as the best context candidate for the unknown context to be used in the recommendation process.

We have implemented an algorithm as a prototype, which is based on a utility scoring method. Given an application context model (as specified in Sect. 3), an end user u and an item j, the system uses context utility function - CUF (1) to measure the utility of item j for user u in relation to each context dimension in the model:

$$\text{item\_util}_{ju} = f_1 * W_1 + \ldots + f_i * W_i + \ldots + f_N * W_N \tag{1}$$

item_util_{ju} is the context utility of item j for user u. It specifies how much the particular context setting satisfies user u when recommended item j. Items with the highest utility are recommended to the user. N is the number of context dimensions in an application model. $f_i * W_i$ represents context dimension i. f_i is the description of context dimension i for user u and item j (in computable format as explained in Sect. 4.1). W_i is the weight for context dimension i: denotes the importance of the context dimension in relation to the other context dimensions (an integer, values 1–5). The weights are defined by the developer based on the importance she would like to denote for each context dimension – different applications may require different weights on context dimensions such as location, temperature, etc.

4.1 Specifying Functionality on Fully Observable Context Parameters

To facilitate the CARS developer in incorporating a context model within a recommendation process, we provide a simple interface by which she can use pseudo code to describe the required functionality to handle the context (Fig. 2). More particularly, for each context dimension in an application context model (i.e. for each f_i, $i = 1 \ldots N$ in (1)), the system supports description of the required functionality to handle the particular context dimension. The description is currently text based and has no effect in system decisions on context; this is planned for future work. Using pseudo code to describe the required functionality on how to handle the context is preferred than using linguistic text based description as it is more descriptive, it is easier for a developer to understand what is needed to be done precisely (especially when the developer is someone other than the context model designer), it avoids ambiguity and it can easier facilitate a system language definition in the future.

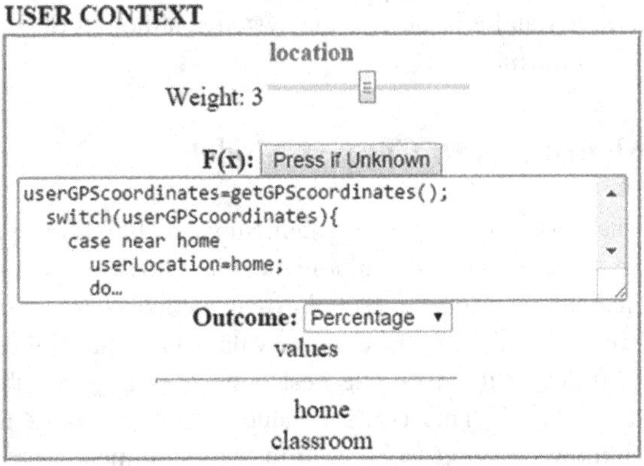

USER CONTEXT

location

Weight: 3

F(x): Press if Unknown

```
userGPScoordinates=getGPScoordinates();
switch(userGPScoordinates){
   case near home
      userLocation=home;
      do…
```

Outcome: Percentage ▾

values

home
classroom

Fig. 2. System interface: CARS developer describes functionality for "location"

The above procedure is feasible for context parameters known at recommendation time (fully observable context), such as the location of the user. As an example suppose the design of a restaurant RS that recommends restaurants nearby the user, hence user location is important. User location can be detected by using GPS coordinates from the user mobile phone. Normally, the developer would specify in her application context model a single context parameter for location, e.g. user location; in our system, she can define context variable name/value pairs so that the context is more descriptive and less ambiguous. Let's suppose she has defined three context variables as one context dimension in her application model: context variables: location:home, location:work, location:other, context dimension: location. In this setting the developer denotes that user location at home and user location at work constitutes important context and should be handled explicitly. On the contrary, user location in other cases other than home or work are not equally important and can be handled collectively. The developer threfore needs to define the functionality for the context dimension location (noted as $f_{location}$ below) and specify the system functionality for each of the three possible context values, e.g.:

```
Function: f_location
   userGPScoordinates=getGPScoordinates();
   switch(userGPScoordinates){
      case near home
         userLocation=home;  do…
      case near work
         userLocation=work;  do…
      default
         userLocation=GPScoordinates;   do…}
```

The above code is clear as to what needs to be done: do specific tasks if user is at home or work, or another task if somewhere else. Note that the system requests from the developer a functionality description for the context dimension $f_{location}$ instead for each of the context variables (e.g. $f_{location:home}$) for simplicity and to avoid repetition.

Describing system functionality as in the above example is valuable when the CARS developer does not want to deal with implementation details at that moment, or has no full knowledge of the platform requirements. Moreover, omitting implementation details that would otherwise be essential for software development saves time and enables the developer not to be precise, as she might not yet be fully aware of other context parameters, their values and their role in the recommendation process. Moreover, it can be the case that the CARS designer is not a software developer and thus cannot code; in such cases ordinary text can be used.

Applying the Context Utility Function. If all context dimensions are known at recommendation time, then Eq. (1) is straightforward: the utility of an item equals to the sum of the context dimension descriptions multiplied by their weights. Of course, context dimension descriptions must return *computable values* so that the total utility of the CUF for each item can be computed. For example, the CUF for the restaurant RS above can utilize context dimension location as follows (the following code is to be appended in function $f_{location}$):

```
For each restaurant in Database Do:
    if dist(restaurant,userLocation)<2km
        return 0.6;
    elseif dist(restaurant,userLocation)>3km
        return 0.1;
    else
        return 0.3;
```

The return values of context dimension descriptions are expressed as the percentage item j fits the particular context dimension for user u. In the above example it is assumed that a walking distance of 2 km is the maximum convenient walking distance to a restaurant for the average user ($f_{location}$=0.6). The utility item_util of each item is computed and the items with the highest utility are recommended to the user.

A variety of CUFs can be specified for each application context model, each of which will differ in context dimension descriptions f_i and/or the assigned weights W_i. These context utility functions can be implemented and incorporated in the RS. The variety of CUFs produced based on the above process are stored in the system and can be shared and reused among CARS developers regardless of application domain.

4.2   Specifying Functionality on Partially Observable Context Parameters

Not all contextual information is known at recommendation time. An example is the context parameter user companion that specifies with whom the user is at a particular time. This contextual information may be important for a movie RS (recommendations on what movie to watch heavily depend on whether the user is with her partner, a friend, a family member or alone), a restaurant RS (a partner's gastronomic preferences can be as important as the user's if they are dining together), etc. User companion cannot be known to the system unless the user explicitly states it or in cases where location information of the person being with the user is known.

To the best of our knowledge in situations of partially observable or unobservable context, the case with systems following a representational view of context is that such context is not defined or used in their context models. On the other hand, systems following an interactional view of context can better cope with unobservable or partially observable context as context may be defined in a more dynamic manner [5].

The interface of the CARS design system includes a button *"Press if Unknown"* in the case of an unknown context dimension such as user companion (Fig. 2). When pressed, it is specified that the particular context dimension is unknown at recommendation time and that user feedback should be utilized for its value to become known.

Applying the Context Utility Function. Suppose that unknown context "user companion" is defined within an application context model as follows: context variables: user_companion:partner, user_companion:friend, user_companion:familyMember; context dimension: user_companion. While a context dimension may be currently unknown, the context-aware recommender system could have utilized this contextual information within the recommendation process, had it been known. For instance, there are no technological means to acquire the value of the context dimension user_companion, but had there be any; the CARS developer would have been able to exploit this information to enrich the recommendation outcome. In the case of context dimension "user companion", the profile of the user's companion u′ can be used in combination with the profile of user u to produce recommendations. This provides added value to the outcomes of a movie RS, a restaurant RS, as well as RS of various domains where user companion is important. For each context variable candidate of an unknown context dimension $f_{unknown}$ the CARS developer is expected to define and implement the desired functionality so that it returns a computable value, as stated in Sect. 4.1. In this manner, each one of the three context variables of context dimension "user companion" can participate in (1) as a context variable candidate of $f_{user\_companion\_unknown}$.

Given a CUF of an application context model (active CUF) and a user u, we define as *Context Utility Function run* (CUF run) the following steps:

i. the computation of $item\_util_{ju}$ for each item j using Eq. (1)
ii. the recommendation of the item(s) with the highest utility score to user u
iii. user u provides feedback on the recommended item, in the form of a rating score, integer from 1 (min) to 10 (max)

Considering an unknown context dimension $f_{unknown}$, in step i. above the available context variable candidates are used as possible values for $f_{unknown}$. One candidate is used for each CUF run, initially in a random manner. After user u provides feedback in step iii, this rating score is used as an evaluation metric for the context variable candidate. In this manner the system evaluates each context variable candidate of an unknown context dimension based on the received user feedback.

4.3   Recommendation of Unknown Context

The system uses a prediction algorithm that recommends to the CARS developer the best context variable candidate for an unknown context dimension to be used in subsequent CUF runs. This method retrieves the user feedback for each CUF run and calculates a score for the context variable candidate. After many CUF runs, the system is able to recommend the context variable candidate with the highest score. The prediction algorithm also considers:

- CUF runs of other CUFs of the same application model: other CUFs may have been defined to facilitate other situations
- CUF runs of CUFs of other application context models

Note that, for using other CUFs than the active one it is necessary that these CUFs include the unknown context variable candidate that is being evaluated as well.

5   Experimental Evaluation

We have evaluated our system by using partially observable context parameters and simulating CUF runs and user feedback.

Experimental Assumptions. For simplicity reasons we have selected an application context model with 1 unknown and 2 known context dimensions: assume an online movie RS that recommends movies to users based on their profile preferences and the context parameters: network capabilities (nc), time of day (td) and user companion (uc). Let the network capabilities and time of day be the 2 known context dimensions, while user companion is unknown. The CUF for item j is:

$$\text{item\_util}_{ju} = f_{nc} * W_{nc} + f_{td} * W_{td} + f_{uc\_unknown} * W_{uc}$$

For simplicity reasons we let $W_{nc} = W_{td} = W_{uc} = 1$. The descriptions of the two known context dimensions f_{nc} and f_{td} have been defined in the system in pseudo code in a similar manner as the location was in Sect. 4.1. We assume these context factors are computable and hence the utility of the recommended item j for user u concerns only the unknown context:

$$\text{item\_util}_{ju} = [\text{some computable values}] + f_{uc\_unknown} \qquad (2)$$

Assume that $f_{uc\_unknown}$ has 7 context variables: user_companion:partner, user_companion:closeFriend, user_companion:socialFriend, user_companion:familyMember, user_companion:colleague, user_companion:otherPerson and user_companion:none. For each of the above context variables the CARS developer defines appropriate functionality in her system so that when a context variable is selected as a candidate for $f_{uc\_unknown}$, the corresponding functionality is executed and a computable value is returned to be used in (1) that depicts the utility of the item regarding the particular context dimension. For instance, if "user_companion:partner" is selected as a candidate, provided that the partner is a user with a profile known to the system, the recommender can combine both user profiles into one extended user profile and calculate the utility of each restaurant for the extended user profile. Similarly, the functionality for the other context variables can be defined.

Based on (2) the following assumption is valid (user feedback assumption): the user feedback on the recommended item j (item with highest $item\_util_{ju}$) corresponds to the current context variable candidate used for the unknown context dimension $f_{uc\_unknown}$. In this manner, user feedback score evaluates context variable candidates, enabling the system subsequently to recommend to the developer the candidate with the highest score.

Running Simulations. We have used a program to simulate CUF runs in order to show that the prediction algorithm is able to discover the preferred context variable and recommend it to the CARS developer to use in subsequent CUF runs. CUF runs are simulated by assigning user feedback scoring on unknown context variable candidates (user feedback assumption and (2)). User feedback reflects user preferences via an integer score from 1–10.

Consider a user that watches movies online and uses the online movie RS to receive movie recommendations. We assume that most of the times the user is with her partner (user_companion:partner) and hence this context setting is often favourable, otherwise a random companion (1 of the 6 remaining context variables). We have defined the following probabilities for this user:

where $P_f(x)$: the probability the user submits feedback score x.

```
if f_uc_unknown equals user_companion:partner
    P_f(1)=P_f(2)=P_f(3)=P_f(4)=0.05;
    P_f(5)=P_f(6)=P_f(7)=P_f(8)=P_f(9)=P_f(10)=0.15;
else
    P_f(1)=P_f(2)=P_f(3)=P_f(4)=P_f(5)=P_f(6)=P_f(7)=P_f(8)=P_f(9)=P_f(10)=0.1;
```

The rationale is that, when the context-aware recommender system randomly selects to use $f_{uc\_unknown}$ = user_companion:partner, the user is more likely to rate higher the recommended movie (selecting a score ranging from 5 to 10) than otherwise. This happens since the particular context variable fits the user's real context, as opposed to any other context variable (since they watch movies together), which will positively affect the recommendation outcome in recommending more suitable movies. Based on various tests with similar probability values and to maintain our argumentation on logical assumptions, we argue that a probability of 15 % instead of 10 % for a high score and a probability of 5 % instead of 10 % for a low score in a preferred context setting are within logical boundaries.

The simulation program executes a number of CUF runs with the above probabilities and stores the user feedback for each run in the system.

6 Results

As stated in Sect. 4.3, the best context variable candidate for an unknown context dimension is recommended by a prediction algorithm that considers CUF runs from various context utility functions and application contexts. Our experiment uses the 7 context variables candidates for the context dimension user_companion presented in Sect. 5. For each of these 7 context variables, the algorithm considers a number of CUF runs, calculates the score for each context variable candidate and recommends the one with the highest score. Assuming that the user prefers movie recommendations computed considering the context variable user_companion:partner (i.e. in the context "user companion is the user's partner"), we consider a valid answer only if the prediction algorithm returns user_companion:partner.

We have conducted three experiments. Initially we have used only CUF runs of the active CUF (Experiment 1, Table 1). Following, we have additionally used CUF runs of other CUFs of the same application model, as well as CUF runs of CUFs of other application context models (Experiment 2, Table 2). At this point note that other than the active CUF mentioned above were defined in the same manner as the active CUF (i.e. based on the user feedback assumption), ensuring that these functions and their application context models utilize the same unknown context dimension user_companion. Finally, we have conducted experiments with random number of CUF runs to simulate more realistic settings (Experiment 3, Table 3).

Table 1. Experiment 1

# CUF runs	Prediction accuracy
200	0.867
175	0.8
100	0.6

Table 2. Experiment 2

# CUF runs	Prediction accuracy
200	1.0
150	0.98
75	0.833
40	0.633

Table 3. Experiment 3

# CUF runs	Prediction accuracy
Random (350–700)	1.0
Random (150–350)	0.9
Random (100–150)	0.833
Random (75–100)	0.8
Random (35–75)	0.633

We argue that a minimum prediction accuracy of 80 % (0.8) is needed, meaning that in a real scenario that a user provides feedback in line with the probabilities of Sect. 5, the system recommends the correct context variable to the CARS developer to use in subsequent CUF runs 8 out of 10 times.

The experiments show that accuracy is enhanced when more CUF runs are used. Experiment 1 needs more than 175 CUF runs to meet the accuracy threshold. This suggests that the context-aware recommender system must provide to the user 175 recommended items within the context specified by the particular CUF, and the user must provide feedback for each of these recommendations for the system to meet the accuracy threshold. Experiments 2 and 3 need considerable less CUF runs to meet the accuracy threshold because the prediction algorithm considers in addition other CUFs than the active one.

Finally, when attempting random CUF runs it is observed that at least 75 CUF runs are needed to meet the accuracy threshold. Moreover, it is noted that for experiments 2 and 3 at least 35 CUF runs are needed for the prediction algorithm to achieve 60 % success. This means that, to be successful more than once out of two times, the context-aware recommender system must provide to the user at least 35 recommended items for feedback.

7   Conclusions and Future Work

The CARS design system presented in this paper aims to provide to CARS developers an efficient and effective way to select and use known, as well as unknown at recommendation time context information for building their own application context models, allowing at the same time for sharing and reuse of context models and information among applications, regardless of the domain they belong to. The novelty of the system relies also in applying a utility-based recommendation algorithm that utilizes these context models by facilitating an interactional approach for partially observable context that incorporates user feedback. The system supports the developer in describing fully observable context, while it relies on user feedback to calculate the utility of a set of possible context values for the partially observable context and recommends the most suitable context value to be used by the system in future recommendation attempts.

Experiments in simulating CUF runs using simple probabilities depicting user feedback patterns have shown that accuracy is enhanced when more CUF runs are used. Less CUF runs are needed if the prediction algorithm considers in addition other context utility functions of the same, as well as other application contexts. In the latter setting, at least 75 CUF runs are needed to meet the accuracy threshold (80 %), while 35 CUF runs are needed for the prediction algorithm to achieve at least 60 % success.

The experimental evaluation described in this work serves as a proof of concept that the CARS design tool, if used in real settings by CARS developers building application context models and context utility functions, can provide valuable recommendations regarding unknown context dimensions. We argue that the assumptions made in this paper are realistic, but fine tuning of the system will be needed in real settings.

As future work we plan to evaluate the system by involving CARS developers, initially people from the university premises. This evaluation will serve as a proof that the system is able to function in real settings and rely on user feedback to recommend partially observable context information to be used within context-aware recommender systems. In addition, we plan to incorporate more recommendation algorithms to the system and experiment in more ways of handling partially observable context, as well as unobservable context. We argue that well known recommendation algorithms can be used within our tool with promising results.

References

1. Adomavicius, G., Tuzhilin, A.: Context-aware recommender systems. In: Ricci, F., Rokach, L., Shapira, B., Kantor, P.B. (eds.) Recommender Systems Handbook, pp. 217–253. Springer, New York (2011)
2. Adomavicius, G., Sankaranarayanan, R., Sen, S., Tuzhilin, A.: Incorporating contextual information in recommender systems using a multidimensional approach. ACM Trans. Inf. Syst. (TOIS) 23(2005), 103–145 (2005)
3. Mettouris, C., Papadopoulos, G.A.: Contextual modelling in context-aware recommender systems: a generic approach. In: Haller, A., Huang, G., Huang, Z., Paik, H.-y., Sheng, Q.Z. (eds.) WISE 2011 and 2012. LNCS, vol. 7652, pp. 41–52. Springer, Heidelberg (2013)
4. Adomavicius, G., Mobasher, B., Ricci, F., Tuzhilin, A.: Context-aware recommender systems. AI Mag. 32(3), 67–80 (2011)
5. Hariri, N., Mobasher, B., Burke, R.: Context adaptation in interactive recommender systems. In: Proceedings of the 8th ACM Conference on Recommender Systems, New York, NY, USA, pp. 41–48 (2014)
6. Dourish, P.: What do we talk about when we talk about context. Pers. Ubiquit. Comput. 8 (1), 19–30 (2004)
7. Anand, S.S., Mobasher, B.: Contextual recommendation. In: Mladenic, D., Semeraro, G., Hotho, A., Berendt, B. (eds.) WebMine 2007. LNCS (LNAI), vol. 4737, pp. 142–160. Springer, Heidelberg (2007)
8. Mettouris, C., Papadopoulos, G.A.: CARS context modelling. In: Proceedings of the 9th International Conference on Knowledge, Information and Creativity Support Systems, KICSS 2014, Limassol, Cyprus, pp. 60–71, 6–8 Nov 2014
9. Costa, A., Guizzardi, R., Guizzardi, G., Filho, J.: CoReS: context-aware, ontology-based recommender system for service recommendation. In: UMICS 2007, 19th International Conference on Advanced Information Systems Engineering (CAISE 2007) (2007)
10. Emrich, A., Chapko, A., Werth, D.: Context-aware recommendations on mobile services: the m:Ciudad approach. In: Meissner, S., Moessner, K., Presser, M., Barnaghi, P. (eds.) EuroSSC 2009. LNCS, vol. 5741, pp. 107–120. Springer, Heidelberg (2009)
11. Moscato, V., Picariello, A., Rinaldi, A.M.: A recommendation strategy based on user behavior in digital ecosystems. In: MEDES 2010 Proceedings of the International Conference on Management of Emergent Digital EcoSystems, p. 25 (2010)
12. Uzun, A., Räck, C., Steinert, F.: Targeting more relevant, contextual recommendations by exploiting domain knowledge. In: HetRec 2010 Proceedings of the 1st International Workshop on Information Heterogeneity and Fusion in Recommender Systems, pp. 57–62 (2010)
13. Rendle, S., Gantner, Z., Freudenthaler, C., Schmidt-Thieme, L.: Fast context-aware recommendations with factorization machines. In: SIGIR 2011 Proceedings of the 34th International ACM SIGIR Conference on Research and Development in Information Retrieval, pp. 635–644 (2011)
14. Rendle, S.: Factorization machines with libFM. ACM Trans. Intell. Syst. Technol. (TIST) 3(3), 57 (2012)
15. Hidasi, B., Tikk, D.: General factorization framework for context-aware recommendations. Data Mining and Knowledge Discovery, 1–30, Springer (2015). doi: 10.1007/s10618-015-0417-y

Unsupervised Indoor Localization with Motion Detection

Yaqian Xu[✉], Linglong Meng, and Klaus David

Chair for Communication Technology (ComTec), University of Kassel,
Wilhelmshöher Allee 73, 34121 Kassel, Germany
{yaqian.xu,comtec}@comtec.eecs.uni-kassel.de,
david@uni-kassel.de

Abstract. Unsupervised indoor localization has received increasing attention in recent years. It enables automatically learning and recognizing the significant locations from Wi-Fi measurements continuously collected from mobile devices in a user's daily life, without requiring data annotation from professional staff or users. However, such systems suffer from continuous Wi-Fi collection, which results in a high power consumption of mobile devices. These problems can be addressed through activating Wi-Fi collection when it is necessary and deactivating Wi-Fi collection when "enough" data is collected. By using the acceleration readings from the embedded accelerometer sensor, a motion detection algorithm is implemented for an unsupervised localization system DCCLA (Density-based Clustering Combined Localization Algorithm). The information of motion states (i.e. a mobile device in motion or not in motion) is then used to automatically activate and deactivate the process of Wi-Fi collection, and thus save power. Tests carried out by different users in real-world scenarios show an improved performance of unsupervised indoor localization, in terms of location accuracy and power consumption.

1 Introduction

The location of a mobile device or a user is one of the most essential pieces of information for emerging location-based services (LBS) and applications. For outdoor localization, Global Positioning System (GPS) has been widely used. For indoor localization, *Wi-Fi fingerprinting* is a promising technique. Wi-Fi fingerprinting first senses the Wi-Fi measurements at desired locations and generates *Wi-Fi fingerprints* for each of these locations. The correlated relationship between Wi-Fi fingerprints and locations is later used to locate mobile devices or users by comparing the current Wi-Fi measurement and learned Wi-Fi fingerprints. The phase of generating the Wi-Fi fingerprints is called the *learning phase*. The phase of determining the current location is called the *positioning phase*.

In the learning phase, the conversion approaches of the Wi-Fi fingerprinting technique usually rely on an extensive site survey with data annotation for the Wi-Fi fingerprint generation. Recently, many research groups focus on generating the Wi-Fi fingerprint without the need of data annotation, i.e., in an unsupervised manner, which is known as unsupervised localization. A typical unsupervised localization system is DCCLA (Density-based Clustering Combined Localization Algorithm) [1]. It can automatically learn and recognize the *significant locations* from Wi-Fi measurements continuously collected in a

© Springer International Publishing Switzerland 2015
H. Christiansen et al. (Eds.): CONTEXT 2015, LNAI 9405, pp. 132–143, 2015.
DOI: 10.1007/978-3-319-25591-0_10

user's daily life, without the requirement of data annotation or an explicit burden on users. Significant locations are locations where a user stays stationary for a while (e.g., at least 10 min).

Given Wi-Fi measurements continuously collected in a user's daily life, we observe that when a user remains stationary in a location, the Wi-Fi measurements collected are similar to each other. Consequentially, the similar Wi-Fi measurements collected at a significant location show a high density. On the other hand, when a user keeps moving, the Wi-Fi measurements are dissimilar to each other, showing a low density. The density differentiation allows DCCLA to discover the significant locations in an unsupervised manner. Once a significant location is discovered, the Wi-Fi measurements collected is used to generate the Wi-Fi fingerprint of the location. As the generation of the Wi-Fi fingerprinting does not require data annotation, the system DCCLA is an unsupervised localization system.

However, since a user often visits the same locations (e.g., home or office) and stays there for a long while, DCCLA suffers from significant power consumption for mobile devices because of the continuous Wi-Fi collection. The problem can be addressed through activating Wi-Fi collection when it is necessary and deactivating Wi-Fi collection when "enough" data is collected. Specifically, the system enables activating Wi-Fi collection when a user stays at a significant location and deactivating Wi-Fi collection when optimal Wi-Fi collection duration is achieved and when the device is motion.

In this paper, we proposed to activate or deactivate the Wi-Fi collection based on the motion information of smartphones. The motion information is detected by using the acceleration readings from the embedded accelerometer sensor. The improved system benefits from saving power by only activating Wi-Fi collection when a user stays at a significant location and the system needs a learning dataset. In addition, we experimentally investigate the optimal Wi-Fi collection duration at a location to trade off the power consumption. The improved unsupervised localization system DCCLA with motion detection is tested by different users in real-world scenarios. The results show an improved performance of unsupervised localization, in terms of localization precision and power consumption.

This paper is organized as follows. In the next section, the works related to unsupervised localization and motion detection are presented. We then introduce the core idea of DCCLA and the motion detection algorithm for DCCLA. The performance of the system without and with motion detection is evaluated in the following section. In the end, the paper gives a conclusion.

2 Related Work

While outdoor localization is well supported by GPS, indoor localization, although attracting much attention, is still a research challenge. Wi-Fi networks provide a potential solution for indoor localization without additional costs of hardware installation. Some commercial software based on Wi-Fi fingerprinting localization has come to the market, including the Mobile Google Map [2], and Skyhook [3]. They usually generate a global database containing Wi-Fi fingerprints, known as a *fingerprint database*, by driving streets in cities and collect the Wi-Fi data at certain locations. However, these systems and applications require time-consuming and labor-intensive site surveys with explicit data annotation.

To eliminate the pre-deployment effort of site surveys, many research groups focus on learning significant locations from data collected in a user's daily life (e.g., GPS, GSM, Wi-Fi, or Bluetooth signals) implicitly. Such approaches do not require manual data annotation.

comMotion [4] is one of the earliest systems for discovering significant locations based on continuously GPS readings. The GPS signal is lost when a user enters a building. If the signal has been lost within a given radius (e.g., 100 meters) on three different occasions, the system infers that the location (building) is significant. The approach from Ashbrook and Starner [5] discovers significant locations, where the GPS readings have a continuous gap of at least 10 min. Such approaches based on GPS reading gaps provide building-level accuracy since the GPS signal is lost within buildings.

BeaconPrint [6] works based on collected Wi-Fi measurements and GSM readings, under the assumption that the user stays in a significant location if the measurements remain fairly *stable* during a pre-defined time window (known as a *stable state*). Once the criterion of a stable state is satisfied, the location is discovered as a significant location. SensLoc [7] is a similar system, but reduces false location detection by exploiting received signal strength (RSS) changes. However, the approaches of discovering significant locations by detecting a stable state are sensitive to the Wi-Fi signal variations and noise during a short time.

In recent years, Density-based clustering [8] has been proposed to address the problems caused by signal variations and noise. It works under the observation that the Wi-Fi measurements at a location, although suffering from signal variations, are quite similar to each other. When a user stays at a location for a while, the location can be discovered based on the high-density of similar measurements. Examples of such systems include ARIEL [9] Place Learning [10], and DCCLA [1]. These systems can automatically learn, and later recognize the significant locations where a user stays for a while. However, such systems suffer from consuming quite some power for continuous Wi-Fi collection.

The embedded accelerometer sensor is utilized to assist localization in some systems by detecting user movements. In most cases, the acceleration data can be used for step counting, displacement estimation or reachability between different areas [11–13], which is further utilized to improve the performance of localization in term of improving the location accuracy.

The accelerometer sensor is also used to save power by determining the moment a procedure of positioning should occur. Shafer, et al. [14] have proposed a strategy that the positioning only occurs when the system detect a user has moved to a new location. However, while such systems save power for positioning when using fingerprint database provided by others (e.g., Google, Skyhook), the power consumption for fingerprint learning is not considered, which is a huge amount when building their own fingerprint database. Different from many previous works, we utilize the accelerometer sensor to save power for both learning and positioning.

3   DCCLA with Motion Detection

DCCLA is an *unsupervised* indoor localization system using the Wi-Fi fingerprinting technique. "Unsupervised" indicates that the Wi-Fi measurements are implicitly and continuously collected in a user's daily life without requiring users' attention or data annotation. By processing the Wi-Fi measurements, the system can automatically discover and learn the significant locations where a user stays for a while (e.g., at least 10 min), and then recognize the locations when the user re-visits them.

A significant location is a location where a user stays stationary for a while (e.g., at least 10 min). The original DCCLA discovers significant locations by analyzing the Wi-Fi continuously collected. When the Wi-Fi measurements show a high-density distribution, a significant location is discovered.

When using an accelerometer sensor, the significant locations can be detected when a mobile device keeping in a stationary state at a location for at least 10 min. Based on motion detection, the unsupervised localization system activates Wi-Fi collection when a mobile device stays in a stationary state, and deactivates Wi-Fi collection when a mobile device stays a moving state.

When "enough" data is collected for place learning in a stationary state, more Wi-Fi collection consumes more power. The system enables deactivating the Wi-Fi collection when the optimal collection duration is achieved.

3.1   DCCLA

The localization procedure of the original DCCLA includes three phases: a collection phase, a learning phase, and a recognition phase. More details are available in the previously published paper [1, 15–17].

Collection phase: The smartphone periodically collects *Wi-Fi measurements* from surrounding APs (Access Points). Wi-Fi measurements consist of a current timestamp, MAC (Medium Access Control) addresses and RSS (Received Signal Strength) values from all detectable APs.

Learning phase: DCCLA performs a density-based clustering algorithm. We define an RSS value from an AP in a Wi-Fi measurement as a point p. A set of points in the learning dataset, whose Euclidean distance (e.g., the absolute value of the difference of two RSS values) to the point p is smaller than a specific distance threshold *Eps*, is the *neighborhood* of p. The density-based clustering algorithm works as follows:

- For a point p, if the number of the neighbourhood of p is equal to or larger than a density threshold *MinPts*, the point p and his neighborhood generate a cluster.
- For a point p, if the number of the neighbourhood of p is smaller than *MinPts*, the point p is regarded as noise.
- If any two clusters contain the same point(s), the two clusters are merged into one cluster.

A set of learned clusters, which belong to different APs related to the same timestamp of a Wi-Fi measurement, are combined to form a Wi-Fi fingerprint of a location.

Recognition phase: As the user visits a location, the current Wi-Fi measurement with n points is compared to the Wi-Fi fingerprints. If at least n-1 points in the current Wi-Fi measurement belong to a Wi-Fi fingerprint, the Wi-Fi fingerprint related to a learned significant location is recognized. Thus, DCCLA can recognize learned locations when the user re-visits them.

3.2   Motion Detection

The accelerometer sensor, embedded in smartphones, can be used to detect a mobile device's/a user's current motion state (i.e., in a stationary state or a moving state). Knowing the motion state, the system enables detecting a significant location and optimizing the Wi-Fi collection duration.

We implement a motion detection algorithm for DCCLA based on the continuous acceleration samples. It works based on the observation that the variation of the acceleration samples is large during a moving state, whereas the variation is small during a stationary state. In other words, a set of successive samples within a *time window* can be used to determine the motion state.

We select 4 s as the time window size. There are two reasons for the selection. First, the Wi-Fi collection frequency in DCCLA is 0.2 Hz. Thus, motion detection occurs at least once between two Wi-Fi collections. Secondly, according to our investigations, the time window of 4 s can tolerate some quick activities within a short period (e.g., to stand up and sit down quickly). The pseudo code of the motion detection algorithm is presented as follows.

```
Input: A set of acceleration samples At; a threshold Ta
Output: the motion state.
1) Label state as stationary.
2) Add the successive acceleration samples in 4 seconds
   into a sample list
3) for each acceleration sample At in the list, do
```
    1) Calculate the standard deviation (SD_{at}) of x_t, y_t, and z_t: $SD_{at} = \sqrt{x_t{}^2 + y_t{}^2 + z_t{}^2}$

```
4) end for
```
5) Calculate the standard deviation SD of the set of SD_{at}: $SD = \sqrt{SD_{at}{}^2 + \cdots + SD_{a(t+4)}{}^2}$

```
6) if the SD is smaller than or equal to Ta, do
```
    1) Label the motion state as *stationary*.
```
7) else, do
```
    1) Label the motion state as *moving*.
```
end.
```

The acceleration samples used in two successive time windows have a 50 % over-lapping that can achieve improved precision than detecting without an overlapping.

In order to determine the optimal sampling frequency, we evaluate the precision of correctly detecting the motion states by using accelerometer readings with different sampling frequencies 5 Hz, 10 Hz, 20 Hz and 32 Hz (32 Hz is a high sampling frequency for activity detection).

Table 1. The precisions of motion detection using different sampling frequencies.

Sampling frequency	5 Hz	10 Hz	20 Hz	32 Hz
Precision	99.53 %	99.53 %	99.76 %	99.77 %

The result in Table 1 shows that the increase of the sampling frequency does not significantly improve the precision of correct detection. With the consideration of power consumption in the next chapter, a low sampling frequency of 5 Hz is optimal for the motion detection.

Based on the motion detection output, the system enables automatically activating and deactivating Wi-Fi collection. The Wi-Fi collection is activated only when the current state is stationary, and the last state is either moving or no state (i.e., the system just starts). It indicates the mobile stays at a fixed location from the moment on. The Wi-Fi collection is deactivated in the following two situations: (1) the current state is moving, and the last state is either moving or no state. It indicates the mobile leaves a fixed location from the moment on. and (2) the Wi-Fi collection duration at a location is longer than the optimal Wi-Fi collection duration. It indicates "enough" data is collected for learning.

4 Experimental Evaluation

The reasons for using an accelerometer sensor in the unsupervised localization system are to save power. As such, we need to evaluate the optimal collection duration and the power consumption without and with motion detection. We have designed three experiments to:

- Determine the optimal collection duration D_{opt}, which provides an optimal perform-ance, when trading off with the power consumption;
- Investigate how much power can be saved with the motion detection algorithm;
- Evaluate the performance of DCCLA with motion detection by different users in real-word scenarios.

We select an office area with adjacent rooms for the investigations. The Wi-Fi meas-urements from available APs in the surrounding are collected with a sampling frequency of 0.2 Hz. The office area is located on the second floor of a three-storey building. The area consists of five office rooms next to each other. The layout of the office area is shown in Fig. 1.

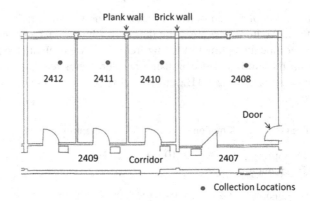

Fig. 1. The office area and the Wi-Fi collection locations, i.e. the positions of smartphones used for the evaluation.

The location accuracy of DCCLA is room-level. The room-level accuracy means the system is likely to be able to locate a mobile device in a room where the mobile device or an occupant is. The room-level accuracy, in the most cases, indicates 3–5 m error distance. DCCLA can learn and recognize a room correctly without having it mistaken for another room, even though the rooms are adjacent.

To evaluate the performance of achieving room-level accuracy, we define a set of evaluations metrics. *Correct* means a smartphone is in a room, and the system recognizes it is in this room. *False* means a smartphone is in a room, but the system recognizes it is in a different room. *Missed* means a smartphone is in a room, but the system does not recognize where it is.

Recognition Precision (RP) is defined as the number of "*Correct*" recognitions divided by the number of recognized attempts. It indicates how well the DCCLA can recognize a room correctly.

$$CP = \frac{\sum Correct}{\sum Correct + \sum False} \times 100\%$$

Response Rate (RR) is defined as the number of "*Correct*"s divided by the number of the total attempts. It indicates how often the Wi-Fi measurements from a given room are correctly recognized.

$$RR = \frac{\sum Correct}{\sum Correct + \sum False + \sum Missed} \times 100\%$$

4.1 Optimal Wi-Fi Collection Duration

In order to determine the optimal Wi-Fi collection duration D_{opt}, we learn the locations (i.e., rooms) using datasets of the following Wi-Fi collation durations D: 5, 10, 15, 20, 30, 45, 60, and 120 min in each room. The recognition phase is performed using the subsequent 60 min of Wi-Fi measurements from each room. The results are summarized in Fig. 2. From previous evaluations of DCCLA without motion detection [1], we deduced that the ideal *MinPts* is approximately one-third of the total number of Wi-Fi measurements. This parameter can be dynamically set when the Wi-Fi collection duration is controlled by the motion detection algorithm.

The results in Fig. 2 show that the optimal Wi-Fi collection duration is between 10 min and 30 min, when observing both *RP* and *RR* performance. Based on the experimental observations, we keep the Wi-Fi collection duration in our experiments between 10 min and 30 min. When a user stays at a location for less than 10 min, the Wi-Fi measurements are discarded. When a user stays at a location for more than 30 min, the Wi-Fi measurement is deactivated to save power, and 30 min of Wi-Fi measurements are used for location learning.

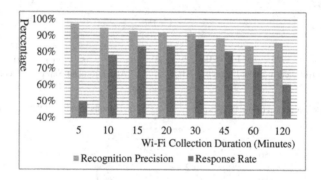

Fig. 2. The performance of recognition precision and response rate with different collection durations (Minutes) (Color figure online).

4.2 Power Consumption

The power consumption of DCCLA with motion detection consists of two parts: the Wi-Fi measurement and the accelerometer sampling. To test the power consumption, we installed the experimental setup as shown in Fig. 3. To measure the power consumption, the experimental setup consists of a combination of hardware and software. The hardware includes a Samsung Galaxy S2 smartphone with a battery of 3.7 V and 1650 mAh capacity, two Peaktech digital-multimeters. The software running on a laptop is to record the voltage and current measurements, respectively. The recording frequency is 2 Hz.

Experiments are carried out in three different settings. In each setting, we take 10 min as measurement duration. We measure each setting three times, respectively.

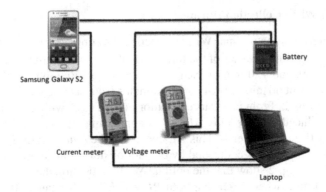

Fig. 3. The experimental setup to measure the power consumption of Wi-Fi measurement and accelerometer sampling.

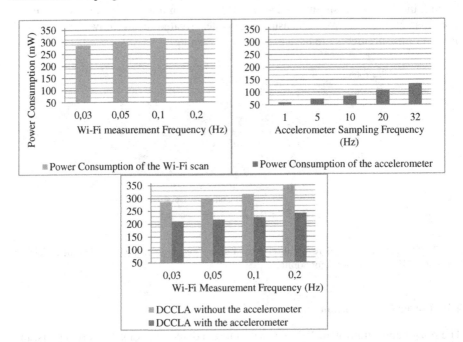

Fig. 4. The power consumption of Wi-Fi measurements (upper right) and accelerometer sampling (upper left), as well as the power consumption of DCCLA without and with motion detection (lower).

Setting 1: A smartphone periodically collected the Wi-Fi measurements with different frequencies of 0.2, 0.1, 0.05, and 0.03 Hz.

Setting 2: A smartphone periodically read the accelerometer samples with different frequencies of 1, 5, 10, 20, 32 Hz.

Setting 3: A smartphone collected Wi-Fi measurements without and with the assistant of accelerometer readings. The Wi-Fi measurement frequency for both systems with and without motion detection is 0.2 Hz. The accelerometer sampling frequency for the system with motion detection is 5 Hz.

The power consumptions for Settings 1, 2 and 3 are shown in Fig. 4. We can observe that the acceleration acquisition consumes 73 mW with a sampling frequency of 5 Hz, whereas it consumes 134 mW with a sampling frequency of 32 Hz. For the Wi-Fi measurement with different measurement durations, the power consumption is at least 280 mW. However, the acceleration acquisition consumes less power than Wi-Fi measurement even with a high accelerometer sampling frequency of 32 Hz. Based on the observations we compare the power consumption of the system without and with the assistant of accelerometer readings. We assume with the assistant of accelerometer readings, the Wi-Fi measurement duration can be reduced to half compared to DCCLA without the assistant of accelerometer readings. In such case, compared with the power consumption of the system without motion detection, the average power consumption for the system with motion detection saves power of almost 30 %.

4.3 Tests in the Real-World Scenario

The real-world tests were carried out in the student activity area as Fig. 5 shows. The area includes a computer pool 2417, a kitchen 2414 and a meeting room 2413, where students spend a part of their day. Smartphones were carried by three different users, whose motion patterns were different.

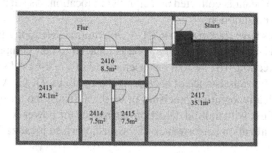

Fig. 5. The student activity area, where the system was tested by three different users.

For each user, the learning dataset in each location was collected in the first 30 min during the stationary state detected based on accelerometer readings. The recognition dataset was the subsequent 15 min from each location after each location is automatically learned.

Figure 6 shows the test results. We can observe that DCCLA without motion detection performs unsupervised localization with an average RP of 78.60 % and an average RR of 68.20 %. DCCLA with motion detection achieves a better performance with an average RP of 99.93 % and an average RR of 64.08 %. The results indicate that DCCLA

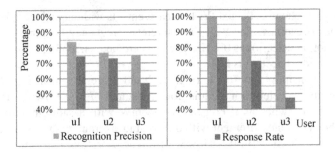

Fig. 6. The comparison of DCCLA performance without (right) and with motion detection tested (left) by three different users.

with motion detection can learn and recognize a room with improved recognition precision, while maintaining the response rate.

5 Conclusion

While the unsupervised localization system DCCLA can automatically learn and recognize the significant locations from the continuously collected Wi-Fi measurements, the system usually suffers from consuming large amounts of power of mobile devices caused by continuous Wi-Fi collection. With motion information (i.e. in a stationary state or in a moving state) determined by acceleration readings, DCCLA with motion detection enables automatically activating Wi-Fi collection when the mobile device stays at a significant location and deactivating Wi-Fi collection when the mobile device keeps moving or "enough" data is collected at a significant location. As such, the system can reduce the power consumption.

Our experiments have shown that DCCLA with motion detection saves almost 30 % power when the Wi-Fi measurement duration is reduced to half compared to DCCLA without motion detection. The system has been tested in a real-world scenario by three different users. With the assistance of accelerometer readings, DCCLA achieves room-level accuracy with a recognition precision of 99.93 %. Thus, the unsupervised localization system DCCLA with motion detection can reduce the power consumption, while maintaining the localization performance of the unsupervised localization system.

Acknowledgments. This work has been [co-]funded by the Social Link Project within the Loewe Program of Excellence in Research, Hessen, Germany.

References

1. Xu, Y., Lau, S.L., Kusber, R., David, K.: DCCLA: autonomous indoor localization using unsupervised Wi-Fi fingerprinting. In: CONTEXT 2013, Annecy, France (2013)
2. Mobile, G.: Google Maps for Android. https://www.google.com/intl/en/mobile/maps/. (Accessed 21 May 2014)
3. "Skyhook," Skyhook Wireless, Inc. (2013) http://www.skyhookwireless.com/. (Accessed 06 Dec 2013)

4. Marmasse, N., Schmandt, C.: A user-centered location model. Pers. Ubiquit. Comput. **6**(5–6), 318–321 (2002)
5. Ashbrook, D., Starner, T.: Learning signicant locations and predicting user movement with GPS. In: The 6th International Symposium on Wearable Computers (2002)
6. Hightower, J., Consolvo, S., LaMarca, A., Smith, I., Hughes, J.: Learning and recognizing the places we go. In: 7th International Conference on Ubiquitous Computing, Venice, Italy (2005)
7. Kim, D., Kim, Y., Estrin, D., Srivastava, M.: SensLoc: sensing everyday places and paths using less energy. In: Proceedings of the 8th ACM Conference on Embedded Networked Sensor Systems (2010)
8. Ester, E., Kriegel, H., Sander, J., Xu, X.: A density-based algorithm for discovering clusters in large spatial databases with noise. In: 2nd International Conference on Knowledge Discovery and Data Mining, Portland, OR, USA (1996)
9. Jiang, Y., Pan, X., Li, K., Lv, Q., Dick, R.P., Hannigan, M., Shang, L.: ARIEL: automatic Wi-Fi based room fingerprinting for indoor localization. In: 14th International Conference on Ubiquitous Computing, Pittsburgh, PA, USA (2012)
10. Dousse, O., Eberle, J., Mertens, M.: Place learning via direct WiFi fingerprint clustering. In: IEEE 13th International Conference on Mobile Data Management, Bengaluru, India (2012)
11. Woodman, O., Harle, R.: Pedestrian localisation for indoor environments. In: UbiComp 2008 Proceedings of the 10th international conference on Ubiquitous computing (2008)
12. Jimenez, A., Seco, F., Prieto, C., Guevara, J.: A comparison of pedestrian dead-reckoning algorithms using a low-cost MEMS IMU. In: IEEE International Symposium on Intelligent Signal Processing, 2009, Budapest (2009)
13. Wu, C., Yang, Z., Liu, Y., Xi, W.: WILL: wireless indoor localization without site survey without site survey. In: INFOCOM, 2012 Proceedings IEEE, Orlando, FL (2012)
14. Shafer, I., Chang, M.L.: Movement detection for power-efficient smartphone WLAN localization. In: Proceedings of the 13th ACM International Conference on Modeling, Analysis, and Simulation of Wireless and Mobile Systems (2010)
15. Xu, Y., Lau, S.L., Kusber, R., David, K.: An experimental investigation of indoor localization by unsupervised Wi-Fi signal clustering. In: Future Network and Mobile Summit, Berlin, Germany (2012)
16. Xu, Y., Kusber, R., David, K.: An enhanced density-based clustering algorithm for the autonomous indoor localization. In: MOBILe Wireless MiddleWARE, Operating Systems and Applications (Mobilware), Bologna, Italy (2013)
17. Lau, S.L., Xu, Y., David, K.: Novel indoor localisation using an unsupervised Wi-Fi signal clustering method. In: Future Network and Mobile Summit, Warsaw, Poland (2011)

Analyzing Real World Scenarios

The Communication Patterns in the Context of Error in an Intensive Care Unit in a Malaysian Hospital

Nik Nailah Binti Abdullah[1(✉)] and Sharil Azlan Ariffin[2]

[1] School of Information Technology, Monash University Malaysia,
Jalan Lagoon Selatan, 47500 Bandar Sunway, Malaysia
nik.nailah@monash.edu
[2] Department of Anaesthesia and Intensive Care,
National Heart Institute of Malaysia,
No 145, Jalan Tun Razak,
50400 Kuala Lumpur, Malaysia
dr.sharil@ijn.com.my

Abstract. This paper describes a study that was carried out in an Intensive Care Unit at a hospital in Malaysia. The objective of our work was to define *what constitutes the context of error* during the ward round practice of patient care management considering how artifacts are used during the communication. Thus the work focused on the analysis of communication patterns. As research method, we have applied the ethnography method and used situated cognition as an analytical perspective to synthesize the communications patterns. In this paper we focus on reporting the empirical analyses of the communication patterns in which errors had occurred. The analyses had highlighted how a clinician team conceptualized information and what majorly constitutes the context of error was that the clinical information on the artifacts were represented 'without a context'.

Keywords: Context and communications · Situated cognition · Medical error · Health information technology · Patient safety

1 Introduction

In the intensive care unit (i.e., ICU) setting, the frequent need for urgent critical and life-saving decision-making potentially creates an environment within which medical errors may happen. In our work we refer to [10] on the definition of medical error, which is defined as the failure of a planned action to be completed as intended or the use of a wrong plan to achieve an aim. Due to the nature of the patient group (i.e., the most critically ill) present within the ICU, the ICU is an environment in which there is less margin for error and less favorable circumstances exist for error recovery [8, 9]. The basic assumption is that error arises within highly complex medical care systems of *people, information systems, workflow, and clinical procedures* and follows a flow of work practice pattern that can be uncovered [6]. Uncovering these patterns of errors would allow clinicians to eliminate or recover from them as soon as possible before the

© Springer International Publishing Switzerland 2015
H. Christiansen et al. (Eds.): CONTEXT 2015, LNAI 9405, pp. 147–158, 2015.
DOI: 10.1007/978-3-319-25591-0_11

errors turn into an actual adverse event[1] [6]. It has been reported that health information technology (HIT) can reduce the risk of serious injury for patients during hospital stays [6]. However, its true potential for *preventing errors* remains only partially realized and, as has been demonstrated in a recent article [7], some systems may even give rise to hazards of their own. Thus, the critical care setting is a uniquely complex field for which computing technology needs to be developed according to novel and often unprecedented design principles [6]. How well the design of the system complements its intended setting and purpose is critically important for patient safety [6]. Patient safety has been defined as "the prevention of harm to patients." Its emphasis is placed on the system of care delivery that: (1) prevents errors; and (2) learns from the errors that do occur [3].

Thus we put forward the following research question - *how can the patient safety element in our use of HIT translate into a design principle?* To begin with, the objective of our research is to study the patterns of *what constitutes context of error, that is, the work system context in which errors occur* during patient care management. By studying the context of error, it will reveal the limitations of the system. A well-defined modeling and representation of *context of error* will reveal specifically how and when artifacts in work practice frequently lead to error. This would enable us to identify safety elements which are required in the design principles that can reduce or eliminate those erroneous situations that may lead to adverse events.

Therefore, we have carried out an ethnography study in an ICU in a studied hospital in Kuala Lumpur, Malaysia for several weeks. Communication exchanges were recorded during the morning ward rounds and analysed using situated cognition theory to synthesize the medical staff communication patterns. The ward rounds involved a clinician team that reviews the patients' cases in the morning to make decisions on the patient care management for the rest of the day.

In this paper we focus on reporting the empirical analyses of the communication patterns in which errors had occurred. The analyses had highlighted how a clinician team conceptualized information. We found that what majorly constitutes the context of error was that the current information on the artifacts were represented 'without a context' (absence for what the patient is being treated for). Thus this paper is organized as follows. We discuss related work, followed by research methods. Then we will illustrate the empirical findings. We summarise the paper with discussion and perspective on future work.

2   Related Work

There are various approaches to studying the use of HIT in improving the patient care management in the ICU and can be mainly divided into two major viewpoints. The first viewpoint looks into the existing clinician work practice to mainly understand the nature of their work practice to derive and/or improve methodologies and theories.

---

[1] Any injury due to medical management, rather than the underlying disease. Example of an injury would be a rash caused by an antibiotic [10].

For example, the work of Abraham et al. [2] looked into conceptualizing a communication model during handoffs in critical care handoffs as communication failure has been reported as the leading cause of medical errors and adverse events. Meanwhile the authors Vimla et al. [8] investigated what are the natural constraints in the ICU environment imposed on error detection and correction during team's decision-making process in patient care management planning.

The second viewpoint looks into the study of the information technology used as part of the work practice in the ICU for the purpose of designing better information system. The work of Thursky and Mahemoff [11] explored the use of user-centered design techniques for developing the requirements for an antibiotic decision support system in ICU. Meanwhile the authors Gibson et al. [5] looked into how general practitioners interact both with their patients and computers with the aim of facilitating the detailed understanding of how GPs use their computers while in consultation with patients.

Our work intertwines both viewpoints. First we would like to understand and capture how the HIT in the studied ICU is used during the clinician team ward rounds. Secondly to uncover what *constitues context of error during the ward rounds.*

2.1 Study Site, Participants and Data Collection

The data was collected at a 30-bedded post-operative adult intensive care unit in a large hospital in Kuala Lumpur. It looks after approximately 2000 admissions per year. A decision support system and paper records were concurrently used for patients care documentation in the unit at the time of study. Three clinicians team from the ICU were included in this study during our 3-day study. Each team consisted of a resident doctor, an intensivist, a nurse, and the ICU clinical director (with once or twice a week participation). This is the common composition of a clinical team participating in the morning handover ward rounds. A total of 10 individuals participated in the 3-day study.

Data was collected during the morning ward rounds, where the patient care planning sessions were done in the ICU. The ward rounds are held twice daily during medical staff shift rotation (0745 and 1800 h) in addition to a more formal morning (0900). The first author followed the team on their formal ward rounds. During these sessions, the clinical team discussed each patient's health status. The resident doctor or the clinical director will lead the ward round to review and decide collaboratively on the patient care management for the day. Each round lasted approximately 3 to 4 h and the first author spent 3 h a day for 3 days consecutively shadowing and observing the clinical teams. Team interactions were video-recorded. Field notes, photographs of artifacts, and interviews with the clinicians and support staff complemented the video recordings. The video interaction records of the morning ward rounds amounted to a total of 9 h.

2.2 Data Transcription Annotation and Coding

We have developed the following steps for relating the clinical communication to the context of error, which was analysed together with the second author.

1. Selected transcriptions were analysed using discourse analysis. An utterance is marked with who is the speaker and listener and the medium (i.e., paper, decision support system) used to mediate their communication, and the topic of their communications [1].
2. The analysed utterances are then applied with a high-level coding of case management and error coding [8] to capture the clinical content. For example, "Klebsiella esbl.....must be carbopenum....carbopenum..where is the carbopenum" is coded as 'information loss'. Details to follow in the next section.
3. Situated cognition theory was applied to synthesize and explain the clinician team communication analysis and patterns [4].

For data coding, we have applied a high-level coding process using codes from the work of Patel et al. [8]. The authors [8] applied an open coding process, where each clinically relevant utterance exhibiting common strategies used during case management was coded for content (e.g., information interpretation). Furthermore, the authors [8] had also developed a coding of errors in communication (including clinical content). If an utterance contained or was related to an error, the authors categorized it as either "generated error", "resolved error" etcs. Please refer to the Appendix for the complete coding definitions. The coding developed however did not look specifically into the *notion of context*. Thus in our work we have incorporated a contextual analysis to relate how artifacts are used during the ward rounds and *how it constitutes part of the work system context in which error had occurred*. The coding were applied to capture the kind of actions that the clinician team formulated in context of errors, and using the contextual analysis to capture how artifacts are used during these contexts.

2.3 Analysis of Context – Conceptualization

In situated cognition [4] context for a person is viewed as a mental construction studied from the notion of *conceptualization* and *contextualization,* and is studied from a moment-by-moment analysis.

Conceptualization is considered from both a social and neuropsychological perspective. In our work, we first focus on the notion of context from the perspective of social psychology. 'Context' from this perspective is explained by conceptualization i.e.: how a person conceptualizes his role considering his situation, and activity - 'What I am Doing Now'. For example, a resident doctor on the ward round is conceptualising what he is doing now: 'making decision whether the patient has a heart failure' in constructing his behavior. From a neuropsychological perspective, the notion of conceptualization involves a composition of higher-order categorization processes at the perceptual-conceptual level that is responsible for our coordinated activity in time. For example, at the perceptual level, a resident doctor at the ICU when he is situated in a context, the way he perceives the context is always through categorizing the details. As an example the resident doctor categorises the details from 'the patient's skin color, blood test results on paper and the nurses speaking about the wounds' in the environment. These categorized details are then given a description or semantic label, as an example 'patient is improving'. At the conceptual-memory level, which is a higher-order categorization process – the details, which are also given descriptions, are then conceptually categorized, as an example 'these are clinical evidence'. Thus, conceptualizing can be viewed as a dynamic process of reconstructing a person's action relating perceptions to higher-level concepts in memory as part of how experiences are formed. Thus physical re-coordination such as taking a paper chart, and reading from it while speaking about is viewed as part of the basis in speaking and comprehending text. The notion on conceptualization is used as a method to synthesize the communication patterns.

3   Results –Communication Analysis

In this paper, we will illustrate the results from our communication analysis of a specific event between the resident and the ICU clinical director in which errors were *generated and unresolved* for about 15 min. The errors (of information) were subsequently revealed and managed appropriately but only after a period of time deliberating and discussing the erroneous points. First we will describe the work setting, followed by the artifacts in use and a brief background of the event. Figures 1 and 2 shows an example of a bedspace at the studied ICU and the mock design of the decision support system correspondingly.

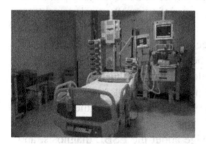

Fig. 1. ICU bedspace.   **Fig. 2.** Mock up of the decision support system

The computer is located on the left hand side at the end of the bed (not shown in Fig. 1), facing away from the bedspace. The system (see Fig. 2) allows users to enter the ICU daily plan, and view the progress of the post-operative vital signs in a time based flow chart. The flow chart shows the reading at different time intervals of the haemodynamic (blood pressure, heart rate, etc.) variables, itemized input and output of the patient (fluids, medications, etc.). The clinical team also refers to an X-ray system, which will include the most recent as well as previous radiological (i.e., X-ray) investigations. Other artifacts in use are the blood gas report, pathology report and microbiology paper record. Figure 3 is an example of a microbiology paper report while Fig. 4 shows an example of the prescription chart.

We give a brief background of the event. The resident doctor had committed an error when referring to the patient record from the decision support system and thus the resident doctor was presented with two contradicting information about his patient. One

Fig. 3. Microbiology report

Fig. 4. Prescription chart

was about the patient in the next room viewed on the decision support system, and another is of the correct patient at that bedspace presented on the paper documents (i.e., microbiology lab result).

We shall refer to the next room patient as patient B, and the correct patient as patient A. Patient B has been diagnosed with an infection "extended-spectrum beta-lactamase (i.e., ESBL) bacteria". Patient A's medical treatment did not indicate that he was diagnosed with the ESBL bacterial infection. Patient A actually had "e-coli bacteria" identified in a healing wound and was being treated with an antibiotic for "corynebacterium bacteria" identified in his sputum. However these facts were not mentioned in the artifacts but instead the drug prescribed for Patient A was mentioned, which was tazocin. For the next 15 to 20 min, the resident doctor and the consultant doctor were engaged both in trying to reconcile the contradictory information by communicating and recalling their clinical knowledge about the ESBL diagnosis, and what's being taught about its medical treatment. Based on their medical knowledge, tazocin is not usually the preferred choice of treatment for an ESBL infection. We show an excerpt of the transcribed communications below:

Resident doctor: "Klebsiella esbl..must be carbapenem, carbpenem, where (looks at the blue folder of drugs prescription) is the carbapenem?"

Resident doctor: "This patient..surgeon request to extubate but still has op-pneumonia..klebsiella pneumonia isolate before //"

ICU clinical director: "//When was the last positive culture?" (continues)...

Applying step 1 (refer to section on Data Transcription and Annotation Coding), the transcribed communications has been annotated with the turn taking, including gestures, artifacts, and the details that the doctors were focused at during the communications. Details could refer to the artifacts, or a concept in the mind (e.g., focused at the detail in the mind about ESBL). Notation such as "//" indicates overlapping of utterances between speaker and listener. The Table 1 below depicts correspondingly at each utterance: (i) the activity-artifact (i.e., paper, computer) referring to which patient's information; (ii) the high-level coding of the actions coded by the categories of the case management, and (iii) if and what are the categories of error that had occurred during the different 'actions'. Situated cognition was applied to give a moment-by-moment account of the communications considering the relationship of details in the artifacts.

Table 1. The annotated utterances applied with case management and error coding.

Ln_n, Spk_x	Utterances	Activity-Artifact	Details	Comment	Case management	Error
1, H	Klebsiella esbl.. must be carbopenem.. where is the carbopenum..	Looked at microbiology report on system (Patient B) Flipping through - Paper chart (Patient A)	Carbopenum (a drug) and ESBL (a type of bacteria)	H verbalising outloud looking for the bacteria culture report and its treatment on the chart pages by pages	Information loss type of failure to follow up	Generated error
2, H	This patient surgeon request to extubate but still has op-pneumonia klebsiella pneumonia isolated before//	Assessing -X-ray and the paper chart (Patient A)	Condition of patient's lung	H refers to historical evidence consisting of the X-ray to look at the lungs, and to the paper chart on reported test results	Information interpretation	Generated error
3,Sh	//when was the last positive culture?	Standing and looking – computer report (Patient A)	ESBL, shadows in the lung field	SH is listening to H's assessment of patient's history	Additional information	N/A
7, Sh	this is 19, this is today	Looking- X-ray (Patient A)	ESBL, shadows in the lung field	Shared representation	Information aggregation	N/A
8, H	slightly better (nodded)	Looking -X-ray (Patient A)	ESBL, shadows in the lung field	Shared representation	Information interpretation	Generated error
9, Sh	not much	Looking -X-ray (Patient A)	ESBL, shadows in the lung field	Shared representation but SH did not agree with the assessment	Information interpretation	Corrected error

The case management coding allowed us to identify the kinds of actions (i.e., information interpretation) that the clinician team formulated in the context of error. At Ln_1, the resident doctor was referring to the microbiology report on the decision support system indicating Patient B had ESBL, at the same time flipping through the prescription chart of Patient A to look for carbopenum, a drug treatment for patients that have ESBL. His action was coded with case management 'information loss' of type 'failure to follow up'. Previously he was inquiring about the patient's A diagnosis, however it was not followed up. Thus he went to look for details of carbopenem that would indicate that the patient A does have ESBL. At this time, it had led to 'error generated' in which the result of his seeking of information was still not being followed up. Thus it has led him *to continue to believe* that the patient A had an ESBL infection.

At Ln_2, the case management was 'information interpretation' to find evidence from the information loss that the patient A does indeed have ESBL (to confirm his diagnosis). Based on the x-ray results, historical records, and the prescription chart he interpreted that the patient A had klebsiella, a bacteria that can have the ESBL property. This led to further 'error generated' because the resident doctor made assumptions that the patient A had a klebsiella infection, which was incorrect still. At Ln_3, the ICU clinical director had applied 'additional information' case management strategy after listening to the resident doctor's interpretation of the patient's A drug treatment and assessing the X-ray. At this moment, the ICU clinical director would like to get past information on how long the infectious disease (i.e., ID) team have isolated the organism, as this would inform him what the patient actually has. From Ln_7 onwards until Ln_9, the resident doctor and the ICU clinical director were both referring to the X-ray, which becomes a shared representation for both. Both of them were focused at analyzing the *details* of the 'shadows in the lungs' that would indicate physiologically that the lungs of patient A have improved.

3.1   Patterns of the Communication Analysis – a Clinician Team's Context

In the previous section we have illustrated how the clinician team formulated actions during a specific event where errors had occurred. Specifically the term *action* in our work encompasses the coordination and communications on the use of the artifacts. In this section, we will discuss the communication patterns that had emerged from our analysis. From the communication analysis, we have found that the "error generated" occurred during two types of case management categories: information loss type 'failure to follow up' (3 times) and information interpretation (2 times), shown in Table 2 below. Thus in total we had identified 5 occasions in which a context of errors had occurred.

Refer to Table 2, we identified the general patterns that has emerged from the category 'information loss' of type failure to follow up:

- The prescription chart (i.e., paper artifact) was used to *get historical information* on the kind of drugs that have been prescribed to the patient, this was followed by;
- X-ray which (i.e., digital X-ray) was used to get clinical evidence that the lung *has physiologically improved.*
- *Microbiology full report on the paper artifact* was used to get the complete results, i.e., full sensitivity report for the microorganism, which is the *reaction of the microorganism with different types of antibiotics.*
- *Microbiology quick report on the decision support system* was used to get the report on *the name of the microorganism that is grown.*

The category 'information interpretation' that is interpreting evidence in hand exhibits the following pattern:

- Digital X-ray which was used to interpret if the lungs had improved and;

Table 2. Type of case management categories in relationship to errors.

Ln$_n$	Categories	Activity-Artifact	Information	Details
1	Information loss	Prescription chart on paper (Patient A)	Shows list of drugs prescribed to patient	Carbopenum (a drug) and ESBL (a type of bacteria)
20	Information loss	Microbiology lab results on paper (Patient A)	Shows results on the microbiology, hematology and biochemistry	Drug (tazocin)
36	Information loss	Patient, ventilator, microbiology lab report on the decision support system (Patient B), X-ray system (Patient A)	Patients demonstrate clinical evidence of health, ventilator, decision support system window showed the microbiology report, X-ray shows lung evidence (improving or not)	Improvements in patients – drug dosage, lungs, clinical evidence (skin, alertness in patient)
2	Information interpretation	X-ray system, prescription chart on paper (Patient A)	X-ray shows lung evidence (improving or not), shows list of drugs prescribed to patient	Drug (tazocin)
8	Information interpretation	X-ray system (Patient A)	X-ray shows lung evidence (improving or not)	Shadows, drugs

- Prescription chart on the paper was used to interpret if the drug treatment have indeed improved the lungs, thus would help clinicians to infer the patient's infectious disease.

The X-ray is used as a 'physiological evidence' that the lungs has improved and the prescription chart is used as an evidence that the drug treatment is indeed the correct one that is improving the lungs.

4   Discussion and Future Work

The objective of our study has been to study *what constitues context of error*. In the previous sections we have illustrated at a moment-by-moment analysis how the clinician team used the artifacts during the context in which error had occurred. To synthesize the conceptualization process, we refer to situated cognition notion [4] of conceptualization. Thus, in this section we will discuss our synthesis of the findings.

The findings had revealed a very complicated work practice interaction among the decision support system, ventilator setting, paper records, and the clinician team clinical knowledge and procedure in conceptualizing *"what is the patient being treated for and is he/she improving?"*. The conceptualisation process demonstrated an act of 'coupling' in the coordinating of using the artifacts *while* speaking about the drugs, and their knowledge on the medical treatment. The coordination here refers to how the actions of *assimilating, interpreting and aggregating* information are used to coordinate the team's decision-making process on the patient's care management.

At the coordination-artifacts coupling level, the clinicians access the information, which was represented on the artifacts as different levels of details to conceptualise their next actions (i.e., what does this information represent and where do I go next to seek evidence?). At the artifacts-speaking coupling level, the details from the artifacts were formed conceptually in the mind (e.g., the patient has EBSL) and the details from the environment were also constructed (e.g., seeing the physical signs on the patient) while speaking. These actions had demonstrated that the clinicians were 'trying to fit' what is being presented as information, and interpreting and aggregating it (model of what is represented in the world) with what they have learned as 'clinical knowledge' and from their experiences on the treatment (model in their mind). This has further revealed that the conceptualisation of the clinicians showed that the formulating actions of *perceiving, interpreting, and aggregating was about contextualising all possible details from the environment on different levels of coupling*. What we refer to as 'all possible details' refers to the action of getting details that can replace the most significant information loss that was 'wh*at has the patient been treated for currently?*'. This information loss was not recorded anywhere on the artifacts.

The synthesis viewed from the perspective of situated cognition highlighted most importantly that the current information was represented 'without a context' (absence for what the patient is being treated for). This has revealed the limitations of the present work system design – and the complexity of the decision making process involved, because of the nature of patient care management in critical care settings. It has also led us to further question - why was the work system (e.g., paper records and artifacts not recording what the patient has) designed in such a way? The initial study of observing context of error had enabled us to highlight the common patterns in the use of artifacts that had led to errors. The observation can lead to design principles that consider safety elements, for example a principle could be that a decision support system must be able to have an alert mechanism. Thus our future work would be conducting longitude studies to obtain a general finding so that the modeling and representing of context of error can be developed.

Acknowledgment. We would like to acknowledge Dr. William J. Clancey at the Florida Institute for Human and Machine Cognition, USA, Dr. Sharifah Suraya Syed Mohd Tahir, and Dr. Suneta Binti Sulaiman, at the Institut Jantung Negara (National Heart Institute of Malaysia) for their constructive comments on the study.

Appendix

Categories of case management coding	Description	Example
Information aggregation	Patient information aggregated by the presenter prior to the its interpretation by the entire team; multiple instances of information aggregation possible depending on the number of ongoing medical issues in the case	"MICU day no.3, she was exubated yesterday. Her problem include mental status, hep C, withdrawal, UTI stage 2 DQ ulcers"
Information interpretation	Patient information interpreted based on the evidence at hand	"Because of her size, I can pretty much guarantee you, what's in there is probably a Bovina"
Information loss	1. Inaccurate recall: Recalled patient information that is inaccurate, where correct information is loss 2. Failure to follow up: Question posed by team member but never addressed in discourse	1. Team member discuss patient having a history of diabetes, when the information available did not show this history 2. Team member asked if patient was passing urine, but this question was never followed up

Categories of error coding	Description
Generated error	1. When the information uttered by a team has something that is incorrect or doubtful 2. Anything that is categorizes as relevant information loss, inaccurate interpretation, of faulty decision making
Corrected error	1. When participants themselves or someone else corrects an error 2. When a mistake is detected and corrective actions are taken 3. When an incorrect interpretation or decision is corrected

References

1. Abdullah, N.N.B., Sharp, H., Honiden, S.: Communication in context: a stimulus-response account of agile team interactions. In: Wang, X., Sillitti, A., Whitworth, E., Martin, A. (eds.) XP 2010. LNBIP, vol. 48, pp. 166–171. Springer, Heidelberg (2010)
2. Abraham, J., Kannampallil, T.G., Patel, V.L.: Bridging gaps in handoffs: a continuity of care based approach. J. Biomed. Inform. **45**(2), 240–254 (2012)
3. Barton, A.: Patient safety and quality: an evidence-based handbook for nurses. AORN J. **90**(4), 601–602 (2009)
4. Clancey, W.J.: Situated Cognition: on Human Knowledge and Computer Representations. Cambridge University Press, Cambridge (1997)
5. Gibson, M., et al.: Multi-tasking in practice: Coordinated activities in the computer supported doctor–patient consultation. Int. J. Med. Inform. **74**(6), 425–436 (2005)
6. Horsky, J., Zhang, J., Patel, V.L.: To err is not entirely human: complex technology and user cognition. J. Biomed. Inform. **38**(4), 264–266 (2005)
7. Koppel, R., et al.: Role of computerized physician order entry systems in facilitating medication errors. JAMA **293**(10), 1197–1203 (2005)
8. Patel, V.L., Alisabeth, L.S., Khalid, F.A.: Error recovery in the wilderness of ICU. In: Patel, V.L., David, R.K., Trevor, C.: Cognitive Informatics in Health and Biomedicine, pp. 91–111. Springer, London (2014)
9. Patel, V.L., Trevor, C.: New perspectives on error in critical care. Curr. Opin. Crit. Care **14**(4), 456–459 (2008)
10. Rothschild, J.M., et al.: The critical care safety study: the incidence and nature of adverse events and serious medical errors in intensive care. Crit. Care Med. **33**(8), 1694–1700 (2005)
11. Thursky, K.A., Michael, M.: User-centered design techniques for a computerised antibiotic decision support system in an intensive care unit. Int. J. Med. Inform. **76**(10), 760–768 (2007)

Understanding the Patient 2.0

Gaining Insight into Patients' Rating Behavior by User-Generated Physician Review Mining

Michaela Geierhos[(✉)], Frederik S. Bäumer, Sabine Schulze, and Caterina Klotz

Heinz Nixdorf Institute, University of Paderborn,
Fürstenallee 11, D-33102 Paderborn, Germany
{michaela.geierhos,fbaeumer,sabine.schulze,caterina.klotz}@hni.upb.de

Abstract. Patients 2.0 increasingly inform themselves about the quality of medical services on physician rating websites. However, little is known about whether the reviews and ratings on these websites truly reflect the quality of services or whether the ratings on these websites are rather influenced by patients' individual rating behavior. Therefore, we investigate more than 790,000 physician reviews on Germany's most used physician rating website jameda.de. Our results show that patients' ratings do not only reflect treatment quality but are also influenced by treatment quality independent factors like age and complaint behavior. Hence, we provide evidence that users should be well aware of user specific rating distortions when intending to make their physician choice based on these ratings.

Keywords: Health 2.0 · Rating behavior · Patient opinion mining on physician rating websites

1 Introduction

Provider rating websites become facilitators for patients in health-related decision-making [10] and hence, it is not surprising that physician rating websites (PRWs) are becoming increasingly popular [6]. A recent study [7] revealed that already more than one third of the Internet users are searching for a physician directly on PRWs. However, "research on these information tools is in its infancy" [11]. So far, only little is known about "the number, distribution, or trend of evaluations on PRWs" [6] as well as about "the content and nature of narrative comments" [7]. Moreover, it is unclear, "whether PRWs have the potential to reflect the quality of care offered by individual health care providers" [3,20]. However, knowledge about these issues is quite important, because PRWs are likely "to influence the image of doctors in society and the self-understanding of both doctors and patients" [20] and could be shown to have a considerable impact on physician choice making [7,8]. We therefore analyzed data from jameda.de, the physician rating portal playing potentially "the most important role in the German physician website movement" [6]. The intention thereby was

© Springer International Publishing Switzerland 2015
H. Christiansen et al. (Eds.): CONTEXT 2015, LNAI 9405, pp. 159–171, 2015.
DOI: 10.1007/978-3-319-25591-0_12

to present the first study on German PRWs analyzing complaint behavior as a treatment quality independent factor, in order to get a more unbiased and comprehensive understanding of patient opinions. We therefore decided to make a proof-of-concept by selecting waiting time for an appointment or in a practice as a prototype for a cause of complaint which is not influenced by the physician's treatment but maybe by age (Subsect. 4.1), health insurance affiliation (Subsect. 4.2) or grading behavior (Subsect. 4.3).

This paper is organized as follows: In Sect. 2, we give an overview of related work. In Sect. 3, the data set (Subsect. 3.1) and the methodology (Subsect. 3.2) of our empirical study are described. Section 4 then presents our results which are evaluated in Sect. 5. Finally, we conclude in Sect. 6 and provide possible directions for future work.

2 Related Work

Since PRWs are becoming more and more popular [9], they have also gained increasing attention in the scientific literature [8]. There are studies dealing with the PRWs themselves. That is, with their quality [3,5,9,21–23], their perception in the scholarly and public debate [2,11] as well as with data protection [17] and ethical issues [11,24] that go along with these websites. Moreover, their influence on the quality of treatment [26] and on doctor-patient relationships [16] have been discussed as well as the question how physicians should deal with physician rating websites [9].

Other research concentrates on profiling users of PRWs. In particular, these studies investigated the motivation for physician rating website usage [2,18] as well as the socio-demographic [1,8,25] and psycho-graphic [25] characteristics of users. Also the health status affiliation of users of PRWs was examined [8,25]. Moreover, evaluation criteria, that are relevant for users when reviewing their physician online, were investigated [7].

Furthermore, the ratings themselves have also been topic of research. On the one hand, there are studies that deal with their overall sentiment [2,9,12,19]. On the other hand, there are other studies that conducted pure numerical analysis of the ratings in order to find out the percentage of physicians that has been rated, the average number of ratings on the PRWs [9], as well as the average amount of assessments per physician [2,12].

However, research on the rating behavior of PRW users remains scarce. Gao et al. [12] conducted an analysis on more than 386,000 ratings from the PRW RateMDS from 2005 to 2010 and investigated what characteristics of physicians (such as specialist field, time of graduation, attended medical school, etc.) increased their likelihood to be rated. However, first of all, this analysis is based on non-actual data. Second, the data set is rather small compared to our one. Third, Gao et al. [12] investigated an U.S. physician rating website and thus no German texts. Fourth, they rather concentrated on the characteristics of the physicians that influence the rating behavior instead of the characteristics of the patients (age, health insurance affiliation, complaint behavior) that affect rating behavior. However, there are also studies that are based on more actual and German data. Emmert and Meier [6] analyzed 127,192 reviews and ratings that were posted on jameda.de in 2012 and Emmert et al. [4] also investigated ratings on jameda.de from 2012 to 2013, whereby the latter only considered ratings on dentists. Both investigations concluded that older patients give better grades than younger patients and that privately health insured patients rate better than statutory insured patients. However, both studies (1) are not based on actual data, (2) rely on a far smaller data set than our analysis and (3) do not consider the influence of complaint behavior on patients' ratings.

Thus, so far, there is no actual investigation of reviews and ratings of German user generated physician reviews and ratings, that considers patients' rating behavior under simultaneous consideration of age, health insurance affiliation and complaint behavior. Besides, the investigated amount of ratings and physicians is unique in the existing research on German PRWs.

3   Empirical Study

3.1   Data Set

Our corpora were created by gathering data from jameda.de on which more than 90 % of German physicians are registered.[1] The data acquisition took place from October 2013 to January 2015 and resulted in a collection of 797,651 reviews on 397,590 physicians that were posted in the time period from January 2009 to January 2015. The distribution of review texts per physician ranges from 0 to 223, whereof less than 2 % are only quantitative reviews with no textual remarks. Most of the reviewers commented on general practitioners (14 %), followed by internists (11 %), dentists (8 %), surgeons (6 %) and psychotherapists (6 %) – to mention only the 5 most frequently reviewed specialties on jameda.de. For example, psychiatrists are rarely reviewed although more than 11,900 are registered on this PRW.

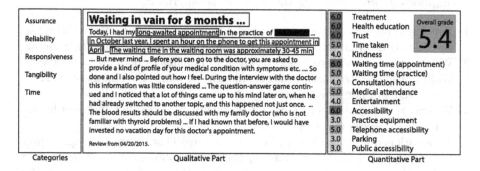

Fig. 1. Sample review on jameda.de translated from German

---

[1] jameda GmbH, "FAQ", http://www.jameda.de/hilfe/?show=user,08/15/2015.

Descriptive metadata, such as age or the type of health insurance, was attached to each review in order to provide classificatory information for a better understanding of the reviewer's background [14]. Each review consists of a qualitative and a quantitative part (cf. Fig. 1). The qualitative part contains the title and the review text. In general, the reviews texts are rather short (51 words on average) and the longest one consists of 348 words. Additionally, around 1.2 % of the collected reviews comprise a response of the particular physician. Next to this qualitative part, the quantitative part contains up to 16 numeric rating categories (e.g. health education, equipment). Each of these categories can be rated by using a grading system (best: 1.0 – worst: 6.0). Besides the grades, the quantitative part also contains metadata (e.g. age, insurance). In total, 60 % of all physician reviews provide details about the statutory health insurance (SHI) or the private medical insurance (PMI). Furthermore, the corpora contain personal information about the physicians such as name, address and specialty.

Before this data was processed, non-physician data records such as veterinarians and pharmacies as well as biased reviews were excluded from the data set. Our data set initially contained reviews on 68 % physicians, 8 % health practitioners, 7 % psychotherapists and 6 % physical therapists. Thanks to previous research [13,14], where we developed an approach to automatically recognize individual inconsistencies in the rating behavior by comparing the sentiment scores of the qualitative part to the grades of the quantitative part of a review [13], we could filter out the biased ones in order to improve data quality.

3.2 Method

First of all, information extraction was performed [15]. That is, we automatically identified and extracted relevant information from the review texts and transformed them into a structured representation (i.e. predefined templates). The patterns for the recognition of sentences containing waiting time information were a priori defined when studying a sample of 200 review texts that was randomly selected from the whole collection. That way, we could make the following assumptions in order to design the right extracting rules:

1. Waiting time in a doctor's office covers the time someone spends sitting in the waiting or even both in and in front of the consulting or treating room.
2. Waiting time in a doctor's office is given in minutes or hours by a reviewer.
3. Waiting times stated in days, weeks, months or years are commonly exaggerations and will not be considered.
4. Waiting time for an appointment is not usually expressed in minutes or hours, but in days, weeks and months.
5. All mentioned waiting times will be rounded up. That means that e.g. 20 min are considered as maximum waiting time in the statement *"I've never been waiting longer than 20 min"* and 4 hours are taken as upper bound from the following utterance *"waiting 2–4 hours at the doctor's office"*.
6. Statements without concrete indications of any waiting time (e.g. *"I was waiting for hours"*) are ignored.

Table 1. Samples of n-grams for waiting time with surrounding context information

#	n-gram	Context trigger(s)
1	*"nach 10 Minuten Wartezeit"*	*"im Wartezimmer"*
	("after a 10 min waiting time")	("in the waiting room")
2	*"noch 50 Minuten gewartet"*	*"trotz Termin"*
	("had still 50 min to wait")	("despite appointment")
3	*"noch nie länger als 20 Minuten"*	*"gewartet"*
	("never more than 20 min")	("waited")
4	*"45 Minuten im Wartezimmer verbracht"*	–
	("spent 45 min in the waiting room")	
5	*"trotz Termin 30 Minuten gewartet"*	–
	("waited 30 min despite appointment")	
6	*"etwa einer Stunde Wartezeit"*	*"Wartezimmer"*, *"Praxis"*
	("about one hour waiting time")	("waiting room", "practice")

In our case, the starting point for the generation of the regular expressions was to go through the physician reviews on jameda.de and look for repetitive patterns (i.e. phrases), containing waiting times. For example, *"... had to wait X hours in the waiting room ..."* or *"... got an appointment already after X weeks"* were phrases occurring quite often in the review texts. The most frequent phrasal patterns (in the length of 3- to 5-grams) were then listed in a candidate list. Since numbers are highly ambiguous when interpreting them without their embedding, we decided to create phrasal patterns considering surrounding context information. Even limited context information such as time units (minutes, hours, days, weeks, months) is not sufficient because the corresponding activity itself is still ignored. Moreover, the phrase structure is essential for the correct information extraction. Since keyword spotting in a window size of, for example, 8 words could even recognize the wrong temporal information for waiting time at a doctor's office as it is shown here: *"Because of $[two_1\ minutes_2\ at_3\ the_4\ doctor's_5\ a_6\ \underline{waiting}_7\ periods_8]$ of up to one hour!"* This could not happen when the predicate-argument structure of a phrase represented in extraction patterns, such as `"waiting period of (up to)? ([\d]{1}) hour(s)?"` and others shown in Table 1 together with so-called context triggers. These are string sequences (n-grams) describing the situation of complaint.

The recognition of phrasal patterns was conducted on sentence level in order to assign the identified waiting times to the right physician(s). That is especially important in cases when more than one physician is mentioned in a review:

> *"Ich bin zu Herrn Dr. X gewechselt, da ich hier nie mehr als 20 Minuten warten musste. Ich kenne andere Ärzte, wo ich bis zu 5 Stunden im Wartezimmer saß."* ("I switched to Dr. X because I never had to wait

Table 2. Grading behavior per age group dependent on waiting time

	< 30 years	30–50 years	> 50 years
Avg. Minutes	75	79	84
Avg. Grade	2.2	2.1	2.0

more than 20 min. I know other doctors where I stayed up to 5 hours in the waiting room.")

From these most frequent n-grams, regular expressions were built which were then applied on the jameda.de review corpus. Depending on the results gathered by these regular expressions, they were refined and again applied on the review texts in order to find out, how to further elaborate them. That is, when we observed for example, that the regular expression "had to wait ([\d]{1,5}) hours"[2] failed to cover a considerable amount of waiting time descriptions because reviewers oftentimes expressed their waiting time in minutes, then we adapted our regular expression. In this case, the regular expression would have been adapted to "had to wait ([\d]{1,5}) [hours| minutes]" combined with a context trigger such as "waiting room". Hence, our information extraction process was conducted by means of iterative adaption of our regular expressions (i.e. bootstrapping). Finally, the waiting times that were expressed in other time units than minutes were converted to minutes (i.e. normalization) in order to make the differently expressed waiting times comparable.

4   Results

The results of the information extraction process (described in the previous section) are presented in the following. After filling the extracted and normalized temporal information in the above mentioned templates for waiting time at a doctor's office and for an appointment per physician, we correlated our findings with the reviewers' age (Subsect. 4.1), health insurance affiliation (Subsect. 4.2) or grading behavior (Subsect. 4.3).

4.1   Age

When considering the rating behavior of the three age groups ("below 30 years", "30 to 50 years" and "above 50 years") not per specialist field[3] but per perceived waiting time (cf. Table 2), it becomes clear that younger patients are more strict than older patients.

While patients below 30 years on average give a grade of 2.2 for a waiting time of 75 min, older patients between 30 and 50 years give a better grade (2.1)

---

[2] [\d]{1,5} denotes that we look for a number with at least one and up to five digits.

[3] In the following, the specialist types "psychiatry" and "psychology" are subsumed under the specialist category "neurology".

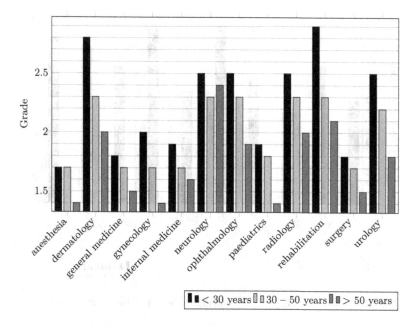

Fig. 2. Average rating per age group by specialist field

even for a longer waiting time (79 min). The patients older than 50 are even less strict, since they give a 2.0 on average for a waiting time of 84 min.

In order to investigate the influence of patients' age on their rating behavior, we looked at the grades given on jameda.de by patients separated in these three age groups. Results showed that younger patients are more critical when rating their doctors than older patients (cf. Fig. 2). The higher the grades on the y-axis are, the more critical the patients are. Only in the case of "anesthesia", patients in the age group "below 30 years" and "30 to 50 years" gave the same grade on average. The second exception is observed in the category "neurology". There, the patients aged between 30 and 50 years rated better than the patients above 50 years. However, the rating difference was just minor. Anyway, we assume that more strict rating behavior of younger patients compared to the rating behavior of older patients is due to the fact that they have – due to less made experiences – higher expectations.

4.2   Health Insurance Affiliation

A further factor that was shown in previous studies to have some influence on patients' rating behavior is the health insurance affiliation. As expected, private patients do not have to wait as long as patients insured by the statutory health system (cf. Fig. 3) and hence give better grades (cf. Fig. 4).

On average, PHI patients have to wait 70 min and give a 1.6 while SHI patients on average have to wait 88 min and give a 2.1.

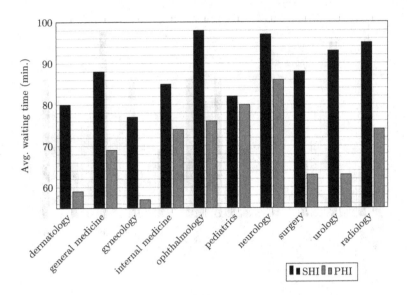

Fig. 3. Average waiting time by specialist field per insurance (in min.)

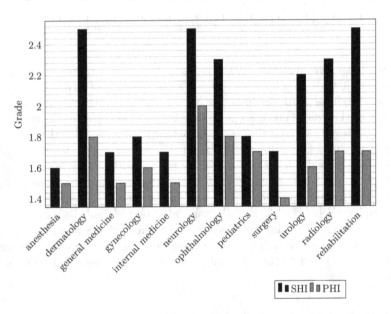

Fig. 4. Average rating by specialist field per insurance

4.3   Grading Behavior

A further aspect we investigated is the influence of individual complaint behavior on patients' ratings. We assumed that variations in (communicated) opinions (i.e. given ratings) do not only depend on the actual perceived treatment but rather also on the individual grading behavior. To prove this, we first grouped the generated waiting time templates per physician by specialty in our data set. Then, we iteratively assigned the corresponding grade for waiting time (appointment) given in the quantitative part of each review to the waiting time extracted from the review text (cf. Fig. 1). After that, we calculated the mean waiting time per grade and per specialty over all reviews.

Table 3. Grading behavior dependent on waiting time

	Grade 1	Grade 2	Grade 3	Grade 4	Grade 5
Dermatology	20 min	43 min	67 min	77 min	84 min
General medicine	30 min	48 min	72 min	83 min	102 min
Gynecology	24 min	49 min	73 min	80 min	96 min
Internal medicine	19 min	55 min	71 min	84 min	95 min
Neurology	49 min	51 min	70 min	95 min	97 min
Ophthalmology	31 min	77 min	74 min	78 min	94 min
Pediatrics	26 min	44 min	70 min	65 min	103 min
Radiology	11 min	47 min	78 min	100 min	101 min
Surgery	25 min	53 min	70 min	87 min	99 min
Urology	27 min	41 min	72 min	81 min	101 min

The results confirm our initial assumption (cf. Table 3). We observed that the same perceived waiting time led to different ratings. For example, a waiting time of 49 min is graded with a 1 (excellent) in case of a neurologist while in case of the gynecologist a waiting time of that length is rated with a grade of 2 (good). This difference in rating behavior becomes even more clear when looking at the waiting times at the ophthalmologists and the dermatologists. While a waiting time of 77 min at the ophthalmologist is graded with 2 (good), it is graded with 4 (acceptable) at the dermatologist.

This shows that the grading behavior is highly context-dependent: On the one hand, patients expect different waiting times for the various specialties and therefore give different grades for the same time period. On the other hand, the waiting times are not monotonically increasing for higher grades in every specialist field (e.g. ophthalmology, pediatrics) which illustrates the subjectivity of patients' grading behavior. In brief, you cannot predict longer waiting times in general when the grades are getting worse (i.e. higher).

5   Evaluation

5.1   Evaluation Track

In order to test the accuracy of our approach, we took a random sample of 200 reviews and manually verified whether the waiting times were correctly extracted from the text. We proved, for example, that the number 45 is recognized as the waiting time and not the 7 in a review text containing the sentence *"I had to wait 45 min with my 7 years old son"*. Furthermore, we checked that, within a review text like *"I had to wait 2 h in the waiting room"*, the temporal information (2 h) is correctly converted into 120 min.

5.2   Evaluation Results

The manual evaluation revealed that our system works quite accurate. Only in 15 of 200 sample reviews, we observed mistakes. This again corresponds to an error rate of 7.5 %. We aimed at covering a great variety of paraphrases expressing waiting time by means of regular expressions. However, errors occurred in some cases. For instance, we had problems with the recognition multi-word time units in the reviews text, which were only extracted in parts and therefore falsely converted: *"I had to wait 1 h and 25 min despite appointment"* was normalized to 60 min (for 1 h) instead of 85 min. Before the first evaluation round we have not considered those complex temporal expression, but now we do.

Furthermore, in other cases, even the human reader may have difficulties in understanding the reviewer's statement: *"Waiting for an appointment 6 weeks waiting at the doctor's 1 h 10 min, should have been waiting additional 15 min."* Since the review is written without interruption, we cannot identify the correct sentence boundaries in the absence of end of sentence points. The misinterpretation of the grammatical rules leads to the extraction of only 15 min. While the error can be simply fixed by adding another regular expression covering this syntactic structure, others are still unsolvable: *"Finally, again a good experience with a spontaneous preferred date. [...] Others expected me to wait 4–5 h"*. This patient is happy because he/she got an appointment with short waiting time but he/she compares this situation to others with longer waiting times. Since our system is not able to differentiate between on-topic and off-topic time statements, it extracts 5 h. This error type occurs three times in the investigated sample of 200 reviews. So this result still leaves room for improvement, but is quite promising.

6   Conclusion and Outlook

The increasing use of physician rating websites has raised concerns whether the information on these sites truly reflect the quality of health care providers. However, this is quite important due to their considerable influence on physician choice and on the image of doctors in society and the self-understanding of both doctors and patients [20]. Therefore, we have investigated – based on 790,000 physician reviews – in how far ratings are influenced by factors that do not have anything to do with health care quality. Our results are in line with [4,6] in terms of that age and health insurance affiliation have an impact on patients' ratings. Furthermore, our analysis revealed that patients' ratings are not only influenced by the received treatment quality itself but also by complaint behavior. Thus, we provide evidence that users should be well aware of user specific distortions of ratings and hence should be careful when making their health care provider decision based on these ratings.

In future research, we therefore plan to investigate (1) what further treatment quality independent factors influence patients' ratings and (2) what evaluation criteria are more important for patients than others when rating their doctors online. The overall goals are to get a better understanding of patients' rating behavior and hence in the long run being able to develop interpretation aid for users when seeking for health care provider information online.

However, also from a technical point of view, there is potential for future research. (1) First of all, since our manual adjustments of the regular expressions is quite time consuming and involves the risk of overfitting and overgeneration, we plan to simultaneously support our information extraction process by machine learning techniques. (2) Second, a further intriguing task for future work would be to create additional domain-specific dictionaries in different languages in order to make our approach multilingual. (3) Third, we could also develop a strategy for dealing with off-topic opinion phrases referring to other experiences with physicians than the one that is on-topic in the review. That is, we have to develop a strategy, for distinguishing automatically which experiences described in the reviews refer to the experience that is to be rated in the reviews, and which descriptions refer to past visits or experiences that the reviewer has just heard from but not made him or herself and hence should not be included in the rating.

Acknowledgments. Partial support for this work was provided by the Ministry of Innovation, Higher Education and Research of North Rhine-Westphalia, Germany.

References

1. Beck, A.J.: Nutzung und Bewertung deutscher Arztbewertungsportale durch Patienten in deutschen Hausarztpraxen. Ph.D. thesis, Universität Ulm (2014)
2. Emmert, M., Gerstner, B., Sander, U., Wambach, V.: Eine Bestandsaufnahme von Bewertungen auf Arztbewertungsportalen am Beispiel des Nürnberger Gesundheitsnetzes Qualität und Effizienz (QuE). Gesundheitsökonomie & Qualitätsmanagement (2013)
3. Emmert, M., Sander, U., Esslinger, A., Maryschok, M., Schöffski, O., et al.: Public reporting in Germany: the content of physician rating websites. Methods Inf. Med. **51**(2), 112 (2012)
4. Emmert, M., Halling, F., Meier, F.: Evaluations of dentists on a German physician rating website: an analysis of the ratings. J. Med. Internet Res. **17**(1), e15 (2015)
5. Emmert, M., Maryschok, M., Eisenreich, S., Schöffski, O.: Arzt-Bewertungsportale im Internet: Geeignet zur Identifikation guter Arztpraxen? Das Gesundheitswesen **71**(4), 218–219 (2009)
6. Emmert, M., Meier, F.: An analysis of online evaluations on a physician rating website: evidence from a German public reporting instrument. J. Med. Internet Res. **15**(8), e157 (2013)

7. Emmert, M., Meier, F., Heider, A.K., Dürr, C., Sander, U.: What do patients say about their physicians? an analysis of 3000 narrative comments posted on a German physician rating website. Health Policy **118**(1), 66–73 (2014)

8. Emmert, M., Meier, F., Pisch, F., Sander, U.: Physician choice making and characteristics associated with using physician-rating websites: cross-sectional study. J. Med. Internet Res. **15**(8), e187 (2013)

9. Emmert, M., Sander, U., Pisch, F.: Eight questions about physician-rating websites: a systematic review. J. Med. Internet Res. **15**(2), e24 (2013)

10. Fischer, S.: Project "Weisse Liste": a German best practice example for online provider ratings in health care. In: Gurtner, S., Soyez, K. (eds.) Challenges and Opportunities in Health Care Management, pp. 339–346. Springer, Switzerland (2015)

11. Fischer, S., Emmert, M.: A review of scientific evidence for public perspectives on online rating websites of healthcare providers. In: Gurtner, S., Soyez, K. (eds.) Challenges and Opportunities in Health Care Management, pp. 279–290. Springer, Switzerland (2015)

12. Gao, G.G., McCullough, J.S., Agarwal, R., Jha, A.K.: A changing landscape of physician quality reporting: analysis of patients online ratings of their physicians over a 5-year period. J. Med. Internet Res. **14**(1), e38 (2012)

13. Geierhos, M., Bäumer, F.S., Schulze, S., Stuß, V.: Filtering reviews by random individual error. In: Ali, M., Kwon, Y.S., Lee, C.-H., Kim, J., Kim, Y. (eds.) IEA/AIE 2015. LNCS, vol. 9101, pp. 305–315. Springer, Heidelberg (2015)

14. Geierhos, M., Bäumer, F.S., Schulze, S., Stuß, V.: "I grade what I get but write what I think". Inconsistency Analysis in Patients' Reviews. In: ECIS 2015 Completed Research Papers, Paper 55. pp. 1–15. AIS (2015)

15. Grishman, R.: Information extraction: techniques and challenges. In: Pazienza, M.T. (ed.) Information Extraction: A Multidisciplinary Approach to an Emerging Information Technology. LNCS, vol. 1299, pp. 10–27. Springer, Heidelberg (1997)

16. Hanauer, D.A., Zheng, K., Singer, D.C., Gebremariam, A., Davis, M.M.: Public awareness, perception, and use of online physician rating sites. JAMA **311**(7), 734–735 (2014)

17. Hennig, S., Etgeton, S.: Arztbewertungen im Internet. Datenschutz und Datensicherheit-DuD **35**(12), 841–845 (2011)

18. Hinz, V., Drevs, F., Wehner, J.: Electronic word of mouth about medical services. Technical report, Hamburg Center for Health Economics (2012)

19. López, A., Detz, A., Ratanawongsa, N., Sarkar, U.: What patients say about their doctors online: a qualitative content analysis. J. Gen. Intern. Med. **27**(6), 685–692 (2012)

20. Reimann, S., Strech, D.: The representation of patient experience and satisfaction in physician rating sites. A criteria-based analysis of english - and german-language sites. BMC Health Serv. Res. **10**(1), 332 (2010)

21. Sander, U., Emmert, M., Grobe, T.: Effektivität Effektivität und Effizienz der Arztsuche mit Arztsuch- und Bewertungsportalen und Google. Das Gesundheitswesen **75**(06), 397–399 (2013)

22. Schaefer, C., Schwarz, S.: Wer findet die besten Ärzte Deutschlands?: Arztbewertungsportale im Internet. Zeitschrift für Evidenz, Fortbildung und Qualität im Gesundheitswesen **104**(7), 572–577 (2010)

23. Strech, D., Reimann, S.: German language physician rating sites. Das Gesundheitswesen **74**(8–9), e61–e67 (2012)

24. Strech, D.: Arztbewertungsportale aus ethischer Perspektive. Eine orientierende Analyse. Zeitschrift für Evidenz, Fortbildung und Qualität im. Gesundheitswesen **104**(8), 674–681 (2010)
25. Terlutter, R., Bidmon, S., Röttl, J.: Who uses physician-rating websites? differences in sociodemographic variables, psychographic variables, and health status of users and nonusers of physician-rating websites. J. Med. Internet Res. **16**(3), e97 (2014)
26. Verhoef, L.M., Van de Belt, T.H., Engelen, L.J., Schoonhoven, L., Kool, R.B.: Social media and rating sites as tools to understanding quality of care a scoping review. J. Med. Internet Res. **16**(2), e56 (2014)

An Analysis Tool for the Contextual Information from Field Experiments on Driving Fatigue

Perrine Ruer[✉], Charles Gouin-Vallerand, Le Zhang, Daniel Lemire,
and Evelyne F. Vallières

LICEF Research Center, Télé-Université du Québec, Montreal, QC, Canada
{pruer,cgouinva,dlemire,evallier}@teluq.ca

Abstract. Elderly drivers will be more present on the road in the next few years. Mobility is fundamental for the elderly because it allows them to maintain an active lifestyle. But the elderly may suffer from cognitive, physical or sensorial decline due to aging. To help them to drive, context-aware systems can assess the status of a driver and warn him or her about hazards. We present a data analysis tool for car driving context information that includes data mining and statistical evaluation algorithms. We applied our system to data collected by sensors into an instrumented vehicle in realistic driving conditions. Results show that our tool is able to store the contextual information collected and to enable an interactive visualization of the data collected. Thanks to this tool, it is easier to share information among the scientists working on the data. Moreover, it makes it convenient to store data in the cloud.

Keywords: Context-awareness · Contextual information · Analytical tool · User modeling · Driving behavior analysis · Road safety

1 Introduction

Aging is a growing phenomenon in almost all countries of the world [1] and the percentage of people 65 and older will more than double between 2010 and 2050 [2]. For instance, there are five millions elderly in Canada and elderly drivers could represent nearly a quarter of the Canadian population of drivers in 2036 [3]. Such a demographic change will have an impact on various aspects of daily life, among them road safety. Indeed, aging has an impact on road safety. Elderly drivers are drivers aged 60 and older. Elderly drivers have their specific safety problems, for instance they have more car crash at intersections. In fact, they have the highest crash rates per vehicle distance of travel [4]. Aging has several consequences for society. Therefore, it is in the best interest to focus on this category of drivers.

In the last decade, there has been a growing interest in intelligent vehicles. A notable initiative on intelligent vehicles was created by the U.S. Department of Transportation with the mission of preventing highway crashes [5]. A range of new technologies allows monitoring and driver assistance, such as automatic speed controls or blind spot monitoring, to prevent motor vehicle accidents. Thus, such vehicles are equipped with multiple sensors that can recognize driver's current activities, and situate the vehicle in

© Springer International Publishing Switzerland 2015
H. Christiansen et al. (Eds.): CONTEXT 2015, LNAI 9405, pp. 172–185, 2015.
DOI: 10.1007/978-3-319-25591-0_13

the environment. Through information provided by these sensors, it is possible to model the context of the users and vehicles; and use these models to assist the drivers and create new kind of driver-car interactions.

Context-aware driving systems can be especially useful to assist elderly drivers by providing adapted assistance that takes into account the driver's interaction capabilities, their cognitive capabilities, the driver habits; while taking into account the cognitive load inferred by such driving assistance [6]. By context-aware system, we mean the ability of a system to capture, model and use specific information about the environment surrounding the system, such as location, time, and user profile [7].

This paper presents a modeling tool for analyzing the driving context. To analyze contextual information, we used a web application and include data mining. The paper is structured as follow. Section 2 presents related works about elderly drivers, context-aware systems and driving behavior model. In Sect. 3, we present and describe our tool. Section 4 reports on the application of our tool to real data. Finally, Sect. 5 concludes the paper and presents perspective and future work.

2   Related Works

Drivers have to respond rapidly to risks with good abilities like attention, perception, motor abilities, information treatment, etc. With aging, some physiological impairment appear and have negative impacts on driving skills [8]. The elderly suffer from cognitive, physical and sensorial decline. The cognitive skills are affected and lead to a longer reaction time, a diminution of attention, and a short memory [9]. Physical abilities are worsening with deterioration of psychomotor skills, development of arthritis, which causes neck problems, or also vulnerability of the body [10]. And sensory functions declines with a diminution of visual acuity, diminution of hearing or even, perceptual ability [9, 11]. These impairments are common among the elderly with normal aging. In addition, with diseases increasing, the elderly take medication that highlights the risk of accidents. Medication alters the driving skills and reduces sensorimotor performance (for example: decreased alertness, impaired vision, etc.) [10]. Therefore, some authors purpose license restrictions to manage elderly driver safety [12]. But driving cessation had adverse negative consequences [10, 11, 13]. It is a stressful experience, which has an impact on quality of life. Elderly drivers want to continue to drive to maintain their independence, even more for elderly drivers living in rural or remote areas. Mobility is fundamental for elderly drivers because it allows them to maintain an active aging. Thus, it is in the best interests of societies to maintain elderly adults driving as long as they can safely do so.

So, driving is a complex and multitask processing that involves good perception and cognition from the driver. It is necessary to have an immediate and appropriate decision while driving. Driving task can be assisted by a context-aware system for vehicle control or vigilance. Context-aware system can detect the status of driver and prevent him about hazard.

Context aware system is defined as a system that uses context to provide relevant information and/or services to the user (relevancy is depending to the user's task) [14]. In 2005, Rakotonirainy [6] specify that context aware system assist the driver to have a safe behavior. These systems could reduce the amount of errors and the likelihood of accident. Indeed, driver errors are the consequences of 90 % of the accidents and most of the accidents occur due to drivers' behavior [15]. Errors would be more prevalent in elderly drivers (or impaired drivers) because of a decreased ability to perceive or quickly interpret the information due to aging [16]. Driving errors are defined as an involuntary deviation from a rule or a norm. Planned actions fail to achieve the desired outcome [17]. Context-aware systems have not been used for applications for elderly drivers [13] although this system can improve vehicle control and prevent accident. Driving behavior is a complex interaction between the driver, the vehicle and the environment. For instance, the driver can have information about his/her physiological state (stress), the environment (traffic) and the car state (speed of traffic) [6]. To achieve an interaction reliable, different types of information are taken with different type of sensors and cameras and provide required information to the driver when it is necessary [18]. Sensors can be categorized into three types: physical sensors (i.e. light sensor, camera, audio sensors, accelerometers, location, temperature sensor, etc.), virtual sensors (i.e. software applications, network event sensors, etc.) and logical location sensors (i.e. combination between physical and virtual sensor) [19]. The best system combines all assistance functions to help the driver and produce good performance [18]. Context aware system's architecture is composed of a direct sensor access to provide sensing data, a middleware infrastructure to present information to users and a context server to manage the information's user and to save context information for a later use [20]. In sum, context aware systems improve driving with understanding the whole driving task (driver, environment and vehicle) and assist driver's decision to reduce road accidents.

But to assist the drivers, such system has to analyze the driver and predict when it is a normal and an abnormal behavior. Indeed, the main actor of the driving activity is the driver. So, there is a need to analyze the driver behavior in the context because the situation impacts on the type of actions [21]. Some cognitive models exist in the literature but they not take into account the context. Indeed, the benefits of context are the explanation of driving behavior and the improvement of the generalizability and reliability of existing driving behavior [21]. It is appropriate to develop behavior models for context aware system. This will help the driver to produce adapted driving actions. Driving behavior models incorporate cognitive state (attention) and behavioral state (motivation, belief or risk assessment). A model is designed for only one particular driving situation like fatigue [21]. Some researchers have developed context aware driving behavior models capable to explain and predict driver's behavior. In 2015, Bhattacharjee and Wankhede [22] develop a context-aware architecture for a driver behavior detection system able to detect four types of driving behavior in real-time driving (normal, fatigued, drunk and reckless driving). Different types of information are collected like speed of the vehicle, yawning angle, steering wheel angle and the vehicle's lane position. To capture static and dynamic aspects of the driver behavior, a dynamic Bayesian network is used. Results show an accurate detection of the abnormal driver's behavior. A context aware system has to integrate driver behavior model. Thus detect driver behavior, it is vital to collect contextual information about the driver, the vehicle and environmental context; then analyze with data mining methods. In the following section, we propose the description of a tool used to analyze the driving and the context, especially for elderly drivers.

3 Tool Description

In the context of our research, we are using an instrumented vehicle, the LiSA (in french: *Laboratoire intelligent de Sécurité Automobile* or Intelligent Laboratory on Automobile Safety), a Nissan Versa 2008 (Fig. 1). Among other, this car is equipped with a data logger AIM Evo4 which can collect the speed, steering movements, acceleration, braking, 3-axis acceleration and GPS location from the car embedded computer. LiSA also include an eye tracking faceLAB 5.0 system, a Microsoft Kinect (with a head tracking software [23]) camera and several other camera to monitor the driver. Finally, all the data is recorded by a computer installed on-board. Thus, LiSA is able to collect a wide range of contextual information.

Fig. 1. The LiSA' instrumented vehicle

To complete the contextual information with specific information on the user profile, we are using surveys during our experimentation to collect sociodemographic variables, such as their age, sex, driving experience, opinions toward safe driving, etc. Moreover, in the context of our last experiment [24], we collect the driver's level of perceived fatigue by using a scale ranging from '0' not at all fatigued to '10' very much fatigued at four times during the experimental session. Each participant was asked his or her level of fatigue just before starting driving (Time 1), after 15 min of driving once the experiment had started (Time 2), after 30 min (Time 3), and when the driving experiment ended (around 45 min) (Time 4). In the case of this experiment, the focus was measuring the level of perceived fatigue and not the physiological level of fatigue. The Table 1 presents an overview of the contextual information collected by LiSA.

Table 1. Contextual information collected by LiSA

Devices	Data collected
AIM Evo4	Time; Distance; Speed; RPM; Gears; Steering wheel rotation; 3-axis acceleration; GPS location (gyro, altitude, latitude, longitude, elevation, etc.)
faceLAB 5.0	Head position and rotation; Eye close (left, right, calibration); Eye gaze (left, right, rotation, calibration); Eye blinking (left, right, number, duration, frequency); PERCLOS; Saccadic eye movements; Pupil movements (left, right, diameter)
Microsoft kinect	Head position and rotation; Blind spot check activity
Surveys	Sociodemographic data; Opinion toward safe driving; Level of perceived fatigue

To analyze these contextual information we built an analysis tool that include data mining and statistical evaluation algorithms that analyze the collected contextual information. The development of such tool was motivated by:

1. Building a development platform to test our algorithms on offline data, which could be later test on real-time data;
2. Getting a tool to stock the contextual information collected during our field experiments for each participants;
3. Building a tool that will allow an easy visualization of the data collected and could be used to make debriefing session with the participants.

Thus, we built this tool using a web application approach by adopting the Node web server technologies, the Javascript language and the MongoDB database. The choice of Node and the Javascript language was motivated by the high adaptability of Javascript language, where we can do server and client side computation. Most data mining and statistical analysis algorithms can be implemented in Javascript easily and developed algorithms can be translated into other languages (e.g. Java or Python) easily. Moreover, there is a huge selection of visualization API available in the community, which provided support for representing our data in graphs or diagrams.

For the database, MongoDB was an evident choice where the noSQL approach, which allows the indexation of data, provided by different sources in different formats (e.g. tabular data). To not loose any information provided by the sensors and their expressivity, we decide to upload the raw data into the MongoDB without any prior transformation. Afterwards, we had several transformations to do on raw data prior to their analysis. For instance, the time frames of the collected data are heterogeneous between sensors, with different sensing rates and time format. To make the data analysis easier, we made an algorithm that transform the data on the most common measure by aggregating the data from sensors with an higher sensing frequency by computing the average values. Another example is with the data provided by the faceLAB about the head position and the eye gaze. To match the cardinal position provided by the faceLAB with the car environment, we made a cardinal (translation) and a scaling transformation (in meters).

However, in the context of the field experiments (see next section), one of the key transformations required to the statistical analysis of the contextual information, such as the driving speed, is the noise reduction. All drivers are varying their speed depending of the driving contexts (obstacle, road topology, etc.). However, even in the case of monotonous driving in flat and straight lanes, the driving speed is varying a lot with several speed spikes at various intensity. Such small variations are not always related to the driver behaviors and can be related to environment and road conditions (Fig. 2) and can therefore be considered as noise. In this example, we are not interested in these small and frequent speed variations but in intentional variation of speeds, that we refer as speed spikes.

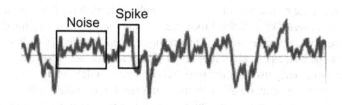

Fig. 2. Example of identified noise and valuable information (spike)

To remove this noises, we implemented a quasi-monotonic segmentation algo-rithm[1] based on Lemire et al. [25] that reduces specific amounts of variation in a signal and keeps the variation in the data that are more significant (i.e. with a significant monotonicity). Intuitively, this algorithm segments a signal into pieces that are "quasi-monotonic" (mostly going up or down). A segment is said to be "quasi-monotonic" if we can approximate the data using a (truly) monotonic function that never deviates from the actual data by more than a small threshold. From a tolerance threshold, we can therefore construct a piecewise monotonic approximation of the original data. The segmentation and the piecewise monotonic approximation can be computed in linear-ithmic time (O(n log n)). Figure 3 presents an example of a speed signal that is processed by the segmentation algorithm, with the original signal and the signal processed with a threshold speed value of 10 km per hour (this threshold can be set by the user). From

---

[1] https://github.com/lemire/MonotoneSegment

this simplified approximation, we are able, for instance, to compute the number of speed spikes and compare it with the level of perceived fatigue, while ignoring variations that are not judged significant.

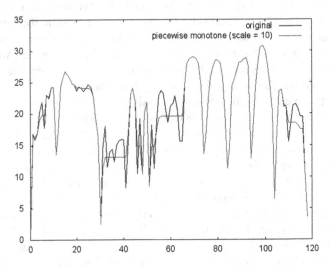

Fig. 3. Speed signals with the segmentation algorithm

Moreover, for the evaluation of the drivers' behavior, we have implemented on the web application a series of statistical tools that enable the evaluation of the contextual information per experiment segments. They compute average speeds, standard deviation and regression, as well as correlation quotient. These results (see next section for more details) were particularly used for the post-experimentation debriefing with the participants. We also developed an interface to link contextual information with geographical locations using the Google map API, while taking into consideration the temporal aspect of these data. This interface allows an easy exploration of the collected data and is used to validate specific events that occurred during the experimentations. This interface can also be used to visualize a participant experimentation through a playback function.

In summary, this contextual information analysis tool represents a multimodal platform that brings several benefits to the researches we conduct on road safety. Thanks to this tool, it is easier to share information among the scientists working on the data and it makes it convenient to store data in the cloud. The used technologies (Node.js and MongoDB) allow an easy implementation of contextual information analysis algorithms and give us the platform to test our algorithm on offline (real) data, before testing them within the technologies we are developing to assist the drivers in their tasks.

4   Results

This section shows the validity and effectiveness of our tool. We used our tool during a pilot evaluation with 20 participants, aged between 56 and 76 years old. The goal of the experiment was to evaluate the intrinsic evaluation/perception of the driving fatigue for elderly drivers and compare it with their driving behavior. During the experimentation, each participant drove around 50 km on a close driving circuit with the LiSA car. During experimentation, field data were collected into the embedded computer of the LiSA then uploaded into the tool following the end of the driving period. During the pilot test, we collected an average size of contextual data for each participant of 29 MB, with 8 MB from the AIM evo4, 18 MB from the faceLAB and 3 MB from our Kinect's software, for 45 min of driving.

The tool enables to visualize the collected data with its graphs (Figs. 4, 5, 6, 7) and maps (Figs. 8, 9) after the experimentation of each participant. As we said in the previous section, the algorithm transforms data on the most common measure. We present the results with a textual description of the graphs for the whole experimentation (Fig. 4), for the segmentation of the speed (Figs. 5, 6), for the blinking frequency (Fig. 7) and for the maps presentation (Figs. 8, 9).

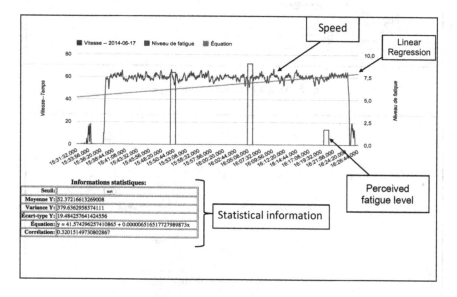

Fig. 4. Interface's explanation

First, for the whole experimentation (Fig. 4), we included speed, the perceived fatigue level and linear regression of the speed during the whole time driving. We add statistics information below the graphic with mean, variance, standard deviation, regression equation and regression correlation.

Then for the speed's attribute, we purpose for the driver a comparison between normal form (Fig. 5) and segmented form (Fig. 6). We suggested four equations (in different colors) every 15 min and for each of them, statistics information (mean, variance, standard deviation, regression, etc.) (Fig. 6).

For the data provided about eye and head movements, Fig. 7 shows an example of blinking frequency with the perceived fatigue level and the regression of blinking frequency during all the experimentation. Likewise for the speed, statistical information are given below the graphic.

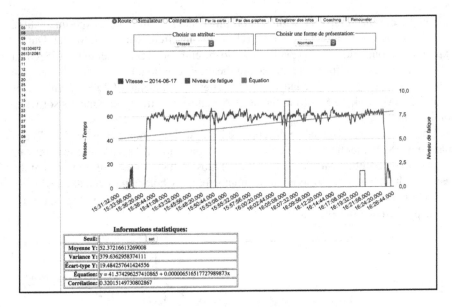

Fig. 5. Graph Interface - Normal speed

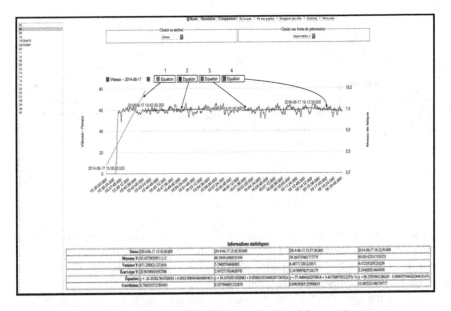

Fig. 6. Graph Interface - Segmentation of the speed

Next, maps form (Fig. 8) gives contextual information with geographical localization using Google maps. It is possible to explore at any time all the information collected by the vehicle. When you click on a particular point of the map, detailed information are given such as time, distance, speed, brake, GPS, etc. (Fig. 9). Similarly, it can be used

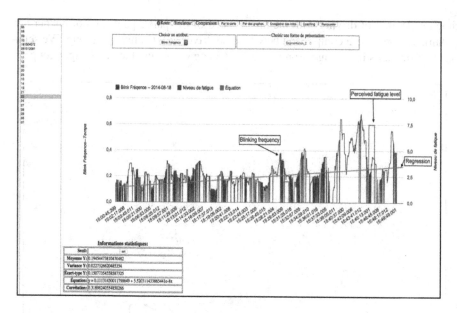

Fig. 7. Blinking frequency

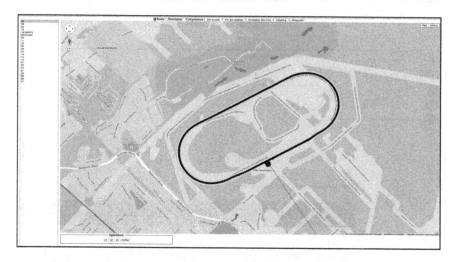

Fig. 8. Maps visualization

to validate a particular event that occurred during the driving (a sudden stop or a speed reduction).

Finally, we propose a debriefing solution to help the driver after driving session (Fig. 10) to visualize their behaviors and talk about their perception of their driving fatigue versus the collected. It also includes participant's information (driving time, number of kilometers, mean speed and localization), curves (participant's speed, the speed limit and the perceived level of fatigue of the driver) and table (divided driving time per 15 min of driving with the mean speed of the participants and of his/her age group). We propose a comparison between the driver and his/her age group. We separated participants in two age groups: the 55–65 age group and the 66–76 age group.

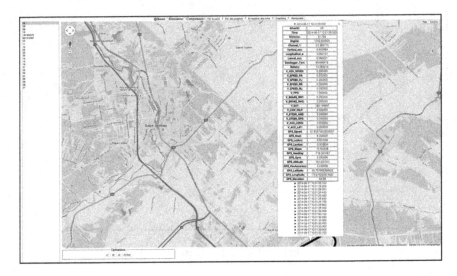

Fig. 9. Maps visualization and information

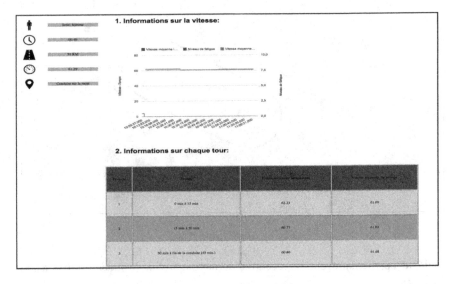

Fig. 10. Debriefing presentation

At the moment, we propose this temporary solution but we still work on that part to improve the feedback. In addition, we will develop a comparison module between real driving condition and simulator driving condition in order to verify ecological validity.

This tool shows promising results. From real data taken during experimentations, the tool allowed an interactive visualization of the collected data and an analysis by diagrams and maps for the driver.

5 Conclusion

In this paper, we presented a data analysis tool for car driving contextual information for elderly drivers. This category of drivers has impairments due to aging. To help them to drive safely, context aware system improve driving with understanding the whole driving task (driver, environment and vehicle) and assist driver's decision. The tool includes a web application approach with Node.js, JavaScript and MongoDB. These one allow an easy implementation of contextual information analysis algorithms and give us the platform to test our algorithm on offline data, before testing them within the technologies we are developing to assist the drivers in their tasks. We applied our system to collect data from the sensors of an instrumented vehicle (the LiSA) in realistic driving conditions. We used our tool during a pilot evaluation with 20 participants, aged between 56 and 76 years old. Results show that our tool is able to store the contextual information collected and to enable an interactive visualization of the data collected for each participant. This contextual information analysis tool represents a multimodal platform that brings several benefits to the researches we conduct on road safety.

Two points are discussed below on the system's outcomes. The first discussion point is to take into account the driving workload with the tool, particularly when the tool has a lot of information like this one. Indeed, car accidents can appear if the system diverts the driver and if it requires too much concentration. Elderly drivers are more likely being distracted by an interface because of the diminution of cognitive abilities due to aging. And elderly have difficulty doing two things simultaneously (for instance: acquire information from the environment and make right decisions about driving) [6, 10]. In addition, too much concentration could induce tool's incomprehension. For an acceptance by the elderly, the interface and outcomes of the context-aware system needs to be affordable and easy to understand to be accepted [6, 10, 13]. Therefore, to improve the tool, we will work on ergonomic characteristics and design aspects directly with elderly drivers.

The second discussion point is in our current experiments the size of the collected data is relatively small; we can see that in our next projects, where the number of participants will increase substantially (100+) as well as the duration, the quantity of collected data will become more important. Moreover, we are planning to add other types of contextual information, such as physiologic data (e.g. heartbeat, EEG), which will even increase the quantity of data to manage and analyze. Our future work in this area will use this platform as an implementation and analysis base of contextual information. Furthermore, it will be interesting to develop an algorithm able to categorized normal and abnormal driver behavior. Indeed, with the analyze tool, that is possible to monitor the behavior and alert the driver when there are change in driver's behavior for example.

Acknowledgments. We want to particularly thank the Canadian Automobile Association (CAA) Foundation, Section of the Quebec Province, for funding the research works behind this paper.

References

1. United Nations, Department of Economic and Social Affairs, Population, Division: World Population Ageing 2013. Technical Report, ST/ESA/SER.A/348 (2013)
2. Sivak, M., Schoettle, B.: Recent changes in the age composition of drivers in 15 countries. Traffic inj. prev. **13**(2), 126–132 (2012)
3. Statistiques Canada. http://www.statcan.gc.ca/pub/91-520-x/91-520-x2010001-fra.pdf
4. Bordeleau, B.: Les collisions mortelles des 65 ans ou plus. Direction des études et des stratégies en sécurité routière. Technical Report, Société de l'assurance automobile du Québec (2007)
5. Emery, L., Srinivasan, G., Bezzina, D., Leblanc, D., Sayer, J., Bogard, S., Pomerleau, D.: Status report on USDOT project – an intelligent vehicle initiative road departure crash warning field operational test. In: 19th International Technical Conference Enhanced Safety of Vehicles, Washington DC (2005)
6. Rakotonirainy, A.: Design of context-aware systems for vehicle using complex systems paradigms. In: Tijus, C., Cambon de Lavalette, B. (eds.) Workshop on safety and context in conjunction with CONTEXT 2005, CEUR Workshop, Paris (2005)
7. Ryan, N., Pascoe, J., Morse, D.: Enhanced reality fieldwork: the context aware archaeological assistant. Bar International Series **750**, 269–274 (1999)
8. Oxley, J., Langford, J., Koppel, S., Charlton, J.: Senior driving longer, smarter, safer: enhancement of an innovative educational and training package for the safe mobility of seniors. Technical Report, Monash University Accident Research Centre, Monash Injury Research Institute (2013)
9. Anstey, K., Wood, J., Lord, S., Walker, J.G.: Cognitive, sensory and physical factors enabling driving safety in older adults. Clin. Psychol. Rev. **25**(1), 45–65 (2005)
10. Eby, D.W., Molnar, L.J.: Has the time come for an older driver vehicle? Technical Report, University of Michigan, Transportation Research Institute (2012)
11. Owsley, C., Mcgwin, G.: Vision and driving. Vis. Res. **50**(23), 2348–2361 (2010)
12. Langford, J., Koppel, S.: Licence restrictions as an under-used strategy in managing older driver safety. Accid. Anal. Prev. **43**(1), 487–493 (2011)
13. Rakotonirainy, A., Steinhardt, D.: In-vehicle technology functional requirements for older drivers. In: The 1st International Conference on Automotive User Interfaces and Interactive Vehicular Applications, pp. 27–33. ACM, New-York (2009)
14. Dey, A.K.: Understanding and using context. Pers. Ubiquit. Comput. **5**(1), 4–7 (2001)
15. Juliussen, E.: Driver assistance system, the road ahead. In: Telematics Research Group Inc. (2011)
16. Van Eslande, P.: Erreurs de conduite et besoins d'aide: une approche accidentologique en ergonomie. Le travail humain **66**, 197–224 (2003)
17. Reason, J., Manstead, A., Stradling, S., Baxter, J., Campbelle, K.: Errors and violations on the roads: a real distinction? Ergonomics **33**(10–11), 1315–1332 (1990)
18. Alghamdi, W., Shakshuki, E., Sheltami, T.R.: Context-Aware driver assistance system. Procedia Comput. Sci. **10**, 785–794 (2012)
19. Indulska, J., Sutton, P.: Location management in pervasive systems. In: The Australasian Information Security Workshop Conference on ACSW Frontiers 2003-vol. 21, pp. 143–151. Australian Computer Society, Inc. Darlinghurst (2003)
20. Baldauf, M., Dustdar, S., Rosenberg, F.: A survey on context-aware systems. Int. J. Ad Hoc Ubiquit. Comput. **2**(4), 263–277 (2007)

21. Rakotonirainy, A., Maire, F.D.: Context-aware driving behavioural model. In: 19th International Technical Conference on the Enhanced Safety of Vehicles (ESV'19). Washington DC (2005)
22. Bhattacharjee, A.A., Wankhede, S.S.: Modeling of driver behaviour recognition and prediction using dynamic bayesian network. Int. J. Eng. Res. Gen. Sci. 3(2), 691–694 (2015)
23. Kedowide, C., Gouin-Vallerand, C., Vallieres, E.F.: Recognizing blind spot check activity with car drivers based on decision tree classifiers. In: 28th AAAI Conference on Artificial Intelligence (AAAI-14). Québec (2014)
24. Vallières, E.F., Ruer, P., Bergeron, J., Mc Duff, P., Gouin-Vallerand, C., Ait-Seddik, K.: Perceived fatigue among aging drivers: an examination of the impact of age and duration of driving time on a simulator. In: SOCIOINT15–2nd International Conference on Education, Social Sciences and Humanities, pp. 314–320. Istanbul (2015)
25. Lemire, D., Brooks, M., Yan, Y.: An optimal linear time algorithm for quasi-monotonic segmentation 1. Int. J. Comput. Math. 86(7), 1093–1104 (2009)

Cognitive Process as a Tool of Tourists' Typology for Rural Destinations

Eva Šimková[✉] and Alena Muzikantová

Faculty of Education, University of Hradec Králové,
Hradec Králové, Czech Republic
{simkoval8,muzikantova.alena}@seznam.cz

Abstract. Cognitive processes for analyses of human typology, motivation and behavior, are also widely applied in tourism for the assessment of psychographic diversity of visitors. One of the primary psychographic methods is Plog's model of psychocenric and allocentric. Identification of clients, knowledge of their needs, motivation factors, expectations, desired destinations and decision making process, are basis for effective planning of activities, and destination marketing. The paper deals with the application of Plog's model in tourism of rural areas in the Czech Republic, including identification of potential problems in practice.

Keywords: Cognitive process · Tourists' typology · Psychocentric · Allocentric · Rural areas

1 Introduction

Tourism represents one of the most important part of global economy. This particular segment of services is also important element of free-time activities, and as such has great impact on life-style of the current generation. That's why tourism represents significant economic, social and cultural phenomenon satisfying human needs for leisure, contact with nature and desires to visit new places, meet other people, human cultures, customs, heritage etc.

Due to the importance of tourism and its contribution to the regional development, impact on the labour market, infrastructure, cultural and sport activities, natural environment, small and medium size enterprises, diversification of activities in the rural areas, competent authorities should create best conditions for its development. Tourism may be beneficial to almost any region and be a source of many benefits. However, as in any other activity it is also tourism that carries along negative externalities and risks, such as environmental damages and social risks. Such negative impacts can eventually lead to the decrease of the quality of tourism activities, tourists' expectations and destination attractiveness.

The recognition of positive and negative impacts of tourism in destination is a critical factor in the decision-making process for all tourism activities. This process is fundamental for regional development so that tourism's sustainability in destinations is widely ensured by competent authorities. Tourism sustainability is also the key aspect in creation of harmonic relations between tourists, local people and the environment.

H. Christiansen et al. (Eds.): CONTEXT 2015, LNAI 9405, pp. 186–198, 2015.
DOI: 10.1007/978-3-319-25591-0_14

In order to do so, destinations must be effectively marketed and managed. Besides basic monitoring (such as situation analysis, SWOT analysis), definition of strategic vision and goals, destination marketing and management includes also definition of visitors (problems of market segmentation) and identification of their needs, desires and motivation factors. Knowledge and understanding of main psychological factors, as well as prediction of tourists' behavior is key aspect for promotion and selling of tourism services, or a destination.

The paper describes individual phases of cognitive process in the analysis of psychographic diversities of visitors, versatility of needs, preferences or motivation. Authors apply one of the basic psychopgraphic methods – the Plog's psychographic model of psychocentric and allocentric.

2 Theoretical Background

2.1 Cognitive Process and Its Application in Tourism

A cognitive process is "collection, restructuring, storage and further application of information by humans, part of which is also cognition, behavioral status, memory, consciousness of quality, quantity and structure of human's outer and inner world" [47, p. 268]. The authors also state that cognitive processes can be influenced by many factors, primarily by individual perception, and former experience in everyday's life, economic, cultural and social features.

Cognitive processes can be widely applied in tourism for their ability to identify and understand tourists' learning processes and then describe psychographic profile. That can be used for the definition of expectations at one hand and experience on the other. The characteristics can be further used for the analysis of psychographic diversity of visitors and their typology. It is the Plog's psychographic model [35], which works with such cognitive features.

The definition above allows for the description of a cognitive process (see Fig. 1), where key elements from tourism psychology perspective are motivation (needs), expectation, decision making, experience, and (dis)satisfaction, including analysis of relations.

The figure indicates one simple thing. It is crucial to understand human needs upon his/her own experience, and which the client receives from particular information (e.g. media), and implication in tourism, such as services (i.e. practical use of tourism services). It is not only motivation, but primarily relations and interconnection of individual features where satisfaction and dissatisfaction retrospectively form the metal picture of a destination, which subsequently drives the client's needs and affect his further motivation and action.

Access to *information* and *knowledge* has gradually gained competition advantage for the entrepreneurs. It is simply because the tourists are able to evaluate information on attractiveness and values of a destination (such as tourist services, destinations' history, nature, crafts, genius loci – sense of place...).

Note: Basic terms applied to field of knowledge in general (e.g. data, information, knowledge, skills, habits etc.) are described in Hošková-Mayerová and Rosická [15].

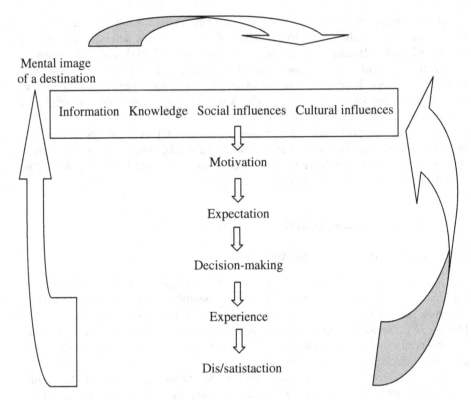

Fig. 1. Description of a cognitive process (Source: authors' own compilation)

Social impacts are represented by the attitude and opinion of an individual. These usually relate to the position and social status and position of such individual, and these are formed upon social interaction. *Cultural impacts* represent ideas, values, knowledge, traditions and morale in a particular community. Cultural impacts are the most influential for an individual. In tourism it is meeting people from different culture, which eventually may bring some problems or even conflicts. As Jakubíková [18] states, cultural differences and respect are among the most important ones.

Palatková [30] states that the *decision making process* starts with the recognition and definition of needs and desires, combination of which then forms his/her *motivation*. Motivation thus represents an internal drive that guides visitor's behavior, e.g. [45]. Decision making has been examined not only in relation to travelling, but also work, sport and other leisure time activities. Motivational research usually incorporates Maslow's hierarchy of needs and the Iso Ahola's motivation model [17].

Maslow's hierarchy of needs says that people are motivated to fulfil basic needs one by one. According to Zelenka and Pásková [47] the Maslow's 5-stage theory on motivation in tourism is extended by cognitive and aesthetic needs. It means that so called low category needs or indispensable needs are extended by cognitive (knowledge and understanding of local life-style, heritage and traditions) and aesthetic needs (such as appreciation and search for beauty and balance) - dispensable needs (so called high category needs – "metaneeds"). It was Macintosh, who upon Maslow's theory specified four groups of motivation factors: physical, cultural, interpersonal, and status and prestige [In 30].

Iso Ahola [17] performed research on a two-dimensional leisure motivation theory in one's personality: escaping dimension (escaping everyday environments, routine, everyday problems, familiar environments, tension, stress and appeal of the new situation), and seeking dimension (seeking some rewards). According to Iso Ahola, both dimensions have personal and interpersonal components. The model covers four aspects of needs according to these main elements (escaping and seeking): escaping personal environment, escaping interpersonal environment, seeking intrinsic personal rewards, seeking intrinsic interpersonal rewards. The push-pull theory is another approach that often complements Iso Ahola's model [6]. Additionally, Crompton and McKay [7] introduced push factors (internal motives) as escaping motives and pull factors (external stimulus) as seeking motives.

Psychographic criteria are basis for market segmentation in strategic marketing. That is often described as STP marketing: Segmenting, Targeting, Positioning. "Market segmentation relates to destination split into homogenous groups (target destination) by their characteristics (age, education, nationality, interests...)" [30, p. 92]. In contemporary tourism market segmentation is done by already mentioned psychographic criteria, such as visitor's motivation. The primary reason is excess of so called hybrid clients, who use tourism services depending on actual situation, i.e. without any planning or pattern.

2.2 Psychographic Aspects of Tourists Segmentation

Psychographic analysis defines personality, consumers' attitudes, needs, interests, activities, their perceptions, preferences, and lifestyle. Psychographic research has been defined e.g. by Wells [46, p. 207] as "quantitative research intended to place consumers on psychological dimensions". Application of psychological factors in understanding and predicting tourist behavior has been widely used both in practice and tourism literature. Some researchers consider psychographic factors very important in tourism in particular (e.g. [1, 43]). Psychographic factors, esp. human values, personalities, travel motivation, preferences for recreational activities, cultural values, lifestyles, are also valuable tools for market segmentation. Psychographic measures enables to identify different types of visitors (tourist segments), and provide more detailed profiles of tourists. Knowledge of psychographic characteristics can help tourism/destination marketers and managers to better understand visitors' behavior and to be more competitive in tourism market and successfully sell destinations and their product to tourists (e.g. [10, 31, 41]).

Some authors value psychographic data as more useful than demographics. They claim that it is because their better description of consumers' differences (e.g. [27, 40]). However in order to better understand consumer market, perceive and predict tourism behavior, psychographic characteristics have been used together with other segmentation criteria, such as demographic, geographic, behavioristic etc. [39].

There are many psychographic studies. Psychographic issues were surveyed by e.g. Shih [43] for using the VALS typology (values, attitudes, lifestyles) to segment tourists in Pennsylvania. Some authors identified tourists' lifestyle characteristics (e.g. [14, 29, 38, 41]). The importance of personality and its usage in tourism is described by e.g. [13, 19, 21, 23, 32].

Yet, there were also studies of theory of motivation to explain visitors' behavior and their destination choice. It was Chon [2], who explored sociopsychological motives (push factors) to explain tourists' needs. Dann [8] analyzed destination attractions (pull factors) for the description of motives for selection of a particular location as a holiday destination. Mayo [27] examined the main reason for travelling to national parks; or to the certain states [41]. Other researchers analyzed travelers' preferences for recreation activities (e.g. [12, 20]).

The importance of the psychographic characteristics was emphasized primarily by Plog [35]. Plog was among the first who described tourists' profile. Plog's model is known as tourism's psychographic typology model for segmenting travelers. In his research "Why destination areas rise and fall in popularity", Plog's psychographic motivation theory classifies tourists according to the traveler's personality into psychocentrics at one side to the allocentrics at the other. *Psychocentrics* are represented by people concerned on their own affairs, i.e. non-adventurous visitors, who often require standard services. They are less active, most often conservative, prefer to travel in groups, participate in common tourist activities and travel to familiar and well-established destinations. On the other hand there are *allocentrics*, who are described as independent tourists seeking for adventure or to experience new things. They are active, motivated by novelty, discovery, participate in varied activities, and meeting new people, and travel to unfamiliar and unique locations. Between these two extremes are *midcentrics*. Midcentrics actually represent majority of tourists. These are tourists relatively flexible in their needs. People who occupy borders with near psychocentrics or near allocentrics are called as *near-psychocentric* and *near-allocentric*. Psychographic distribution of visitors, as described above, forms Gaussian curve.

We can now summarize the main points and define general characteristics of psychocentrics (labeled as dependables) and allocentrics (labeled as venturers) [36, 37]:

Characteristics of Psychocentrics/Dependables:

- less adventuresome and less exploring
- cautious and conservative in their daily lives
- restrictive in spending income
- prefer popular, well-known brands of consumer products
- seek well known and overdeveloped destinations
- high tendency towards revisit destination if they are satisfied etc.

Characteristics of Allocentrics/Venturers:

- curious about and want to explore the world around them
- make decisions quickly and easily
- spend income more readily

- like to choose new products rather than sticking with popular brands
- seek new experiences to enrich their daily lives
- seek new destinations before others have discovered etc.

Plog in his further research [36] states, that also destinations can be classified in the same continuum as tourists. E.g. Tibet, Nepal, Amazon, Vietnam etc. represent allocentric destinations (also called as venturer), Atlantic City, Myrtle Beach, Orlando beach resorts etc. are examples of the psychocentric destinations (dependable). Plog also determined the preference of travellers for visiting the same type of destinations that reflect their personal psychographic characteristics. As Litvin and Smith state, this is very important especially for marketers, who should understand the value of aligning destinations with psychographic characteristics and behavior of visitors [26]. According to Pásková [34] better infrastructure in a destination usually attracts higher number of conservative visitors. Such destination, however, has usually lost its authenticity, character and genius loci. Plog's model was also tested and revised by some researchers (e.g. [24, 44]).

We can mention also Cohen [5] and his typology of visitors by physical attractiveness of destinations and required safety levels. Cohen defined four basic types of visitors: "drifters" are tourists who do not want to be associated with the tourism industry (similar to Plog's allocentric), "explorers" arrange trips by themselves and stays in contacts with residents, but seeks standard services, "individual mass tourists" usually seeks help from tour operators, while "organized mass tourists" are less adventurous, usually let others do all the work, and requires high quality services.

Yet, there are even newer researches that apply the Plog's model in practice, e.g.:

- The relationships among psychographic factors such as human personality, cultural values, lifestyle, motivation for travelling, and preference for doing leisure activities, was studied by Reisinger and Mavondo [39]. They analyzed psychographic profiles among students who wanted to travel;
- Research of the critical view of the use of typologies in tourism planning [4];
- Litvin [24] analyzed the Plog's model on an "ideal destination" and "most recent destinations";
- Merritt [28] analyzed on the basis of Plog's model association between preferred recreation activities of different generations of travelers and their psychographic profiles;
- George, Henthorne and Williams [9] decomposed Plog's model and identified five smaller bell shaped curves presenting five tourist personas;
- Park and Jang [33] tested Plog's model and verified whether this model is flawed by defining travelers in terms of a "static" category;
- Research exploring the relationship between the Big Five Factors (such as openness to experience, conscientiousness, extraversion, agreeableness, and neuroticism) of personality and travel personality [19];
- Litvin and Smith [26] tested effectiveness of the Plog's model as a possible predictor of travel behavior and discussed practical marketing and destination management implications.

2.3   Critical Assessment of Practical Application of Plog's Psychographic Model

As already mentioned Plog was among the first to use the psychographic concept of tourists' profile in understanding tourists motivation of travel destinations by personality. Now, this model is extensively used. The main point of Plog's psychographics model is that travelers' personality characteristics are the key determinants of their travel preferences. Plog's model of allocentricity and psychocentricity is valuable tool for understanding tourists' behavior and marketing segmentation. Identifying different types of visitors can help marketers to attract tourists to their destination.

However, Plog's categorization has also gained some criticism. It was for the following negative issues:

- model can't predict travel behavior with different visitors motivation (e.g. [11, 16]),
- Chon and Sparrowe [3] consider model as impractical for the use in tourism,
- Litvin [25] criticized model for ignoring political factors (esp. government control) and effective management of a destination,
- Li and Cai [22] demonstrated Plog's personality types as weak predictors of travel behavior,
- Park and Jang [33] verified that Plog's model isn't "static", but its psychographic segmentation may change quite rapidly, esp. if the level of satisfaction is high,
- Litvin and Smith [26, p. 6] found that "regardless of a destination type, the destinaton's visitor count is likely to be heavily dominated by midcentric visitors".

Another failure of the Plog's model and other psychografic models are in their definition of motivation to travel, such as defined by Seaton and Bennett [42]:

- Tourism is a heterogeneous entity with many activities and services, where destination may have minor importance than activities themselves (such as walking, cycling, rock climbing, via ferrata);
- Sometimes it is difficult to get to know the primary reason for selection of such product (such as due to the reluctance of a respondent);
- Changes in the motivation driver, such as cultural and/or social drivers;
- Drivers for travelling are often in contrary each other, i.e. relaxation and rest at one hand and desire for exploration;
- Motivation can be fed up in many ways;
- There are two main levels of motivation, i.e. primary (exploration, relaxation), and secondary (safety during travelling, safety at the destination) etc.

3   Materials and Methods

The paper deals with typology of tourists upon pre-set psychographic characteristics. The main objective of the paper is, using the Plog's psychographic model [35], to define the type of tourists visit rural areas in the Czech Republic and also identify potential problems in its application in practice.

In our survey we have tested this basic research question: "What type of tourists is motivated for visiting rural areas?" On this basis we stated this hypothesis: "Rural areas are visited predominantly by allocentric travelers".

The research has been based on a survey performed at the beginning of 2015. The survey was done by direct questioning of 150 respondents (students from the University of Hradec Králové). In order to determine the psychographic tourists' profile the Plog's psychographic model was used.

Using main characteristics of allocentric and psychocentric, we have selected five items. Also, because of our personal interest in rural tourism and development of rural areas in particular, we have focused on main characteristics of tourists in rural areas as an example of an allocentric destination. Research results are presented in the next chapter.

4   Analysis of Psychographic Differences Among Tourists

Our questionnaire consisted of five primary items (see Table 1). Additionally, there were three questions on personal characteristics. Since rural areas represent allocentric type of destinations (according to Plog [36]), we have focused our questionnaire so that it corresponded with general characteristics of allocentrics.

Table 1. Sample of items used in psychographic distribution of tourists (Source: authors' own compilation using [36, 37])

Items for psychographic tourist profile	1	2	3	4	5
I prefer individual approach (I am independent tourist)					
I always seek meeting new people and environment					
I seek for authenticity/genius loci					
I prefer new, unexplored activities					
I am an adventurous traveler					

Each item was evaluated in the scale from 5 – highest score, to 1 – lowest score. Then, using statistical calculation we have got final score of all respondents. This number was put into Table 2.

Distribution of respondents based upon their summed score to psychographic items is presented in Figs. 2 and 3.

Table 3 presents research results in comparison with Plog's model. According to our survey results, midcentric tourists with 75 % prevail among the respondents. Psychocentrics reached 8 % of all the respondents and remaining 17 % were classified as allocentrics. These results approximately correspond with Plog's projected split.

We analyzed types of tourists profile for rural areas that belong to allocentric type of destinations. Results of our survey are opposite to Plog's philosophy [36] about tourists' preference to visit such destinations that reflect their personal psychographic

Table 2. Distribution of respondents upon summed score (Source: authors' own compilation)

Score	5	6	7	8	9	10	11	12	13	14	15	16	17	18	19	20	21	22	23	24	25	Total
Number of respondents	0	0	0	1	3	2	5	8	18	11	26	14	22	14	11	5	5	1	3	1	0	150
Tourists profile (%)	Psychocentrics							Midcentrics							Allocentrics							100 %
	8 %							75 %							17 %							

Note: Summed scores 5–11 characterize psychocentrics, scores 12–18 characterize midcentrics, scores 19–25 characterize allocentrics.

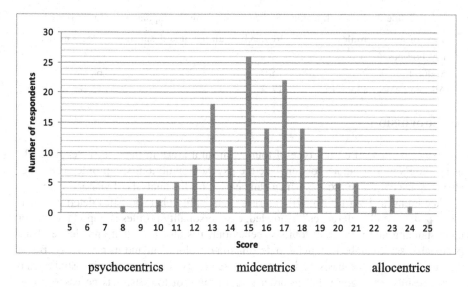

psychocentrics   midcentrics   allocentrics

Fig. 2. Distribution of respondents - column chart (Source: authors' own compilation)

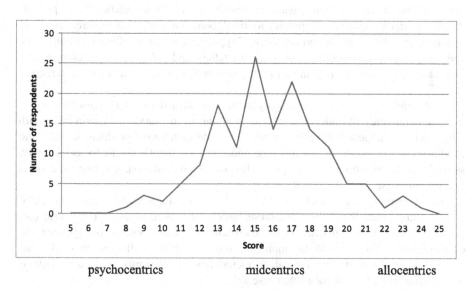

psychocentrics   midcentrics   allocentrics

Fig. 3. Distribution of respondents – line chart (Source: authors' own compilation)

characteristics. Our hypothesis therefore was not confirmed. On the other hand, our results correspond with the survey of Litvin and Smith [26], who stated that regardless of the destination type, midcentric visitors heavily dominate among visitors.

Table 3. Research results (Source: authors' own compilation using [26])

Type of tourist profile	Research result for rural destinations	Plog's model
Psychocentric/near psychocentric	8 %	17 %
Midcentric	75 %	68 %
Allocentric/near allocentric	17 %	15 %

5 Conclusion

The paper is focused on the type of tourists in rural areas in the Czech Republic. By using Plog's model the authors analyze the psychographic tourists' profile. The paper also presents cognitive process which can be used for the analysis of visitors' typology.

According to Plog [36, p. 16] "tourists' personality characteristics determine their travel patterns and preferences". Knowledge of personality profiles is important tool for understanding and predicting tourists' behaviour. It's also true, that knowledge of the personality scale can be beneficial to destination development and its popularity. But at the same time on the basis of broad literature review of psychographic studies and survey results we suggest that in order to get profile of tourists, it is necessary to do more than analyze travel-related psychographic data. The analysis must be much broader. Analysts must also search for demographic, geographic and other cognitive items. Identification of all elements of cognitive process, as well as effective management and marketing can ensure understanding of these relationships, provide detailed tourists' profile, and predict tourists' behavior and their preferences. Only system approach to identification of tourists' typology can ensure effective, competitive and comprehensive formulation of tourism strategy and policy. It is therefore a very important tool for tourism planning of the sustainable development of the particular region.

At the end of the paper it's necessary to mention other findings. Our research results indicate dominancy of midcentric travelers (similar to the survey of Litvin and Smith [26]), and also suggest that it is necessary to provide such tourist products, services and conditions in a destination, that meet the midcentric segment's needs: focus on safety, some level of comfort and average quality of accommodation, catering, and other tourism services.

This, however, should not be generalized. In many rural areas (that can be specified as allocentric type) there are destinations which characterized as psychocentric (dependable) destinations, are often visited by wealthy tourists or visitors of psychocentric characteristics. These could be mountain resorts, western villages, geoparks etc. Finding destinations having typical characteristics of allocentric (venturer) type is prerequisite and stimulus for further research.

Although our research brought interesting findings, we have to mention that we have worked only with a readily available sample of people. Those were students of the University of Hradec Králové. Not all these students, however, eventually proved being tourism oriented. Our sample of respondents was thus not fully comparable with "average" tourists. That's why the authors prepare additional, more extensive research. That will be realized directly in tourist destinations and cover various types of tourists for better findings of their personality profiles.

References

1. Abbey, J.: Does life-style profiling work? J. Travel Res. **18**(1), 8–14 (1979)
2. Chon, K.: Understanding recreational traveler's motivation, attitude and satisfaction. Tourist Rev. **44**(1), 3–7 (1989)
3. Chon, K.H., Sparrowe, R.T.: Welcome to Hospitality: An Introduction, 2nd edn. South-Western Publishing, Cincinnati (2000)
4. Coccossis, H., Constantoglou, M.E.: The use of typologies in tourism planning: problems and conflicts. In: Coccossis, H., Psycharis, Y. (eds.) Regional Analysis and Policy, pp. 273–275. Physica-Verlag, Heidelberg (2006)
5. Cohen, E.: Toward a sociology of international tourism. Soc. Res. **39**(1), 164–182 (1972)
6. Crompton, J.L.: Motivations for pleasure vacation. Ann. Tourism Res. **6**(4), 408–424 (1979)
7. Crompton, J.L., McKay, S.L.: Motives of visitors attending festival events. Ann. Tourism Res. **24**(2), 425–439 (1997)
8. Dann, G.: Tourist motivation: an appraisal. Ann. Tourism Res. **8**(2), 187–194 (1981)
9. George, B.P., Henthorne, T.L., Williams, A.J.: The internal structure of destination visitation model and implications for image management. Rev. Turismo Patrimonio Cult. **11**(3), 47–53 (2013)
10. Gladwell, N.: A psychographic and sociodemographic analysis of state park inn users. J. Travel Res. **26**, 15–20 (1990)
11. Goeldner, C.R., Ritchie, J.R.B.: Tourism: Principles, Practices, Philosophies, 9th edn. Wiley, New York (2003)
12. Graham, J., Wall, G.: American visitors to Canada: a study in market segmentation. J. Travel Res. **16**(3), 21–24 (1978)
13. Gretzel, U., Mitsche, N., Hwang, Y.H., Fesenmaier, D.R.: Tell me who you are and I will tell you where to go: use of travel personalities in destination, recommendation systems. Inf. Technol. Tourism **7**, 3–12 (2004)
14. Holman, R.H.: A values and lifestyles perspective on human behavior. In: Pitts, R.E., Woodside, A.G. (eds.) Personal Values and Consumer Psychology, pp. 35–52. Lexington Publishing Company, Lexington (1984)
15. Hošková-Mayerová, Š., Rosická, Z.: Programmed learning. Procedia Soc. Behav. Sci. **31**, 782–787 (2012)
16. Hudson, S.: Consumer behavior related to tourism. In: Pizam, A., Mansfield, Y. (eds.) Consumer Behavior in Travel and Tourism, pp. 7–32. The Haworth Hospitality Press, New York (1999)
17. Iso Ahola, S.E.: Towards a social psychological theory of tourism motivation: a rejoinder. Ann. Tourism Res. **9**(2), 256–262 (1982)
18. Jakubíková, D.: Marketing in Tourism. Grada Publishing, Praha (2012)
19. Jani, D.: Relating travel personality to Big Five Factors of personality. Tourism **62**(4), 347–359 (2014)
20. Kim, K.Y., Jogaratnam, G.: Activity preferences of Asian international and domestic American university students: an alternative basis for segmentation. J. Vacation Mark. **9**(3), 260–270 (2003)
21. Leung, R., Law, R.: A review of personality research in the tourism and hospitality context. J. Travel Tourism Mark. **27**(5), 439–459 (2010)
22. Li, M., Cai, L.A.: The effects of personal values on travel motivation and behavioral intention. J. Travel Res. **51**(4), 473–487 (2012)
23. Lin, Y., Kerstetter, D., Nawijn, J., Mitas, O.: Changes in emotions and their interactions with personality in a vacation context. Tourism Manag. **40**, 416–424 (2014)

24. Litvin, S.W.: Revisiting Plog's model of allocentricity and psychocentricity... one more time. Cornell Hotel Restaurant Adm. Q. **47**(3), 245–253 (2006)
25. Litvin, S.W.: Apocalypse not: a commentary on destination maturation models. J. Vacation Mark. **16**(2), 157–162 (2010)
26. Litvin, S.W., Smith, W.W.: A new perspective on the Plog psychographic system. J. Vacation Mark. **16**(4), 1–9 (2015)
27. Mayo, E.: Tourism and the national park: a psychographic and attitudinal study. J. Travel Res. **14**(1), 14–17 (1975)
28. Merritt, R.T.: Destination Recreation: A Generational Exploration of Psychographic Characteristics related to Vacation Recreation Activity Preferences, p. 112 (2013). http://thescholarship.ecu.edu/bitstream/handle/10342/1760/Merritt_ecu_0600M_10888.pdf?sequence=1
29. Mitchell, A.: The Nine American Lifestyles. Macmillan, New York (1983)
30. Palatková, M.: Marketing Strategy of Tourism Destination. Grada Publishing, Praha (2006)
31. Palatková, M.: Destination Marketing Management. Grada Publishing, Praha (2011)
32. Park, S., Tussyadiah, I.P., Mazanec, J.A., Fesenmaier, D.R.: Travel personae of American pleasure travelers: a network analysis. J. Travel Tourism Mark. **27**(8), 797–811 (2010)
33. Park, J., Jang, S.: Psychographics: statistic or dynamic? Int. J. Tourism Res. **16**(4), 351–354 (2014)
34. Pásková, M.: Tourism sustainability. Gaudeamus, Hradec Králové (2014)
35. Plog, S.C.: Why destination areas rise and fall in popularity. Cornell Hotel Restaurant Adm. Q. **14**(4), 55–58 (1974)
36. Plog, S.C.: Why destination areas rise and fall in popularity. an update of a Cornell Quarterly classic. Cornell Hotel Restaurant Adm. Q. **42**(3), 13–24 (2001)
37. Plog, S.C.: The power of psychographics and the concept of venturesomeness. J. Travel Res. **40**(3), 244–251 (2002)
38. Perreault, W., Darden, D., Darden, W.: A psychographics classification of vacation life-style. J. Leisure Res. **9**(1), 208–223 (1977)
39. Reisinger, Y., Mavondo, F.: Modeling psychographic profiles: a study of the US and Austalian student travel market. J. Hospitality Tourism Res. **28**(1), 44–65 (2004)
40. Ryel, R., Grasse, T.: Marketing ecotourism: attracting the elusive ecotourist. In: Whelan, T. (ed.) Nature Tourism: Managing for the Environment, pp. 164–186. Island Press, Washington, DC (1991)
41. Schewe, C., Calantone, R.: Psychographics segmentation of tourists. J. Travel Res. **16**(3), 14–20 (1978)
42. Seaton, A.V., Bennett, M.M.: The Marketing of Tourism Products: Concepts, Issues and Cases. International Thomson Business Press, London (1996)
43. Shih, D.: VALS as a tool of tourism market research: the Pennsylvania experience. J. Travel Res. **24**(4), 2–11 (1986)
44. Smith, S.L.J.: A test of Plog's allocentric/psychocentric model: evidence from seven nations. J. Travel Res. **28**(4), 40–43 (1990)
45. Strnadová, V., Voborník, P.: Tools to develop quality of life for technical thinkers in various generation periods. In: International Conference on Society for Engineering and Education (CISPEE 2013), Porto. Portugal, pp. 23–30 (2013)
46. Wells, W.D.: Psychographics: a critical review. J. Mark. Res. **12**, 196–213 (1975)
47. Zelenka, J., Pásková, M.: Tourism – Dictionary. Linde, Praha (2012)

Data Analysis and Context

Multi-domain Adapted Machine Translation Using Unsupervised Text Clustering

Lars Bungum and Björn Gambäck[✉]

Department of Computer and Information Science,
Norwegian University of Science and Technology,
Sem Sælands vei 7–9, 7094 Trondheim, Norway
{larsbun,gamback}@idi.ntnu.no
http://www.idi.ntnu.no

Abstract. Domain Adaptation in Machine Translation means to take a machine translation system that is restricted to work in a specific context and to enable the system to translate text from a different domain. The paper presents a two-step domain adaptation strategy, by first making use of unlabeled training material through an unsupervised algorithm, the Self-Organizing Map, to create auxiliary language models, and then to include these models dynamically in a machine translation pipeline.

Keywords: (Statistical) Machine Translation · Domain adaptation · Self-Organizing Maps · Hierarchical clustering · Unstructured data

1 Introduction

Intuitively we reply "it depends" when asked to translate a word into our native language. Indeed the meaning of a word depends, but on what? Research into Machine Translation (MT) is partially about making these dependencies tangible and accounted for. The dependencies in "it depends" can be seen as constraints on an MT system. Kay (1980) formulated these constraints as *quality, automation, general purpose* and *computational efficiency*. While no system has been even close to solving the unconstrained problem, Machine Translation is successful in scenarios where some of these requirements are relaxed. Restricting the system to certain contexts, by only allowing the input text to belong to a specific *domain* means relaxing the *general purpose* constraint. Domain Adaptation (DA) in turn means adapting an MT system catered for a general or specific purpose into a system capable of translating text from a different domain.

Faced with data from new domains and contexts, MT systems have problems, to a varying degree. Reasonably so, as some domains are closer than others, even if it can be hard to specify exactly what separates them: as language is constantly changing and evolving, it is difficult to draw boundaries between domains. When the focus of Machine Translation shifted from rule-based to data-driven (and later hybrid) approaches in the wake of the landmark IBM Models

© Springer International Publishing Switzerland 2015
H. Christiansen et al. (Eds.): CONTEXT 2015, LNAI 9405, pp. 201–213, 2015.
DOI: 10.1007/978-3-319-25591-0_15

(Brown et al. 1990), Domain Adaptation received more attention as a particular problem. In data-driven approaches, training material is a scarce resource; hence all available data is used, without necessarily being adapted to any specific domain.

Plank (2011) and Jiang (2008) argued that annotating more data in the new domain is a theoretical, but not practical strategy, thus stating the goal of Domain Adaptation as making systems usable in new contexts without needing to annotate new data or to specifically exploit labeled data from one domain to train classifiers on a new domain. Jiang (2008) also stressed a major point when she observed that context and domain adaptation is a general problem, and that techniques developed for Domain Adaptation are applicable to most classification tasks where the distributions of the training and the test sets differ.

Here we aim to explore how a Machine Translation system could (a) adapt to multiple domains without knowing the domain of the input document, and (b) choose the appropriate domain on the basis of an entire input document (and not only one string). This would presuppose a system that can make use of, in principle, unlimited unsorted data, and furthermore could pre-process the data in such a way that it could be used online. Building on an idea from Moore and Lewis (2010), this entails addressing the question of how to segment a large text corpus into sections that can be used for Domain Adaptation. More specifically, the present work uses only segments of the total corpus on which a Self-Organizing Map is drawn, but selects an auxiliary Language Model based on the perplexity of the *input* document (i.e., roughly its level of ambiguity).

The rest of the paper is organised as follows: Sect. 2 discusses the background and related work on Domain Adaptation; Sect. 3 then introduces the methods and data sets used in the present work. In Sect. 4, Self-Organizing Maps are used to segment text corpora into smaller portions using various hierarchical clustering algorithms; these sub-corpora are then utilized in a Machine Translation pipeline. Section 5 sums up the discussion and concludes.

2   Background and Related Work

The discussions in Theoretical Linguistics on text types have mostly been using the term *sublanguages* for specific types of texts within one language. In the Machine Translation area, the term *language domain* is more frequent; however, there is no clear definition of what a domain is. Price (1990) talks about domains as the specificity of a sub-language. According to Kittredge and Lehrberger (1982) sublanguages vary in terms of: subject matter; lexical, syntactic and semantic properties; grammar rules; frequency of specific constructions (as idioms); text structure; and use of special symbols.

Karlgren (1993) pointed out the problematic sides of the mathematical readings of the terms 'sub' and 'super' in that sublanguages often will exhibit properties that are not present in the perceived *superlanguage*, the general language from which a sublanguage is divided. He instead proposed the term *register*, borrowed from Sociolinguistics. Registers are defined as varieties of language according to use, in contrast to varieties according to speaker or geographical location. This definition is therefore more process-oriented, to be understood as properties of the speaker directly.

In Bungum and Gambäck (2011), we gave an account of domain adaptation with a special emphasis on shared knowledge with cognitive sciences. There we observed how much work is relating to an original domain and a new domain, where a system suited for the former is adapted into fitting the latter. In the present work we are prototyping a system that will adapt a Machine Translation system simultaneously to many different domains and contexts. This means that there is a need to translate on the document level, that is, to adapt a general MT system according to the document properties.

Early attempts at adapting Statistical Machine Translation (SMT) models went into adapting the Language Model (LM), inspired by language model adaptation work in speech recognition (Iyer et al. 1997); Rosenfeld (1996). Bellegarda (2004) gave a review of the work on Statistical Language Model Adaptation from the perspectives of Automatic Speech Recognition (ASR) at the beginning of the Big Data era, poignantly making the case for Domain Adaptation by pointing to both that Language Models are very brittle across domains (Rosenfeld 2000) and that a small (2 million words) corpus can be better in perplexity terms than a large (140 million words) for a specific ASR task.

Building on Information Retrieval (IR) methods, Mahajan et al. (1999) used cosine similarity between document vectors to interpolate a general LM to a domain specific model computed on the retrieved documents, in what they called a Dynamic Language Model, where an auxiliary LM is built depending on the documents similar to the input document. Eck et al. (2004) also used IR techniques, retrieving the most similar documents and sentences from a large collection. They explored two aspects of adaptation: the best amount of documents to retrieve and the best unit of retrieval, noting that sentence level retrieval performed best in terms of perplexity reduction, but with a weak correlation between improvement of the LM and the overall SMT task. While this was not strictly speaking Domain Adaptation, as all the test data was of the same kind, the ideas and methods would still be applicable to different text types. IR-inspired strategies have also been applied by others, although with limited success: Wang et al. (2014) took an edit-distance approach based on normalized Levenshtein (1966) similarity, while Lu et al. (2007) added a log-linear interpolation of *sub-models*, i. e., models built on data from each of the domains.

Moore and Lewis (2010) proposed a method of selecting relevant sentences from an out-domain corpus to build an auxiliary Language Model. They experimented on the Europarl and Gigaword corpora, with the former defined as in-domain text. The method selected sentences from the Gigaword corpus by comparing the difference in Cross-Entropy of an LM built on the English-French part of Europarl to one based on a random selection of sentences from Gigaword. The Gigaword sentences were then split into eight equal portions, ranked after this difference, from which new LMs were built. The method was compared to other selection criteria, such as ranking the sentences only on the perplexity score from Europarl or scoring each Gigaword sentence on the log-likelihood of testing the Europarl corpus on unigram models built with and without this sentence, as well as a random selection of Gigaword sentences. The method based on Cross-Entropy difference had a lower Perplexity than all the other methods until all data was added, but more importantly, using a only small portion of the data gave a lower perplexity than adding the whole corpus. Axelrod et al. (2011) performed similar extraction of *pseudo-in-domain* sentences based on Cross-Entropy measures; pseudo because they are similar, but not identical to the in-domain data. These extracted corpora were used to train smaller domain-adapted translation models that performed better for the target context than a model built on the entire material, and improved even more in combination. Discarding as much as 99 % of an original, general-purpose corpus, Axelrod et al. (2011) achieved better results with the pseudo-in-domain model.

There have been some notable efforts of injecting Domain Adaptation in a full Statistical Machine Translation pipeline. Carpuat et al. (2012) performed experiments on Phrase-Sense Disambiguation, a discriminative approach to MT where the correct Phrase is chosen according to a number of criteria at decode-time. Carpuat and Wu (2007) discuss how the method can be integrated into an SMT framework by means of dynamic phrase tables. However, the approach yielded no improvements in their experiments. The Moses SMT toolkit (Hoang et al. 2007) was used by Louis and Webber (2014) with cached language models to store domain specific information, and by Sennrich (2011) to build "dynamic" phrase tables, after an addition to a limited in-domain parallel corpus, and to combine these with Moses' log-linear decoding procedure. Going further, Sennrich et al. (2013) investigated the idea of using unsupervised methods to classify text, combining them with mixture models to perform Domain Adaptation.

3   Methods and Data

We have formulated a Domain Adaptation problem that requires two steps. First an outward-looking approach making use of additional, unlabelled training material, and second a way to include this dynamically in a Machine Translation pipeline. An unsupervised algorithm, the Self-Organizing Map (SOM) is employed to induce structure from an unstructured body of text and to create auxiliary Language Models for SMT decoding. The number of clusters to be created from the SOM is decided through bottom-up hierarchical agglomerative clustering. The method provides n separate clusters of text, from which standard n-gram Language Models are built to be used as auxiliary LMs in Domain Adaptation. In this way, the number of auxiliary text corpora (and later LMs) are determined by the clustering algorithm, enabling Domain Adaptation into the *available* domains, which is why an unsupervised method was chosen.

3.1   Data

Two data collections have been used in this work. The first, the SdeWac corpus (Faa and Eckart 2013) consists of 10,830 unorganized German web documents,

9,056 of which were included here, comprising 8.1 G of text. *Unstructured* text corpora with distinct features were extracted from SdeWac to create clusters for later use in Domain Adaptation. Each document in the collection was vectorized with Scikit-learn (Pedregosa et al. 2011), which includes several different vectorizers, on word and character level. A unigram TF-IDF (term frequency-inverse document frequency) vectorizer was mostly used in these experiments. Some tests were also run combining it with n-gram frequencies and on bigrams.

The second data collection was created from standard Statistical Machine Translation training and test corpora. A baseline SMT model was trained on Europarl version 7 (Koehn 2005) and News Commentary (Tiedemann 2012), containing news text and commentaries from the Project Syndicate. Two datasets were used for development (for tuning weights), the EMEA and Subtitles corpora, while a held-out Newstest corpus was used for testing (evaluation).

3.2 Self-Organizing Map as Clustering Device

Self-Organizing Maps are created with an unsupervised algorithm that effectively visualizes underlying structure of complex data in a 2D environment, in that respect a dimensionality reduction. Pioneering research into computer simulations of self-organizing systems was conducted at the Helsinki University of Technology (Kohonen 1982) with Kohonen et al. (1996) employing it to self-organize usenet (newsgroup) data, Kohonen et al. (2000) to cluster patent data, and Lagus et al. (2004) the Encyclopedia Britannica.

The underlying idea has its basis in neuroscience and assumes that unordered inputs can be mapped to a "topological order" through a self-organizing process. The process starts with an array of randomly initialized nodes according to some topology, each node represented by a vector of the same dimension as the training data. During training, input samples are compared to the vectors, and the node whose vector is closest to the input is considered winner (Best-Matching Unit). The winner's weights, as well as those of the neighboring nodes, are updated to be more similar to the input sample. The degree to which the weights are updated, both for the winning node and its neighborhood, is parameterized. The process is repeated for a (often pre-defined) number of iterations. As the iterations progress, the learning rate (the degree at which the nodes update) diminishes according to a parameterized function. The resulting Self-Organizing Map gives insights into the relations between the input samples: similar samples should attach to adjacent nodes, hence *self-organizing* into the same area.

The areas in the Self-Organizing Map are not delimited. In order to transform them into actual clusters, it is necessary to perform clustering on the map, represented as a grid of vectors. To cluster these vectors, a *metric* to determine similarity between clusters is chosen, alongside a *method* to select which data points the distances are measured between. According to these two facilities, the nodes are merged successively from the bottom up. Some clustering methods compute distances between nodes based on raw observations (node vectors), but more commonly a distance metrics, such as Euclidean or Chebychev distance, is used to construe a *Distance* Matrix containing the distances between nodes in the grid. Based on these distances, a *Linkage* Matrix is computed, that contains the binary links between nodes. This can be represented with a dendrogram.

Clustering Methods. Hierarchical agglomerative clustering algorithms are algorithms that successively merge clusters. Whereas a top-down approach needs metrics to split clusters, a bottom-up approach needs criteria to merge clusters. The number of clusters that come out as a result from the whole clustering procedure is determined by choosing the height of the tree. At its minimum it is only one cluster, and the maximum is the number of nodes in the tree; in our experiments the number of nodes in the Self-Organizing Map.

The number of clusters can either be specified directly, or chosen according to some criteria. There is no analytical way to determine the best number of clusters (ultimately it depends on the task clustering for). Salvador and Chan (2004) present several methods to determine the number of clusters automatically, such as information theoretic methods that quantify the fit of cluster members to their centroids and optimize the number of clusters on this, or the elbow/knee method, that finds the point before merging two clusters that will add the most to the distance between clusters (i. e., the point where the marginal gain of adding another cluster is the lowest). Determining the second derivative of the points is one way to find the knee, and we have used this method in our experiments.

The clustering method, that is, the algorithm that measures the distance between two clusters, is a choice of what points to compare distances between. The methods that have been explored in this work are *single-linkage, average distance, complete linkage, weighted linkage, Ward clustering, centroid clustering,* and *median.* In single-linkage clustering, the distance between two clusters is determined as the distance between the two points in the clusters that are closest. On the other extreme, complete linkage uses the maximum distance between two points in the clusters, and the average method the average distance between the points in the two clusters. The weighted method takes into account the children of a cluster as well, using the average of the distance between the two clusters that formed cluster A and B as the distance between the two.

Evaluation. Evaluating the unsupervised clustering process is two-fold; first the clustering can be evaluated *intrinsically* according to properties of the mathematical relation between the nodes, or *extrinsically* according to the performance gain or loss on the task for which the clustering was done. For extrinsic evaluation, we applied the corpora resulting from the above clustering method and created Language Models from them; models that were then used as auxiliary LMs in a Statistical Machine Translation system.

For intrinsic evaluation we have used the Silhouette Coefficient, which is a measure of the clusters' separation and how compact they are. Following Han (2005, pp. 489–490), the Silhouette is calculated by first determining $a(o)$, the average distance between an object o $[o \in C_i (1 \leq i \leq k)]$ and all other objects in the cluster C to which o belongs: $a(o) = \sum_{o' \in C_i, o \neq o'} \text{dist}(o, o')/(|C_i| - 1)$, and then $b(o)$, the minimum average distance from the object to all other clusters to which it does *not* belong: $b(o) = \min_{C_j : 1 \leq j \leq k; j \neq i} \sum_{o' \in C_j} \text{dist}(o, o')/|C_j|$. Then the Silhouette Coefficient is determined by

$$s(o) = \frac{b(o) - a(o)}{max\{a(o), b(o)\}} \tag{1}$$

The value of the Silhouette Coefficient will range between $[0, 1]$. A higher number indicates a better clustering. The term $a(o)$ is an indication of *compactness*. A lower value indicates a more compact cluster, i. e., low distance between the members inside a cluster, whereas $b(o)$ describes the separation from other clusters. The larger it is, the more separated the cluster is from the others. This means that when the Silhouette Coefficient reaches 1, the clusters are compact [low $a(o)$] and far away from the other clusters [high $b(o)$]. It would also be possible to optimize the number of clusters based on this score.

4   Domain Adapted Machine Translation

The strategy chosen for domain adapted Machine Translation is to first segment large corpora with a Self-Organizing Map and then utilize language models built on the basis of these corpus segments in a Statistical Machine Translation system. The first phase is conducted off-line, whereas the employment of the language models is done on-line, while decoding an SMT model given input sentences.

When a document is input for translation, it is matched against the n created Language Models, ranked after perplexity. The Moses SMT system (Koehn et al. 2007) allows for the use of user-defined features in its log-linear model. The LM with the lowest perplexity is selected by a new Moses feature providing additional information for the decoder. This setup creates a platform in which a system can do adaption to multiple domains, as the additional feature in the SMT decoding phase can select the most appropriate auxiliary LM on-the-fly. The Language Models where built with the KENLM toolkit (Heafield 2011), which was also used to read the LMs when decoding the SMT models.

4.1   Clustering Experiments

A quadratic layout of nodes in a Self-Organizing Map was used in the experiments, and for each sample the comparisons to all the different vectors in the node were done in parallel, as was the updating of the nodes in the next stage. Each SOM was configured with a separate configuration file, where the number of iterations, the initial size of the neighborhood (given as a radius in Euclidean space), and the initial learning rate were specified. In this file, details about the vectorization of the document collections could also be entered.

After a pre-defined maximum number of iterations was reached, the last run of the input samples was kept, which left similar samples with the same winning node. An agglomerative clustering algorithm was then run on the nodes to cluster them by similarity. The algorithm chose a cut-off point in the clustering according to the relative change in the similarity measure, as described above.

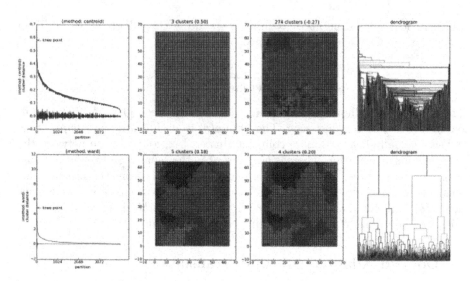

Fig. 1. Hierarchical clustering (centroid and Ward) for a 64 × 64 SOM.

In Fig. 1, the detailed clustering of the SdeWac corpus using the first and second knee-points is shown. The knee-points are in the first column, the resulting clusterings in the 2nd and 3rd columns, and the corresponding dendrogram in the 4th. The two methods *centroid* and *Ward* exemplified here illustrate the differences in results between methods. Ward's method generally resulted in more evenly sized clusters and was used as a basis for the further experiments. The color codings come from Principal Component Analysis (PCA) decomposition of the node vectors into four dimensions. These plots do not, however, show the size of the clusters as some nodes will contain more samples than others. This can be illustrated with heat maps, but those are left out here for space reasons.

The creation of Self-Organizing Maps is subject to many variables, such as grid size, vectorization methods, and distance metrics. We conducted several clustering experiments on the SdeWac corpus with different SOM sizes, summarized in Table 1. The table shows the Silhouette Coefficients for the partitionings, with the number of clusters in each in parentheses. The two sets of experiments come from two different vectorizations of the SdeWac data. In the first

Table 1. Silhouette scores for corpora clustered with SOM.

Grid size	8 × 8	16 × 16	32 × 32	64 × 64
1st Part. (#)	0.33 (2)	0.46 (2)	0.35 (2)	0.31 (2)
2nd Part. (#)	0.23 (4)	0.31 (10)	0.31 (4)	0.29 (4)
1st Part. (#)	0.39 (2)	0.41 (7)	0.37 (3)	0.33 (2)
2nd Part. (#)	0.30 (3)	0.34 (11)	0.32 (8)	0.32 (4)

experiments, the TF-IDF for the n-grams up to 7 were computed with a fixed vocabulary from the entire corpus, but on a subset of 1,000 documents (dimensionality 157,053). In the second set of experiments, the vectorization process had the same cut-off points, but the n-grams were computed only from that part of the corpus (dimensionality 53,016).

We then went on to use one of these experiments for Domain Adaptation, presented in the next section.

4.2 Using Self-Organized Clusters for Domain Adaptation

A Statistical Machine Translation model consists of a Translation Model and a Language Model, that are decoded to provide the optimal Target Language string given the input. The term *auxiliary Language Model* denotes an additional LM used to aid this decoding. We have shown above how the text collections are divided into sub-corpora, through a Self-Organizing Map and an Agglomerative Clustering procedure. A selection method among the sub-corpora is necessary in order to test our hypothesis that building an Language Model over the most relevant sub-corpus will give nearly as good results as using the entire text.

With training data available in the target domain, the perplexity of the Target Language version of the development set can be used to rank the auxiliary LMs. In the absence of such data, a two-pass solution is possible, i. e., that a document is translated with the baseline Machine Translation first, and the perplexity of this translation is used to rank the LMs. This is the method we have applied in this work. Other possible selection criteria include matching a vectorization of the input document against the Self-Organizing Map used as a basis for clustering, using the cluster to whose SOM node it matched at closest. Such matching can be done on sentence, document or collection level.

An extra Language Model can be included in a Machine Translation system in a number of ways; concatenating the text corpus with the original LM, using an additional LM in Moses' configuration file or interpolating Language Models directly. When using such features, their individual weights can be optimized using the Minimum Error Rate Training (Bertoldi et al. 2009) training procedure. When using prepared datasets for Domain Adaptation research, it is reasonable to treat the entire sub-part of the corpus belonging to the same text type as one, a given premise in how they are developed. Then all sentences can be tested in the SMT system against the auxiliary LM found most relevant. However, we also developed a separate Moses feature that uses a specific LM for each input sentence, depending on which of them was closest according the selection criterion mentioned above. Concretely, this information was looked up in a file that was compiled from labeling individual sentences with the most relevant sub-corpus (chosen as described in the previous paragraph).

4.3 Domain Adaptation Experiments

As a baseline system, we used the data presented in Sect. 3. An SMT model was created from parliamentary text, and three domain specific corpora were used

Table 2. Perplexity scores including and excluding OOV words. Low scores in bold.

LM	EMEA	Subs	News		LM	EMEA	Subs	News
1	**25,269**	281	893		1	12,501	182	470
2	41,386	374	1030		2	13,764	242	510
3	25,494	**261**	**744**		3	12,566	**167**	**406**
4	30,320	556	1030		4	**11,260**	331	494

| (a) including | (b) excluding |

Table 3. Domain Adaptation results on English-German text across different LM configurations. Highest score in bold, LM with lowest perplexity in italics.

LM	BLEU	Meteor		LM	BLEU	Meteor		LM	BLEU	Meteor
LM1	*19.28*	*0.3492*		LM1	12.85	0.3332		LM1	**10.53**	0.2704
LM2	17.97	0.3358		LM2	12.47	0.3366		LM2	9.93	0.2646
LM3	18.55	0.3418		LM3	*13.05*	*0.3367*		LM3	*10.01*	*0.2670*
LM4	18.43	0.3420		LM4	12.14	0.3304		LM4	8.88	0.2551
No Aux.	16.82	0.3347		No Aux.	11.29	0.3236		No Aux.	8.48	0.2563
Total	**19.29**	0.3479		Total	**13.09**	0.3366		Total	10.47	0.2668

| (a) EMEA | (b) News | (c) Subtitles |

for testing: medical text (*EMEA*), TV subtitles text, and a news corpus. The auxiliary LMs came from building a SOM trained on the SdeWac corpus that after Hierarchical Agglomerative Clustering with the Ward method resulted in four corpora, from which Language Models were built. In order to test which LM was the most relevant for each of the input documents, perplexity was measured on a) the translation of each test corpus provided by the parallel corpora and b) the translation of the test set provided by the generic SMT model.

The Ward clustering, that is, the number of clusters from the dendrogram, from the first partitioning (knee-point) was used in further experiments into Domain Adaptation. This meant three clusters for this data, and accordingly three text corpora and Language Models. The next step was to measure the perplexity of the three *in-domain* corpora, to determine which auxiliary LM was closest. The results are presented in Table 2. The *EMEA* corpus selected LM1, whereas the *Subs* and the *News* corpora selected LM3 according to this criterion.

In Table 3 the results from using these Language Models in the full Machine Translation pipeline are shown, using BLEU and Meteor evaluation. For the EMEA and News corpora, the LM with the lowest perplexity also gave the highest gain in MT performance of the auxiliary LMs, for the Subtitles corpus it came second. The results are also compared to not using any auxiliary LM and using all the text to build one large Language Model.

5   Discussion and Conclusion

The work presented here introduces a solution to the Domain Adaptation problem that looks for external sources to increase the available training data rather than exploiting some limited source of *in-domain data in refined ways*.

The purpose of the Language Model in a Machine Translation pipeline is to provide *fluency*, by selecting translation candidates that are likely members of the target language. The famous Firthian truism *"You shall know a word by the company it keeps"* speaks to understanding a word from its context. By mapping input documents to close-matching Language Models, we try to provide this context for arbitrary input documents.

However, it is not always obvious how to distinguish text domains from each other, where to draw the line between them and according to what dimensions (such as writing style, topic, author or target age groups) to separate them. An unsupervised approach to segmentation of a vast data source is a way of enabling a Machine Translation system to respond to various input domains also along such dimensions. The proposed Domain Adaptation method includes a step to establish which of the auxiliary Language Models is the most relevant for the given input document, and thereby enabling context adapted simultaneous translation of documents of different types.

The strategy builds on a Self-Organizing Map (SOM) approach to finding relations within unorganized data, a strategy which has also been successfully utilized in other application areas. Results indicate that it is possible to attain improvements equal to—or even better than—those stemming from adding the whole available text collection by using the most relevant part, also by unsupervised methods.

More research is necessary to establish the feasibility of the Self-Organizing Map to extracting relations useful for extra training material for Statistical Machine Translation models, notably into self-organizing parallel corpora.

References

Axelrod, A., He, X., Gao, J.: Domain adaptation via pseudo in-domain data selection. In: Proceedings of the Conference on Empirical Methods in Natural Language Processing, EMNLP 2011, Edinburgh, UK, pp. 355–362. Association for Computational Linguistics (2011)

Bellegarda, J.R.: Statistical language model adaptation: review and perspectives. Speech Commun. **42**, 93–108 (2004)

Bertoldi, N., Haddow, B., Fouet, J.-B.: Improved minimum error rate training in moses. Prague Bull. Math. Linguist. **91**, 7–16 (2009)

Brown, P.F., Cocke, J., Pietra, S.A.D., Pietra, V.J.D., Jelinek, F., Lafferty, J.D., Mercer, R.L., Roossin, P.S.: A statistical approach to machine translation. Comput. Linguist. **16**(2), 79–85 (1990)

Bungum, L., Gambäck, B.: A survey of domain adaptation in machine translation: towards a refinement of domain space. In: Proceedings of the India-Norway Workshop on Web Concepts and Technologies, Trondheim, Norway. Tapir Academic Press (2011)

Carpuat, M., Wu, D.: How phrase sense disambiguation outperforms word sense disambiguation for statistical machine translation. In: Proceedings of the 11th Conference on Theoretical and Methodological Issues in Machine Translation, pp. 43–52, September 2007

Carpuat, M., III, H.D., Fraser, A., Quirk, C., Braune, F., Clifton, A., Irvine, A., Jagarlamudi, J., Morgan, J., Razmara, M., Tamchyna, A., Henry, K., Rudinger, R.: Domain adaptation in machine translation: final report. In: 2012 Johns Hopkins Summer Workshop Final Report. Johns Hopkins University (2012)

Eck, M., Vogel, S., Waibel, A.: Language model adaptation for statistical machine translation based on information retrieval. In: Proceedings of the 4th International Conference on Language Resources and Evaluation, Lisbon, Portugal, May 2004, pp. 327–330. ELRA (2004)

Faaß, G., Eckart, K.: SdeWaC - a corpus of parsable sentences from the web. In: Biemann, C., Zesch, T., Gurevych, I. (eds.) GSCL 2013. LNCS, vol. 8105, pp. 61–68. Springer, Heidelberg (2013)

Han, J.: Data Mining: Concepts and Techniques. Morgan Kaufmann Publishers Inc., San Francisco (2005)

Heafield, K.: KenLM: faster and smaller language model queries. In: Proceedings of the Sixth Workshop on Statistical Machine Translation, Edinburgh, Scotland, July 2011, pp. 187–197. ACL (2011)

Hoang, H., Birch, A., Callison-burch, C., Zens, R., Aachen, R., Constantin, A., Federico, M., Bertoldi, N., Dyer, C., Cowan, B., Shen, W., Moran, C., Bojar, O.: Moses: open source toolkit for statistical machine translation. In: ACL, Prague, Czech Republic, June 2007, pp. 177–180. Association for Computational Linguistics (2007)

Iyer, R., Ostendorf, M., Gish, H.: Using out-of-domain data to improve in-domain language models. IEEE Signal Process. Lett. 4, 221–223 (1997)

Jiang, J.: Domain adaptation in natural language processing. University of Illinois at Urbana-Champaign (2008)

Karlgren, J.: Sublanguages and registers - a note on terminology. Interact. Comput. 5(3), 348–350 (1993)

Kay, M.: The proper place of men and machines in language translation. Technical Report CSL-80-11, Xerox Palo Alto Research Center, Palo Alto, California (1980)

Kittredge, R., Lehrberger, J. (eds.): Sublanguage: Studies of Language in Restricted Semantic Domains. W. de Gruyter, Berlin, New York (1982)

Koehn, P., Hoang, H., Birch, A., Callison-Burch, C., Federico, M., Bertoldi, N., Cowan, B., Shen, W., Moran, C., Zens, R., Dyer, C., Bojar, O., Constantin, A., Herbst, E.: Moses: open source toolkit for statistical machine translation. In: Proceedings of the 45th Annual Meeting of the ACL on Interactive Poster and Demonstration Sessions, ACL 2007, Stroudsburg, PA, USA, pp. 177–180. Association for Computational Linguistics (2007)

Koehn, P.: Europarl: a parallel corpus for statistical machine translation. In: Conference Proceedings: The Tenth Machine Translation Summit, Phuket, Thailand, pp. 79–86. AAMT (2005)

Kohonen, T., Kaski, S., Lagus, K., Honkela, T.: Very large two-level SOM for the browsing of newsgroups. In: von der Malsburg, C., von Seelen, W., Vorbrüggen, J.C., Sendhoff, B. (eds.) Artificial Neural Networks — ICANN 96. LNCS, vol. 1112, pp. 269–274. Springer, Berlin (1996)

Kohonen, T., Kaski, S., Lagus, K., Salojrvi, J., Paatero, V., Saarela, A.: Organization of a massive document collection. IEEE Trans. Neural Netw. Spec. Issue Neural Netw. Data Min. Knowl. Discov. 11(3), 574–585 (2000)

Kohonen, T.: Self-organized formation of topologically correct feature maps. Biol. Cybern. **43**(1), 59–69 (1982)

Lagus, K., Kaski, S., Kohonen, T.: Mining massive document collections by the WEB-SOM method. Inf. Sci. **163**(1–3), 135–156 (2004)

Levenshtein, V.I.: Binary codes capable of correcting deletions, insertions and reversals. Sov. Phys. Dokl. **10**(8), 707–710 (1966)

Louis, A., Webber, B.: Structured and unstructured cache models for SMT domain adaptation. In: Shuly Wintner, I., Stefan Riezler, G., Sharon Goldwater, U., (eds.) Proceedings of the 14th Conference of the European Chapter of the Association for Computational Linguistics, vol. 2: Short Papers, Gothenburg, Sweden, April 2014. Association for Computational Linguistics (2014)

Lu, Y., Huang, J., Liu, Q.: Improving statistical machine translation performance by training data selection and optimization. In: Proceedings of the 2007 Joint Conference on Empirical Methods in Natural Language Processing and Computational Natural Language Learning (EMNLP-CoNLL), Prague, Czech Republic, June 2007, pp. 343–350. Association for Computational Linguistics (2007)

Mahajan, M., Beeferman, D., Huang, X.D.: Improved topic-dependent language modeling using information retrieval techniques. In: ICASSP (1999)

Moore, R.C., Lewis, W.: Intelligent selection of language model training data. In: Proceedings of the 48th Annual Meeting of the Association for Computational Linguistics, volume Short papers, Uppsala, Sweden, July 2010, pp. 220–224. ACL (2010)

Pedregosa, F., Varoquaux, G., Gramfort, A., Michel, V., Thirion, B., Grisel, O., Blondel, M., Prettenhofer, P., Weiss, R., Dubourg, V., Vanderplas, J., Passos, A., Cournapeau, D., Brucher, M., Perrot, M., Duchesnay, É.: Scikit-learn: machine learning in Python. J. Mach. Learn. Res. **12**(1), 2825–2830 (2011)

Plank, B.: Domain adaptation for parsing. Ph.D. thesis, University of Groningen (2011)

Price, P.J.: Evaluation of spoken language systems: the ATIS domain. In: Proceedings of the Workshop on Speech and Natural Language, HLT 1990, Stroudsburg, PA, USA, pp. 91–95. Association for Computational Linguistics (1990)

Rosenfeld, R.: A maximum entropy approach to adaptive statistical language modeling. Comput. Speech Lang. **10**, 187–228 (1996)

Rosenfeld, R.: Two decades of statistical language modeling: where do we go from here. Proc. IEEE **88**, 1270–1278 (2000)

Salvador, S., Chan, P.: Determining the number of clusters/segments in hierarchical clustering/segmentation algorithms. In: ICTAI, pp. 576–584. IEEE Computer Society (2004)

Sennrich, R., Schwenk, H., Aransa, W.: A multi-domain translation model framework for statistical machine translation. In: ACL (1), pp. 832–840. The Association for Computer Linguistics (2013)

Sennrich, R.: Combining multi-engine machine translation and online learning through dynamic phrase tables. In: EAMT 2011: The 15th Annual Conference of the European Association for Machine Translation, Leuven, Belgium, May 2011. European Association for Machine Translation (2011)

Tiedemann, J.: Parallel data, tools and interfaces in OPUS. In: Chair, N.C.C., Choukri, K., Declerck, T., Dogan, M.U., Maegaard, B., Mariani, J., Odijk, J., Piperidis, S. (eds.) Proceedings of the Eight International Conference on Language Resources and Evaluation (LREC 2012), Istanbul, Turkey, May 2012. European Language Resources Association (ELRA) (2012)

Wang, L., Wong, D.F., Chao, L.S., Lu, Y., Xing, J.: A systematic comparison of data selection criteria for smt domain adaptation. Sc. World J. **2014**(1) (2014)

Identifying Context Information in Datasets

Georgia M. Kapitsaki$^{(\boxtimes)}$, Giouliana Kalaitzidou, Christos Mettouris,
Achilleas P. Achilleos, and George A. Papadopoulos

Department of Computer Science, University of Cyprus,
1 University Avenue, Nicosia, Cyprus
{gkapi,gkalai01,mettour,achilleas,george}@cs.ucy.ac.cy

Abstract. Datasets are used in various applications assisting in performing reasoning and grouping actions on available data (e.g., clustering, classification, recommendations). Such sources of information may contain aspects relevant to context. In order to use to the fullest this context and draw useful conclusions, it is vital to have intelligent techniques that understand which portions of the dataset are relevant to context and what kind of context they represent. In this work we address the above issue by proposing a context extraction technique from existing datasets. We present a process that maps the given data of a dataset to a specific context concept. The prototype of our work is evaluated through an initial collection of datasets collected from various online sources.

Keywords: Context extraction · Dataset · Context matchmaking

1 Introduction

Context-awareness is an area that has gained tremendous interest from the research community in the latest years targeting in many cases pervasive and mobile computing systems. Adaptive, personalized services that take into account location as well as other user-related data predicate the existence of well-formed information with respect to users environment, referred to as contextual information or context. Although context is in many cases mapped to location information many definitions of context that include its different aspects apart from location, such as weather conditions, user profile information, device and netwotk connectivity, can be found in the literature [8,20]. However, the most popular definition is given by Dey and Abowd [1]: *Context is any information that can be used to characterize the situation of an entity. An entity is a person, place, or object that is considered relevant to the interaction between a user and an application, including the user and applications themselves.* The techniques that enable the exploitation of contextual information are generally known as *context-handling* techniques, while the use of context to provide relevant information and/or services to the user, where relevancy depends on the users task, is known as *context-awareness*.

Various applications, where context is utilized for various purposes, can be found and these are mainly mobile applications and the recently emerged

© Springer International Publishing Switzerland 2015
H. Christiansen et al. (Eds.): CONTEXT 2015, LNAI 9405, pp. 214–225, 2015.
DOI: 10.1007/978-3-319-25591-0_16

Context-Aware Recommender Systems (CARS) that utilize context-information to provide better recommendations to end-users [4]. Context plays also a vital role in mobile context-aware applications for users on-the-move, where user surroundings and current activities are used for offering a personalized experience to users [16].

In order to be able to utilize effectively the available context information for given applications, context identification is necessary. in the framework of this work we define context identification as *the process of identifying which information constitutes context information and which not*. This process should also include more information on the kind of context addressed, i.e., one should also indicate whether specific context information refers to location data, data connected to the user, data connected to hardware devices used, etc.

In this work we address the above by introducing a process that assists in making sense of context data hidden in given datasets. Specifically, we propose a context extraction technique from an existing dataset as input source. We present a process that classifies the existing information in a given dataset as a specific context concept. The context concepts are based on a context taxonomy that we introduce for this purpose, although any context model can be used instead [6]. The prototype of our work is evaluated through an initial set of datasets containing various files that we have collected from online sources and research works, such as the one used in the multi-agent system for the care of elderly people living at home on their own [13]. This evaluation serves as proof-of-concept for the usefulness and the effectiveness of the proposed approach.

To the best of our knowledge, context extraction from datasets is a process that has not been adequately studied before. We are currently handling this task in the following way: appropriate string comparisons are performed on the dataset feature names with a context taxonomy as a reference, arguing in this manner whether a particular feature could be used as a context element by a context-aware application or not. At a second phase, a similar process is followed for the feature values, but in this case concepts with a wider sense are compared against the feature values using a lexical database that contains information on such relations between words. Although our approach is simple, we argue that it can serve as a first step towards context information extraction from datasets that can potentially enhance context-aware applications, such as Context-Aware Recommender Systems in incorporating the extracted context in their recommendation method. Identifying which information constitutes context can be a useful asset for making better use of the available data.

The rest of the paper is structured as follows. Section 2 presents the area of context modeling in recommeder systems giving at the same time a brief overview of related work on context modeling and identification. Section 3 presents our main contribution and the extraction steps proposed along with implementation details. Section 4 is dedicated to the presentation of the evaluation of our work using online datasets and to a discussion on the obtained results. Finally, Sect. 5 concludes the paper.

2   Motivation and Related Work

Recommender systems use a variety of filtering techniques and recommendation methods to provide personalized recommendations to their users. The information used is mostly retrieved from the user profile, from user's usage history, as well as from item-related information. However, these traditional recommender systems use limited or none contextual information to produce recommendations. Instead, they only focus on two dimensions: the user and the items (also called two-dimensional recommenders: *Users* and *Items* are used in order to produce *Ratings*), excluding other contextual data that could be used in the recommendation process, such as the day/time, with whom the user is with, weather conditions, etc. A typical dataset of such recommender systems includes information on the user (user ID), information on the item (item ID, item features, item price, availability, etc.) and ratings of users on items. Datasets that do not include context information are known to be two dimensional.

On the contrary, Context-Aware Recommender Systems focus on using contextual information to enhance recommendations combining *Users*, *Items* and *Context* to construct *Ratings* [3,4]. The goal is to enhance their datasets with context information so that CARS produce better, enhanced and more personalized recommendations. Context information was first utilized into the recommendation process by Adomavicius et al. by proposing three approaches: the Pre-filtering approach, the Post-filtering approach and the Multidimensional Contextual Modeling approach [3].

Based on the research work of Adomavicius and Tuzhilin [4], context can be used in two ways for producing recommendations, i) the *Recommendation via Context-Driven querying and search*, where systems use contextual information from the environment (e.g., location), the user (user profile) and the system to retrieve the most relevant items to recommend (ubiquitous and location-based systems such as systems that recommend restaurants and POIs (Points of Interest) in the user's proximity), and ii) the *Recommendation via Contextual preference elicitation and estimation* [2], where systems focus on modeling user preferences by using various methods, e.g., observing the user while interacting with a system or by receiving appropriate feedback from the user regarding the recommendations. In this paper we are dealing with the second method of incorporating the context for producing recommendations, which is used by Context-Aware Recommender Systems. CARS do not use two-dimensional datasets as with traditional recommenders; rather they face the challenge of utilizing multidimensional, context enriched datasets that include additional contextual dimensions besides 'users' and 'items [2]. For more information on the two ways for producing recommendations the reader may refer to previous works [2,3].

Based on the above, we argue that including and recognizing possible context elements in a given dataset to be later utilized in the recommendation process in order to produce multidimensional, context-aware recommendations by Context-Aware Recommender Systems is a useful and important process. Any Context-Aware Recommender System that uses datasets in combination with sophisticated recommendation methods such as those met in traditional recommender systems to produce recommendations should be able to produce better results in cases where the datasets used are enriched with context information. The validity of the above statement is supported by the fact that Adomavicius and Tuzhilin were the first to prove that using contextual information in CARS (from context enriched datasets) indeed enhances the recommendation process [2,3] incorporating. This observation is also validated and supported by numerous works in the CARS research literature [2,4,5,10,15,21,22].

Moreover, another class of systems to be benefited by context-enriched datasets is the Ubiquitous Context-Aware Recommender Systems class (Ubi-CARS) [18]. UbiCARS constitute a subset of Ubiquitous Recommender Systems [18] and utilize both ways of using context mentioned above: *Recommendation via Context-Driven querying and search* to enable the provision of recommendations on location via mobile devices (e.g., identification of near-by products), as well as *Recommendation via Contextual preference elicitation and estimation* by using context-enriched (multidimensional) datasets and context-aware recommendation techniques and methods as CARS do. UbiCARS systems can be used for in-situ products recommendations and will potentially be able to provide better recommendations than common Ubiquitous Recommender Systems, since, besides utilizing the surrounding context as Ubiquitous Recommender Systems do, they also consider the multidimensional context enriched datasets (as used by CARS) in their recommendation process.

Many other related works on context have addressed the issue of context modeling with the main motivation of using context in specific applications. Context modeling is relevant also in the framework of the current work, since it can provide the structure for representing the extracted context information [6]. Since we are focusing on CARS in this work, we are not presenting in detail context modeling techniques from other domains.

3   The Context Extraction Process

3.1   Analysis Steps

The proposed context extraction process is shown in Fig. 1. The elements of a given dataset are indicated as features with a given name and value using the terminology of machine learning. The usual case is for the first row in a separate dataset file to contain the feature names with the following rows containing the records with specific values for each feature.

The context identification process is divided into two distinct phases that operate on different level on the dataset files:

– *Phase 1 - Feature names matcher*: The feature matchmaking phase classifies a specific feature in the dataset as either a context or non-context value. This phase also specifies the context category the feature belongs to based on the introduced context taxonomy (e.g., location, user, etc.).
– *Phase 2 - Feature values matcher*: Feature values are examined in this phase. These values are also classified as context or non-context with an indication of the context category.

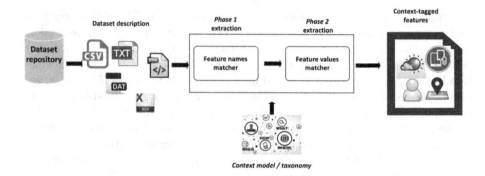

Fig. 1. The proposed context identification process.

The feature names (i.e., column names in the dataset files) correspond to the main terms whose values are contained in the dataset. These names may correspond to context-relevant elements. For the identification of such context elements and for matching purposes different string matching algorithms can be exploited. In our work the following string matching algorithms are used in the first phase of feature name matching:

- The Jaro Winkler string distance calculation algorithm [9].
- Our WordNet distance similarity algorithm introduced in a previous work [14]. Wordnet is a lexical database that retrieves similar concepts to the input word given [23]. The algorithm considers the type of connection between the examined terms. Connections of the type of same words (same), synonyms (syn), meronyms (mer), hypernyms (hyper) and related terms (rel) are considered in the following equation:

$$\sigma(n_x, n_y) = l \times \sigma_{same}(n_x, n_y) + p \times \sigma_{syn}(n_x, n_y) + q \times \sigma_{mer}(n_x, n_y)$$

$$+ r \times \sigma_{hyper}(n_x, n_y) + t \times \sigma_{rel}(n_x, n_y)$$

The constants l, p, q, r and t express the importance of each similarity level retrieved through Wordnet. At most one of the operands in the similarity calculation will evaluate to 1.0. We have used the following values for each weight: l=1.0, p=0.7, q=0.2, r=0.2, t=0.0. Using these values 1.0 is returned only if the terms compared are exactly the same. If the terms are synonyms, then a similarity score of 0.7 is assigned.

Other algorithms that could ba considered in phase 1 can be found in the WordNet similarity algorithm from the xssm[1] (XML Schema Similarity Mapping) library in Java that returns a score between terms using a preprocessed WordNet and corpus data, and similarity based on n-grams [24].

Regarding the second phase of feature value matchmaking is performed on a different level than the first phase. Instead of using string similarity algorithms,

---

[1] https://code.google.com/p/xssm/.

the connection of terms in WordNet is exploited. Specifically, the hypernyms of a given term are examined for potential matching to a specific term of the context taxonomy. If an adequate hypernym is found for the feature values, then the respective feature name is considered context-relevant. For instance, if different city names (e.g., Athens, London, New York, Moscow) are indicated as feature values, these can be matched to the same hypernym synset in WordNet *town* giving an indication of a feature representing location information. Since hypernyms can reach terms in different levels, e.g., in WordNet *town* is a hypernym of *Athens* and *municipality* is a hypernym of *town*; hence, *municipality* is connected with *Athens* through 2 levels, we have selected an appropriate value for level. Based on the conducted experiments, 2 was chosen as a plausible value.

Note that this second phase considers only the first 100 entries in the dataset file. Since many datasets contain huge numbers of entries (100,000 or more), we have observed that this is an appropriate number of entries for drawing useful conclusions. Examining more rows would only add to the processing time without improving the results of the process.

Conclusions for the final characterization of a term as context-relevant or not are drawn by combining the results of the two distinct phases. Specifically, in the first phase a matching is considered succesful if the similarity algorithm returns a value higher than 0.8, whereas in the second phase this is considered for cases, where an exact hypernym of level 2 (or lower) is found. However, the significance of each phase is not the same. Phase 2 provides less accuracy, since as we observed in many dataset files only number indications are given in feature values (e.g., user or item ID numbers, year expressed in a number etc.). These cannot assist in drawing conclusions on the meaning of the values and in such cases (i.e., feature values in numeric format) phase 2 is not applied on the dataset files. Also in cases, where features values are in text format, the results of phase 1 are considered more relevant for the final results using the following weights:

$$\sigma(n_x, n_y) = 0.8 \times \sigma_{phase1}(n_x, n_y) + 0.2 \times \sigma_{phase2}(n_x, n_y)$$

Note also that important information can be found in the README files of the dataset descriptions. These are, however, not considered in the current state of our work due to the large heterogeneity of such files.

3.2   Context Model

As aforementioned a variety of context models tailored to specific domains can be found in the literature [6]. The context categorization employed in the current work has resulted from our study conducted on the state-of-the-art on context-aware systems including context-aware ubiquitous and location-aware systems, as well as context-aware recommender systems. This context categorization includes the most important context elements we have retrieved during our research. For the resulting context model captured in a taxonomy of 3 levels (main context category, subcategories, and subcategory items) in Table 1 we have used the system database from our previous work [17], as well as related literature on context models [2,4,5,10,15,21,22]. The system database of the CARS

Table 1. Context taxonomy introduced.

Context category	Context subcategory	Context terms in subcategory
User	Profile	Name, age, gender, companion, marital status, children, research interest, social role, expertise, goal, experience, employment status, education preferences, contacts, payment info
	Activity	People nearby
Environment	Weather	Temperature, humidity, Celsius, Fahrenheit, rain possibility
	Other	Season, lighting, noise level, traffic conditions
Time		Date, year, month, day, hours, minutes, seconds, timestamp, timezone
Location	Address	Street, road, city, town, municipality, prefecture, country, post code
	GPS	GPS coordinates, Latitude, longitude
System		Battery level, computing platform, bandwidth, network connectivity, communication cost, nearby resources

Context Modeling System [17] includes context models presented and used by research works in the field of context-aware recommender systems, as well as context models built by developers and experts on context-aware development at the university premises.

Please note that our aim was to build a context categorization that would be generic enough to facilitate a wide range of context-aware applications, and that our system is developed so that it can use other context categorizations as well, provided that these are given in the appropriate format.

3.3   Implementation Tools

The proposed context extraction process has been implemented in Java with the assistance of different libraries: e.g., Apache Commond CSV[2] for the parsing of the CSV files and the JWNL[3] Java WordNet Library. Appropriate implemenations of the aforementioned string matching algorithms in Java were also used.

4   Evaluation and Discussion

4.1   Testing Set and Experiments

We have collected a number of datasets from various web sources. Each dataset is composed of one or more CSV files, whereas README files with more

---

[2] https://commons.apache.org/proper/commons-csv//.
[3] http://sourceforge.net/projects/jwordnet/.

Table 2. Datasets employed in the evaluation.

Dataset #	Application use	Number of dataset files	Source	Number of context features
1	Activity Recognition in Home Setting	4	[19]	1
2	Activity Book recommendations	3	[28]	3
3	Travel recommendations	1	[27]	5
4	Social networks: Facebook	1	*online*[a]	3
5	Social networks: Delicious	7	[7]	17
6	Social networks: last.fm	6	[7]	9
7	Social networks: MovieLens	12	[7]	24
8	Microblog spamming detection	4	*online*[b]	8
9	Wireless sensor network	4	[25]	8
10	Climate data	1	[12]	3
11	Texting zone locations	1	*online*[c]	6

[a]https://github.com/ManuelB/facebook-recommender-demo/blob/master/src/main/resources/DemoFriendsLikes.csv
[b]https://archive.ics.uci.edu/ml/datasets/microblogPCU
[c]https://catalog.data.gov/dataset/texting-zone-locations

information on the dataset use are provided in some cases. Note that, as afore-mentioned, although these README files may contain useful information for the dataset, they are neglected in the current prototype implementation of the context extraction tool due to their diversity and use of free text language that renders uniform processing impossible.

The datasets and their domains are depicted in Table 2. Note that the major-ity of files in the datasets contains headers with the feature names used. For cases, where these headers were missing, they were added by our context extractor in order to facilitate the processing phases. All datasets used are available from previous works or can be found online as indicated in the table.

4.2   Main Results and Discussion

In order to measure the results of our approach we have used the following three metrics from the information retrieval field [11,26]:

$$precision = \frac{\#correctMatchesReturned}{\#totalCorrectMatches}$$

$$recall = \frac{\#correctMatchesReturned}{\#totalMatchesReturned}$$

$$f-measure = \frac{2 \times precision \times recall}{precision + recall}$$

Some feature names and some terms in the context categories consist of more than one words. For those cases each word is examined independently for matching. If a match for any of the words is found, then the feature name or the context term respectively is considered a match as a whole. Further study of the consideration of n-grams instead of unigrams could improve the matching results.

Note that a returned match is considered correct, if the correct context category is also indicated. If only the characterizaton as context is correct, then the result is not considered correct. We have measured the above for the following cases: *Consideration only of Phase 1 results*, and *Consideration of results from both Phases*. For each of the above cases the two aforementioned string similarity algorithms were employed for the first phase (i.e., Jaro Winkler distance, our custom WordNet distance similarity algorithm).

An example of values returned for the third examined dataset of *Travel recommendations* consisting of data from TripAdvisor is shown in Table 3. The results were the same for both examined algorithms and the second phase would not add any additional information on the existing observations: using the WordNet hypernyms only *hotel city* feature values were recognized as relevant to location. This feature contained values of cities in the United States, such as Houston, Los Angeles, Oklahoma City and Boston. This is a case of good context extraction, where all relevant terms have been tagged as context-relevant but only one term has been placed in a wrong category (i.e., *user timezone* has been placed under *user* instead of *time*). Note that the terms in parenthesis in the table correspond to the context category, if it exists.

The summary of results with average precision, recall and f-measure values for all datasets in the testing set are depicted in Fig. 2. These initial results indicate that the proposed approach assists in making sense of context data that may be included in the features of the dataset files. Recall and precision values reach 0.9 in many cases. Overall phase 2 does not improve the accuracy of the results, due to the rare cases of encountering feature values in text format. For this reason, we have also selected a small weight for the similarity score of phase 2. Concerning the accuracy of phase 1 better results are observed for the Jaro

Table 3. Context identification results for the Travel recommendations dataset.

Algorithm	Terms matched as context by our approach	# of terms matched as as context correctly	Context terms in dataset	Precision	Recall
Jaro-Wrinkler (Ph1,1+2), our WordNet (Ph11+2)	user(:user), user state(:user), user timezone(:user), hotel city(:location), hotel timezone(:time)	4	user(:user), user state(:user), user timezone(:time), hotel city(:location), hotel timezone(:time)	0.8	0.8

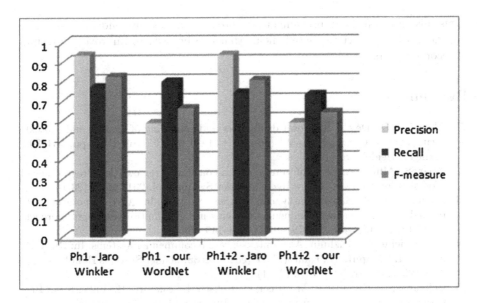

Fig. 2. Main evaluation results.

Winkler similarity algorithm that provides a higher number of terms macthed as context-relevant in comparison to the dedicated WordNet algorithm.

In general, the accuracy of the results is high, but even higher values could have been achieved. This is attributed mainly to the choice of names given to the features of datasets and also to errors in the spelling of the features names given by the dataset creators (e.g., *temperature* is spelled as *tepmrature* in the Wireless sensor network dataset). Since no common terminology exists, dataset creators are using terms that best suit their needs without conforming to any guidelines. A similar problem appears with the use of abbreviations (e.g., for state names of the United States). The utilization of the README file might assist in improving the matching results alleviating the above problems, as well as stemming or stopword removal preprocessing actions that were not employed in the framework of our work.

5 Conclusions

In this paper we have presented our work on context information identification from dataset files. We have defined a process of two phases for matching features in the datasets with context elements from a given context taxonomy. This extraction process can be a useful tool for context-aware application development and context-aware recommender systems, since it can point out context elements from huge amounts of data. We have also performed an initial evaluation of our process using a number of datasets from different application domains.

As future work we would like to utilize the results of our process for assisting software engineers in the creation of context-aware applications. We intend to focus this effort on the improvement of context-aware recommenders by suggesting the most appropriate context fields that can be used to improve the results of recommenders.

References

1. Abowd, G.D., Dey, A.K.: Towards a better understanding of context and context-awareness. In: Gellersen, H.-W. (ed.) HUC 1999. LNCS, vol. 1707, pp. 304–307. Springer, Heidelberg (1999)
2. Aciar, S.: Mining context information from consumers reviews. In: Proceedings of Workshop on Context-Aware Recommender System, vol. 201. ACM (2010)
3. Adomavicius, G., Sankaranarayanan, R., Sen, S., Tuzhilin, A.: Incorporating contextual information in recommender systems using a multidimensional approach. ACM Trans. Inf. Syst. (TOIS) **23**(1), 103–145 (2005)
4. Adomavicius, G., Tuzhilin, A.: Context-aware recommender systems. In: Ricci, F., Rokach, L., Shapira, B., Kantor, P.B. (eds.) Recommender Systems Handbook, pp. 217–253. Springer, New York (2011)
5. Baltrunas, L., Kaminskas, M., Ricci, F., Rokach, L., Shapira, B., Luke, K.H.: Best usage context prediction for music tracks. In: Proceedings of the 2nd Workshop on Context Aware Recommender Systems (2010)
6. Bettini, C., Brdiczka, O., Henricksen, K., Indulska, J., Nicklas, D., Ranganathan, A., Riboni, D.: A survey of context modelling and reasoning techniques. Pervasive Mob. Comput. **6**(2), 161–180 (2010)
7. Cantador, I., Brusilovsky, P., Kuflik, T.: 2nd workshop on information heterogeneity and fusion in recommender systems (hetrec 2011). In: Proceedings of the 5th ACM conference on Recommender systems. RecSys 2011. ACM, New York (2011)
8. Chen, G., Kotz, D., et al.: A survey of context-aware mobile computing research. Technical Report TR2000-381, Department of Computer Science, Dartmouth College (2000)
9. Cohen, W., Ravikumar, P., Fienberg, S.: A comparison of string metrics for matching names and records. In: KDD Workshop on Data Cleaning and Object Consolidation. vol. 3, pp. 73–78 (2003)
10. Domingues, M.A., Jorge, A.M., Soares, C.: Using contextual information as virtual items on top-n recommender systems. arXiv preprint arXiv:1111.2948 (2011)
11. Marianne, H., Mathieu, L., Clémentine, N., Jean-Rémy, F.: Metamodel matching for automatic model transformation generation. In: Ober, I., Uhl, A., Völter, M., Bruel, J.-M., Czarnecki, K. (eds.) MODELS 2008. LNCS, vol. 5301, pp. 326–340. Springer, Heidelberg (2008)
12. Hansen, J., Sato, M., Ruedy, R., Lo, K., Lea, D.W., Medina-Elizade, M.: Global temperature change. Proc. Nat. Acad. Sci. **103**(39), 14288–14293 (2006)
13. Kaluža, B., Mirchevska, V., Dovgan, E., Luštrek, M., Gams, M.: An agent-based approach to care in independent living. In: de Ruyter, B., Wichert, R., Keyson, D.V., Markopoulos, P., Streitz, N., Divitini, M., Georgantas, N., Mana Gomez, A. (eds.) AmI 2010. LNCS, vol. 6439, pp. 177–186. Springer, Heidelberg (2010)
14. Kapitsaki, G.M., Achilleos, A.P.: Model matching for web services on context dependencies. In: Proceedings of the 14th International Conference on Information Integration and Web-based Applications & Services, pp. 45–53. ACM (2012)
15. Lombardi, S., Anand, S.S., Gorgoglione, M.: Context and customer behaviour in recommendation (2009)

16. Lovett, T., O'Neill, E. (eds.): Mobile Context Awareness. Springer, London (2012)
17. Mettouris, C., Papadopoulos, G.A.: Cars context modelling (2014)
18. Mettouris, C., Papadopoulos, G.A.: Ubiquitous recommender systems. Computing **96**(3), 223–257 (2014)
19. Munguia Tapia, E.: Activity recognition in the home setting using simple and ubiquitous sensors. Ph.D. thesis, Massachusetts Institute of Technology (2003)
20. Schmidt, A., Beigl, M., Gellersen, H.W.: There is more to context than location. Comput. Graph. **23**(6), 893–901 (1999)
21. Sielis, G.A., Mettouris, C., Papadopoulos, G.A., Tzanavari, A., Dols, R.M., Siebers, Q.: A context aware recommender system for creativity support tools. J. UCS **17**(12), 1743–1763 (2011)
22. Sielis, G.A., Mettouris, C., Tzanavari, A., Papadopoulos, G.A.: Context-aware recommendations using topic maps technology for the enhancement of the creativity process. In: Educational Recommender Systems and Technologies: Practices and Challenges: Practices and Challenges, p. 43 (2011)
23. Stark, M.M., Riesenfeld, R.F.: Wordnet: an electronic lexical database. In: Proceedings of 11th Eurographics Workshop on Rendering. MIT Press (1998)
24. Suen, C.Y.: n-gram statistics for natural language understanding and text processing. IEEE Trans. PAMI-Pattern Anal. Mach. Intell. **1**(2), 164–172 (1979)
25. Suthaharan, S., Alzahrani, M., Rajasegarar, S., Leckie, C., Palaniswami, M.: Labelled data collection for anomaly detection in wireless sensor networks. In: 2010 Sixth International Conference on Intelligent Sensors, Sensor Networks and Information Processing (ISSNIP), pp. 269–274. IEEE (2010)
26. Heinze, T., Voigt, K.: Metamodel matching based on planar graph edit distance. In: Gogolla, M., Tratt, L. (eds.) ICMT 2010. LNCS, vol. 6142, pp. 245–259. Springer, Heidelberg (2010)
27. Zheng, Y., Burke, R., Mobasher, B.: Differential context relaxation for context-aware travel recommendation. In: Lops, P., Huemer, C. (eds.) EC-Web 2012. LNBIP, vol. 123, pp. 88–99. Springer, Heidelberg (2012)
28. Ziegler, C.N., McNee, S.M., Konstan, J.A., Lausen, G.: Improving recommendation lists through topic diversification. In: Proceedings of the 14th international conference on World Wide Web, pp. 22–32. ACM (2005)

Warehousing Complex Archaeological Objects

Aybüke Öztürk[1,2]($\boxtimes$), Louis Eyango[2], Sylvie Yona Waksman[2],
Stéphane Lallich[1], and Jérôme Darmont[1]

[1] Laboratoire ERIC, Université de Lyon,
5 avenue Pierre Mendès France, 69676 Bron Cedex, France
[2] Laboratoire Archéométrie et Archéologie, Université de Lyon,
7 rue Raulin, 69365 Lyon Cedex 7, France
{aybuke.ozturk,stephane.lallich,jerome.darmont}@univ-lyon2.fr,
{louis.eyango,yona.waksman}@mom.fr

Abstract. Data organization is a difficult and essential component in cultural heritage applications. Over the years, a great amount of archaeological ceramic data have been created and processed by various methods and devices. Such ceramic data are stored in databases that concur to increase the amount of available information rapidly. However, such databases typically focus on one type of ceramic descriptors, e.g., qualitative textual descriptions, petrographic or chemical analysis results, and do not interoperate. Thus, research involving archaeological ceramics cannot easily take advantage of combining all these types of information.

In this application paper, we introduce an evolution of the Ceramom database that includes text descriptors of archaeological features, chemical analysis results, and various images, including petrographic and fabric images. To illustrate what new analyses are permitted by such a database, we source it to a data warehouse and present a sample online analysis processing (OLAP) scenario to gain deep understanding of ceramic context.

Keywords: Archaeology · Archaeometry · Ceramics · Complex objects · Databases · Data warehouses · OLAP

1 Introduction

Archaeology is a branch of humanities that investigates past societies. It includes the study of material culture left behind by past human populations, especially pottery, which is seen as the most common archaeological material, providing with information on many aspects including chronology, trade, and technology. The form and decoration of pottery changed over time, which makes it a potential chronological marker, and its circulation is an indication of exchanges and trade. Another interesting fact about pottery is that, once a pottery was broken, it could not be recycled, unlike iron or glass, for instance. Thence, potteries have remained to exist until today. Therefore, it is one of the most important archaeological material to help reconstruct past civilizations.

H. Christiansen et al. (Eds.): CONTEXT 2015, LNAI 9405, pp. 226–239, 2015.
DOI: 10.1007/978-3-319-25591-0_17

In recent years, the use of digital systems and tools in archaeological studies developed rapidly. Using web-based databases, geographical coordinates, digital mapping, and digital photography is becoming popular. Scientific developments and statistical techniques have further contributed to the analysis of archaeological materials. Nowadays, digital systems are needed by archaeologists to study a variety of archaeological information and to share them. Digital systems also give an opportunity to study different aspects of the same objects or categories of objects.

Archaeological ceramics[1] can be described in different ways, by archaeologists, museum curators or archaeological scientists, e.g., through chemical, mineralogical, and petrographic analyses. In addition, ceramics can be used to determine contextual relationships, which help to highlight archaeologically meaningful data from the mass of individual data. In other words, exploiting ceramic data allows to discover patterns that are only visible in larger and more distributed ceramic samples than can be collected about any single ceramic. In archaeology, core data are highly contextual. Thence, ceramics and their properties can help to obtain comprehensive knowledge about technological, cultural, and geographical information. Such information may also contribute to understand the period and provenance of ceramics.

However, information is globally very heterogeneous. Databases have different file formats, access protocols, and use various query languages. There is no standardized terminology, especially in terms of the description of ceramic materials and their properties. Moreover, databases have a strong focus by and large. For instance, in Lyon archaeometric studies carried out on ceramics [1,2] led to the development of the Ceramom database [3]. In Ceramom, whose development began in the late 1970's, ceramics were until recently mainly described by their chemical composition together with a text summary of archaeological information. Eventually, databases little interoperate, most being offline and the others only providing a web interface, but no API (Application Programming Interface). Thus, combining various information about archaeological objects, such as textual, numerical, and graphical documents, which would allow powerful computer analyses, is at best an intricate task as of today.

Thus, in this application paper, we introduce the new Ceramom database, which models previously little-exploited textual descriptions of ceramic samples and includes image descriptions (technical drawings, photos, etc.) as well. This new database aims to be the basis of powerful analyses, such as OLAP and data mining, which should integrate various points of view on ceramic objects, e.g., text descriptors, chemical analysis results, technical drawings, binocular images resulting from fabric analyses, and petrographic images resulting from petrographic analyses to be able to learn deep contextual information from all the bits and pieces of ceramic. To this aim, we use the new Ceramom database to source a data warehouse, which is original in its storage of data that are not only numerical.

---

[1] In this paper, we use both "pottery" and "ceramic" to designate all the range of categories of these archaeological objects.

The remainder of this paper is organized as follows. Section 2 presents a selection of ceramic database projects, including Ceramom. Section 3 further details the new Ceramom database. Section 4 deals with remodeling this database as a multidimensional, data warehouse schema, presents a sample OLAP analysis scenario and discusses issues in ceramic data analysis. Finally, we conclude this paper and hint at future research in Sect. 5.

2   Ceramic Database Projects

In recent years, several databases were created to highlight different perspectives in pottery research. These databases have different types of contents, depending on the aspects of ceramics studies they focus on. Moreover, specific formats may be implied based on different contents, e.g., numbers for chemical analyses or text and/or images for petrographic analyses.

Databases usually have a main type of contents, and may focus on specific categories of ceramics, time periods or regions. Databases can be either publicly available online or not. Additionally, interface features may also be available, such as interactive maps, interactive 2D or 3D views, or statistical tools, the latter being of particular interest in the context of our research. Table 1 lists a selection of databases that we consider representative of this diversity of contents, formats, statuses and features, with some indications on their specificity. In Table 1, primary content is indicated by X, secondary content by x, and occasional content by (x).

The *Levantine Ceramics Project* (LCP), directed by Boston University, proposes an archaeological database focusing on ceramic wares produced in the Levant, from the Neolithic to the Ottoman periods [4]. It mainly includes archaeological data (typological, chronological, and geographical), but also provides with fabric and petrographic data. The format of LCP data is either text or image. LCP is an open, interactive internet resource. *Roman Amphorae: a digital resource*, proposed by the University of Southampton, provides an online introductory resource for the study of Roman amphorae, based on a rich corpus of archaeological information together with petrographic and fabric data [5]. *POTSHERD* is a collection regarding pottery from the Roman period (1st cent. BC – 5th cent. AD) in Britain and Western Europe, including distribution maps and links to complementary resources [6]. The *Worcestershire On-line Ceramic Database* was designed to make available the complete pottery fabric and form type series for Worcestershire, from the Neolithic to the early post-medieval period [7].

Other types of databases are more centered on either archaeological data or petrographic and fabric data. The *National Roman Fabric Reference Collection* (NRFRC) is the online version of a reference book providing detailed and standardized fabric descriptions of Roman wares found in Britain [8]. *FACEM* focuses on fabric data of Greek, Punic, and Roman pottery in the Southern Central Mediterranean area [9]. FACEM includes interactive maps and allows downloading detailed information. *Petrodatabase* is a petrographic relational database proposing interactive maps [10]. *ICERAMM* focuses on medieval and modern ceramics in western and northern France, Belgium, and Switzerland [11]. *PECL* is a project of encyclopedia for ceramics of the Mediterranean and sub-Saharian region of all periods, including detailed archaeological contexts information [12]. *ASCSA.net* presents archaeological objects and contexts from the excavations of the American School of Classical Studies at Athens in the Athenian Agora and in Corinth [13].

There are several image databases that are designed for a larger audience, using digital representations of ceramics. One of them, *Sgraffito in 3D*, proposes 3D reconstructions of late medieval pottery collection from the Museum Boijmans Van Beuningen (Rotterdam, the Netherlands) [14].

Yet other databases focus on chemical data. The *CeraDAT project*, developed by the Demokritos National Centre of Science Research (Athens, Greece), is a prototype relational database including interactive maps and focusing on the Aegean and the wider Eastern Mediterranean Region [15]. The *Archaeometry Laboratory Database* of the MURR laboratory (Missouri, USA) presents chemical analysis of ceramic artifacts from many regions, including Northern, Central and Southern America, and the Mediterranean [16]. It also gives access to "historical" chemical databases, such as the Berkeley laboratory's. Archaeological information of MURR data is presented as a bibliography. The Archaeometry Research group at the University of Fribourg (Switzerland) has established several reference groups with the chemical composition of ancient ceramics from Switzerland, Italy, France, and Germany [17]. Archaeological information is also presented as a bibliography.

Eventually, Lyon's *Ceramom database* used to be mainly a chemical database [1,2], including only limited archaeological information. The new Ceramom 3.0 model we detail in Sect. 3 has been recentered on ceramic objects, and enriched in archaeological and multimedia contents. It covers all periods. Ceramom is not available online yet, but it will be soon at the following address [3]. Table 1 clearly shows that Ceramom 3.0 is one of the most comprehensive database model for ceramic data, including enriched archaeological (geographical and graphical descriptions) and archaeometric (numerical values of various analysis results) data.

3   The Ceramom Database

Data may be qualified as complex if they are [18]:

- multiformat, i.e., represented in various formats (databases, texts, images, sounds, videos);
- and/or multistructure, i.e., diversely structured (relational databases, XML documents repository);
- and/or multisource, i.e., originating from several different sources (distributed databases, the Web);

Table 1. Ceramic database features

Ceramic databases	Database type			Fabric	Data type			Features		
	Archaeological	Chemical	Petrographic		Textual	Numerical	Multimedia	Online	Structured	Stat tools
LCP	X	x	x	x	x		x	x	x	
Roman Amphorae	X	X	X	x	x		x	x	x	
POTSHERD	X			x	x		x	x		
Worcestershire Ceramics	X	(x)	X	x	x		x	x	x	
NRFRC	(x)	X	X	x	x		x	(x)		
FACEM	x			X	x		x	x	x	
Petrodatabase	(x)	X	x	x	x		x	(x)	x	
ICERAMM	X			x	x		x	x		
PECL	X	(x)	(x)	x	x		x	x	x	
ASCSA	X				x		x	x	x	
Sgraffito in 3D	x				x		x	x		
CeraDAT	(x)	X			x	x			x	x
MURR	(x)	X			x	x		x		downloadable
Fribourg	(x)	X			x	x		x		
Ceramom 2.0	x	X	(x)		x	x			x	x
Ceramom 3.0	X	X	x	x	x	x	x	x	x	x

– and/or multimodal, i.e., described through several channels or points of view (radiographies and audio diagnosis of a physician, data expressed in different scales or languages);

– and/or multiversion, i.e., changing in terms of definition or value (temporal databases, periodical surveys).

The Ceramom database was designed from the requirements of ceramic specialists for recording, using, and analysing data generated by different techniques. Thence, it stores complex data. Mineralogical and petrographical data with extensive definitions are indeed combined with rich location data. In addition to these data, graphical documents that are necessary to complement ceramic information are added in the new model, such as drawings and images of ceramic samples with variety of references. According to this model, the Ceramom database is centered on pottery samples, which are described by geographical features, several analyses, and various descriptions (Fig. 1). Each of these packages is further detailed in the following subsections. All conceptual models are depicted as UML class diagrams [19].

3.1 Geography

Figure 2 displays the Geography package of the data model. The LOCATION class connects geolocation data to STORAGE OUTSIDE

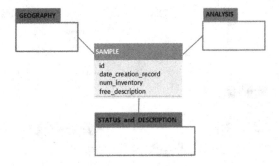

Fig. 1. Ceramom conceptual schema: global view

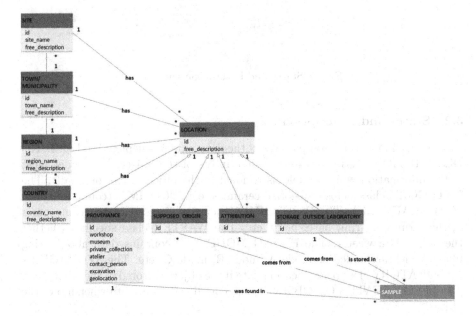

Fig. 2. Geography package

LABORATORY, PRO-VENANCE, SUPPOSED ORIGIN, and ATTRIBUT-ION classes. The STORAGE OUTSIDE LABORATORY class represents where the object is physically stored.

The PROVENANCE class bears information regarding the location data where the object was found. The SUPPOSED ORIGIN class provides a supposed origin before analysis. The ATTRIBUTION class indicates where the object was demonstrated to come from by after analysis. Finally, the SITE, TOWN/MUNICIPALITY, REGION, and COUNTRY classes represent a hierarchy of LOCATIONS.

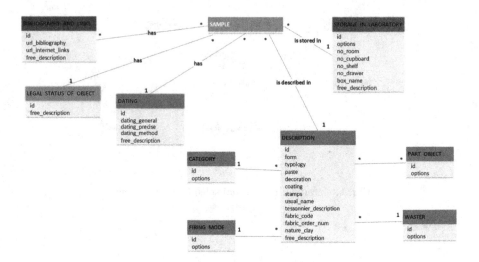

Fig. 3. Status and Description package

3.2 Status and Description

The Status and Description package of the data model is depicted in Fig. 3. The DESCRIPTION class bears textual descriptors of the object. It includes most of the information which enable us to identify the object archaeologically. The CATEGORY class helps categorize ceramics, e.g., "COMM". (common ware), "CARREAU" (tile). The PART OBJECT class collects data regarding parts of the form, e.g., rim, base, body, etc. The WASTER class specifies whether the object is a waster. The FIRING MODE class contains data about firing mode, which are coded as mode A, mode B, mode C, etc. The STORAGE IN LABORATORY class bears location data if the object is stored in the laboratory. The LEGAL STATUS OF OBJECT class contains data about ownership of the object.

3.3 Analysis

Figure 4 depicts the Analysis package of the data model. Analysis results of a sample are represented in class ANALYSIS RESULT. Each analysis corresponds to a series of separate records, each of which contains an individual measure. This is due to the fact that samples may be analyzed by several techniques, such as chemical or petrographical analyses. A given sample may also be analyzed several times using the same technique, but with different parameters. For examples, as a larger number of chemical elements were determined since the 1970's, when a same sample was re-analyzed a given chemical element (e.g. aluminium or calcium) would be assigned several values of concentration.

The DIFFRACTION, DILATO, PETRO, CHEMISTRY, BINO, SEM, and additional analyses classes bear data regarding diffraction, dilatometry,

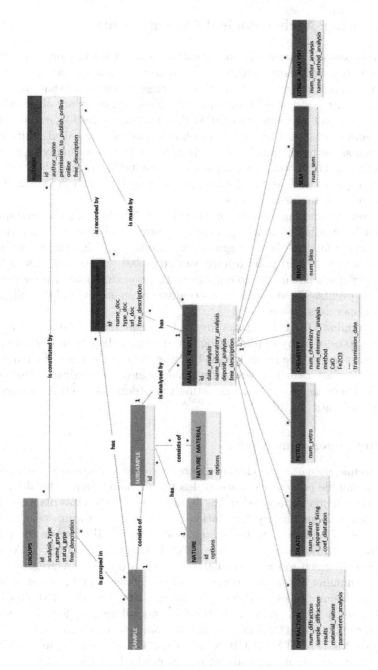

Fig. 4. Analysis package

petrography, chemistry, binocular microscopy, scanning electron microscopy, and additional, miscellaneous analyses, respectively.

4   Exploring Archaeological Ceramic Data

Data warehouses are actually databases with a specific model tailored for efficient OLAP analyses. In a data warehouse, the observed data are called facts, e.g., sales in a business context. They are characterized by measures that are usually numerical, e.g., quantities sold and amounts of money. Facts are observed with respect to different analysis axes called dimensions, e.g., sold products, store location, and sale date. Thus, data warehouse schemas are called multidimensional schemas, or more casually star schemas, for facts are usually represented in the center of the model, with dimensions gravitating around. Star schemas help answer queries such as "total sales revenue of each product in Lyon in 2014", to go on with our business example.

Moreover, dimensions may be organized in hierarchies, e.g., a time dimension could be subdivided into day, month, quarter, and year. Such a structure helps observing facts at different granularity levels, e.g., "dezooming" from one quarter of a year to said year to have a more global (aggregate) view of sales, or "zooming" from one month to one day in this month to have a more detailed view of sale events. These operations actually correspond to OLAP's rollup and drill-down operators, respectively.

Thus, to allow OLAP navigation in the Ceramom data, we must select facts to observe, axes of analysis (dimensions) and import data from Ceramom into the data warehouse. The result is called a cube (hypercube when the number of dimensions is greater than 3), where dimension values are coordinates that define a fact cell. There are a couple of interesting works done using OLAP analysis on archaeological data [20,21].

4.1   Multidimensional Model

In this sample scenario, we choose to observe chemical dosages with respect to ceramic sample provenance, dating, description, and groups. Our data warehouse's star schema is provided in Fig. 5, again as a UML class diagram. Facts are modeled as a quaternary association-class connected to dimension classes. To make use of numerical values for analyses, the SAMPLE class from Fig. 1 is combined with the Analysis package (Fig. 4) into the SAMPLE ANALYSES class in Fig. 5, which models our analysis facts. In our case, aggregates (summaries) are number of sample, average number of sample, and number of analyses, etc. Dimension classes are PROVENANCE, GROUPS, DESCRIPTION, and DATING, which are the same as in the Ceramom database (Figs. 2, 3 and 4). Moreover, the LOCATION class individually connects to all classes in the SITE, TOWN/MUNICIPALITY, REGION, and COUNTRY hierarchy to still allow a connection in case of missing value at one hierarchy level (Sect. 4.3).

4.2   OLAPing Archaeological Ceramic Data

Once part of Ceramom data are multidimentionally remodeled, OLAP analyses can be done. OLAP actually helps interactively navigate the data warehouse,

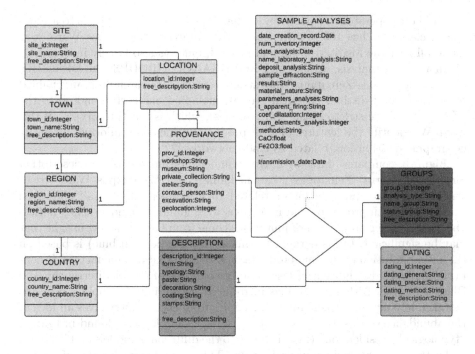

Fig. 5. Chemical data warehouse's multidimensional schema

e.g., to discover outliers or hidden patterns. We use Pentaho Business Analytics[2], a suite of open source business intelligence, as our OLAP engine. Pentaho features a user console that is a web-based design environment. The console helps to visualize and navigate hypercubes, which are created from the data warehouse with the help of the schema-workbench tool.

As an example of OLAP analysis, let us examine the contents of a specific chemical group coming from the GROUP class, and compare them to the initial typological classification from the DESCRIPTION class. Samples within a given chemical group belong to the same pottery production, i.e., they share the same origin. They usually come from several excavations (PROVENANCE class), and their circulation and corresponding fluxes provide insight into past contacts between populations and trade networks. When different workshops manufactured similar wares, classified under the same typology, chemical analysis "sorts out" the different productions and enables archaeologists and historians to better understand economic trends and cultural influences.

In a first analysis, successive rollups help aggregate PROVENANCE data at the country level to achieve a coarser view of data. We take interest in ceramics of the Byzantine period (Medieval period in the DATING class) called "Zeuxippus Ware" and we select the DATING, DESCRIPTION (typology), and PROVENANCE (country) dimensions and count "Zeuxippus Ware" occurrences (OLAP

---

[2] http://www.pentaho.com.

slice and dice operators, respectively). "Zeuxippus Ware" corresponds to a typological class that has 163 occurrences in the database. "Zeuxippus Ware" was found all over the Mediterranean and beyond, but was also largely imitated.

In a second analysis, we slice on PROVENANCE, GROUPS, and DESCRIPTION, and dice on "Zeuxippus Ware stricto sensu". A research program enabled to define several distinct chemical groups, including one corresponding to the "Zeuxippus Ware stricto sensu" (87 samples), which is the "prototype" of this ware. We identify the features of each production, including information on its geographic distribution, related to trade networks [22].

Figure 6 compares data with the number of samples whose description includes the term "Zeuxippus", i.e., including both "Zeuxippus Ware stricto sensu" and wares imitating it or related to it typologically. In all countries, examples of both prototype and imitations were found. It is nonetheless noticeable that a larger proportion of imitations come from Greece, a new insight that may be significant. In histogram, "chemical classification" (in blue) is based on the actual diffusion of ceramic products, it is related to economic factors. It confirms the large distribution of this ware in countries of the Mediterranean and Black sea areas. Although the bias introduced by the initial sampling needs to be taken into account, the number of samples from each country gives an idea of the abundance of this ware, e.g., only very few examples were found in France. "typological classification" (in red) refers to the diffusion of models and fashions, and is thus more related to cultural factors. This example somehow simulates the comparison of data obtained on the same categories of objects from two databases, focusing each on another aspect of these wares. It shows the discrepancies, but also the added value that may be obtained when connecting information. It also shows how OLAP analysis may contribute to the understanding of economic and cultural relationships at the Byzantine period, thanks to its ability to bring ceramic information back into a wider context.

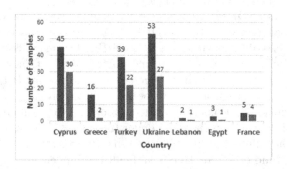

Fig. 6. Distribution of samples by country, for descriptions including the term "Zeuxippus" (in red, "typological classification"), and in chemical group "Zeuxippus Ware stricto sensu" (in blue, "chemical classification") (Color figure online).

4.3   Issues in Archaeological Ceramic Data Analysis

We have been confronted to a couple of major challenges before OLAP analyses and when performing OLAP analyses onto Ceramom data. First, we encountered a classical problem in databases, i.e., missing values. In our example scenario, there is location information for provenance studies, but in practice, some information, i.e., site, town, region or country might be missing in the database. For example, some samples relate to Sudak, Ukraine and Acre, Israel, with no archaeological site reference. This is why we complement the geographical hierarchy with direct associations from provenance to site, town, region, and country. Some of this missing information (town to region to country relationships) shall be found in external sources, though.

Moreover, data about archaeological ceramics mostly consists of textual and numerical data. On one hand, textual data include information about the characteristics of ceramics, such as form, chronological information, etc. On the other hand, numerical data include information produced by various analyses performed on the ceramic material. Both can be warehoused. However, while classical OLAP provides a good tool for analyzing numerical data (through aggregation functions such as sum, average, minimum, maximum, etc.), it is not very convenient for textual data, whose individual values can only be counted. Thus, in order to gain in-depth knowledge about ceramics, there is a crucial need to better take textual data into account in OLAP.

5   Conclusion and Perspectives

Designing comprehensive archaeological databases/tools is a challenge because of many reasons. Various dimensions should be integrated from distant databases that describe the same categories of objects in a complementary way. Thus, different point of views and parameters should be combined coming from different, heterogeneous databases.

In this paper, we first survey representative ceramic databases and show there is no single comprehensive resource for studying ceramic materials. Then, we propose a new conceptual data model for archaeological ceramics. We believe this model could be a good starting point to help ceramic databases interoperate in the future. Moreover, we illustrate how ceramic data can source a data warehouse to perform OLAP. Such analyses help navigate and observe data from different perspectives, thus enabling archaeological researchers with better insights on their data. The main benefit of the proposed approach is to identify hidden patterns and possibly unexpected values, especially only visible in larger and more distributed ceramic samples, in order to contextualize information and help building up knowledge of past societies [23,24].

Our first perspective is to include textual and graphical descriptions of ceramic samples in the warehouse to do OLAP analyses. Some research already address non-numerical data integration into data warehouses, e.g., by performing some preprocessing on text before storage into the data warehouse. Actually, integration of such data might not be in textual format [25–27].

Moreover, data mining could also be combined to OLAP to complement data navigation with automatic pattern or structure discovery. For example, clustering techniques could help enhance current statistical tools for categorizing ceramics based on textual, numerical, and graphical information. Each context allows to create a clusterer. The challenge will be to manage the collaboration between the different clusterers either for producing a consensus clustering or to explain and annotate a given clustering or to identify strong forms.

In the longer run, we finally aim to build smart links between ceramic databases in order to achieve interoperability between Ceramom and partner databases, and allow cross-analyses to run over a database federation.

Acknowledgements. This project is supported by the Rhône Alpes Region's ARC 5: "Cultures, Sciences, Sociétés et Médiations" through a PhD grant to A. Öztürk, who gratefully acknowledges this support. We also sincerely thank the Archaeological Ceramics team of the "Archéométrie et Archéologie" Laboratory (CNRS UMR 5138), and especially C. Brun, for its contribution to the design of the Ceramom model.

References

1. Picon, M.: L'analyse Chimique des Céramiques: Bilan et Perspectives. In: Archeometria della Ceramica. Problemi di Metodo, Atti 8e Simposio Internazionale della Ceramica, Bologna, pp. 3–26 (1993)
2. Waksman, S.Y.: Etudes de Provenance de Céramiques. In: Dillmann. P., Bellot-Gurlet. L (dir.) Circulation et Provenance des Matériaux dans les Sociétés Anciennes, pp. 195–216 (2014)
3. CNRS UMR 5138, Université de Lyon: Ceramom, forthcoming. http://www.arar.mom.fr/ceramomdatabase/. Accessed June 2015
4. Boston University: Levantine Ceramic Project. http://www.levantineceramics.org/. Accessed June 2015
5. University of Southampton: Roman Amphorae: A Digital Resource. http://archaeologydataservice.ac.uk/archives/view/amphora_ahrb_2005/index.cfm?CFID=573996&CFTOKEN=39545028. Accessed June 2015
6. Tyers, P.A.: POTSHERD: Atlas of Roman Pottery. http://potsherd.net/atlas/potsherd. Accessed June 2015
7. Worcestershire Archive and Archaeology Service: Worcestershire On-line Ceramic Database. http://pottery.rigorka.net/. Accessed June 2015
8. Tomber, R., Dore, J.: The National Roman Fabric Reference Collection: A Handbook. Museum of London Archaeology Service, London (1998)
9. University of Vienna: FACEM. http://facem.at/. Accessed 2011
10. Quinn, P., Rout, D., Stringer, L., Alexander, T., Armstrong, A., Olmstead, S.: Petrodatabase: an on-line database for thin section ceramic petrography. J. Archaeol. Sci. **38**(9), 2491–2496 (2011). Academic Press
11. Husi, P.: ICERAMM Network. http://iceramm.univ-tours.fr/index.php. Accessed 2014
12. CNRS USR 3125, Aix Marseille Université: Prototype D'Encyclopédie Céramologique en Ligne (PECL). http://www.cnrs.fr/inshs/recherche/pecl.htm. Accessed 2012

13. The American School of Classical Studies at Athens: ASCSA Digital Collections. http://www.ascsa.net/research?v=default. Accessed June 2015
14. Museum Boijmans Van Beuningen, Sgraffito in 3D. http://www.sgraffito-in-3d. com/en/. Accessed June 2015
15. Demokritos, National Centre of Scientific Research, ceraDAT. http://www.ims. demokritos.gr/ceradat/. Accessed June 2015
16. Missouri University: Archaeometry Database. http://archaeometry.missouri.edu/ datasets/uman/index.html. Accessed 2015
17. University of Fribourg: Reference Groups with the Chemical Composition of Ancient Ceramics. http://www.unifr.ch/geoscience/mineralogy/archmet/index. php?page=746. Accessed 2015
18. Darmont, J., Boussaid, O., Ralaivao, J.C., Aouiche, K.: An architecture framework for complex data warehouses. In: 7th International Conference on Enterprise Information Systems, pp. 370–373. INSTICC, Setubal (2005)
19. Object Management Group: Unified Modeling Language™ (UML) Resource Page. http://www.uml.org. Accessed June 2015
20. Sun, Q., Xu, Q., Li, Q.: Multidimensional Analysis of distributed data warehouse of antiquity information. Open Cybern. Systemics J. **9**, 55–61 (2015)
21. Musco, S., Salvatori, A., Mazzei, M., D'Agostini, C.: Lapis Pallens: Integrated Research on Ancient Roman Quarries of Red Tuff of Aniene River Known as Latomie di Salone (ROME), p. 49 (2011)
22. Waksman, S.Y., François, V.: Vers une Redéfinition Typologique et Analytique des Céramiques Byzantines du Type Zeuxippus Ware. Bulletin de Correspondance Hellénique, 128–129,2.1, pp. 629–724 (2004–2005)
23. Kansa, E.C.: A community approach to data integration: authorship and building meaningful links across diverse archaeological data sets. Geosphere **1**, 97–109 (2005)
24. Kansa, E.C., Kansa, S.W., Burton, M.M., Stankowski, C.: Googling the grey: open data, web services, and semantics. Archaeologies **6**(2), 301–326 (2010)
25. Tournier, R., Tournier, R., Pujolle, G., Pujolle, G., Teste, O., Teste, O., Ravat, F., Ravat, F., Zurfluh, G., Zurfluh, G.: Multidimensional database design from document-centric XML documents. In: Dayal, U., Dayal, U., Cuzzocrea, A., Cuzzocrea, A. (eds.) DaWaK 2011. LNCS, vol. 6862, pp. 51–65. Springer, Heidelberg (2011)
26. Darmont, J., Olivier, E.: A complex data warehouse for personalized, anticipative medicine. In: 17th Information Resources Management Association International Conference, pp. 685–687. Idea Group Publishing, Hershey (2006)
27. Aknouche, R., Asfari, O., Bentayeb, F., Boussaid, O.: Decisional architecture for text warehousing: ETL-text process and multidimensional model TWM. In: 19th International Conference on Management of Data, pp. 101–104. Computer Society of India, Mumbai (2013)

Context Models and Multi-Agent Systems

A Contextual Model of Turns for Group Work

Kimberly García[✉] and Patrick Brézillon

University Pierre and Marie Curie (UPMC), Paris, France
{Kimberly.Garcia,Patrick.Brezillon}@lip6.fr

Abstract. The Contextual-Graphs formalism has been conceived to represent task realizations in the way they are actually performed. The objective is to provide decision makers with a clear panorama of the different ways a task can be realized (i.e. practices), and the implications of choosing one way or another. The Contextual-Graphs formalism has been successfully used in many fields, such as medicine, biology, and transports, for representing task realizations involving a single actor. In this paper we explore the representation of a group task by analyzing the paper-submission example, from which a turns mechanism is proposed as a way to adapt the Contextual-Graphs formalism to support this type of tasks. Moreover, the types of interaction among actors involved in group task realizations are studied in detail based on a set of definitions introduced in this paper. We claim that the real understanding of group task realizations will not just help decision makers, but will also provide groupware designer with real requirements for building successful applications.

Keywords: Contextual model for CSCW · Turns mechanism · Cooperation · Collaboration · Coordination

1 Introduction

Cooperation and collaboration are two ambiguous notions that have different meanings across domains, and sometimes from one author to another. The difference between these two concepts seems related to the sharing of the goal in the interaction. Collaboration means to "work together", i.e. a joint development of a negotiated and consensual solution, thus it is a task realization in different ways. Cooperation means to "operate together", it is a negotiated division of the task realization among participants (a common goal but autonomous actions) and a pooling by the assembling of each subtask realization in a linear way, each participant has to handle a definite part of the shared task realization. Coordination is the technical organization of the different elements of a task realization to enable participants to work together effectively according to a plan. If coordination has a relatively well-accepted definition across disciplines and types of approaches, such as, cognitive (AI, psychology, etc.) or technology (CSCW, interface design, etc.) [2,7,9], it is not the case of collaboration and

© Springer International Publishing Switzerland 2015
H. Christiansen et al. (Eds.): CONTEXT 2015, LNAI 9405, pp. 243–256, 2015.
DOI: 10.1007/978-3-319-25591-0_18

cooperation. In this paper, we retain that collaboration refers to work together while cooperation is to operate together.

We are interested in bringing the concept of context to group tasks as the relevant knowledge, regarding the task, that each actor possess and shares with the rest of the group, as this would help understanding the way the job gets actually done, which will bring benefits to decision makers and to groupware designers. Commonly, in the technology research field, context is restricted to a set of physical measurements that denotes the state of the environment (e.g., location, and temperature). Our study is based on the Contextual-Graphs formalism [5], which offers a uniform representation of elements of knowledge, reasoning, and context. This formalism has been used by Fan et al. [8] as a contextual complement to scientific workflows in virtual screening, demonstrating that it is more important to model a task realization than its theoretical model. However, the Contextual-Graphs formalism was conceived for task realizations involving a single actor. Therefore, we want to extend it to support group tasks, which represents a different challenge, as each actor involved realizes a set of subtasks in an ordered way (e.g., an actor is able to perform a subtask after another actor has completed theirs), such an actor also needs to share with the rest of the group those elements that are considered important for the group task realizations (i.e. shared context). The interaction between actors evolves, since it can start with superficial communication, but within the task realization development, it could become a strong engagement of the actors. In order to represent such an interaction, we have explored different ways of modeling the example of the paper submission in a contextual model based on the Contextual-Graphs formalism. From this analysis, it was found out that it is necessary to create a mechanism that establishes turns between actors. During each turn, an actor realizes a set of subtasks, and updates the shared context so the next actor understands the state of the task and is able to contribute to the group task.

This paper is organised as follows. In Sect. 2, the Contextual-Graphs formalism as a foundation of the paper is described. Then, in Sect. 3, a series of definitions regarding the levels of interaction among people working together are presented. In Sect. 4, an analysis of the common known paper submission process [1] is introduced in order to find a suitable way to represent group tasks through the Contextual-Graphs formalism. After exploring some options, we propose a mechanism for turns to manage the flow of this type of group tasks in Sect. 5. Then, in Sect. 6, the practice tree implementation as a way to facilitate the visualization of the exchange of turns is introduced. Finally, the conclusions are presented in Sect. 7.

2   On the Contextual-Graphs Formalism

Pomerol and Brézillon [10] define context as the sum of: (1) the contextual knowledge, which is all the knowledge relevant for a person in a given decision problem, (2) the external knowledge, corresponding to the rest of the knowledge that is not important for the current situation, and (3) the proceduralized context, which is a part of the contextual knowledge that becomes important at a specific step of the decision problem. Based on this context definition, Brézillon [4] introduces the Contextual-Graphs (CxG) formalism for obtaining a uniform representation of elements of knowledge, reasoning and context. Thus, a contextual graph represents the realization of a task, each path is a practice developed by an actor in a particular context for realizing the task. A contextual graph represents the accumulated experience of one or several actors realizing the same task.

The elements of a contextual graph are: actions, contextual elements, activities and parallel action groupings. An action is the building block of contextual graphs. A contextual element is a pair of nodes, a contextual and a recombination one; the former has one input and N outputs (branches) corresponding to the N instantiations of the contextual element; the latter is [N, 1] and represents the moment at which the instantiation of the contextual element does not matter anymore. An activity is a graph by itself that is identified by actors because it appears in the same way in different problem solving processes. An activity is defined in terms of the actor, situation, task and a set of actions. Finally, a temporal branching for action grouping expresses the fact that several set of actions can be realized in parallel or in a sequential way, no matter the order.

A contextual element corresponds to an element of the nature that must be analyzed. The value taken by a contextual element when the focus is on it, its instantiation is considered as long as the situation is under analysis. The proceduralized context evolves dynamically during a practice development by addition (at the contextual node) or removal (at the recombination node) of a contextual element during the progress of the focus. Moreover, for group tasks, the shared context is formed by persistent (known by all the actors) contextual elements introduced by an actor, and eventually accepted by others after negotiation. Thus, contextual elements can be used as a way to manage turns among actors involved in a group task, as their instantiation would denote whose turn is next. The working context corresponds to all the contextual elements of a Contextual Graph and their instantiations [3].

3   Levels of Interaction

In the Computer Supported Cooperative Work (CSCW) research field, the group awareness need has been identified since early works as a requirement for applications supporting non-collocated teams [6], since variables such as presence, availability and activity of each member of the team need to be known to coordinate activities and achieve collaboration. Thus, many applications incorporated the availability and presence modes (e.g., busy, online, available, away), to let others users know if the interaction they require could be possible at a specific moment. Then, with the evolution of the field, the creation of applications for supporting not only non-collocated collaboration, but also collocated one, and the increasing attention to the Ubiquitous Computing field [12], arose a new way of thinking about context, since current authors have embraced and developed the view of context as any variable that characterize a situation [11]. Nowadays, the goal is to customize the response of an application based on the environment that surrounds the user and the user themselves. Thus, the most popular variables considered in a group application are presence, availability, activity, location, time and interests. This narrowed vision of context in CSCW is due to researchers concern on building applications limited to physical and environmental features. However, group tasks can be studied at another level, by focusing on human interaction and not just in the applications technical aspects.

We propose a deeper analysis of the interaction among users involved in a group task. Thus, in Table 1, we illustrate through a set of definitions that there is a large spectrum of collaboration/cooperation, which always involves a degree of coordination. In Table 1 we present six levels in this spectrum, namely: the task level, the task realization level, the actor level, the activity level, the team level, and the planning level. For each level we identify the actors' commitment degree, then we propose a visual guide for each level of interaction, in which rectangles depict tasks (i.e., a task can consist of several actions) performed by a single participant at a specific moment. Finally, in the last column of the table, we use well-known collaborative applications as examples to illustrate that although they are all known as collaborative applications, they are targeted to provide support to one or several levels of interaction.

As mentioned before, most works in the CSCW field are focused on building applications, which is probably the reason why an agreement about the collaboration and cooperation concepts has not been reached yet. However, our objective is to provide support at a lower level, i.e. by understanding the way people actually realize tasks together. In Table 1, we see that it is important to differentiate the level of interaction among actors realizing a task. Since, the over simplification of using the term collaboration and cooperation to refer to any task group leads to lose the essence of the interactions, in which the shared context, the type of turns, and the degree of collaboration/cooperation varies significantly from one situation to another. Our conceptual framework allows to capture some contextual features related to the management of turns that otherwise would not be visible.

4   The Paper Submission Example as a Group Task

The submission of a scientific paper to a journal is a well-known process in the research community. In short, an author submits a paper to the editor of a journal. The editor may either accept the paper for reviewing, or reject it due to mismatch between the scope of the journal and the topics covered in the paper. An accepted paper is sent for evaluation to (at least) two reviewers. The reviewers read the submitted paper and provide their feedback before a deadline assigned by the editor. The reviewing process is an individual task that the reviewer performs based on his personal knowledge, expertise and point of view. Once the reviews are received, the editor makes a decision by comparing the reviewers' evaluations. This process can be long, so the editor must evaluate their options and time constraints in order to decide the tactics to choose. If

Table 1. Interaction between two actors' tasks

Level of interaction	Actors engagement	Applications corresponding to the type of interaction
1. Task level. Actors have independent tasks, but one task can start only after the completion of another. Interaction is limited to actors' tasks coordination.		An instructor using a learning platform, such as Blackboard, is able to upload an assignment for their students. The instructor will be able to give feedback and grade the student's work. It is a series of tasks that require coordination at all times.
2. Task realization level. Several actors realize (in parallel or at different times) the same task. Practices are accumulated in the same space. This "experience sharing" leads to direct or indirect collaboration.		A wiki is created by a user and fed by others in a synchronous or asynchronous way. Entries are shared in a common space by people in an indirect collaboration, except if someone edits their entry to contribute to the content.
3. Actor level. A helping actor is responsible for an activity in the task realization of another actor. It is a weak cooperation since the global goal belongs to a unique actor. Thus, a quick change of turns is required when the helping actor is involved in the task realization.		Question and answer websites and mailing lists help people to solve technical problems. By posting a question a user receives multiple answers from experts whose interest is just in helping others.
4. Activity level. Actors have independent tasks (and independent objectives) in a joint activity. In some cases, they could be competing but also cooperating to accomplish a common activity. Several actors' turns are activated at the same time, but in order to complete the joint activity, a coordination of turns is required at specific moments.		Crowdsourcing for software development often leads to this type of interaction. In a project, each participant has to realize a precise task as part of the common general task. The client is constantly making new requests. Thus, at specific moments, developers have to work together to perform the joint task, but as their personal goal is to maintain a good reputation in order to be hired again by the client, developers could compete to present the best ideas and work.

(Continued)

Table 1. *(Continued)*

Level of interaction	Actors engagement	Applications corresponding to the type of interaction
5. Team level. The focus is on the team task realization. There is a strong cooperation because, even when the focus is on a given actor at one moment, the others must coordinate their activities according to the actor "in focus". The shared context concerns all the team members who update it frequently. Interaction is dynamic, and turns are fast.		An application for collaborative edition of documents, such as Google Docs, supports this level of interaction. All members of a team working on a document at the same time coordinate their activities to edit a paragraph. Thus, the actor in focus changes constantly, increasing the need of dynamism of the shared context.
6. Planning level. Interaction between actors is not direct, a manager is responsible for the actor's interactions in their task realizations. The manager needs information about the schedules and actor's plans. It is a lose collaboration that could lead to cooperation once the team is constituted. The manager's turn can be paused while waiting for the termination of each actor's turn.		Google Agenda is an application supporting this level of interaction, by using it, a user is able to create an event, add comments and documents related to it. In case the event is shared and users are accepted to use the agenda, any of those users can provide feedback to interact with the people involved in it.

the paper is not rejected, the editor could: (a) conditionally accept the paper by demanding to the authors an improved version, which is verified by the editor or the reviewers, or (b) accept the paper with the minimal suggested changes. If the final editor's decision is to reject the paper, the author is notified with reviewers' comments. Otherwise, after receiving the new version, the editor sends the accepted submission to the publisher for publication. The paper submission process requires interaction among different actors with different roles: author, editor, reviewers and publisher. We choose to model the actors' interaction in the Contextual-Graphs formalism.

4.1 An Actor's Task as Part of Another Actor's Task

Figure 1 shows a fragment of a contextual graph centered in an actor's vision: the editor, who communicates to other actors whose tasks are embedded in the editor's graph. In Fig. 1 the activities 3 and 5 correspond to the reviewers'

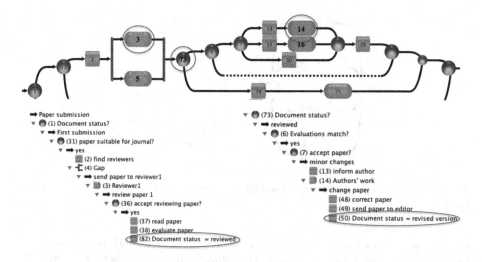

Fig. 1. A partial view of an actor's task inside of another actor's task

tasks, while activities 14 and 16 are author's tasks. In this contextual graph, "Document status" represents the common goal as a shared contextual element. "Document status" may be changed by one actor and then used by another. In this case, when a reviewer has finished or decides not to be involved in the task realization, he will change the "Document status". Figure 1 shows in contextual element 73, that in case the reviewers have evaluated the paper, the instantiation of "Document status" becomes "reviewed" and the editor will be able to continue the decision process, by comparing the evaluations. In case a reviewer has refused to evaluate the paper, the Document status stays "to be reviewed", so the editor needs to find a new evaluator. The introduction of this shared contextual element brings some advantages, as any person could easily understand the task realization flow without having to know the general protocol, they just need to understand the common goal. Furthermore, the working context is always denoting the current status of the process, making easy to obtain information. However, this representation is sequential, since the editor's job is paused until the reviewers change the status of the shared context.

4.2   One Branch for Each Actor's Task Realization

Another way to represent interaction between the editor and reviewers as a working group is to consider all actors' task realization at the same level. In Fig. 2, each branch of the graph corresponds to an actor's activity. The branches correspond (from top to bottom) to the activities of the author, the editor, the reviewer, and the publisher. Thus, an actor is selected by the instantiation of the contextual element.

This representation becomes difficult to follow when a change of actors is needed, since the Contextual-Graphs formalism is not adapted to model group

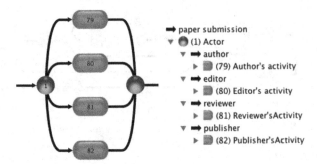

Fig. 2. A partial view of the paper submission problem presenting an actor in each branch

tasks. Thus, elements for guiding the representation should be introduced. Figure 3 shows the extended view of the editor and reviewer activities. Specifically, in action 8 the editor's task is to inform a reviewer about their new assignment. Once the request is sent, the next turn concerns the reviewer. This is realized by reapplying the global contextual graph of Fig. 3 in the new working context with reviewer as the actor in the focus. This representation, understandable in this well-known example, could become rapidly difficult to follow when jumping from one branch to another, without completing any branch. The representation is also inefficient, as it is difficult to know important information about the submission, such as its current status: Is the paper in the reviewers' hands? Has the editor received the evaluations but has not yet made a decision? Is there a conflict between reviewers? The sequence fashion of the representation and the lack of shared contextual elements make it really difficult to notice information rapidly.

4.3 Lesson Learned

Modelling a group task realization supposes the management of several actor's viewpoints in order to be able to follow the development of the task. In the framework of the Contextual-Graphs formalism, a contextual graph is the representation of a task realization. Thus, it is possible to represent the interaction between actors as an activity representing an actor's task realization as a clearly defined part of another task realization, as in Fig. 1. However, it is not efficient to use this approach when two actors interact several times during the group task realization, as it supposes that an actor sending a request waits for the completion of the recipient's task. It is an over-simplification of a group effort. The representation of an actor's activity on each branch of a Contextual Graph is not efficient either for a task involving more than two actors, as it is contradictory with the definition of a contextual element with exclusive instantiations. Moreover, many task realizations could get mixed in a single contextual graph, violating this philosophy, which states that a contextual graph represents a single task realization. Thus, a particular mechanism for turns management in addition

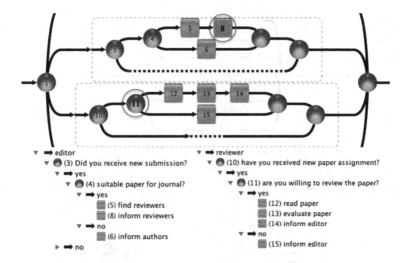

▼ ➡ editor
  ▼ ● (3) Did you receive new submission?
    ▼ ➡ yes
      ▼ ● (4) suitable paper for journal?
        ▼ ➡ yes
          ▤ (5) find reviewers
          ▤ (8) inform reviewers
        ▼ ➡ no
          ▤ (6) inform authors
  ▶ ➡ no

▼ ➡ reviewer
  ▼ ● (10) have you received new paper assignment?
    ▼ ➡ yes
      ▼ ● (11) are you willing to review the paper?
        ▼ ➡ yes
          ▤ (12) read paper
          ▤ (13) evaluate paper
          ▤ (14) inform editor
        ▼ ➡ no
          ▤ (15) inform editor

Fig. 3. A partial view of the branches representing the editor and reviewer's task realization

to the use of the shared context between the actors needs to be considered. The shared context will play the role of a virtual working context of the turns mechanism with contextual elements coming from the two working contexts associated with the task realization of the two actors.

5   The Turns Mechanism

To address the limits pointed out in the previous sections, we propose to manage turns with a specific contextual element. Figure 4 shows a "meta graph" for the paper submission example. This graph does not correspond to any actor's particular view, but to the management of the turns between actors. The different actor's views (e.g. the editor and the reviewer) are represented in individual contextual graphs shown in the activities. This representation introduces some reserved words to denote the name of the contextual elements in charge of the sequence of turns among actors. Such words are MANAGER, RECIPIENT and SENDER. The MANAGER is the actor responsible for the task realization at hand (the actor on focus); the RECIPIENT denotes the actor whose turn is next (the next actor on focus); and the SENDER is the actor whose turn is just ending (the actor releasing the focus). Thus, at the completion of the current task, the MANAGER informs the correct RECIPIENT, who in turn, will realize one task or another depending on the SENDER and the shared context. At the beginning of the graph in Fig. 4 the MANAGER becomes the last assigned RECIPIENT, thus the value of the contextual element 1 is known, and the corresponding branch is selected. By the end of a turn, the SENDER takes the value of the current MANAGER, since the next actor in turn needs to know the SENDER. Thus, a cycle is created through these contextual elements by instantiating the

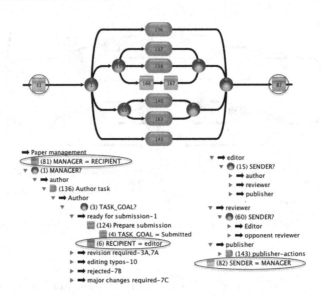

Fig. 4. A partial view of the Turns Mechanism model

initial working context of the current turn with the final working context of the previous turn (i.e. MANAGER = RECIPIENT).

This representation preserves and enhances the use of shared contextual elements. The semantic of these contextual elements are task dependent, and can adopt different values from one actor to another. The idea is to give a common ground to all the actors, for them to be able to know the current situation of the task realization and perform the corresponding task when their turn comes. A shared contextual element is instantiated by one actor and used by another. The TASK_GOAL is a reserved contextual element shared among the actors corresponding to the status of the document (e.g., submitted, reviewed, etc.). Figure 5 shows an example of changing turns. As shown in Fig. 5 at the reception of a new paper, the editor can decide if the paper is suitable for the journal or not. In case the paper is suitable, the corresponding actions are done in the editor's side, then the contextual element TASK_GOAL is instantiated to "to be reviewed" and the RECIPIENT is assigned as "reviewer"; which means that the turn of the editor has ended for now, and the focus should be placed on the reviewer, shown in Fig. 5 as a change from activity 137 to 140. Once the branch corresponding to this specific interaction between the author and the editor has been completed, the SENDER takes the value of "editor" in action 82 on Fig. 5, creating a loop in the reading of the graph, since the MANAGER instantiation will change and the next actor in focus should be found. In this case the focus will be on the reviewer (following the legends of Fig. 5) who detects that the editor has sent a paper to be evaluated, since the TASK_GOAL shared contextual element has been changed in the previous turn to "to be reviewed". The reviewer will change the TASK_GOAL instantiation to either accept or refuse

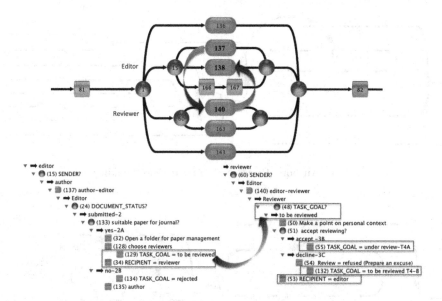

Fig. 5. A partial view of the Turns Mechanism shared context

reviewing the paper, and the RECIPIENT of the focus will be assigned to the editor. Thus, the loop in the reading of the diagram continues, this time from activity 140 to 138. Such a loop is broken when the RECIPIENT is equal to NIL (i.e. it is the end of the overall task).

6   The Practice Tree for Identifying Turns

The Contextual-Graphs software includes a Practice Tree View that helps decision makers to quickly identify the possible practices to develop, considering the information they possess. This view is also important for group tasks, since each tree branch corresponds to a turn. Figure 6 shows the practice tree of the contextual graph in Fig. 5 followed by the practice tree of the actor's activity 137. In the deployed view of activity 137 in Fig. 6, it should be noticed that a decision maker might need to have gathered a lot of information in order to choose the paths containing more contextual elements.

Figure 7 presents the author's practice tree, which is encapsulated in activity 136 in the general graph. Each branch finishes in a change of turns (e.g. author to editor, and author to publisher), meaning that the focus of attention moves from the editor to another actor. Branch 1 of Fig. 7 corresponds to the beginning of the group task, since the author sends a paper to the editor of the journal. The level of interaction in branch 1 is at "task level", regarding the types of interaction presented in Table 1, since the editor's turn is activated until a new paper submission has been received, the author instantiates the TASK_GOAL contextual element to "Submitted" just before their turn is over. The second branch

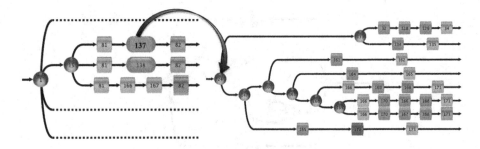

Fig. 6. Author's Practice Tree for decision making

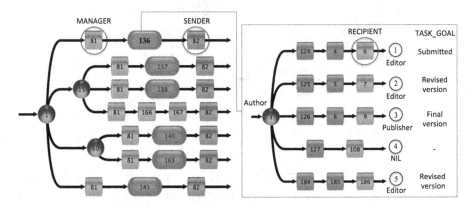

Fig. 7. Author's Practice Tree

corresponds to the author finishing his turn by instantiating the TASK_GOAL contextual element to "Revised version" and passing the focus of attention to the editor. Here the interaction is at the "activity level", since the editor task is different to the author, but they communicate to reach agreements regarding the submitted paper, e.g. the author attaches a document explaining the changes made to the revised version. Once the submission is accepted, on branch 3, the author and the publisher interact at the "team level", since they both work on the document to create the last version of the paper. When the author finishes correcting the typos, he instantiates the TASK_GOAL contextual element to "final version". Branch 4 does not create any further interaction, since it corresponds to the case in which the paper has not been accepted. Finally, branch 5 is really similar to branch 2, because it corresponds to the case in which the author has been asked by the editor to change the paper. The difference with branch 2 is that this time, the editor asks for a complete new version, not just some small changes. Thus, the interaction is at "activity level" and the TASK_GOAL contextual element is instantiated to "revised version". Although not all interaction levels presented in Table 1 can be found in the paper submission example, it is important to do not forget about them, since they are present in different types of group tasks, that have lead to the development of divers applications, such as the ones mentioned in Table 1.

By analyzing the levels of interaction detected in the paper submission group task, we see that, as the actors become more engaged in the group task realizations, the need of just coordination among them, evolves to collaboration and then to cooperation. In branch 1 of Fig. 7, the interaction between the author and editor requires coordination, as the editor tasks starts after receiving a new paper. Not presented in the author's interactions, but following the flow of the submission example, once the editor has sent the paper for reviewing, the interaction between the two reviewers denotes collaboration, as they both realize the same task in their own way. However, the interaction among the editor and the reviewers is at "planning level". Moreover, in Fig. 7, branches 2 and 5, the collaboration evolves to cooperation between the author and the editor, as their common goal is to obtain a new paper version, but each one has a different task to realize. While the author is in charge of providing a new paper version, the editor should answer the author's questions. Finally, once the paper has been accepted for publishing, the author and the publisher engage into a strong cooperation, as they both work in the same document to get it ready for publishing.

Beyond the Paper Submission Example. The paper-submission example is a task that involves several actors constantly interacting to reach a goal. However, group tasks are not limited to this type of team configuration, in which a leader coordinates the group actions and the shared context is reduced. Shorter and more dynamic group tasks (e.g. brainstorming sessions), as long as those including two groups working together at specific times might have several leaders coordinating the task realization of their own group. Thus, further discussion can be held regarding the passing of turns among groups, and the efficiency on the quick management of the shared context modifications.

7 Conclusions

Discussions on collaboration and cooperation must be considered in a large spectrum, going from a superficial approach focused on the built of applications, as found in the literature, to an approach aiming to understand the way people really work together, as proposed in this paper.

There is a need to extend the scope of the research on collaboration and cooperation upstream (experts working jointly to develop a contextual model satisfying everybody) as well as downstream (the contextual model as a medium of communication between experts). This means that it looks difficult to obtain a unique way to create contextual graphs for representing group task realizations. However, this paper explore some approaches. Through the analysis of the paper-submission example, we have notice the importance of the shared context, since it is a way to make compatible an actor's viewpoint with the rest of the group, because each actor is an expert in their own domain. Furthermore, the shared context, along with some reserved words, conform the turns mechanism we propose. Such mechanism helps experts to use the Contextual-Graphs formalism to model their group tasks and to analyze the type of interaction produced in each turn. Moreover, the understanding of a group interaction, will also help experts to make decisions regarding the way the task are realized, and application designers would be able to easily spot the real requirements for supporting a specific group tasks.

Acknowledgment. We thank the CONACyT (Consejo Nacional de Ciencia y Tecnología) for funding Kimberly García's post-doctoral fellowship at LIP6, UPMC.

References

1. Baresi, L., Garzotto F., Paolini, P.: Extending UML for modeling web applications. In: 34th Annual Hawaii International Conference on System Sciences, IEEE (2001)
2. Bedwell, W.L., Wildman, J.L., DiazGranados, D., Salazar, M., Kramer, W.S., Salas, E.: Collaboration at work: an integrative multilevel conceptualization. Hum. Resour. Manag. Rev. **22**(2), 128–145 (2012). Elsevier Science
3. Brezillon, P.: Contextualization of scientific workflows. In: Christiansen, H., Kofod-Petersen, A., Schmidtke, H.R., Coventry, K.R., Beigl, M., Roth-Berghofer, T.R. (eds.) CONTEXT 2011. LNCS, vol. 6967, pp. 40–53. Springer, Heidelberg (2011)
4. Brézillon, P.: Context modeling: task model and practice model. In: Richardson, D.C., Kokinov, B., Roth-Berghofer, T.R., Vieu, L. (eds.) CONTEXT 2007. LNCS (LNAI), vol. 4635, pp. 122–135. Springer, Heidelberg (2007)
5. Brézillon, P.: Task-realization models in contextual graphs. In: Dey, A.K., Leake, D.B., Turner, R., Kokinov, B. (eds.) CONTEXT 2005. LNCS (LNAI), vol. 3554, pp. 55–68. Springer, Heidelberg (2005)
6. Dourish, P., Bly, S.: Portholes: supporting awareness in distributed work groups. In: Human Factors in Computer Systems CHI 1992, California, USA, May 1992
7. Elmarzouqi, N., Garcia, E., Lapayre, J.C.: CSCW from coordination to collaboration. In: Shen, W., Yong, J., Yang, Y., Barthès, J.-P.A., Luo, J. (eds.) CSCWD 2007. LNCS, vol. 5236, pp. 87–98. Springer, Heidelberg (2007)
8. Fan, X., Zhang, R., Li, L., Brézillon, P.: Contextualizing workflow in cooperative design. In: 15th International Conference on Computer Supported Cooperative Work in Design (CSCWD-11), pp. 17–22. IEEE, Lausanne, Switzerland, June 2011
9. Gulati, R., Wohlgezogen, F., Zhelyazkov, P.: The two facets of collaboration: cooperation and coordination in strategic alliances. Acad. Manag. Ann. **6**, 531–583 (2012)
10. Pomerol, J.-C., Brézillon, P.: Dynamics between contextual knowledge and proceduralized context. In: Bouquet, P., Serafini, L., Brézillon, P., Benercetti, M., Castellani, F. (eds.) CONTEXT 1999. LNCS (LNAI), vol. 1688, pp. 284–295. Springer, Heidelberg (1999)
11. Salber, D., Dey, A.K., Abowd, G.D.: The context toolkit: aiding the development of context-enabled applications. In: The SIGCHI Conference on Human Factors in Computing Systems (CHI 1999), pp. 434–441. ACM, NY, USA, New York (1999)
12. Weiser, M.: The computer for the 21st century. Sci. Am. **265**(3), 94–104 (1991)

Representing and Communicating Context
in Multiagent Systems

Sonia Rode and Roy M. Turner[✉]

School of Computing and Information Science,
University of Maine, Orono, ME 04469, USA
{sonia.rode,rturner}@maine.edu

Abstract. Context-aware agents operating in a cooperative multiagent system (MAS) can benefit from establishing a shared view of their context, since this increases coherence and consistency in the system's behavior. To this end, agents must share contextual knowledge with each other. In our prior work on context-mediated behavior, agents used frame-based contextual schemas (c-schemas) to explicitly represent and reason about context. While an expressively rich approach, the lack of formal structure poses problems for mutual understanding of c-schemas among agents in a MAS. As we are interested MASs with heterogeneous agents, not only will agents represent c-schemas in idiosyncratic ways, but the set of c-schemas known by each agent will differ. In this paper we propose a new, related representation of contextual knowledge using description logic and a shared ontology, and we present a technique for communicating contextual knowledge while respecting bandwidth limitations.

Keywords: Multiagent systems · Communicating contextual knowledge · Context representation

1 Introduction

Multiagent systems (MASs) are groups of intelligent agents that interact, usually to carry out a set of goals. They are of interest for a variety of tasks, from autonomous exploration to data collection (e.g., [6]) to e-commerce. A MAS may be cooperative, in which case the agents work together to achieve common goals, or the individual agents may be self-interested and each work to satisfy their own goals, which may or may not align with some global set of goals.

While a great deal of work has been done on the problem of ensuring that individual agents behave appropriately for their context, much of it reported in this conference series, much less has focused on agents working in a MAS. However, context-appropriate behavior is just as important for an agent in a MAS, and it is important that the MAS as a whole behaves appropriately for its context.

The problem is more difficult than for a single agent. The contexts of individual agents now *always* include other agents, which may be unpredictable to

© Springer International Publishing Switzerland 2015
H. Christiansen et al. (Eds.): CONTEXT 2015, LNAI 9405, pp. 257–270, 2015.
DOI: 10.1007/978-3-319-25591-0_19

some extent and which are themselves behaving in ways influenced by their own contexts. There is also the opportunity for agents to gather information from others to better understand their context, but at the cost of added complexity, effort, and time.

While we could focus only on agent-level context recognition and hope that globally-appropriate behavior will emerge from the interactions of the agents, this suffers from problems similar to agent-level control of planning in multi-agent systems, primarily a lack of global coherence. Instead, the MAS, or at least a subset of its agents, should attempt to share information about the context to arrive at a shared, more complete "partial global context" (to borrow from partial global planning [7]). The individual agents, as well as any control mechanisms for the MAS as a whole, can then take the shared context into account when determining how to achieve individual goals, organize the agents, assign tasks, and coordinate agents' actions.

The ability to reason and communicate about the context implies that the context is explicitly represented. It also implies that there is a *message protocol* for communicating about the context, and that there is a way for different agents to represent contexts and agree on the meaning of contextual knowledge. This implies that there exists a representation language for contextual knowledge as well as a shared ontology that the agents can refer to for terms' meanings.

In earlier work [18], we described an approach to multiagent context-appropriate behavior that we called distributed context-mediated behavior, which we refer to here as multiagent context-mediated behavior, or MASCon.[1] This approach relies on agents communicating about their perceived context, knowledge, goals, and percepts in order to arrive at a representation of the MAS's global context (the context representation, or CoRe). The process involves context representation, local context assessment, communication, and a distributed assessment of the global context.

This paper focuses on the communication aspects of MASCon, including how contextual knowledge is represented to facilitate communication (and reasoning) about context. We focus first on representation, and describe the description logic-based representation of contexts and contextual knowledge, an ontology for contextual knowledge, and how that knowledge is represented as c-schemas. The c-schemas themselves can be viewed collectively as forming a kind of ontology for contexts. We then discuss context-related communication in MASCon. Part of this involves a message protocol for communicating about context that attempts to minimize bandwidth needed, (synergistic with any data compression that might be used) which is a key concern for some domains (e.g., a MAS consisting of underwater vehicles). The other part is deciding what to communicate about the context in order for agents to arrive at a shared understanding.

---

[1] The name seems appropriate, since just as a mascon is a concentration of mass that affects (e.g.) a satellite's orbit, our approach relies on a concentration of contextual knowledge to affect a MAS's behavior.

2   Representation Language

Agents need a shared language and ontology in order to communicate about anything, not just their context, and there has been much work on both in artificial intelligence, especially in the area of multiagent systems. Unfortunately, most work on knowledge representation and ontologies has not focused on contextual knowledge per se, but rather on domain and problem-solving (e.g., planning) knowledge. This is largely because context has seldom been considered as a first-order concept, but instead has been treated implicitly.

Context *has* been considered a first-order concept in some formal logic work, especially in the context community (e.g., [4,8,12]), and in some non-logical approaches (e.g., [3,9]). Our previous work has also addressed this by creating explicit representations for contexts (c-schemas) and for the contextual knowledge they must contain [17].

Unfortunately, our prior work lacked a formal representation language, and the ontology and semantics were idiosyncratic to each project and somewhat ad hoc. This is problematic if agents are to interact with others that may not have the same designers or reasoning mechanisms, such as would be the case in some open multiagent systems (e.g., autonomous oceanographic sampling networks [6]). In addition, the representation was frame-based, which has some beneficial properties, especially knowledge clustering, but for which there are no really good, widely-accepted reasoning mechanisms as there are, say, for formal logic.

What is needed, then, is an ontology for contextual knowledge and a way to represent it for communication that has a well-defined, formal basis, for which there are tractable reasoning mechanisms, and that is amenable to being related to a shared ontology.

For these reasons, we are now basing our contextual knowledge representation on *description logic* [1], a widely-used formalism in multiagent systems and the semantic web [2]. There has been other work on representing contextual knowledge as description logic, for example the work of Wang et al. [19], which was based on the Web Ontology Language (OWL). However, since their representation of context does not include guidance for behavior, it is not sufficient for our purposes.

We assume that most readers will have some familiarity with description logic (DL), and only a quick overview is presented here to allow others to understand terms in the rest of the paper. DL is a set of languages based around the idea of sets of individuals, restrictions on set membership, operators, and subsumption. A description of a set of individuals is termed a *concept*, for example, AUV (autonomous underwater vehicle). Concepts are viewed as having *roles* that can be used to restrict the individuals that are members of the set; for example, (AND AUV (SOME hasColor Yellow))[2] would denote the set of yellow AUVs. This example also shows an operator, conjunction, and the existential quantifier.

---

[2] Sometimes written AUV ⊓ ∃ hasColor.Yellow.

Determining *subsumption* is the primary inference type in DL: if A and B are concepts (i.e., descriptions of sets), then A is said to subsume B if $B \subseteq A$. For example, the atomic concept AUV subsumes the more restricted description above for yellow AUVs. Although relatively straightforward, for some description logics, subsumption checking is intractable in the worst case.

The particular DL we use in our work is a version of the language L1 as specified by Teege [15], which allows concept union, concept intersection, existential role restriction, minimum cardinality role restriction, and role composition. Axioms can be defined using these operators and used as concept definitions. The operators provide sufficient expressive power for our purposes. The L1 language has the property of *structural subsumption*, which means that the subsumption test on concepts always reduces to subsumption tests of single clauses. A clause is a description that cannot be further decomposed into a conjunction. The reason we require structural subsumption is to allow for an efficient algorithm for communicating contexts, which will be discussed in detail later.[3]

The version of L1 we use adds datatypes, equivalent to the way one of the Web Ontology Language (OWL) variants, OWL-DL [13], uses them. Datatype roles are permitted, which are similar to regular roles but with data values (e.g., integers, strings, etc.) as opposed to concepts as the fillers. This does not interfere with the use of structural subsumption in our algorithm. For our purposes, datatype roles are treated like regular roles, and we consider that a data value "subsumes" another whenever the two values are of the same type.

A concept such as Yellow or AUV is an *atomic concept*. These concepts live in an *ontology*, an isa hierarchy that directly shows the set–subset relationships between the concepts. Figure 1 shows a portion of the ontology we use in this project, for example (with subtrees not shown for some concepts).

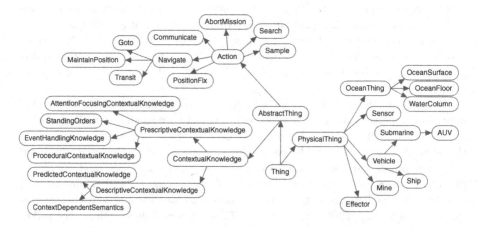

Fig. 1. A portion of the ontology. Thing is the top-level concept. Some subtrees are hidden to save space.

---

[3] Structural subsumption is weaker than logical subsumption [1].

All concept definitions as well as all axioms in DL belong to what is called the reasoner's *terminological box*, or TBox. In the case of this project, all agents will have some common knowledge in their TBoxes, that is, their shared knowledge, including the ontology, will mostly reside there.

3 Contextual Knowledge

There are two main concerns when explicitly representing knowledge about contexts: representing the contexts themselves, and representing the kinds of contextual knowledge they can contain.

MASCon, like all of our projects based on context-mediated behavior, represents contexts as knowledge structures called *contextual schemas*, or c-schemas. In the past, these have been frame-like structures with roles defining the kinds of knowledge being represented: knowledge for handling unanticipated events, for modifying goal priorities, and so forth. In this project, we have largely done away with this frame-like nature. Instead, c-schemas are primarily containers for description logic statements (concept descriptions and axioms) that apply in the represented context (cf. Guha's [10] microtheories).

A c-schema contains several types of knowledge about the represented context, each of which is best represented as a type of knowledge in its own right, i.e., a concept in the ontology. This allows the reasoner to easily determine what the knowledge is and how it is meant to be applied in the situation. Some researchers in the pervasive computing community (e.g., [19]) have also developed ontologies for context, but the kinds of contextual information used in that community tend to be only a subset of what is needed for context-sensitive behavior in agent-based and multiagent based systems.

We can broadly classify the needed contextual knowledge as being either *descriptive* or *prescriptive*. The former describes the features expected in the context, that is, the features of situations that are instances of the context. This knowledge is used in assessing the context, and it can be used to generate predictions about unseen features of the situation and to help understand newly-seen features. Part of descriptive knowledge is also any context-dependent semantics, for example, what a fuzzy logic or description logic concept might mean in the context that is different from its normal meaning. Prescriptive knowledge tells the agent how to behave in the context. There are several kinds: knowledge about goals and their context-appropriate priority, ways to achieve goals in the context, how to recognize and handle unanticipated events, and how to set non-goal-based behavioral parameters appropriately (e.g., sonar status, recommended depth envelope, etc.). In the case of multiagent systems, actions would also include such things as how/when/what to communicate, how to organize the group of agents, authority relationships, if any, and so forth.

Each kind of contextual knowledge is present in the agents' shared ontology, as shown in Fig. 1 (StandingOrder represents behavioral parameter settings). This allows agents to communicate about the contents of c-schemas without the problem of being misunderstood. The definitions of concepts in the ontology contain not just their name, but also their roles and their definitions in

PredictedContextFeatures:
  (expectsPresenceOf *some* AbstractThing) *or*
  (expectsPresenceOf *some* PhysicalThing)
ContextDependentSemantics:
  (hasFuzzyFeature *some*
    (AbstractThing *and*
      (hasFuzzyMembershipFunction *some* FuzzyMembershipFunction)))
StandingOrders:
  (hasActivePeriod *some* ContextLifeCycle)*and*
  (hasOperationalSetting *some*
    ((SelfState *and* (hasAdvisedLowerBound *some* Number)) *or*
    (SelfState *and* (hasAdvisedUpperBound *some* Number)) *or*
    (SelfState *and* (hasAdvisedValue *some* ValuePartition))))
EventHandlingKnowledge:
  (handlesEvent *some* Event) *and* (hasImportance *some* DefaultValuePartition)
  and (respondsWithAction *some* Action)
AttentionFocusingKnowledge:
  (definesGoal *some* AbstractThing) *and*
  (hasCost *some* DefaultValuePartition) *and*
  (hasDegreeExpected *some* DefaultValuePartition) *and*
  (hasImportance *some* DefaultValuePartition) *and* (isAchievedBy *some* Action)
ProceduralKnowledge:
  (definesAction *some* Action)

Fig. 2. Some definitions of contextual knowledge concepts

terms of other concepts. Figure 2 shows some of the definitions for our contextual knowledge.

The actual concept descriptions in a c-schema will make use of these concepts; for example, the description:

```
(AND EventHandlingKnowledge (SOME handlesEvent PowerFailure)
                            (SOME hasImportance High)
                            (SOME respondsWithAction AbortMission))
```

describes a piece of event-handling knowledge telling an AUV that is suffering a power failure that the event is very important and can best be handled in this context by aborting the mission.

Within a c-schema, a piece of contextual knowledge (an assertion) is associated with a name that is unique across all of the agent's knowledge. For shared ("prototype") c-schemas that are part of the MAS's common knowledge, all agents know these unique names. We require this to reduce bandwidth via our message protocol (see below).

In addition to a name, each assertion can also have associated metadata that is not part of the DL description. This is useful for knowledge that would be inconvenient or impossible to represent using the description logic in use. For example, we would like each concept within a c-schema to have an associated certainty factor (CF) representing the agent's certainty that the concept occurs or is relevant to the context; this is used by MASCon in context assessment, making predictions, etc. However, if the CFs were represented as part of the assertion, then they would be taken into account during subsumption, causing subsumption that would otherwise succeed to instead fail due solely to differing CFs. Consequently, CFs are represented as metadata.

Figure 3 shows an example of part of a contextual schema in our approach, in this case, one that represents being in the context of performing a sampling mission. The names of the pieces of contextual knowledge are unimportant for our purposes, but note that each piece has a description and a piece of metadata, the certainty factor. The c-schema predicts (or matches) that the mission area is large (BSM-1) and that the agent has a sampling mission active (BSM-2). What "large" (Broad) means is also defined in terms of a fuzzy membership function (BSM-3). This is one way in which the semantics of terms used by the agent are context-dependent in our approach. The c-schema also suggests a behavioral parameter setting that is appropriate for the context (a "standing order"), i.e., that obstacle sensitivity should be High (BSM-4). The portion shown also contains some event-handling knowledge about sensor failures as well as some action information.

```
BSM-1(0.60): (AND PredictiveContextualFeature
                  (SOME expectsPresenceOf
                        (AND MissionArea (SOME hasFuzzyValue Broad))))
BSM-2(0.34): (AND PredictiveContextualFeature
                  (SOME expectsPresenceOf (AND SamplingMission
                                              (SOME hasSamplingTarget Thing))))
BSM-3(0.41): (AND ContextDependentSemantics
                  (SOME hasFuzzyFeature
                        (AND Broad
                             (SOME hasFuzzyMembershipFunction
                                   (AND ShoulderFunction
                                        (SOME hasLocalMinAt (AND Float 5.0 mi^2))
                                        (SOME hasLocalMaxAt
                                              (AND Float 20.0 mi^2)))))))
BSM-4(0.59): (AND StandingOrder
                  (SOME hasActivePeriod DuringContext)
                  (SOME hasOperationalSetting
                        (AND ObstacleSensitivity (SOME hasAdvisedValue High))))
SM-5(0.064): (AND EventHandlingKnowledge
                  (SOME handlesEvent SensorFailure)
                  (SOME hasImportance High)
                  (SOME respondsWithAction
                        (AND TransferData (SOME hasObject
                                               (AND Sensor
                                                    (SOME hasStatus FailureImminent)))
                                          (SOME toObject MAS))))
BSM-7(0.55): (AND EventHandlingKnowledge
                  (SOME handlesEvent SamplingComplete)
                  (SOME hasImportance High)
                  (SOME respondsWithAction Transit))
BSM-8(0.74): (AND ProceduralKnowledge
                  (SOME definesAction
                        (AND Transit (SOME hasIndex Survey))))
```

Fig. 3. Part of the contextual schema BroadSamplingMissionCtx (Format: label(CF): description)

Conceptually, contextual schemas themselves are part of the ontology. Each c-schema describes a concept corresponding to a context, or set of situations. C-schemas exist within generalization/specialization hierarchies in much the same way concepts are related an ontology. For example, the context "in a harbor" is a generalization of "in Portsmouth Harbor". In addition, agents share knowledge about c-schemas representing prototype contexts.

However, c-schemas are different than other concept descriptions. First, they contain a significant amount of knowledge that is not in the form of DL roles (e.g., metadata, information to index other related c-schemas, etc.). Second, although they do have specialization/generalization relationships, these can represent other c-schemas that they were derived from or that they can be found from (in memory) instead of true subclass relationships. Third, unlike an ontology, both the set of c-schemas and their relationships to one another are expected to change relatively frequently as an agent experiences new situations that lead to creating new c-schemas or modifying existing ones. And fourth, unlike ontology concepts, many c-schemas will be idiosyncratic to particular agents, since not all agents will experience the same contexts as they operate.

For these reasons, we treat c-schemas differently than other parts of an agent's ontology. An agent has a separate c-schema memory that changes over time as it gains experience. The schema memory is assumed to organize c-schemas in generalization/specialization hierarchies that can be traversed based on features of the situation to find c-schema(s) matching the situation [17,18]. Such a memory is essentially a set of dynamic discrimination networks that change based on the memory's contents. As an agent gains experience, it will create and store new c-schemas in this memory, and it may learn new connections between existing c-schemas. Shared prototype c-schemas are considered fixed across the agents, corresponding in some ways to a shared "upper ontology". However, how c-schemas are indexed in agents' memory and any idiosyncratic c-schemas derived from the prototypes will likely differ from agent to agent.

4   Communicating About Context

In MASCon, agents communicate about their context during distributed context assessment as well as when deciding what the context means in terms of their behavior. If only the prototype contexts were considered, communication would be trivial: just an identifier for the shared c-schema would need to be sent. However, it is more likely over time, as agents learn new contexts, that agents will each believe that the current situation is an instance of one of their own known idiosyncratic contexts.

We do not want the agents just to send the complete contents of such c-schemas to others. One reason is bandwidth. In many domains of interest, bandwidth is quite limited; for example, in the underwater vehicle domain, maximum bandwidth is on the order of 60 kbit/s or less [14]. Consequently, saving bandwidth is critical for MASs operating in those environments. The second reason is a matter of focus. If all information is sent, then the other agent has to try to match a large amount of knowledge against all of its own c-schemas; if we can provide some commonality, then the receiver can focus immediately on its own c-schema and the differences between it and what was sent.

In our approach, an agent makes use of its shared contextual knowledge, represented as prototypical c-schemas, to communicate only what is needed to allow the recipient to regenerate the idiosyncratic contexts from its own prototypical contextual schemas. This is reminiscent of earlier work on the agent communication language COLA [16], which also was concerned with limiting bandwidth by appeal to shared knowledge. While at the current time we are focused on communicating contextual knowledge, our approach is not incompatible with a COLA-based communication system.

There are two major problems to be addressed for agents communicating about contextual knowledge. The first is what *message protocol* to use when exchanging messages. The second is determining what to send.

4.1 Message Protocol

MASCon's message protocol focuses on the different kinds of relationships between knowledge in an idiosyncratic context and in prototype ancestors. Communicating about an idiosyncratic context will require multiple messages, since the prototypical context will have to be identified, then differences from it will need to be communicated. Which message types are sent is determined by the algorithms described in the next section.

Figure 4 shows the grammar for our message protocol. Strings on the right-hand side of `<ck-code>` are abbreviations for our six types of contextual knowledge: predicted context features, context-dependent semantics, standing orders,

```
        MESSAGE ::= ALL-MSG | ALL-EX-MSG | SOME-MSG | NEW-MSG |
                    MOD-MSG | START | END
          START ::= "CTX"
            END ::= "CTX-END"
        ALL-MSG ::= "ALL" <ck-type>? <proto> <new-cf>*
     ALL-EX-MSG ::= "ALL" <ck-type>? <proto> "EXCEPT" <ck-code>+
                    <new-cf>*
       SOME-MSG ::= SOME <ck-code>+ <new-cf>*
        NEW-MSG ::= "NEW" <full-ck-descrip> <new-cf>
        MOD-MSG ::= "MODIFIES" <ck-code> <dl-difference>+ <new-cf>?
      <ck-type> ::= "PCF" | "CDS" | "SO" | "EHK" | "AFK" | "PK"
        <proto> ::= a name of a prototypical context
       <new-cf> ::= float
      <ck-code> ::= name of contextual knowledge item in prototypical
                    context
<full-ck-descrip> ::= a KRSS (this is a DL syntax) description that
                    is-a ContextualKnowledge
<dl-difference> ::= a KRSS description
```

Fig. 4. The message protocol

```
;; AbortMission goal has high importance, has medium cost,
;; and is achieved by action Abort:
rm-afk-1(0.93): (AND (SOME definesGoal AbortMission)
                     (SOME hasImportance High)
                     (SOME hasCost Medium)
                     (SOME isAchievedBy Abort))
;; AbortMission has medium importance, has medium cost, and is
;; achieved by aborting to sea floor}:
sm-afk-1(0.89): (AND (SOME definesGoal AbortMission)
                     (SOME hasImportance Medium)
                     (SOME hasCost Medium)
                     (SOME isAchievedBy
                     (AND Abort (SOME hasObject SeaFloor))))
```

Fig. 5. Two partially-matching pieces of contextual knowledge

event-handling knowledge, attention-focusing knowledge, and procedural knowledge.

When an idiosyncratic context K contains some of the same contextual knowledge as a prototype ancestor (shared) c-schema P, an ALL-MSG, ALL-EX-MSG, or SOME-MSG is used. Suppose P has three pieces of event-handling knowledge with identifiers $p1$, $p2$, and $p3$. Then the message

ALL EHK P 0.9 0.8 0.7

that K has all the event-handling knowledge from P. The float values are the certainty factors for the three pieces of knowledge in K. The ordering of the certainty factors in the message correspond to the order in which the event-handling knowledge is found in P. In contrast, the message

ALL EHK P EXCEPT p2 0.9 0.7

means that K includes all the event-handling knowledge from P except $p2$. We can refer to the pieces of knowledge by name, as we can rely on their order, since the representations of the prototypes are known to both the sender and receiver as shared knowledge. Alternatively, the message

SOME p1 p3 0.9 0.7

has the same meaning as the ALL-EX message, and is the better choice in this case because it is shorter.

Up to this point, we have been concerned with the case in which K may not match P exactly, but some or most of its corresponding contextual knowledge does match. However, corresponding pieces of contextual knowledge may only partially match in many cases. This case is handled by the MOD-MSG message type.

The two descriptions of AttentionFocusingKnowledge shown in Fig. 5 have some role restrictions in common and some that differ. If we suppose that rm-afk-1 belongs to a prototypical c-schema, we can use a MOD-MSG to express sm-afk-1. The <dl-difference>+ part of a MOD-MSG message will describe the differences between rm-afk-1 and sm-afk-1 and allow for the agent receiving the message to reconstruct sm-afk-1.

A NEW-MSG message, which sends a verbatim description of knowledge, is used as a last resort when none of the other message types can capture a piece of knowledge. The circumstances in which a NEW-MSG is used are outlined in the next section.

4.2   Deciding What to Send

We have devised an algorithm by which agents can generate a set of messages to completely describe an idiosyncratic context. The algorithm, which assumes an agent can correctly retrieve the prototypical ancestor(s) of an idiosyncratic context, is divided into two phases. The first phase generates message types ALL-MSG, ALL-EX-MSG, and SOME-MSG, which are the message types that deal with direct matches between idiosyncratic and prototypical knowledge. The second phase generates message types MOD-MSG and NEW-MSG, which cover all the rest. Both parts of the algorithm require a DL reasoning engine.

Direct Match Algorithm. Let C_I be an idiosyncratic c-schema and C_P be the set of its prototype ancestors, and let ck_i refer to a piece of knowledge in C_I. To find direct matches for contextual knowledge in C_I, the DL reasoner is used to look for concept synonyms in the combined knowledge of C_P for each ck_i.

Our goal is to partition the set of directly matched ck_i into subsets, where each subset is covered by a single message from the types ALL-MSG, ALL-EX-MSG, or SOME-MSG, such that the total number of bytes of the messages is minimized. Unfortunately, this is an exponential-time problem. Consequently, we use a greedy algorithm to approximate this partitioning. The algorithm repeatedly loops through the six contextual knowledge types until all the direct matches have been handled by a message. It finds the prototype context with the most unhandled direct matches of the current knowledge type, and uses this prototype context as a reference point to create a message covering this knowledge. Thus at each iteration we pick a subset and produce a message for it. The messages that cover each subset in the partition can then be sent, with their certainty factors based on the idiosyncratic context C_I.

Computing Differences. In addition to messages for the directly-matching contextual knowledge, the agent must send messages for the rest of the knowledge in the idiosyncratic c-schema C_I. These messages will have to be of types MOD-MSG or NEW-MSG, each covering a single piece of knowledge ck_i. Let $M_I = \{ck_i|ck_i$ has no direct matches$\}$. The algorithm iterates through each $ck_i \in M_I$, determining which message type to use.

The algorithm attempts first to use a MOD-MSG message, since that should be shorter than using a NEW-MSG message. To do this, it checks to see if there is a piece of knowledge ck_p from C_P that can be used as a point of reference for ck_i. The differences between ck_p and ck_i are then expressed in the <dl-difference>+ portion of the MOD-MSG message.

Our algorithm for computing differences between ck_i and ck_p makes use of the DL *subtraction operation* [15]. If B and A are two DL concepts and A subsumes B, then the subtraction (difference) operation, $B - A$, gives a new concept such that each piece of information in B that is present in A is removed. For DLs with structural subsumption, a simple implementation of this operator is possible.

In our case, we can find the differences even if the one concept does not subsume the other, since we know that the two are both of the same Contextual Knowledge type: they contain the same restriction clauses and only differ in the role ranges for the clauses. Consider again the two pieces of knowledge shown in Fig. 5, where `rm-afk-1` corresponds to ck_p and `sm-afk-1` is ck_i. The MOD-MSG message expressing `sm-afk-1` is

```
MODIFIES rm-afk-1 (SOME hasImportance Medium)
  (AND Abort
    (SOME hasObject seaFloor)) 0.89
```

Rules outline how the recipient can reconstruct `sm-afk-1` based on the message contents. Let d be a description in the `<dl-difference>`+ portion of a MOD-MSG M, p be the description of the `<ck-code>` in M, and `type(p)` be the ContextualKnowledge subclass to which p belongs. Then d is interpreted as follows:

1. If d is a restriction on role r, a required role of `type(p)`, d is meant to replace the restriction on role r found in p.
2. If d is a restriction on any other role, it is meant to be ANDed to p.
3. If d is a conjunction containing a primitive base concept, let g be a superclass of d such that g is found in p. Then d is meant to replace each instance of g in p.
4. If d is of any other form, it is an error.

Sometimes it is not possible to create a MOD-MSG message for two pieces of differing contextual knowledge, for example in the case where the only Contextual Knowledge type in common is ContextualKnowlege itself. The rules for reconstructing a MOD-MSG allow successful creation of a MOD-MSG referencing any ck_p that is the same ContextualKnowledge bottom-level type as ck_i. However in some cases all possible ck_p for ck_i will have no subsuming role ranges for the required roles. In this case the MOD-MSG produced will contain all the conjoined clauses from ck_i, which is essentially the same as what is contained in a NEW-MSG. In this case, the NEW-MSG is shorter than using MOD-MSG and referencing a piece of prototypical knowledge.

5   Evaluation

The message generation algorithms have been implemented and tested using a randomly generated c-schema hierarchy. This was done to avoid accruing the extensive domain knowledge of the AUV domain needed to create a set of realistic contexts, which is not the focus of this work. Instead, well-structured contexts were generated based on the ontology.

The context-generation mechanism is implemented in Common Lisp and uses the reasoner RACER [11] for DL inference. The program first reads the description of the ontology, stored as an OWL ontology file. To create a complete c-schema, several pieces of each ContextualKnowledge type are generated and combined. To create a piece of contextual knowledge, for each role in its definition, a suitable concept is chosen as the value. If there are unfilled roles in that concept, then the process continues recursively.

A small number of c-schemas were created to serve as prototype contexts and placed in a context hierarchy. Idiosyncratic contexts were then generated by choosing prototypes as parents, then combining and randomly modifying their knowledge. Modifications included replacing concepts with their siblings or descendants and adding role restrictions.

For a preliminary evaluation, twenty idiosyncratic c-schemas were generated from four prototype c-schemas. The message generation procedure was run on each idiosyncratic c-schema, and the resulting messages were processed by a message receiver procedure that implemented the rules for interpreting messages. The c-schema produced by the receiver procedure matched the original c-schema each time. Further evaluation will be aimed at comparing the number of bytes of the generated messages to the minimum number of bytes required to encode the c-schema to determine the effectiveness of bandwidth-reducing heuristics.

6   Conclusion and Future Work

This paper describes an approach to context representation and communication to support multiagent context assessment by allowing agents to share their individually-known contexts with each other in an efficient manner. An ontology for contextual knowledge, represented using description logic, has been developed, and the representation of contexts themselves as c-schemas is also very much like an ontology. A message protocol and algorithms to support its use have been developed to allow an agent to decide which pieces of contextual knowledge it needs to send and how to send them.

The work reported is at an early stage, and so at this point, evaluation has been limited. The next step is to perform much more extensive evaluation to determine strengths/weaknesses of the approach and to quantify the efficiency of context communication in this approach.

Beyond representation and communication in MASCon, we are working on the problem of how agents can negotiate to come to an agreement on their shared context. To arrive at a consensus, agents must be able to evaluate others' knowledge based on their own. Our DL representation facilitates many possible techniques for an agent to compare pieces of knowledge. After receiving a contextual knowledge message, an agent can look for concept ancestors and descendants in its own evoked c-schemas, where ancestors represent more general and descendants represent more specialized knowledge. This can help an agent determine what aspects of the received knowledge it agrees with. An agent can also find the least common subsumer [5] of two knowledge descriptions, which finds the largest set of commonalities between two descriptions. In addition to techniques for evaluating knowledge, we are extending the message protocol to include message types for negotiation. Messages will be added for agreeing and disagreeing about received contextual knowledge and reasons for disagreement.

References

1. Baader, F.: The Description Logic Handbook: Theory, Implementation, and Applications. Cambridge University Press, Cambridge (2003)
2. Berners-Lee, T., Hendler, J., Lassila, O.: The Semantic Web. Sci. Am. **284**(5), 28–37 (2001)
3. Brezillon, P., Pasquier, L., Pomerol, J.C.: Reasoning with contextual graphs. Eur. J. Oper. Res. **136**(2), 290–298 (2002)
4. Buvač, S.: Quantificational logic of context. In: Working Notes of the IJCAI 1995 Workshop on Modelling Context in Knowledge Representation and Reasoning (1995)
5. Cohen, W.W., Borgida, A., Hirsh, H.: Computing least common subsumers in description logics. In: AAAI, pp. 754–760 (1992)
6. Curtin, T., Bellingham, J., Catipovic, J., Webb, D.: Autonomous oceanographic sampling networks. Oceanography **6**(3), 86–94 (1993)
7. Durfee, E.H., Lesser, V.R.: Using partial global plans to coordinate distributed problem solvers. In: IJCAI, pp. 875–883 (1987)
8. Giunchiglia, F.: Contextual reasoning. Epistemologia **16**, 345–364 (1993)
9. Gonzalez, A.J., Stensrud, B.S., Barrett, G.: Formalizing context-based reasoning: a modeling paradigm for representing tactical human behavior. Int. J. Intell. Syst. **23**(7), 822–847 (2008)
10. Guha, R.: Contexts: a formalization and some applications. Ph.D. thesis, Stanford University (1991)
11. Haarslev, V., Möller, R.: RACER system description. In: Leitsch, A., Nipkow, T., Goré, R.P. (eds.) IJCAR 2001. LNCS (LNAI), vol. 2083, pp. 701–705. Springer, Heidelberg (2001)
12. McCarthy, J.: Notes on formalizing context. In: IJCAI, pp. 555–560 (1993)
13. McGuinness, D.L., Van Harmelen, F.: OWL web ontology language overview. Technical report, W3C, W3C Recommendation, February 2004. www.w3.org/TR/owl-features
14. Song, H., Hodgkiss, W.: Efficient use of bandwidth for underwater acoustic communication. J. Acoust. Soc. Am. **134**(2), 905–908 (2013)
15. Teege, G.: Making the difference: a subtraction operation for description logics. KR **94**, 540–550 (1994)
16. Turner, E.H., Chappell, S.G., Valcourt, S.A., Dempsey, M.J.: COLA: a language to support communication between multiple cooperating vehicles. In: Proceedings of the Symposium on AUV Technology (AUV 1994), pp. 309–316. IEEE (1994)
17. Turner, R.M.: Context-mediated behavior. In: Brézillon, P., Gonzalez, A. (eds.) Context in Computing: A Cross-Disciplinary Approach for Modeling the Real World Through Contextual Reasoning, pp. 523–540. Springer, New York (2014)
18. Turner, R.M., Rode, S., Gagne, D.: Toward distributed context-mediated behavior for multiagent systems. In: Brézillon, P., Blackburn, P., Dapoigny, R. (eds.) CONTEXT 2013. LNCS, vol. 8175, pp. 222–234. Springer, Heidelberg (2013)
19. Wang, X.H., Zhang, D.Q., Gu, T., Pung, H.K.: Ontology based context modeling and reasoning using OWL. In: Proceedings of the Second IEEE Annual Conference on Pervasive Computing and Communications, pp. 18–22 (2004)

Using Contextual Knowledge for Trust Strategy Selection

Larry Whitsel[1](✉) and Roy M. Turner[2]

[1] University of Maine at Augusta, Bangor, ME 04401, USA
larry.whitsel@maine.edu
[2] School of Computing and Information Science, University of Maine,
Orono, ME 04469, USA
rturner@maine.edu

Abstract. In open multiagent systems in which some of the agents may be self-interested, it is vital that an agent be able to make trust decisions about its peers to determine which of them may be untrustworthy and how to behave in response. The agent's context, including the environment, observed and inferred actions and motives of other agents, and properties of the MAS as a whole, is critical to making an informed trust decision and especially to choosing a strategy for taking actions. However, most prior work in the area has ignored context or only treated it implicitly. In this paper, we present an implemented approach that explicitly represents the agent's context, informed by known contexts, and that uses that contextual knowledge to select the best strategy, even in the presence of untrustworthy agents.

Keywords: Multiagent systems · Trust decisions · Self-interested agents · Context representation · Miscreant agents

1 Introduction

An open multiagent system (OMAS) is a multiagent system in which agents can come and go and may not all be under the control of the same entity. An agent in an OMAS that contains self-interested agents must be able to identify others in the society that are likely to act in a way inimical to its own interests, to the society's interests, or both. We refer to these as *miscreant agents*. Once a miscreant has been identified, the agent must then decide which actions to take when interacting with it to protect and advance its own interests.

The problem of identifying a miscreant agent is essentially a *trust decision*: for each other agent in the system, can we trust that agent? We use *trust* here the general sense of Castelfranchi and Tan [3] to mean both that the agent is trustworthy in the normal sense and that it is capable of performing whatever action we are concerned with.

Unfortunately, trust decisions are made difficult by several factors. Self-interested agents will usually have utility functions for their actions that are

© Springer International Publishing Switzerland 2015
H. Christiansen et al. (Eds.): CONTEXT 2015, LNAI 9405, pp. 271–284, 2015.
DOI: 10.1007/978-3-319-25591-0_20

hidden or otherwise unpredictable, and so an agent will need to observe their behavior over time to detect patterns indicating their trustworthiness. In addition, an agent's view of the current situation will be incomplete, since in most domains the environment will be only partially-observable. Consequently, the agent will need to infer what its observations may have missed, including others' actions and their effects on others.

Determining how to behave in the presence of miscreants is also difficult. We call an agent's mapping (implicit or explicit) from particular configurations of the situation to actions its *strategy*. An agent will have multiple possible strategies to choose from, but not all will be appropriate for the context. In particular, the strategy in use will depend on the presence and kind of miscreant behavior.

Consequently, knowledge about the current context must be used to determine which strategy to select at any given time. As the context changes—for example, as miscreants come and go in the OMAS—the agent should use its contextual knowledge to select strategies to automatically tune its behavior in response.

Research on trust decisions has generally been concerned with identifying what we call miscreants by focusing primarily on the other agent. For example, socio-cognitive approaches [5], or that of Mao [10], used game theoretic or psychological attribution theory to either respond directly to or to infer an agent's motives. Reputation-based approaches (see, e.g., [12]) decide to trust an agent based on its past observed behavior. Machine learning approaches (e.g., [7]) have attempted to learn strategies based on a set of interactions with another agent.

Few if any approaches, however, have focused on using the context, including the environment, the goals of the agent (and the MAS), and observed properties of other agents, including their reputation, to infer the presence of miscreants and how to respond to them. Yet doing so allows better assessment of the true situation, and it can allow previously-successful strategies to be immediately identified and used.

In our approach, we use explicitly-represented contextual knowledge to help an agent identify and respond to miscreant agents. Known contexts—classes of situations—are represented in knowledge structures called *contextual schemas* (c-schemas) [17]. Features of the current situation, including environmental features, current goals, and possible indications of miscreant behavior, are used to select one or more matching c-schemas, essentially identifying the current context. The c-schema(s) then suggest appropriate strategies for the situation.

The approach is called CATS (Context-based Agent Trust System), and it has been implemented and tested in a toy domain (Liar's Dice) that is a reasonable surrogate for many real-world open multiagent system domains [18,19]. In this paper, we focus on how contextual knowledge is represented and used in CATS.

In the remainder of the paper, we will first discuss context and its representation, then how an agent using the CATS approach assesses and manages its context. This makes use of inferences about properties of agents based directly on their own actions as well as on their interactions with others (society-level analysis of motives [18]). We then briefly touch upon our evaluation of CATS via experiments in a simulated MAS.

2   Context and Context Representation

We view a *context* as a kind of situation, where a situation is itself the sum of all observed and unobserved features of the environment, the agent, other agents, etc. Contexts are classes of situations that have some implication for how an agent should behave [17]. For example, the context of an agent operating in an OMAS with a pair of collusive miscreant agents would likely encompass many different situations: playing bridge, participating in an auction, and so forth, where such agents were present. The utility of recognizing a particular situation as an instance of a known context is that it allows the agent to focus on the salient features and to select appropriate behavior.

Contexts are related to one another. One context can be a specialization of another, for example, or it can be a blend of several other known contexts. For example, an agent engaged in an Internet bandwidth auction where there are collusive agents present could consider its context as composed of other contexts such as participating in an auction, negotiating about bandwidth, and operating in the presence of collusive agents. In our approach, if there is something about the composite context that has important implications for the agent's behavior, then it will be remembered; otherwise, the agent will blend the components when it is encountered again.

As we have argued elsewhere (e.g., [17]), an agent should explicitly represent contexts it knows about rather than spreading contextual knowledge across its knowledge base (e.g., as rule antecedents or plan preconditions). Context representations can be a means of bundling facts and assertions about the world (cf. [6]), allowing all relevant information to be retrieved at once about the context, as well as facilitating knowledge acquisition and learning. In addition, explicit representations allow an agent to commit to what its context is, then automatically behave appropriately until the context changes, thus saving reasoning effort.

We represent contexts as *contextual schemas* (c-schemas) [17]. C-schemas both limit the scope of reasoning and bundle together related knowledge about (in CATS) trust decisions and appropriate behavior. Previous work on trust has used the agent's context only to limit scope, e.g., turning a dynamic situation into a static one by limiting the agent's predictions or decisions to one case and ignoring the rest. In contrast, we allow multiple contexts to be recognized at once and represented by multiple c-schemas as well as make use of context as a bundling mechanism by adding additional knowledge to our assessment of other agents' actions to help fill in missing knowledge.

CATS relies on a library of c-schemas from which the appropriate one(s) can be found based on the situation. In other work [9,17], c-schemas have been organized in a dynamic conceptual memory (e.g., [8]). Here, we make no commitment as to how c-schemas are stored; we simply assume that they can be retrieved as needed based on the situation's features.

We are not concerned in CATS with many of the kinds of information usually stored in c-schemas (e.g., context-dependent semantics, event-handling information, etc.). Instead, we are more concerned with strategies, that is, mappings from

```
(defrule in-wary
  (is-me ?me ?cf)
  (game-round (bidder-is ?x)(certainty ?cf0))
  (or (and
          (claim (claim-type ally)(source ?x)(target ?y)(certainty ?cf1))
          (claim (claim-type enemy)(source ?me)(target ?y)(certainty ?cf1)))
      (system-state (state-name global-alert)(certainty ?cf1))
      (claim (claim-type enemy)(source ?me)(target ?x)(certainty ?cf1))
      (or (belief-strength (belief trustworthy) (source ?x)
          (strength very-low)(certainty ?cf1))
          (belief-strength (belief trustworthy) (source ?x)
          (strength low)(certainty ?cf1)))
      (claim (claim-type cheating)(target ?x)(certainty ?cf1))
      (claim (claim-type enemy)(source ?x)(target ?me)(certainty ?cf1))
      (belief-strength (belief bid-aggressive) (source ?x) (target ?me)
          (strength very-high)(certainty ?cf1))
      (belief-strength (belief bid-aggressive) (source ?x) (target ?me)
          (strength high)(certainty ?cf1)))
  =>
  (assert (in-context wary ?cf1)))
```

Fig. 1. A rule from the contextual schema in-wary.

situations to actions. C-schemas in this project are associated with a strategy that is appropriate for all situations that are instances of the context represented. Thus, contexts in this project are delimited based on changes in strategy.

At the present time, to populate our agent's c-schema library, we rely on the system designer. In an ideal system, the agent would learn its own c-schemas, either by modifying those initially given to it or creating them de novo. This would require the agent to track the success of strategies in different situations and group together similar situations that share strategies successfully.

Showing a complete c-schema would require more space than can be allotted in this paper, but an example of one production rule that acts as descriptive knowledge for a c-schema called in-wary is shown in Fig. 1. In this case our agent is making use of an embedded CLIPS [13] rule-based system to implement its context manager.

3   Context Assessment and Management

Representing context is half the problem. The other is identifying the context the agent is in and managing the corresponding representations.

In CATS, we break this into four processes or modules: tracking the situation to produce inferences useful for recognizing the context (Track); society-level analysis of motives to produce additional interagent-based inferences (Analyze); recognizing the context and managing the contextual knowledge (Contextualize); and selection and execution of context-appropriate strategies by the agent (Execute).

Figure 2 shows an overview of our approach, the major modules discussed below, and some of the data paths between them.

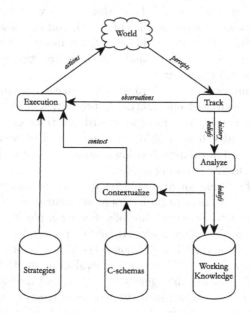

Fig. 2. Overview of the reasoner's architecture

3.1 Tracking the Situation

The Track module is primarily concerned with making non-society-level inferences about the situation that can help identify the context. It uses the agent's percepts and the history of those percepts over time to produce beliefs about the current situation that can then be used by the other CATS modules. This is reminiscent of some robotics software architectures (e.g., [2]), where raw, objective sensor data is aggregated and processed into subjective, symbolic, agent-centric inferences for use by (e.g.) situation assessment.

We are primarily interested here in how an agent can differentiate between agents it can trust and those who are miscreants. Thus, Track is concerned with establishing the values of *trust-warranting properties* (e.g. [1]) of others. For this, we look to humans for inspiration.

Humans, being social animals, are greatly concerned with recognizing deception by others; it is even possible that we have evolved specialized "cheater-detection" circuits in our brains [4]. However, even for humans, detecting deception is a difficult task. It is even more difficult in environments with electronic-mediated communications [15], which of course has implications for trust decisions by artificial agents.

Humans use the perceived intent of others' actions to help decide about trusting them. To do this, we rely on a combination of observable trust-warranting properties ("manifesta"), a history of previous interactions, and beliefs we have about the motivations of others [1,3]. From these, we develop beliefs about other, hidden trust-warranting properties ("krypta" [1]). Humans also use others' reputations to make trust decisions, as do some agent-based systems (for a review, see [14]). Others' reputations can be maintained by an agent itself or by relying on a (hopefully trusted) third party.

Both of these sources of trust-warranting information are mirrored in CATS' Track module. One source of information is properties of the problem domain that are easily observable, for example, board positions or moves in a game domain or bidders and values of bids in an auction. The value of this kind of immediately-available information has low *durability*, that is, it decays rapidly over time, and it is highly domain-specific.

Another kind of information available has to do with properties of the OMAS itself, such as protocols used, organizational structures in effect, and rules of behavior. These tend to be more durable. For example, the speed at which system rules or protocols change is constrained by the need to disseminate them to all participating agents. Although knowledge of how another agent interacts with the system itself may not be readily generalizable to other OMASs, it can give significant indications of the agent's intentions and trustworthiness.

Track produces beliefs about the world and about trust-warranting properties for each other agent in the society. We represent beliefs as tuples $< P, O, V, C >$, where P is the property that is the subject of the belief, O is the object(s) or agent(s), V is the value of the property for O, and C is a confidence value in the range $[-1, 1]$. O can consist of more than one object. For example, the belief that agent A has goal G might be represented as $< \mathtt{hasGoal}, (A, G), 0.735 >$.

Track adjusts its beliefs, including confidence in existing beliefs, as the situation changes. It takes what may be called a "frequentist" approach to belief confidence revision: the more observations that are made that support a belief, the more confident Track becomes about that belief. Knowledge about the durability of a property, as discussed above, factors into this as well.

Interagent interactions are another source of information important for making trust decisions. This is the purview of the Analyze module and its society-level analysis of motives.

3.2   Society-Level Analysis of Motives

An important source of trust-warranting information is how another agent interacts with others in the OMAS. By observing this, an agent can infer possibly-hidden motives. This is called *society-level analysis of motives* (SLAM, introduced earlier in relation to collusion detection [18]), which is done by the Analyze module.

Analyze depends on the beliefs created by Track about domain, system, and agent properties. It uses knowledge about how agent actions and their results can affect others to create beliefs about relationships between agents. It also takes into consideration exogenous events and actions such as messages between the agents, system-wide information messages, agent entry or exit, and so forth. The beliefs are then used by Contextualize as it determines what the current context is. Analyze also makes use of contextual information from Contextualize to change which knowledge it uses to create beliefs. This circularity is unavoidable, since the meaning of agent interactions is itself context-dependent.

Analyze examines each action taken by another agent to determine what it might mean to other agents, including itself. For example, an agent may take an action that indicates that it is unusually aggressive, that it is similar to another, or that provides disproportionate benefits to some other agent (possibly indicating collusion).

Occasionally, any agent's actions may seem to indicate miscreant behavior, for example, by going against our agent's interests or disproportionately benefiting some other agent. However, it is the *pattern* of an agent's behavior over time that is most important in making a trust decision, as well as any sudden unexpected changes in the pattern, for example, if agent A always takes actions that benefit B more than itself, or if an aggressive agent suddenly begins to act benignly toward another.

As an example, suppose our agent is engaged in a series of "English", or outcry, auctions with other agents. Suppose that one of the other agents always bids aggressively, yet never wins the bid, thus bidding up the final price. Its actions have a cost to itself, i.e., the actual cost of participating and the potential cost incurred if it should win an auction. However, the benefits, except if it wins, go to another agent in the system: the seller. Our Analyze module would regard this as grounds for the hypothesis that there is an unsanctioned coalition (i.e., collusion) between that agent and the seller, even though no explicit collusive actions or messages have been observed. It would consequently treat the other agents as miscreants.

Analyze accrues evidence about factors indicating relationships by keeping a weighted count of the occurrences (or non-occurrences) of the factors, then, during each of its decision cycles, calculating the mean and standard deviation for all observations. Bayesian inference is then used to infer a new belief state from the prior one, and the probabilities are used to assign a symbolic value to the degree of belief, from very-weak to very-strong. Rules are then used to combine these individual beliefs into beliefs about relationships.

For example, in our auction example, we inferred that one of the agents bid aggressively, had actions that disproportionately benefited the auctioneer, and seldom won a bid. The combination of these factors would allow Analyze to believe the hypothesis that the agent was a shill for the auctioneer and, hence, a miscreant, with a confidence value derived from the probabilities determined for each factor involved.

Once Analyze has inferred an agent's motives and its relationships to others, it remembers these as a *durable hypotheses* about the agent. A durable hypothesis is different from other beliefs in that it persists from one decision cycle to another, whereas others are generated from historical and perceptual information at the start of each decision cycle. The duration of hypotheses will vary based on the stream of more transient beliefs and on a decay process that the Analysis module uses to reduce the confidence in the hypothesis over time. We have found it useful to have multiple classes of durable hypotheses, each with its own decay rate. The more significant a hypothesis is to our assessment of the situation, the longer we wish to consider it, but still at some point the agent forgets an old hypothesis if beliefs cease to support it.

Based on its durable hypotheses and other beliefs, the Contextualize module then can determine the current context.

3.3 Context Management

The Contextualize, or context management, module uses the beliefs and durable hypotheses from Track and Analyze to determine what the current context is and to manage the application of associated contextual knowledge.

As an example of context management, consider this scenario. On a cold winter day, you are sitting at your desk at work. A man enters the building wearing a ski mask. Suddenly, you hear the sound of sirens outside, and the man steps quickly into another room and closes the door. Here, the data is straightforward. We have two human actors, three locations (outside, the entryway, and the other room), and at least three objects (the desk, the ski mask, and the siren). Why does the siren beginning to sound result in a change in your perception of the other person's trustworthiness?

In our approach, the explanation is that you perceive a change in the context. Prior to the sound of the siren, the observation that a man had arrived wearing a ski mask was evaluated in the context of cold weather; our agent would have formed the hypothesis that the other person had donned the mask as a response to local conditions. When the siren sounds and the other agent takes an action to conceal himself, the agent's idea of what the context is and what the man's motivations are would change, and its trust in the agent should drop. Whereas our prior context-motivated strategy may have caused us to continue working on a paper, after the context change, a new "act warily" strategy may suggest the action "call police".

Contextualize has four major tasks. First, it must find contextual schemas matching the current situation. Second, it must *activate* the appropriate c-schemas from this set and deactivate any currently-active c-schemas that no longer match the situation. Third, it must select a strategy appropriate for the current context and make that available to the Execute module. And fourth, it must update the knowledge used by the other modules to reflect what is known or predicted about the context based on the knowledge from active c-schemas.

C-schema Retrieval. Contextualize has to find c-schemas that match the current situation. In previous work, we relied on a content-addressable "dynamic" schema memory for this (e.g., [8,9]) of the kind sometimes used in case-based reasoning. However, we do not require such a mechanism. As long as an appropriate set of c-schemas can be found that match the current situation, CATS is agnostic as to the mechanism.

In our approach, we refer (after [11]) to c-schema retrieval as *evoking* a set of c-schemas based on the situation's features. In CATS, the primary features used are the beliefs about the world and interagent interactions produced by Track and Analyze. At the evoking stage, we are not overly concerned with the degree of match between a contextual schema and the situation; it is enough that some beliefs "bring to mind" (evoke) the c-schema with some level of confidence (based on the belief's confidences).

Contextualize continually watches the situation to detect context changes that should bring to mind new c-schemas; or, to better phrase it, it is constantly looking for c-schemas that now match the situation, which could indicate a change in context. If it detects beliefs predicted by a c-schema not currently in use, than that c-schema is considered. In addition, c-schemas in CATS explicitly list the "boundary conditions" for entering or leaving the represented context. This is used by Contextualize as well to change the set of c-schemas under consideration.

There may be situations in which no specific c-schema can be identified. For this eventuality, CATS has a "default" c-schema that is always applicable, albeit with low confidence.[1] Thus, when no other, more-specific context is recognized, this c-schema would provide a default strategy; otherwise, it is "overruled" by more specific c-schemas during activation/deactivation.

C-schema Activation/Deactivation. Retrieving contexts is only part of the problem of context assessment, since not all will be good matches. For example, being involved in an Internet auction for bandwidth might remind an agent of auctions, auctions involving some of the agents that are present now, and Internet auctions; only the latter two would be good matches, since the first is just a generalization of the latter.

In other work, we use a differential diagnosis process to determine which of the evoked c-schemas should be used as the representation of the current context [17]. In this work, we are more focused on representing and using contextual knowledge, and so we use a correspondingly simpler diagnostic process. Contextualize uses a version of MYCIN's certainty factor combination technique [16] to assess the match between the observed/inferred situation and what is predicted by the agent's c-schema. This includes the boundary rules mentioned previously that c-schemas may contain that, when matched to the situation, change Contextualize's belief in what the current context is. The c-schema or c-schemas that this process selects as matching the current situation are then *activated* to represent the current context.

Some c-schemas that had matched the situation may be found to be no longer appropriate as the situation changes. Contextualize will notice this during its diagnostic process. In addition, it can use any boundary rules in the old c-schema that suggest the context is no longer in effect. Contextualize then deactivates such c-schemas.

---

[1] In other projects, the corresponding c-schema would automatically be retrieved from memory when no more-specialized ones were found.

Strategy Selection. Once it has assessed the context, Contextualize must find the best strategy for the current context and pass that along to Execute. This is done using knowledge provided by the c-schemas representing the current context, i.e., by using their suggestions for strategies appropriate for the current context.

Currently the results of our contextual reasoning is the selection of a single strategy from those suggested by the active c-schemas. This may be a weakness if a truly novel context occurs in the society, since our a priori c-schemas may not respond well to the unanticipated context. In the future, we will look at how to mitigate this problem by combining strategies from multiple c-schemas, each of which may capture some aspect of the novel context, to create a new strategy that will work better than any existing one.

Once Execute receives a strategy suggestion from Contextualize, it can use this strategy until the situation changes enough for Contextualize to send it a new strategy.

Propagate Contextual Knowledge. In addition to suggesting a strategy to Execute, Contextualize is also responsible for propagating other contextual knowledge to the agent's modules. C-schemas provide the agent with declarative and procedural knowledge. This includes knowledge the other modules use to do their jobs, for example, to create beliefs about the state of the world and about agent motives. When the context changes, such knowledge from active c-schemas needs to be activated, and knowledge unique to exiting c-schemas should be deactivated. If there are conflicts between knowledge from different c-schemas, then Contextualize chooses from among them based on its confidence in the c-schema as a fit for the situation.[2]

There are several kinds of knowledge that can be activated from c-schemas. For example, predictions about the world (e.g., a particular miscreant is present) may need to be activated by establishing a belief or a durable hypothesis. Existing beliefs may also be revised by changing their confidence values. Diagnostic rules (e.g., boundary rules from a c-schema) may be added to aid future context recognition. Other rules may be activated to be used by Track and Analyze to help them make decisions that are appropriate to the group. This will cause them to make inferences that are automatically appropriate for the context.

To illustrate how this might work, consider an agent that has been a long term member of an OMAS. Also existing in the same system is Agent X, for whom we have a long history of observations about its action. We have labeled Agent X as trustworthy, and therefore use a cooperative strategy when dealing with it. During the course of observing activity in the system we notice that Agent X is treated well by Agent Z, but that Agent X makes inaccurate, negative reputation reports about Agent Z, in effect slandering Agent Z. Our agent would generate a number of hypotheses about this situation, e.g.: (1) Agent X seeks to undermine Agent Z; (2) Agent Z's reputation is lower than it should be; and (3) Agent X is not as trustworthy as we thought. These hypotheses, together or separately, should cause Contextualize to recognize a new context and recommend a new

---

[2] Other work in our lab looks at how to merge such conflicting contextual knowledge.

strategy the agent should use. As a result of context recognition, the agent should revise beliefs it holds to reflect the new hypotheses and beliefs about the effects of Agent X's actions. First, it might add a new assumption about the relationship of Agents X and Z. This would entail asserting a new durable hypothesis. Second, it should lower its confidence in any beliefs about the reputation of Agent Z, since it suspects that reputation has been compromised by Agent X's slander. Since beliefs are generated anew at the start of each decision cycle, simply updating the facts for the current cycle is not sufficient. Instead, Contextualize needs to modify Track's and Analyze's working knowledge so that they make the right decisions. Finally, it should discard its old durable hypothesis that Agent X is a trustworthy agent and prevent it from arising again by modifying the knowledge upon which Analyze makes its decisions, for example, by asserting that Agent X tends to slander Agent Z.

4   Empirical Evaluation

To determine if adding context-based reasoning provides an improvement over general-purpose strategies, we constructed an agent using the techniques described in this paper. The agent interacted with other agents using our Liar's Dice (a poker-like dice game) test framework [19]. Each of the other agents employed a single general-purpose strategy (although possibly different for each agent), while our context-based agent chose which of the strategies to use based on its contextual reasoning. We tested the agent in several different test scenarios with both mixed-strategy and homogeneous-strategy societies. In some tests we introduced miscreant agents that employed a collusive strategy, while other cases did not employ miscreants.

For each test scenario we conducted a session of 36,000 games. With this number of games, there is a > 0.99 probability that the rarest roll (five sixes) will occur in the session. We can therefore expect that all possible hands will be experienced.

There were 25 distinct cases based on: whether the CATS agent was present or replaced by an agent following a random (*coinflip*) strategy; the presence of collusive agents (i.e., miscreants); and other context changes, such as the entry of a non-collusive agent or a system message announcing a change of state.

We measured the *success* of each agent, meaning the improved performance in the games, as well as its *correctness*, the percentage of correct decisions. We were also concerned with the *strength* of the strategy used by the agent, measured as $(w - l)/(w + l)$, where w and l are the number of wins and losses, respectively.

The results showed a statistically significant ($p < 0.05$) improvement in both correctness and success for our context-aware agent compared to other agents in the society when the test scenario included a miscreant agent.[3] In all mixed-strategy experiments, the CATS agent displayed the highest level of correctness among all agents. In all but one mixed-strategy experiment, the context-aware

---

[3] See [19] for complete details, including the statistical analysis.

agent was also the most successful agent in the society. In homogeneous-strategy society experiments, the context-aware agent performed well with or without the presence of miscreant agents, with two exceptions. Without miscreants present, the context-aware agent was equivalent to other members of the homogeneous society of stochastic agents, and it was inferior to other agents in the case of a homogeneous society composed of altruistic agents. Barring these two special cases, the CATS agent showed better success and correctness in all cases, with statistically significant improvement when miscreant agents were part of the society. The improvement in strength of CATS over the second best strategy was > 2.6 for all, while the improvement in correctness over the second best strategy was > 3.32.

In experiments in which we inserted miscreant behaviors, the CATS agent maintained or slightly increased its correctness, while the correctness of other dynamic strategies (i.e., those that changed their actions based on conditions) were reduced by significant amounts ($p \leq 0.05$). Since the CATS agent uses only the strategies available to the other agents in the society, this indicates that its correctness is the result of its ability to better recognize the situation occurring in the society and respond to it by choosing a better strategy. This argument is further supported by the experiments in which no miscreant behavior was present. In these experiments, the CATS agent continued to exhibit comparable levels of correctness, but, since there were no miscreants to affect the other agents, their performance was improved. In the absence of miscreant behavior, but in the presence of context change, our approach continued to outperform other strategies, but the effect was not statistically significant. in terms of strength, the CATS agent was the same or slightly better than the others in the presence of miscreants ($p < 0.05$), but in experiments in which there were no miscreants, there was no significant difference in performance.

The results show that a context-aware agent enjoys an advantage over other agents in environments that include context change, and that the agent's ability to detect miscreant behaviors provides a statistically significant improvement in performance. The CATS agent's ability to detect miscreants allows it to choose the best strategy, rather than being locked into a single strategy, as were the other agents.

We believe that the size of the effect of context-based reasoning will vary based on the domain and on the amount of tuning put into the creation of the c-schemas. In the future we expect to be able to confirm and quantify this relationship.

5   Conclusion

We have described how a context-aware agent employing CATS can make better trust decisions, resulting in better performance with respect to its goals. In our approach, an agent not only tracks environmental factors and the actions of other agents, it also performs a society-level analysis of motives to uncover hidden relationships and motives agents may have. All of these are indicators of what the current context really is. Once the context is assessed, then contextual schemas representing that context provide suggestions for an appropriate strategy, e.g., to deal with miscreants, as well as other knowledge the agent can use in the context. We have implemented and tested this approach in a toy domain that has many characteristics of real-world open multiagent system domains. Experiments support the conclusion that context-aware strategy selection is beneficial in an OMAS. The approach was effective in two ways: (1) allowing an agent to make a correct decision about whether or not to trust another agent; and (2) maintaining good overall performance with respect to accomplishing the agent's own goals.

The results of these experiments are very encouraging and provide several further paths of inquiry, some of which we have mentioned in this paper.

References

1. Bacharach, M., Gambetta, D.: Trust as type detection. In: Castelfranchi, C., Tan, Y.-H. (eds.) Trust and Deception in Virtual Societies, pp. 1–26. Springer, Netherlands (2001)
2. Blidberg, D.R., Chappell, S.G.: Guidance and control architecture for the EAVE vehicle. IEEE J. Oceanic Eng. (OE) **11**(4), 449–461 (1986)
3. Castelfranchi, C., Tan, Y.: Introduction: why trust and deception are essential for virtual societies. In: Castelfranchi, C., Tan, Y. (eds.) Trust and Deception in Virtual Societies, pp. 17–31. Kluwer, Dordrecht (2001)
4. Cosmides, L., Tooby, J., Fiddick, L., Bryant, G.: Detecting cheaters. Trends Cogn. Sci. **9**, 505–506 (2005)
5. Falcone, R., Castelfranchi, C.: Social trust: a cognitive approach. In: Castelfranchi, C., Tan, Y. (eds.) Trust and Deception in Virtual Societies, pp. 55–90. Kluwer Academic Publishers, Berlin (2001)
6. Guha, R.: Contexts: A Formalization and Some Applications. Ph.D. thesis, Stanford University, 1991. Technical report STAN-CS-91-1399-Thesis, and MCC Technical Report Number ACT-CYC-423-91 (1991)
7. Iszuierdo, L., Izquierdo, S.: Dynamics of the Bush-Mosteller learning algorithm in 2x2 games. In: Weber, C., Elshaw, M., Mayer, N. (eds.) Reinforcement Learning Theory and Applications, p. 424. I-Tech Education and Publishing, Vienna (2008)
8. Kolodner, J.L.: Retrieval and Organizational Strategies in Conceptual Memory. Lawrence Erlbaum Associates, Hillsdale (1984)
9. Lawton, J.H., Turner, R.M., Turner, E.H.: A unified long-term memory system. In: Branting, L.K., Bergmann, R., Althoff, K.-D. (eds.) ICCBR 1999. LNCS (LNAI), vol. 1650, p. 188. Springer, Heidelberg (1999)
10. Mao, W.: Modeling social causality and social judgment in multi-agent interactions. Ph.D. thesis, University of Southern California (2006)
11. Miller, R.A., Pople, H.E., Myers, J.D.: INTERNIST-1, an experimental computer-based diagnostic consultant for general internal medicine. N. Engl. J. Med. **307**, 468–476 (1982)
12. Mui, L., Mohtashemi, M., Halberstadt, A.: Notions of reputation in multi-agents systems: a review. In: Proceedings of the First International Joint Conference on Autonomous Agents and Multiagent Systems: Part 1. pp. 280–287. AAMAS 2002, ACM, New York (2002). http://doi.acm.org/10.1145/544741.544807
13. Riley, G.: CLIPS - a tool for building expert systems. On the World Wide Web at http://www.ghg.net/clips/CLIPS.html (Accessed 16 Dec 2004)

14. Sabater, J., Sierra, C.: Review on computational trust and reputation models. Artif. Intell. Rev. **24**(1), 33–60 (2005)
15. Santos, Jr., E., Johnson, Jr., G.: Toward detecting deception in intelligent systems. In: Defense and Security, pp. 130–141. International Society for Optics and Photonics (2004)
16. Shortliffe, E.H.: Computer-Based Medical Consultations: MYCIN. Elsevier, New York (1976)
17. Turner, R.M.: Context-mediated behavior. In: Brézillon, P., Gonzalez, A. (eds.) Context in Computing: A Cross-Disciplinary Approach for Modeling the Real World Through Contextual Reasoning, pp. 523–540. Springer, New York (2014)
18. Turner, R., Whitsel, L.: A context-based approach to detecting miscreant behavior and collusion in open multiagent systems. In: Kofod-Petersen, A., Beigl, M., Christiansen, H., Schmidtke, H.R., Coventry, K.R., Roth-Berghofer, T.R. (eds.) CONTEXT 2011. LNCS, vol. 6967, pp. 300–306. Springer, Heidelberg (2011)
19. Whitsel, L.T.: A context-based approach to detecting miscreant agent behavior in open multagent systems. Ph.D. thesis, School of Computing and Information Science, University of Maine, University of Maine, Orono, ME 04469 USA, December 2013

Semantics and Philosophy

Descriptive Indexicals, Propositional Attitudes and the Double Role of Context

Katarzyna Kijania-Placek[(✉)]

Institute of Philosophy, Jagiellonian University, Krakow, Poland
katarzyna.kijania-placek@uj.edu.pl

Abstract. This paper offers an account of some uses of indexicals in the context of propositional attitude ascriptions, i.e. reports that concern the cognitive relations people bring to bear on propositions. While the contribution of indexicals to the truth conditions of an utterance is usually singular, their interpretation is general in the case of so called descriptive uses. I will propose an interpretation of the descriptive uses of indexicals via a mechanism of descriptive anaphora and apply this mechanism to the case of attitude ascriptions. I will emphasize the role of context both in the suppression of the default referential reading of the indexical, as well as in the reconstruction of the relevant interpretation of the whole utterance.

Keywords: Descriptive indexicals · Propositional attitudes · Descriptive anaphora · Linguistic context · Extralinguistic context · Suppressive role of context · Constructive role of context

1 Introduction

Indexicals are typically considered as vehicles of direct reference. Some contexts of propositional attitude ascriptions make it clear, however, that the singular mode of presentation deployed by an ascriber cannot be attributed to the ascribee. An example has been given by Nunberg in [42]:

(1) The Founders invested me with the sole responsibility for appointing Supreme Court justices.
uttered by George H.W. Bush in 1992

Existing accounts of propositional attitudes (deploying the transparent/ opaque, *de re/de dicto* or similar distinctions; see for example [1,5,18,30,50,51, 54,56,57]) seem to imply that by uttering such a sentence George H. W. Bush might be committed to the absurd claim that the Founders had *de re* thoughts about himself.[1] My aim in this paper is to offer an account of the reconstruction

---

[1] In treating (1) as a proposition attitude ascription I assume the propositional analysis of intensional verbs (see [31,36,50] and the references therein. Moltmann does not subscribe to the propositional analysis of intensional verbs). My arguments, however, are independent of this analysis and could be reformulated as arguments against an obligatory referential interpretation of indexicals in the scope of intentional verbs in general.

© Springer International Publishing Switzerland 2015
H. Christiansen et al. (Eds.): CONTEXT 2015, LNAI 9405, pp. 287–301, 2015.
DOI: 10.1007/978-3-319-25591-0_21

of the proposition expressed by the original utterance, reported in (1), in terms of a descriptive interpretation of indexicals that would not have such unintuitive consequences. I will characterize the double role that context plays in this reconstruction.

2   Descriptive Indexicals

Descriptive uses of indexicals are uses where indexical utterances express general propositions (see [6,8,9,14,15,17,19–26,41–43,46,47,53,55,60]), like in the following example ([41,55]):

(2) He is usually an Italian, but this time they thought it wise to elect a Pole.
uttered by someone gesturing towards John Paul II, as he delivers a speech with a Polish accent shortly after his election

One expresses here not a singular proposition about John Paul II, but a general one concerning all popes. Because 'usually' is a quantifier that requires a range of values to quantify over, and because 'he' in its standard interpretation provides just one object, there is a tension in this sentence which triggers the search for an alternative interpretation. The tension is not caused by the fact that John Paul II himself is the possible referent but it is a tension between the generality of the quantifier and the singularity of the indexical in its default interpretation. The tension would be present regardless of who the referent was. As a result, the pronoun's basic referential function is suppressed.

2.1   The Mechanism of Descriptive Anaphora

I postulate that the alternative – general – interpretation is obtained by a process I call descriptive anaphora. Via the descriptive anaphoric mechanism, an indexical expression inherits its semantic value from its antecedent. However, in contrast to classic anaphora, that antecedent stems from an extra-linguistic context: it is an object identified through the linguistic meaning of the pronoun (in the case of pure indexicals) or by demonstration (for demonstratives). In a communication context, those objects serve as a means of expressing content and, as such, they acquire semantic properties.[2] The object is used as a pointer to a property corresponding to it in a contextually salient manner. That property contributes to the general proposition. What is important is that the property is not a referent for the pronoun. The structure of the general proposition is determined by a binary quantifier and the property which is retrieved from the context serves as a context set that limits the domain of quantification of the

---

[2] Compare Frege's treatment of objects as means of expressing content (e.g. [11–13, 22,25,29,45]). Also, Nunberg wrote about indexicals that "this is the characteristic and most remarkable feature of these expressions. They enable us to turn the context itself into an auxiliary means of expression, so that contextual features are made to serve as pointers to the content of the utterance." [42, pp. 19–20].

quantifier. Typically, as in the case of (2), it is that quantifier which triggered the mechanism of descriptive anaphora in the first place but the quantifier does not have to be overt (compare [22, 24, 26]).

In the case of (2), the mechanism of descriptive anaphora is triggered by the inconsistency between the indexical and the quantifier and John Paul II is the demonstrated antecedent. His salient property of 'being a pope' serves as the context set for the binary quantifier 'usually' – USUALLY$_x(\phi(x), \psi(x))$ – interpreted in accordance with the generalized quantifiers theory (see [3, 38]).[3] In this case, the general proposition becomes:

$$\text{USUALLY}_x(\text{POPE}(x), \text{ITALIAN}(x))$$

with the usual truth conditions for the (generalized) majority quantifier (see [3]):

$$\mathfrak{M}^{gi} \models \text{USUALLY}_x(\phi(x), \psi(x)) \quad \text{iff} \quad |\phi^{\mathfrak{M}gi} \cap \psi^{\mathfrak{M}gi}| > |\phi^{\mathfrak{M}gi} \backslash \psi^{\mathfrak{M}gi}|,$$

where g is an assignment and i is a context.[4]

The descriptive use of an indexical is not its basic use. The process of descriptive anaphora is triggered by the inadequacy of its basic uses - (classically) anaphoric, deictic or deferred.

2.2   Three Types of Descriptive Uses of Indexicals

Not all cases of descriptive uses of indexicals are triggered by an inconsistency between an indexical and a quantifier (i.e. Type I).[5] I distinguish two other types of descriptive uses of indexicals, introduced below. They differ only in what triggers the search for an alternative interpretation, but the mechanism of the interpretation is the same: we are looking for a salient property that is in correspondence with the object demonstrated and the property serves as a context set that limits the domain of quantification for the quantifier that provides the structure to the general proposition expressed. Those quantifiers need not be overt.

---

[3] I use SMALLCAPS font style for formal counterparts of natural language quantifiers and predicates.

[4] In what follows $\mathfrak{M}$ is a model, g is an assignment of objects from the domain of the model to individual variables, i is a context, $\models$ is a satisfaction relation obtaining between a sentence (or an open formula) and a model and context, under an assignment; ϕ and ψ are open formulas, $|A|$ signifies the cardinality of the set A, and $\phi^{\mathfrak{M}gi}$ is the interpretation of formula ϕ in model $\mathfrak{M}$ and context i under assignment g.

[5] Other examples of descriptive uses of indexicals of Type I were given by Nunberg [41–43] Recanati [53, 55], Bezuidenhout [4], Stokke [60], Hunter [16, 17], Elbourne [9], Galery [15] and Kijania-Placek [22]. The present author is responsible for the typology.

Type II – Unavailability of Basic Interpretations. The unavailability of a referent may be a result of a physical absence in the context of the utterance of anybody/anything that fits the constraints of the linguistic meaning of the expression deployed. Consider Schiffer's example [58]. On seeing a giant footprint in the sand someone exclaims:

(3) He must be a giant.

Here the potential deictic referent is not present in the context of the utterance, and, since (3) is a conversation starter, there is no linguistic antecedent for the pronoun. Because the speaker has no particular male individual in mind, the deferred interpretation is not an option either (compare [22,42]).[6] The descriptive interpretation is thus considered due to the failure of other interpretations.[7]

Yet, the mechanism of descriptive anaphora works in the same way as in the cases discussed in the previous section (in cases of Type I): we search the context for a salient property that is in correspondence with the object demonstrated (in this case – the footprint). The property may have (and in the case of (3) does have) the structure of a relation, whose one relatum is fixed by the demonstrated object. While this (possibly many argument) relation is salient in the context, it is the resulting property (relation with one open argument) that contributes to the general proposition. In the case of (3), the property is 'being somebody who left this (demonstrated) footprint'. The sentence does not contain an overt quantifier which constrains the structure of the proposition expressed, but a

---

[6] In the case of deferred reference, the resulting proposition is singular, involving the rigid attribution of a property to the actual (deferred) referent, while in (3) the expressed proposition is rather a general one, such as 'The man who left this footprint, whoever he is, must be a giant' (see [58] and footnote 7 below). If somebody else had left this footprint, the speculation would have concerned the other person and not the original one. Arguably, there is some confusion regarding the concept of deferred reference, and sometimes it is used inclusively for both cases when a contribution of the indexical is singular and when it is general. For my arguments to go through, it suffices that the referential reading of deferred reference is excluded in the case of (3). On such an inclusive interpretation of deferred reference, descriptive anaphora may be considered an elucidation of the general reading of deferred reference. I have supported the need for drawing the distinction between deferred reference and descriptive interpretation more extensively in [22,25]. My thanks to the referee who pointed out the need to clarify this point.

[7] The point that it is the lack of the referent in the context that triggers the descriptive reading was made by Bezuidenhout [4, p. 401]: "[I]t is precisely because the listener is unable to think of the referent in an identifying way in the context (i.e. because the listener is unable to track an individual in the context) that the listener understands the speaker to have used the indexical attributively". Bezuidenhout is using the term 'attributively' for what I here call 'descriptively'. Other examples were given by Loar [34], Nunberg [40–42], Recanati [53], Galery [15] and Kijania-Placek [22].

covert binary quantifier 'the' is reconstructed from the context[8] and the structure of the proposition expressed is the following:[9]

$$\text{THE}_x(\text{MALE-WHO-LEFT-THIS-FOOTPRINT}(x), \text{GIANT}(x))$$

– 'The man who left this footprint (whoever he is) is (must be) a giant', where 'the', interpreted as a generalized quantifier, has the following truth conditions:

$$\mathfrak{M}^{gi} \models \text{THE}_x(\phi(x), \psi(x)) \quad \text{iff} \quad |\phi^{\mathfrak{M}gi}| = 1 \text{ and } \phi^{\mathfrak{M}gi} \subset \psi^{\mathfrak{M}gi}|$$

But in the case of propositional attitude ascriptions containing indexicals in their that-clauses, there are additionally situations when the conventional referent of the indexical is present in the context of the ascription but, for reasons obvious from the context, could not have been present during the reported utterance. In such cases it seems to be obvious that there is a discrepancy between the mode of presentation used by the ascribee (singular) and the mode of presentation of the ascriber (general). I will return to these kinds of cases in Sect. 4.

Type III – Irrelevance. Sometimes, however, descriptive anaphora is triggered by the blatant irrelevance of the referential interpretation, because of the incompatibility of that interpretation with the goal of the utterance or due to its obvious triviality or falsity. An interesting example was again given by Nunberg in [42]: A doctor during a lecture points to his own chest and states:

(4) When a person is shot here, we can usually conclude that it was not suicide.

Here 'usually' does not quantify over persons being shot – one person may be shot several times during his or her life – but over events of shooting at a chest.[10] Yet, because it is not excluded that somebody might be shot several times during one event, there is no semantic inconsistency between the referential reading

---

[8] For the role of context in the selection of the covert quantifier see Sect. 3.2 below.

[9] I have ignored 'must' in this analysis. An anonymous referee suggested, however, that assuming the descriptive anaphora interpretation of 'he' we would expect that it could interfere scopally with 'must' giving a possible reading in which 'must' takes wide scope, so that the sentence reads 'It must be the case that there is a unique person who left this footprint and is a giant'. According to the referee such a reading does not seem to be available. I agree with this opinion as long as we interpret 'must' as a metaphysical modality. But that would amount to attributing metaphysical necessity, which on both narrow and wide scope readings is counterintuitive. Rather, 'must' in (3) is either evidential (Nunberg, p.c.) or should be interpreted as an epistemic necessity, which on the wide scope reading of the modal gives 'For all I know, there is a unique person who left this footprint and is a giant'. I find this reading highly accessible.

[10] In this example, the quantifier 'usually' scopes out of the consequent, which is legitimate due to its binary character (see footnote 12 below), the conservativity of generalized quantifiers (i.e. QAB iff $QA(A \cap B)$, see [61]) and the anaphoricity of 'it' in the consequent. Thanks to an anonymous referee for making me clarify this point.

of 'here' and the quantifier 'usually'. It is thus not a case of a descriptive use of an indexical of type I. But, anyhow, we are not concerned with the particular place demonstrated, because it is obvious for the addressee that the speaker is demonstrating a chest which has not been shot at in the relevant way pertaining to murder. After all, for something to be considered suicide, there must be a case of death. The referential interpretation – with 'here' referring to the chest of this particular speaker – would therefore give a blatantly false, if at all comprehensible, and thus irrelevant proposition.[11] In typical cases of this type such as (5) [42], and unlike in (4), the indexical is embedded under modal operators (see [17]):

(5) If the Democrats had won the last few presidential elections, we might have been liberals.
   said by Supreme Court Justice O'Connor

I will not be concerned with descriptive uses of indexicals of type III in this paper.

3   The Double Role of Context

In the case of descriptive uses of indexicals, context plays a role both in suppressing the default referential reading of the indexical, as well as in constructing the relevant interpretation of an utterance.

3.1   The Suppressive Role of Context

In typical cases, descriptive anaphora is triggered by the use of quantifying words such as traditionally, always, or usually in contexts in which they quantify over the same kind of entities that the indexicals refer to. In such contexts the generality of the quantifiers clashes with the singularity of the default referential reading of indexicals. Whether there is a clash is, however, a pragmatic matter, as it depends on the domain of quantification of the quantifier, which for most adverbs of quantification is not given as part of the semantics of the word. If 'usually' quantified over periods of time or events – like in 'He usually spends his holidays in Rome' – there would be no conflict between 'usually' and 'he'.[12] Since in the case of descriptive uses of indexicals of type I it is the conflict between the generality of the quantifier and the singularity of the indexical which results in suppressing the referential reading of the indexical, both linguistic and extralinguistic context play a role here. The domain of quantification is dependent mainly

---

[11] For the analysis of this and the following example via the mechanism of descriptive anaphora see [22]. Other examples were given by Nunberg [40–42], Recanati [53], Bezuidenhout [4], Powell [46], Borg [6], Elbourne [9], Galery [14,15], Hunter [17], and Kijania-Placek [22–25].

[12] Compare [33]. A binary structure is standardly postulated for adverbial quantification, regardless of its explicit structure. Thus 'He usually goes on holiday to Rome' would be analyzed as 'Usually, if he goes on holiday, he goes to Rome' (see [32,49]).

on what is predicated of the objects quantified over (linguistic context) but in some cases relies as well on such extra-linguistic features of context as world knowledge.[13] For example in (2) – in contrast to 'He usually spends his holidays in Rome' – a relatively static property is attributed to the subject, a property which typically does not change with time, but changes from person to person. And it is the attribution of this property that is one of the factors that determines the domain of people as the domain of quantification in (2), leading to the suppression of the referential interpretation of 'he' and thus to the descriptive interpretation of the pronoun.

In the case of descriptive uses of indexicals of type II, linguistic context only plays a negative role in excluding classically anaphoric interpretation when no linguistic antecedent is present and most of the suppressing work is dependent on extra-linguistic context. We know from the extralinguistic context if a suitable (i.e. complying with the linguistic meaning of an indexical) referent is available and if other requirements of potential basic interpretations (such as having a particular object in mind, for deferred interpretation) are fulfilled. Their non-fulfillment suppresses the referential interpretation. The triggering factors of the descriptive uses of indexicals of type III are solely dependent on the extra-linguistic context, as, by definition, a general interpretation is caused there by the pragmatic irrelevance of the (available) referential one.

3.2 The Constructive Role of Context

The extra-linguistic context figures prominently in the construction of the general interpretation of the indexical. In all cases it is the property salient in the context that is the propositional contribution of the indexical term.

But the relevance of the context does not end there. The structure of the general proposition expressed is provided by a binary quantifier and the quantifier is not always overt. Both linguistic and extra-linguistic context play a role in reconstructing covert quantifiers. In example (3), repeated from above,

(3) He must be a giant.

the sentence does not contain an overt quantifier which constrains the structure of the proposition expressed, but as with the use of bare plurals for the expression of a quantified sentence, we reconstruct a covert binary quantifier. It will usually be a universal quantifier or the definite description – but which quantifier in particular is the relevant one is a contextual matter and depends mainly on what is predicated of the objects quantified over (compare [7,28]). An analogy with bare plurals can be illustrative here: 'Mice are mammals' is interpreted by a universal quantifier, while 'Mice will come out of this hole if you wait long enough' – by an existential quantifier. The same plural noun is here quantified universally or existentially, depending on what is predicated of mice.[14] In the

---

[13] This last dependence is exemplified by the discussion of (3) in below.

[14] "the source of the existential quantifier is not the determiner of the NP, but rather what is being predicated of it at the time" ([7, p. 451]).

case of (3) the type of the quantifier – the definite description – is dictated by the predicated properties of 'having left a footprint' (a property retrieved from the extra-linguistic context) and 'being a giant' (linguistic context), which typically are the properties of just one individual (world knowledge). As a result, the structure of the proposition expressed is the following:

$$\text{THE}_x(\text{MALE-WHO-LEFT-THIS-FOOTPRINT}(x), \text{GIANT}(x))$$

– 'The man who left this footprint is a giant.

On the other hand, indexicals interpreted descriptively seem to be semantically numberless in the sense that they do not provide a clue as to what kind of quantifier should be used in providing the structure of the general proposition.[15] In (3) it was a definite description, but in (6), another example of Nunberg's [42]:

(6) He is always the last one to know,

it would be the universal quantifier, even though in both the pronoun was the same.[16] The descriptive reading of (6) is triggered only if we interpret this utterance as a proverb – this is usually uttered in the out of earshot presence of the relevant husband – and thus it is the context of the proverb that triggers the universal interpretation 'Every husband is always the last one to know', where 'husband' is the property retrieved from the context via the mechanism of descriptive anaphora.[17] Since (6) may in exceptional situations be interpreted as

---

[15] On the semantical numberlessness of pronouns see [39].

[16] An anonymous referee proposed an alternative analysis of (6) – 'In every extra marital affair situation/event x, the husband in x is the last to know' – where, according to the proposal, the indexical is analysed as a definite rather than a universal quantifier and suggested that it might be possible to analyse all of the descriptive uses of indexicals as definites. I agree that this analysis is possible for (6), but its adequacy comes from the dependence of the definite quantifier on the universal one. As a result, semantically we still quantify over all husbands of unfaithful wives in all extra marital affairs. Since both 'always' and 'every' are universal quantifiers, the analysis proposed in this paper and the one proposed by the referee are equivalent. Nonetheless, the strategy of analysing all of the descriptive uses of indexicals as definites would not work for examples such as 'Today is always the biggest party day of the year', uttered on New Years Eve (adopted from Nunberg [43]), where we universally quantify over days that are New Years Eves.

[17] Importantly, the quantifier 'always' quantifies here over events – affairs of the wives of the relevant husbands – and thus it does not give the structure to the whole proposition, but appears in the second argument of the quantifier:

$$\text{EVERY}_x(\text{HUSBAND}(x), \text{ALWAYS}_y(\text{AFFAIR-OF-WIFE-OF}(y, x),$$
$$\text{LAST-ONE-TO-KNOW-OF}(x, y)))$$

– '(Every) husband is always the last one to know (about his wife's affair)'.
The truth conditions of both the quantifier 'every' as well as of the quantifier 'always' are those of a universal quantifier:

concerning just the person demonstrated (not necessarily a husband), the fact that (6) is interpreted as a proverb is a purely contextual matter. The context of a proverb supplies the generality requirement, which makes the referential interpretation trivial.

I conclude that both linguistic and extra-linguistic context play crucial and quite specific roles in triggering, as well as in constructing, the descriptive interpretation of indexicals.

4 Propositional Attitudes

In most accounts of propositional attitude ascriptions it is assumed that attitude ascriptions that contain indexicals are *de re* ascriptions (see [57]), i.e. such that the mode of presentation of the referent of the indexical does not affect the truth conditions of the belief report and we are usually not told how the subject of the attitude thinks about the referent.[18] Thus in contrast to belief reports such as (7) ([10]):

(7) John believes that the winner will go to Hong Kong,

which are ambiguous between the wide scope ('There is somebody who is the winner and of whom John believes that he will go to Hong Kong') and the narrow scope readings ('John believes that there is somebody who is the winner and who will go to Hong Kong'),[19] the sentence (8):

(8) John believes that you will go to Hong Kong,

is supposed to have only one – *de re* – reading, the thesis of which can also be expressed by saying that indexicals always take the wide scope in the context of proposition attitude verbs or that indexicals are 'open to exportation' within the that-clause ([35]; compare [54]).

Because the exercised mode of presentation (the mode used by the reporter; see [53]) is referential due to the referentially of the indexical, there is an object the attitude is about, given by the context of the report. Since the believer is

---

$$\mathfrak{M}^{gi} \models \text{EVERY/ALWAYS}_x(\phi(x), \psi(x)) \quad \text{iff} \quad \phi^{\mathfrak{M}gi} \subset \psi^{\mathfrak{M}gi}.$$

For details of this construction as well as for the analysis of indexicals used in proverbs as descriptively used indexicals see [26].

[18] The notions used in the literature for describing the behavior of indexicals in attitude contexts include '*de re*', 'relational' ([50,54]) or 'transparent' ([51,53,54]). In [54] Recanati argues that the *de re*/*de dicto* distinction should not be confused with the relational/notional distinction. I will not go into detail here, as they do not affect my argument. In [54] he also uses the terms transparent/opaque in a slightly different way, than in [53]. Again, these distinctions will not be relevant for the point I am going to make.

[19] According to Recanati [53] belief ascriptions that contain definite descriptions are in fact ambiguous in many-ways, but this complication will not be relevant for the case of indexicals, so I will ignore it here.

usually not a part of the context of the utterance of the report, according to Recanati [53] it seems that "there is no reason to suppose that the mode of presentation in question is also a constituent of the believer's thought" (p. 400). Yet, Recanati admits, the believer may be part of the context, as cases of self-ascriptions of belief testify, so we must leave room for transparent interpretations [in the sense of being about a particular and identified object] in which a specific mode of presentation – that supplied by the linguistic meaning of the indexical – is ascribed to the believer. Such readings would not, however, be general but they would contain the object referred to together with the mode of presentation of the object ([53]). Also Balaguer argues in [2] for the necessity of including the linguistic meaning of the indexical in the content of the reported belief in cases of what he calls the essential uses of indexicals, which are reminiscent of Prior's "Thanks goodness it's over" examples ([48]; see [44]). In [52] Recanati concedes, persuaded by Morgan [37], that indexicals also admit readings in which the ascribed mode of presentation is not the linguistic meaning of the indexical but some other mode of presentation of the referent – for example a visual mode of presentation – that is supplied by the context. All these cases remain non-general, however, because "even on the opaque reading of a belief sentence in which a singular term occurs, reference is made to some particular individual" ([54, p. 132]). Thus, from the fact that an indexical has been used in the that-clause of an attitude ascription we cannot infer that no mode of presentation of the referent is ascribed to the believer. But, Recanati maintains, we can infer that the ascribed mode of presentation is singular:

Singularity of the Ascribed Belief
An indexical within the that-clause of an attitude ascription indicates a singular mode of presentation of the referent in the ascribed belief.

The implied singularity allows for an explanation of the intuitions of Kripke concerning Sosa's example [59]:

"[A] spy and his accomplice see through a window how an investigator finds some incriminating evidence in the spy's footlocker. The accomplice could very naturally say 'He knows that you are a spy now. You must escape.' In fact, and so far as the accomplice knows, the investigator does not know the spy, and knows practically nothing about him: the footlocker had been searched only as part of a general investigation of the base. What the investigator knows is ⌜the owner of the footlocker is a spy⌝ " [59, p. 891].

Kripke, commenting on this example in [30], finds Sosa's intuitions pertaining to the appropriateness of the accomplice's remarks 'strange', since, according to Kripke "Sosa's accomplice obviously would not say, 'he knows that you are a spy now,' though he might say, 'watch out, they may soon find out that you are a spy, once they find out who owns the footlocker." [30, p. 340] Here I agree with Kripke against Sosa. What makes this report inappropriate is the implied singularity of the ascribed mode of presentation of the spy. 'He knows that you are a spy now' suggests that the investigator has a *de re* knowledge about the spy, while they only know *de dicto* that the owner of the footlocker is a spy. This has practical consequences. If it was only the reporter who saw the scene and reported it to the spy, the latter, assuming the police know of him (*de re*) that he is a spy might have undertaken a decision of immediate and risky flight in a situation, which in fact left some (minimal) time for preparation. Such a result might, of course, be intended by the reporter.[20] Paraphrasing Richard we might put it as follows: "an ascription is true provided [it] ascribes belief in a proposition which is believed and the ascription doesn't imply anything false about what pictures are held by the believer" [57, p. 446].[21]

While Sosa's case is misleading, which means at least that it allows for a singular interpretation of the ascribed belief,[22] there are situations in which it is obvious from the context that the reported belief could not have been a singular one. In such cases it seems that the indexical used in the that-cause is just exercised but not attributed, contrary to Recanati's thesis of the singularity of the ascribed belief. This would happen when the referent of the indexical is present in the context of the ascription but, for reasons obvious from the context, could not have been present during the reported utterance. Example (1), repeated here, is a case in point:

---

[20] Examples of intentional alteration of expressions used in the report and the consequences thereof were discussed by Bonomi [5], Aloni [1] and King [27]; see footnote 22.

[21] This is a citation from Richard, but the paraphrasing aspect stems from the fact that I do not explain the technical meaning Richard assigns to the notion of a 'picture' in his theory and instead I intend the notion to be understood in a non-technical, common-sense way.

[22] A similar example was given by King in [27]: "Suppose Glenn believes all politicians are corrupt. [...] Glenn has never met or heard of Bob [the mayor of San Diego]. [...] Glenn's boss is throwing a party as a fundraiser for a charity. [...] [He] tells Glenn and the other employees to look over the guest list, which includes Bob, and tell him if anyone corrupt is on it. The boss is adamant that should anyone fail to tell him about someone they believe to be corrupt, they will be fired. Glenn and the others look over the list and no one says anything. Alan, a conniving coworker of Glenn's who is always trying to get Glenn in trouble and who knows both Glenn's views on politicians and Bob's profession, says to Glenn's boss at the party pointing at Bob: 'I am surprised Glenn didn't say anything: Glenn believes he is corrupt.' This seems false in the new context, as would 'There is someone at this party Glenn believes to be corrupt' or any other such *de re* ascription concerning Bob to the effect that Glenn believes him to be corrupt." But the important difference between both this and Sosa'a examples and the example that follows – (1) – is that unlike in (1), the hearer may interpret the reporter as ascribing a *de re* believe about the relevant subject to the believer. The inappropriateness of the remark relies upon this possibility and in King's example it even constitutes the intended outcome. Since referential interpretation is available and relevant, descriptive interpretation is thus not triggered in Sosa's and King's examples (and they were not intended as descriptive by these authors).

(1) The Founders invested me with the sole responsibility for appointing Supreme Court justices.
uttered by George H. W. Bush in 1992

In this case it is obvious that the Founders could not have had *de re* thoughts about George H. W. Bush and the hearer, if aware of the fact, does not interpret the president as claiming so much. Additionally, the hearer is able to reconstruct the reported general belief by relying on the mechanism of descriptive anaphora.

On the descriptive anaphora interpretation, George H. W. Bush (the person, not the name) is the extra-linguistic antecedent of this token of 'he' and points to his silent property of 'being the president of the United States' ('US-president' for short). The quantifier that gives the structure to this general proposition is the binary universal quantifier and the property obtained from the context serves as its context set. As a result, we obtain the following structure of the original declaration (RASCJ is short for 'having been given the responsibility for appointing Supreme Court justices'):

$$\text{EVERY}_x(\text{US-PRESIDENT}(x),\ \text{RASCJ}(x))$$

– 'Every president of the United States has been given the responsibility for appointing Supreme Court justices', which seems to be the intended interpretation of the reported belief.

Another example is the following (see [42]):

(9) According to all the textbooks, you often get in trouble with that move.
uttered by a chess teacher giving an introductory lesson to a student who has just played 4. N x P ...

The authors of textbooks are unlikely to know the present player and have *de re* attitudes towards him, not to mention this particular move of his. (9) contains an indexical 'you' and a demonstrative 'that move'; both receive descriptive interpretation.[23] 'Often' is a quantifier that in this context quantifies over events of type X (4. N x P ...) and the property delineating the type is supplied via descriptive anaphora by this particular move. This property serves as the context set for 'often', while the property of 'being a player who has played a move of type X (4. N x P ...)', supplied by the referent of 'you', serves as the context set for the covert quantifier 'the', dependent on 'often':

$$\text{OFTEN}_x(\text{CHESS-MOVE-OF-TYPE-X}(x),\ \text{THE}_y(\text{PERSON-WHO-MAKES}(y, x),$$
$$\text{GET-IN-TROUBLE-WITH}(y, x))),$$

– 'The person who makes a move of type X (4. N x P ...) often gets in trouble with that move'. This captures the reported belief.[24]

---

[23] For reasons why 'that move' in this example should not be interpreted as deferred reference to a kind and for details of the analysis see [22].

[24] Other examples of this kind were given by Recanati [53], and Bezuidenhout [4].

The unavailability of the referent in the context, which is a characteristic trigger for the descriptive use of indexicals of type II, may thus be the result of the physical absence of a suitable referent in the context of utterance,[25] or – as it is the case in some attitude ascriptions – may come about due to the impossibility of the presence of the referent of the expression used in the report in the context of the reported belief. Crucially, for the mechanism of descriptive anaphora to work, it is required that the referent of the indexical in question exemplified, or was taken to exemplify,[26] the property that is essential for the reported belief.

5 Conclusion

I hope to have shown that, contrary to the prevailing view, the use of an indexical in attitude ascription does not guarantee the singularity of the ascribed mode of presentation of the relevant object. Additionally, the indexical plays a double role in some cases – the referential role in the exercised mode and the descriptive role in the ascribed mode. In such cases it is possible to reconstruct the reported belief via the mechanism of descriptive anaphora that, I claim, is operative in all cases of descriptive uses of indexicals.

Acknowledgements. This work has been partly supported by the (Polish) National Science Centre 2013/09/B/HS1/02013 grant. I would like to thank my three anonymous referees for their comments which have helped me to improve this paper.

References

1. Aloni, M.: A formal treatment of the pragmatics of questions and attitudes. Linguist. Philos. **28**(5), 505–539 (2005)
2. Balaguer, M.: Indexical propositions and *de re* belief ascriptions. Synthese **146**(3), 325–355 (2005)
3. Barwise, J., Cooper, R.: Generalized quantifiers and natural language. Linguist. Philos. **4**, 159–219 (1981)
4. Bezuidenhout, A.: Pragmatics, semantic underdetermination and the referential/attributive distinction. Mind **106**, 375–409 (1997)
5. Bonomi, A.: Transparency and specificity in intensional contexts. In: Leonardi, P., Santambrogio, M. (eds.) On Quine: New Essays, pp. 164–185. Cambridge University Press, Cambridge (1995)
6. Borg, E.: Pointing at Jack, talking about Jill: Understanding deferred uses of demonstratives and pronouns. Mind Lang. **17**, 489–512 (2002)
7. Carlson, G.N.: A unified analysis of the English bare plural. Linguist. Philos. **1**, 413–456 (1977)
8. Elbourne, P.: Situations and Individuals. MIT Press, Cambridge (2005)

---

[25] Deferred and (classically) anaphoric interpretations must be excluded as well, see Sect. 2.2.

[26] For a discussion of the weakening of the requirements of the factivity of the relevant property see [22].

9. Elbourne, P.: Demonstratives as individual concepts. Linguist. Philos. **31**, 409–466 (2008)
10. Fauconnier, G.: Mental Spaces. MIT Press, Cambridge (1985)
11. Frege, G.: Über Sinn und Bedeutung. Z. Philosophie Philosophische Kritik **100**, 25–50 (1892)
12. Frege, G.: Logik (1897, unpublished). In: Hermes, H., Kambartel, F., Kaulbach, F. (eds.) Posthumous Writings, pp. 126–151. University of Chicago Press, Chicago (1979). (transl.)
13. Frege, G.: Der Gedanke. Beiträge zur Philosophie des deutschen Idealismus **I**, 58–77 (1918)
14. Galery, T.N.: Singular content and deferred uses of indexicals. UCL Working Papers in Linguistics, vol. 20, pp. 157–201 (2008)
15. Galery, T.N.: Descriptive pronouns revisited. The semantics and pragmatics of identification based descriptive interpretations. Ph.D. thesis, University College London (2012)
16. Hunter, J.: Structured contexts and anaphoric dependencies. Philos. Stud. **168**, 35–58 (2014)
17. Hunter, J.: Presuppositional indexicals. Ph.D. thesis, The University of Texas at Austin (2010)
18. Kaplan, D.: Quantifying in. Synthese **19**, 178–214 (1968)
19. Kijania-Placek, K.: Descriptive indexicals and the referential/attributive distinction. In: Peliš, M. (ed.) Logica Yearbook 2009, pp. 121–130. College Publications, London (2010)
20. Kijania-Placek, K.: He is usually an Italian, but he isn't. Organon F **3**, 226–234 (2011)
21. Kijania-Placek, K.: Deferred reference and descriptive indexicals. Mixed cases. In: Stalmaszczyk, P. (ed.) Philosophical and Formal Approaches to Linguistic Analysis, pp. 241–261. Ontos Verlag, Frankfurt (2012)
22. Kijania-Placek, K.: Pochwała okazjonalności. Analiza deskryptywnych użyć wyrażeń okazjonalnych (Praise of indexicality. An analysis of descriptive uses of indexicals, in Polish). Semper, Warsaw (2012)
23. Kijania-Placek, K.: Situation semantics, time and descriptive indexicals. In: Stalmaszczyk, P. (ed.) Semantics and Beyond: Philosophical and Linguistic Inquiries, pp. 127–148. Walter de Gruyter, Berlin (2014)
24. Kijania-Placek, K.: Descriptive indexicals and epistemic modality. forthcoming in Topoi, doi:10.1007/s11245-015-9340-5
25. Kijania-Placek, K.: Descriptive indexicals, deferred reference, and anaphora, under review
26. Kijania-Placek, K.: Indexicals and names in proverbs, in preparation
27. King, J.: Singular thought, Russellianism and mental files, forthcoming
28. Kratzer, A.: Stage-level and individual-level predicates. In: Carlson, G., Pelletier, F.J. (eds.) The Generic Book, pp. 125–175. Chicago University Press, Chicago (1995)
29. Kripke, S.A.: Frege's theory of sense and reference: Some exegetical notes. Theoria **74**, 181–218 (2008)
30. Kripke, S.A.: Unrestricted exportation and some morals for the philosophy. In: Philosophy Troubles, pp. 322–350. Oxford University Press, Oxford (2011)
31. Larson, R.: The grammar of intensionality. In: Peter, G., Preyer, G. (eds.) Logical Form and Language, pp. 228–262. Clarendon Press, Oxford (2002)
32. Lewis, D.: Adverbs of quantification. In: Keenan, E. (ed.) Formal Semantics of Natural Language, pp. 3–15. Cambridge University Press, Cambridge (1975)

33. Lewis, D.: On the Plurality of Worlds. Blackwell, Oxford (1986)
34. Loar, B.: The semantics of singular terms. Philos. Stud. **30**, 353–377 (1976)
35. McKay, T., Nelson, M.: Propositional attitude reports. In: Zalta, E.N. (ed.) The Stanford Encyclopedia of Philosophy. Spring 2014 edn. (2014). http://plato.stanford.edu/archives/spr2014/entries/prop-attitude-reports/
36. Moltmann, F.: Intensional verbs and quantifiers. Nat. Lang. Semant. **5**(1), 1–52 (1997)
37. Morgan, D.: First person thinking. Ph.D. thesis, University of Oxford (2011)
38. Mostowski, A.: On a generalization of quantifiers. Fundamenta Math. **XLIV**, 12–36 (1957)
39. Neale, S.: Descriptions. MIT Press, Cambridge (1990)
40. Nunberg, G.: Indexicality in context. Presented at the colloquium on "Philosophy, the human sciences, and the study of cognition," Cerisy-la-Salle, France, 14–20 June 1990 (1990)
41. Nunberg, G.: Two kinds of indexicality. In: Barker, C., Dowty, D. (eds.) The Proceedings from the Second Conference on Semantics and Linguistic Theory: SALT II, pp. 283–302. Department of Linguistics, Ohio State University (1992)
42. Nunberg, G.: Indexicality and deixis. Linguist. Philos. **16**, 1–43 (1993)
43. Nunberg, G.: Descriptive indexicals and indexical descriptions. In: Reimer, M., Bezuidenhout, A. (eds.) Descriptions and Beyond, pp. 261–279. Clarendon Press, Oxford (2004)
44. Perry, J.: The problem of the essential indexical. Nous **13**, 3–21 (1979)
45. Poller, O.: Wyrażenia okazjonalne jako wyrażenia funkcyjne w semantyce Gottloba Fregego (Indexicals as functional expressions in Frege's semantics, in Polish). Diametros **17**, 1–29 (2008). Published under the name Volha Kukushkina
46. Powell, G.: The deferred interpretation of indexicals and proper names. UCL Working Papers in Linguistics, vol. 10, pp. 1–32 (1998)
47. Powell, G.: Language, thought and reference. Ph.D. thesis, UCL (2003)
48. Prior, A.N.: Thank goodness that's over. Philosophy **34**, 12–17 (1959)
49. Quine, W.V.O.: Elementary Logic. Ginn and Company, New York (1941)
50. Quine, W.V.O.: Quantifiers and propositional attitudes. J. Philos. **53**, 177–187 (1956)
51. Quine, W.V.O.: Word and Object. MIT Press, Cambridge (1960)
52. Recanati, F.: Mental Files. Oxford University Press, Oxford (2012)
53. Recanati, F.: Direct Reference: From Language to Thought. Blackwell, Oxford (1993)
54. Recanati, F.: Oratio Obliqua, Oratio Recta: An Essay on Metarepresentation. MIT Press, Cambridge (2000)
55. Recanati, F.: Deixis and anaphora. In: Szabó, Z.G. (ed.) Semantics versus Pragmatics, pp. 286–316. Oxford University Press, Oxford (2005)
56. Recanati, F., Crimmins, M.: Quasi-singular propositions: the semantics of belief reports. Proc. Aristotelian Soc., Supplementary Volumes **69**, 175–193, 195–209 (1995)
57. Richard, M.: Direct reference and ascriptions of belief. J. Philos. Logic **12**, 425–452 (1983)
58. Schiffer, S.: Indexicals and the theory of reference. Synthese **49**, 43–100 (1981)
59. Sosa, E.: Propositional attitudes de dicto and de re. J. Philos. **67**, 883–896 (1970)
60. Stokke, A.: Indexicality and presupposition. Explorations beyond truth-conditional information. Ph.D. thesis, University of St Andrews (2010)
61. Westerståhl, D.: Quantifiers in formal and natural languages. In: Gabbay, D., Guenthner, F. (eds.) Handbook of Philosophical Logic, vol. IV, pp. 1–131. D. Reidel Publishing Company, Dordrecht (1989)

Epistemic Contextualism: An Inconsistent Account for the Semantics of "Know"?

Stefano Leardi[(✉)] and Nicla Vassallo

DAFIST, School of Humanities, University of Genova,
Via Balbi 4, 16126 Genoa, Italy
stefano.leardi.91@gmail.com, nicla.vassallo@unige.it

Abstract. The contextualistic account for the semantic behaviour of the term "know" - a position labelled as "epistemic contextualism" - combined with the widely accepted idea that "know" is a factive verb seems to lead to a very unpleasant conclusion: epistemic contextualism is inconsistent. In Sect. 1 we first examine some aspects of the epistemological meaning of the contextualist semantics of "know", then in Sect. 2 we sketch the problem which leads to the supposed inconsistency of epistemic contextualism and in Sect. 3 we analyse some solutions that have been proposed to solve the problem which are, in our view, unsatisfactory. In Sect. 4 we present our attempt of solution.

Keywords: Knowledge · Epistemic contextualism · Context · Semantics of knowledge ascriptions · Semantics of "know"

1 Contextualizing "Know": An Epistemological Point of View

Epistemic Contextualism (therefore: EC) is one of the latest landing of contemporary epistemology. Proposed in many different readings, in its more common form EC claims that the meaning of the term "know" and that of the propositions which attribute or deny knowledge - as "*S* knows that *p*" and "*S* doesn't know that *p*" - depends upon the context in which those expressions are uttered. According to the core thesis of EC, the term "know" and the propositions that contain it show the same semantic behaviour that characterizes indexical terms or predicates as "large", "rich" or "tall".[1] As propositions such as "I'm here" or "My destination is near", also an expression like "*S* knows that *p*", if closed off by its declaration context, doesn't express a complete and clearly determinable meaning; the practical interests and purposes of the subject who attributes knowledge, as well as the goals of the conversation in which he is involved, are in fact responsible for the arrangement of that set of conditions - *i.e.* the epistemic standard, - which details how strong must be the epistemic position of a subject *S* related to a proposition *p* in a certain context *C*, so that a knowledge ascription as "*S* knows that *p*" comes out to be true in *C*.

As an example,[2] suppose that Thomas is at the King's Cross railway station of London, waiting for a train to Cambridge and wondering whether the train will arrive at

---

[1] DeRose (2009) pp. 166–174, Davis (2013), Kompa (2014).
[2] This is an adaptation of Cohen's airport example. Cfr. Cohen (1999).

© Springer International Publishing Switzerland 2015
H. Christiansen et al. (Eds.): CONTEXT 2015, LNAI 9405, pp. 302–315, 2015.
DOI: 10.1007/978-3-319-25591-0_22

destination before 10.30 am. Beside him, a passer-by asks to Lucy - who's waiting on the platform too, - if she knows the arrival time of the train to the station of Cambridge; Lucy checks her itinerary and answers: "Yes, I know. The train will arrive at 9.30 am". Thomas, who has heard the conversation, wonders if the itinerary checked by Lucy is reliable; getting to Cambridge before 10.30 am is very important for him - at that time he have an appointment of which he cannot be late, - and since the itinerary could be misprinted, Thomas prefers not to attribute knowledge to Lucy and decides to ask for information at the tickets office.

Suppose that the arrival time of the train is 9.30 am. Now, the epistemic position of Thomas is the same of Lucy - they both share the same information: The itinerary's data. On the contrary, what distinguish the two are their respective practical interests, purposes and needs. Knowing the precise arrival time of the train is very important for Thomas and so he prefers not to attribute knowledge to Lucy; he needs something more than the itinerary's data to believe in the proposition "The train will arrive at 9.30 am at the station". On the other hand, Lucy, who doesn't have any particular reason to wonder whether the train will arrive before a certain hour and doesn't have any doubt about the reliability of her itinerary, attributes knowledge to herself about the arrival time of the train. We then have two subjects in the same epistemic position, but with different practical interests and, as it has been forecast by EC, the truth conditions of the proposition "Lucy knows the arrival time of the train" change depending upon the context of utterance. In the case of Lucy the epistemic standard is low, her itinerary is sufficient to her for attributing knowledge to herself and then in this context the proposition "Lucy knows the arrival time of the train" is true. While in the context of Thomas the epistemic standard is significantly higher, and therefore in this case the proposition "Lucy knows the arrival time of the train" is false. Then, in both cases the practical interests of the subjects have established the epistemic standard, i.e. that set of condition which, if satisfied, makes true a knowledge attribution.

EC then looks in first place as a semantic theory, which grounding on the habits of the speakers of the ordinary language, proposes to understand the meaning of the expressions that attribute or deny knowledge as we have said. Even if the formulation of EC is probably mainly due to epistemological reasons - as defeating skepticism or overtaking invariantism - the plausibility of its premises is justified not on considerations about normativity, but on evidences supplied by strengthened and spread linguistic habits exhibited by the speakers of the ordinary language. According to its supporters, the ability of EC to give a sound account for those cases - as the our set at the station, - which show a very elastic way of understanding the meaning of "know" is the main argument in favour of the theory.[3]

If then the role of the semantic aspects of EC is quite clear, more complex is their relation with the epistemic aspects of the theory. The epistemological meaning of the hypothesis about the context sensitivity of "know" is in fact controversial.

Keith DeRose supported the neutrality of EC - understood as a semantic theory, - in respect to its various possible epistemological interpretations.[4] For example, it is

---

[3] DeRose (2009) pp. 47–79.

[4] *Ivi.* p. 21.

possible to be a contextualist and to endorse - at the justification level - foundationalism, coherentism or even an another intermediate option between the two, because EC would be compatible with both and with their possible intermediate versions. Or perhaps we could also built a contextualist theory which is compatible with different versions of the theory of the relevant alternatives or the theory of the conclusive reasons. But even if EC can be understood in many ways, it's still not very clear how we should regard, by a peculiar epistemological point of view, a knowledge attribution as "*S* knows that *p*" when is considered true in a certain context. For DeRose, when a proposition as "Lodewyk knows that the pheasant's head is blue" is considered true in a certain context - and then when Lodewyk's epistemic position satisfies the standard which is at stake in his knowledge attributor context, - it seems that we can say that Lodewyk "counts as knowing" the proposition at issue.[5] The use of this quite ambiguous expression is probably due to the fact that for EC there isn't knowledge simpliciter, but a subject always knows only relatively to a specific set standard. For EC - at least in the view that DeRose seems to endorse, - the truth of a knowledge attribution as "*S* knows that *p*" shouldn't suggest that *S* "posses" knowledge; knowledge indeed cannot be characterized as a more or less broad set of proposition "in possess" of the epistemic subject, but it should be described as a condition or as a status of a subject. In this way, the epistemic subject, to whom now is attributed knowledge in a context, now is denied in another, doesn't risk to see his knowledge vanishing and reappearing suddenly - a phenomenon this which seems very unattractive as well as implausible. To change is just the relation between his epistemic position and the epistemic standard that varies depending on the context of attribution. At the same time two different subjects could attribute and deny knowledge to Lodewyk about a certain proposition *p*, but this would not affect his epistemic position; unware of the evaluations expressed upon him, Lodewyk could in fact quietly still attributing or denying knowledge to himself on the basis of his own epistemic standard.

Compared to DeRose, Stewart Cohen seems to be more interested in the epistemic aspects of EC than in the semantic ones. As a contextualist he endorses the idea that knowledge is in part determined by the social context[6] and supports one of the main claims of EC - i.e. that it's possible that two speakers attribute and deny knowledge at the same time to a subject about the same proposition *p* without any contradiction. Wondering then which context has to be considered to evaluate the epistemic performance of a subject, Cohen gives a first answer: Is the one of the social group which the subject belongs to. However, from this first answer follows a significant consequence: If the relevant context is the one of the social group which the subject belongs to, then we find ourselves with an "indefinite number of concepts of knowledge":

> Is "knowledge" then ambiguous between various concepts each based on a different standard? This would entail an indefinite number of concepts of knowledge. It would also entail that, were our reasoning powers to improve or decline, our concept of knowledge would change.[7]

---

[5] *Ivi.* p. 187.

[6] Cohen (1987).

[7] *Ivi.* p. 15.

So Cohen introduces attributor contextualism:

A better way to view matters is to suppose that attributions (or denials) of knowledge are indexical or context sensitive. The standards that apply are determined by the context of attribution.[8]

But even so, the problem isn't solved. Now we have an indefinite number of contexts of attribution that determines an indefinite number of epistemic standards which defines as many meanings for "know" and concepts of knowledge. In this article and in its later works, however, Cohen seems to have lost interest in this question; what seems pivotal for him is perhaps the epistemic agility that is granted by EC, an agility which is necessary to EC to propose interesting solutions to well-known epistemological problems. Cohen seems then to endorse a reading of EC which is alike the one of DeRose: When a subject S is an object of a knowledge attribution about a proposition p, in the context of the knowledge attributor, S counts as a knower of p.[9]

The admission of an "indefinite number of concepts of knowledge" seems a result hardly evadible for EC, which is then forced to concede some room to relativism - but how much is debated, - to support one of its main thesis - i.e. that it's possible that two speakers attribute and deny knowledge at the same time to a subject about the same proposition p without any contradiction. A thesis that certainly represents one of the finest and more interesting aspect of EC, to which contextualists should not renounce, otherwise they would fall in those problems in which invariantism[10] incurs. The classical analysis of knowledge and many of the theories that had tried to complete it have been often formulated exactly according to the principles of invariantism, however, the rigidity of this approach has the unpleasant consequence to create unattractive asymmetries. A too exigent set of conditions, for example, if by one hand let us to safeguard the value of knowledge - attributing it only at strict conditions, - on the other risks to condemn our ordinary knowledge claims. At the same time, if we would grant our knowledge in everyday context we would be compelled to define weaker conditions, but if deciding whether a subject knows or not would be very important to us, then it seems difficult that we would be satisfied by so weak criterions. However, EC is not a valuable theory only because it overtakes this kind of worries; its premises let contextualists to propose interesting solutions for well-known epistemological problems as the one of lottery or the Gettier cases, and moreover, EC developed an original argument against skepticism which makes our ordinary knowledge claims compatible with the exercise of radical skeptic doubts.[11]

2 Is Epistemic Contextualism Inconsistent? the Problem in a Nutshell

Clearly, the contextualist solutions have been - and still are, - hotly debated. The main part of the critics are focused on the semantic aspects of EC (The account proposed by EC for the semantics of "know" really depicts the linguistic behaviour of the speaker of

---

[8] *Ibid.*

[9] Cohen (1999), (2005).

[10] Unger (1984). According to invariantism there is only one epistemic standard for knowledge.

[11] Lewis (1996), Cohen (1998), DeRose (1995).

the ordinary language? Is "know" really a context-sensitive term?), while others wonder about the epistemological relevance of the theory (Semantic contextualism is a sufficient ground to formulate an adequate epistemological contextualism?).

Lesser broad is the debate about a serious problem that seems to afflicts EC and that questions the consistence of the theory itself: the factivity problem[12] (Therefore: FP). Let's recall our example set at the station, we have:

(1). Thomas doesn't know$_H$ that p.
(2). Lucy knows$_L$ that p.

In prose: "Thomas doesn't know that p" is true in his high standard context while "Lucy knows that p" is true in her low standard context; with p: "The train will arrive at destination at 9.30".

According to EC, even if we are in the more demanding context of Thomas the proposition (2) is still true in the context of Lucy; as we have seen in the Sect. 1 knowledge cannot vanish and despite the evaluations of the knowledge attributor Lucy's epistemic position is still the same. Therefore Thomas couldn't be considered a proper contextualist if he would not give an account for (2) in his own context:[13]

(3). Thomas knows$_H$ that (2).

But from (3) we can derive a very unpleasant contradiction if we combine that proposition with two epistemological principles that the contextualist endorses; the factivity principle (F), according to which some verbs - as "to know", - implies truth:

(F). S knows that $p \rightarrow p$.

and the closure principle (C) which claims that the knowledge of a subject can be extended to that proposition which are entailed from the ones that he yet knows:[14]

(C). If S knows that p and S knows that $(p \rightarrow q)$, then S knows that q.

Now, because for (F) proposition (2) implies p:

(2$_F$). Lucy knows$_L$ that $p \rightarrow p$.

and because Thomas knows$_H$ that (2):

(3$_C$). Thomas knows$_H$ that ⟨Lucy knows$_L$ that p⟩ and that ⟨Lucy knows$_L$ that $p \rightarrow p$⟩.

for (C) and (3) we obtain:

(4). Thomas knows$_H$ that p

Which contradicts (1). EC seems to be inconsistent.

---

[12] For a specific analysis of FP and its variations see Williamson (2001), Brendel (2005), (2014), Wright (2005), Kallestrup (2005), Steup (2005) and Baumann (2008).

[13] Brendel (2005), p. 47.

[14] Luper (2006)"The Epistemic Closure Principle" *The Stanford Encyclopedia of Philosophy* (Spring 2006 Edition), Edward N. Zalta (ed.).

3 Hypothesis for a Consistent Contextualism

According to Wolfgang Freitag and Alexander Dinges[15] FP arises only because EC is misunderstood: for them contextualist is not committed to (2).

Freitag argues that the sufficient and necessary condition for the arise of KP is:

(α). $\exists p \in B$: $(C \rightarrow [p \wedge \neg K_X(S, t, p)])$

In prose: Exist a proposition p which belongs to the set of empirical propositions B such that there is a theory C that implies that p is true, but that p cannot be known by a subject S at a moment t in a certain context X. Clearly, if EC satisfies (α) or not depends upon the way in which we construe our theory,[16] however, according to Freitag we shouldn't ever allow that (α) is fulfilled because if so "we would have to make the empirical claim that {I} a certain proposition p is true and that {II} S doesn't know$_X$ that p at t".[17] We may think that EC implies {II} because of skepticism, which is generally conceded by contextualist;[18] nevertheless, according to Freitag skepticism isn't part of EC, which if not entails skepticism then doesn't entail {II} neither. According to Peter Baumann[19] {I} derives from the factivity principle, from (2) - that EC seems to concede - and from the claim of EC according to which in ordinary context our knowledge attributions are generally true. But for Freitag Bauman's interpretation of EC is not correct. If EC would be committed to empirical claims like (2), then it would perilously depend upon contingent facts: e.g. if (2) would come out false EC would be refuted. For Freitag knowledge in ordinary contexts is expected, but is not part of EC theory. EC is then not committed neither to (1) and (2), so FP appears to be solved. However, two remarks seems to cause troubles to Freitag argument. First, we could agree that EC doesn't entail skepticism, but EC claims that there are high standard - not skeptic, - contexts in which happen that S doesn't know that p, which it was the case that S knew in a lower standard context. This kind of situations seems a proper part of EC theory, then {II} seems justified: indeed there are contexts in which happen that S doesn't know a proposition that he know in others, less demanding contexts. Freitag could perhaps object that specific attributions - as (2) - or denials - as (1) - of knowledge are not parts of EC, and here we come to the second remark: maybe propositions as (1) or (2) are not proper part of EC theory, but it could happen that the contextualist finds himself in a practical situations as the one described in Sect. 2; if so, he couldn't simply reject (1) or (2). Freitag's solution perhaps saves EC from a general point of view, but seems to not have any practical application: the contextualist still have to deal with a theory that doesn't survive when is put to the test.

---

[15] Freitag (2011), Dinges (2014).

[16] Freitag suggests that we could formulate EC reducing it to just its anti-skeptical form:
(EC$_S$). $\neg \forall x, y \in X, \forall S \in G, \forall t \in T, \forall p \in B$: $\wedge[K_X(S, t, p) \longleftrightarrow K_y(S, t, p)]$ which wouldn't suffer of KP. However this solution looks unattractive to Freitag because contextualist should aims to a more complex and articulate theory. Freitag (2011) p. 281.

[17] *Ibid.*

[18] See footnote 11.

[19] Baumann (2008).

Dinges proposes to refute (2) because it seems to not follow from any premise of EC. Proposition (2) doesn't follow from the anti-skeptic claim of EC - according to which our ordinary knowledge attributions are generally true, - because claiming that doesn't mean endorsing the truth of any particular empirical proposition. For example, we could say that the major part of the tickets of a lottery will lose even if we don't know which tickets will lose in particular; in the same way we could understand the anti-skeptical claim of EC: the contextualist knows that the ordinary knowledge attributions are generally true, even if he doesn't know of any particular true knowledge attribution. Even the contextualist anti-skeptical argument[20] seems to not entailing the truth of any particular proposition: the contextualist argues that we cannot show that a proposition as "S knows that p" is false in an ordinary context by showing that it is false in a skeptical one, but that doesn't entail the truth of p. Again, if we could concede to Dinges that (2) is not a proper part of EC, it seems that FP in its practical form isn't solved. Dinges considers this aspect of the problem, but according to him the argument of FP should be refute because appears to be based on a "a tricky logical issue".[21] To show that, Dinges proposes an argument which has very implausible conclusions, and which has a structure that looks like the one of FP: let's suppose to assign different properties to a predicate depending on the time in which it is uttered. "Know" would then express different relations depending on the time of utterance: e.g. if S has forgot something, there would propositions that he $know_{(past)}$ but that he doesn't $know_{(now)}$. So, according to Dinges, S would say that he $know_{(past)}$ that p but that he doesn't $know_{(now)}$ that p. But for the knowledge norm of assertion (KNA)[22] S should $know_{(now)}$ that he $know_{(past)}$ that p, but if so, for factivity and closure we could derive that S $know_{(now)}$ that p, which deny our assumption that S has forgot that p. The structure of this argument looks very similar to the one of FP, however, even if we obtain an implausible conclusion this doesn't mean that it is the structure of the argument that need to be rejected; the problem could in fact depends upon the premises of the argument, but Dinges doesn't propose any reason to exclude this alternative. Refuting (2) then doesn't appear a suitable strategy to overtake FP, because doing so means to solve the problem from a general point of view, but the practical inconsistency remains: EC is still a not working theory.

According to Anthony Brueckner and Christopher Buford[23] EC shouldn't concede "asymmetrical" knowledge attributions like (3), which therefore should be refuted. When the contextualist is pondering about his theory he should limit himself to say something like "It's possible that there are two context C_1 and C_2 such that a proposition as "S know that p" is true in a context but false in the other". Now let's consider FP in his practical form; for the example of Sect. 2 we have that (1): Thomas doesn't $know_H$ that p and that (2): Lucy $knows_L$ that p. According to Brueckner and Buford, if we - sharing the same context C_H of Thomas, - would tell him that Lucy is in a low

---

[20] For EC reply to skepticism: DeRose (1995), Rysiew (2011) "Epistemic Contextualism", *The Stanford Encyclopedia of Philosophy* (Winter 2011 Edition), Edward N. Zalta (ed.).

[21] Dinges (2014) p. 3550, footnote 20.

[22] According to KNA an utterance of p is appropriate only if the speaker knows that p.

[23] Brueckner and Buford (2009).

standard context C_L, that she satisfies the standard at stake in C_L and that she attributes herself knowledge about p saying "I know that p", Thomas shouldn't endorse (3), but his answer should be:

> Well, it sounds as Lucy is in a position to be saying something true via uttering his "knowledge"-sentence, given his wimpy context C_L and ordinary evidence. So I know that the conditions for the truth of "Lucy knows$_L$ that p" are satisfied up to the "truth condition", i.e. the condition that p is the case. However, to know that "Lucy knows$_L$ that p" is true in C_L, I must know whether p is the case [...]. But I have just told you that I do not know p; [...] "Lucy knows$_L$ that p" is not true in our context C_H.[24]

In virtue of his lack of knowledge about p Thomas is not in the position to utter (3), which is therefore to be considered an illegitimate step. But if so, we wonder if we can still speaking about a proper EC. As Brueckner and Buford recognize,[25] their solution saddles EC with a 'stability problem': the contextualist thesis that in a low standard context "S knows that p" is true cannot be known anymore in a more demanding context. According to Baumann, it's highly controversial whether EC "can live with the above mentioned statability limitation",[26] it's then doubtful if Brueckner and Buford's solution can be considered as a progress of any kind.

Martin Montminy and Wes Skolits[27] proposes to understand (3) not as a proper assertion, but as a weak one, which is a kind of illocutionary act which includes conjectures, guesses and hypothesis. As the two authors underline, this is the typical way that philosophers adopt when they propose or defend their views. Weak assertions are not governed by the KNA: a weak assertion that p is epistemically appropriate if the speaker have some evidence that p - the number of evidence required depends upon the strength of the assertion. But if (3) is a weak assertion, then when the contextualist utters (3) it seems that he is not properly recognizing (2) - i.e. that Lucy know$_L$ that p. Perhaps he would say something like "It seems that Lucy knows that p in her own context", but therefore the epistemic status of (2) is indeterminate. This solution appears to meet worries analogous to the ones that the solution of Brueckner and Buford meets; there is a knot that needs to be unravelled: for Thomas - which is in the high standard context, - Lucy knows or not that p? Montminy and Skolits' solution seems then to suggest to Thomas to endorse and answer analogous to the one which Brueckner and Buford have proposed, something as "I know that the conditions for the truth of ⟨Lucy knows$_L$ that p⟩ are satisfied up to the truth condition, but I cannot proper say that she knows that p", but we have already seen what kind of troubles that answer involves.

Peter Baumann[28] doubts about the plausibility of the disquotation principle that is involved in FP - which according to him is: (D) ["p" is true $\rightarrow$ p] - because the supposed context-sensitivity of "know" wouldn't allow us to apply (D) and infer from (2) and (3) that (4). Of course it would be a trouble for EC to deny any kind of

---

[24] *Ivi.* pp. 434–435. This quote has been adapted to our exposition of the FP.

[25] *Ivi.* pp. 436, 437.

[26] Baumann (2010), p. 88.

[27] Montminy and Skolits (2014).

[28] Baumann (2008).

disquotation principle, so Baumann proposes that we should formulate a principle which would make explicit the context-sensitivity of "know" and that would understand "'knowledge' as referring [...] to a ternary relation between a person, a proposition and a standard".[29] We should then formulate a contextualist friendly disquotation principle as:

(D$_{EC}$). An utterance of "S knows that p" in a context C$_X$ is true $\rightarrow$ S knows$_X$ that p

However, if we consider (3) - "Thomas knows$_H$ that Lucy knows$_L$ that p" - and we apply factivity, closure and D$_{EC}$ we can still obtain (4) - i.e. "Thomas know$_H$ that p" - which leads to the contradiction. But according to Baumann, the warrant that Thomas needs to know Lucy's epistemic performance is not the same that he needs to know what Lucy knows; Thomas could have very sophisticated knowledge about the rough nature of Lucy's knowledge of, e.g. the average weight of an hippopotamus, but by no means it follow that he has sophisticated knowledge about the very same thing. For Baumann there is a certain failure of transmission of warrant:

(T). S_1 has warrant for knowledge$_H$ that S_2 knows$_L$ that p $\rightarrow$ S_1 has warrant for knowledge (at some level, but not necessarily for knowledge$_H$) that p

We should then endorse a new principle of closure as:

(C$_{EC}$).   For all subjects S, propositions p and q, and knowledge relations know$_A$, there is a knowledge relation know$_B$ (where know$_A$ is not more demanding than know$_B$) such that:
[S knows$_A$ that $p \wedge$ S knows$_A$ that $(p \rightarrow q)$] $\rightarrow$ S knows$_B$ that q.

Because of (T) and the new principle of closure (C$_{EC}$) we cannot derive (4) any-more; at least we could obtain (4$_B$): "Thomas know$_L$ that p", which doesn't deny (1). According to Baumann in fact, it could be the case that, considered the factivity principle (F): [S knows$_X$ that $p \rightarrow p$], Thomas would know$_H$ the antecedent of (F) - i.e. that Lucy knows$_L$ that p - but he wouldn't know$_H$ the consequent. However, Baumann's solution is controversial: If we still endorse a factivity principle as (F), as Baumann suggests, then in his context Thomas should infer from (2) - "Lucy knows$_L$ that p" - that p is true *simpliciter*; the truth of p is in fact one of the necessary conditions for the truth of (3), a condition that Baumann doesn't seem to refute. So, as Montminy and Skolits have noticed,[30] Thomas would find himself saying something as (5): "I know$_H$ that p is true, even if I don't know$_H$ that p", which is quite odd; bizarre propositions as "While I do not know that I'm a bodiless brain in a vat, I do know that I have hands" or "Even though I don't know that these are not well dis-guised mules, I know that they are zebras"[31] should be avoided. Even Baumann solution then seems unattractive.

---

[29] *Ivi.* p. 589.
[30] Montminy and Skolits (2014), p. 325.
[31] *Ibid.*

4   Solving the Factivity Problem: A Further Attempt

We have seen that denying specific knowledge attributions doesn't seem to be a good strategy to solve FP because it turns out to be an impassable path when practical cases are considered. Denying (3) as Brueckner and Buford suggest however, has proved to be a dangerous solution as well: It seems to solve FP, but saddles EC with the limitation of the stability problem. On the other hand, the solution developed by Baumann involves odd consequences which every theory about knowledge-attributing sentences - not only EC, - should avoid. Nevertheless, Bauman's solution seems to be the only one that really attempts to preserve the possibility for the contextualist to know - in his high standard context, - that the knowledge attribution made in lower standard contexts are true, a possibility that, according to Baumann, seems to be pivotal for EC:[32]

> What is the attraction of contextualism if one cannot (at least as a contextualist) coherently say (or think) that knowledge attributions made in a lower context are in fact true? [...] The kind of contextualism that results would be a very much weakened one and not very attractive.[33]

Therefore, a formulation of EC which could be called "robust" and "attractive" appears to be compelled not only to avoid the contradiction of the FP, but also to pursue this goal without denying the possibility mentioned above. To achieve this purpose and to explore the features of this reading of EC let's consider an example: Suppose that Thomas, an amateur ethologist, is attending a lesson about primates in a natural reserve. During the lesson, Thomas notices an animal on a tree that appears to be a chimpanzee, so he say

(a). "The animal on that tree is a chimpanzee!"

The ethology professor - who knows that the animal is a chimpanzee, - asks to Thomas: "How do you know that?" and Thomas, aware that his only answer could be

(b). "I saw many images of chimpanzees and that animal looks as one of them"

and that however this too generic answer could not satisfy the professor, prefers to reply: "No, I don't know that the animal on that tree is a chimpanzee". Now imagine to say to Thomas that Lucy - who is in the natural reserve for a safari, - has said that (a) speaking of the same ape seen by Thomas, that she has justified her claim saying that (b) and that her trip mates, satisfied with her explanation, have attributed knowledge to her: What kind of answer should we suggest to Thomas, who is a contextualist? Does Lucy know (a) according to the standard at stake in *her* context?

A first useful remark is to remember that, according to EC, the truth conditions of an expression as "S knows that p" can be defined only considering the characteristics of the knowledge attributor's context - i.e. the practical interests and purposes of the attributor as well as the goals of the conversation in which he is involved; those characteristics are in fact responsible for the setting of the epistemic standard, and

---

[32] On this point Brendel seems to agree with Baumann, see Brendel (2005) pp. 45–47.

[33] Baumann (2008) p. 583.

therefore for the definition of the truth conditions of the knowledge-attributing or denying sentences. We can then deduce that - at least according to her trip mates, - Lucy is well positioned enough in respect to the standard at stake in the safari context, and therefore that, in that context, she counts as a knowing that (a). To clarify this point we could imagine the evaluations made by Lucy's trip mates. Indeed, if it's up to the knowledge attributor defining the conditions at which an expression as "S knows that p" is true, we could also imagine that, when the attributor is evaluating the epistemic performance of a subject, it's up to him saying something as, in our case: "(i) Lucy believe that p, (ii) her belief is justified with enough good reasons and (iii) p is true"; but then, if it's the knowledge attributor the one who decides if the truth conditions are satisfied or not, then, even (iii), the truth condition, should be understood according to a contextualistic point of view. Indeed, before that the attributor could say that

(c). "S knows that p"

he should claim that

(d). "I know that p is true"

knowing that p is true is in fact a supposed required condition to know that (c); however, (d) is a knowledge-attributing sentence and then, according to EC, it should be evaluated considering the practical interests of the knowledge attributor. A clue, this one, that seems to suggest that it would be coherent for EC to argue that it's up to the knowledge attributor deciding if p is "true enough"[34] or "reasonably true" according to his purposes and practical interest - and not true *simpliciter*. After all, when the epistemic performance of a subject is evaluated, the judgement of the knowledge attributor always grounds on some specific epistemic basis. Imagine to evaluate if a subject S knows a proposition p: our epistemic custom - which appears to be mainly concerned with the practical aspects of the knowledge attributing practice, - would suggest to define a reasonable perimeter for our evaluation; e.g. if an ordinary epistemic standard would be at stake, we would admit many truths that in a more demanding context we would not assume. In a skeptical context, for example, Thomas would probably deny knowledge to himself about (a) since the possibility that an evil demon is deceiving him would be salient; nevertheless, in an ordinary context he would ignore the evil demon possibility and would smoothly attribute knowledge to himself. Indeed, in such a context certain error possibilities would be ignored, but also certain propositions would be assumed as true and certain methods to catch truth would be approved. Suppose that Thomas is at the zoo, looking to an animal that looks like a chimpanzee in a pen beside a banner that says 'chimpanzees': in this context, looking to the animal and considering that it quite exactly resembles to a chimpanzee could be enough for Thomas to admit that the animal in the pen is in fact a chimpanzee. Clearly someone could ask to Thomas: "How can you be sure that the animal in the pen is a chimpanzee?" and then he could reply making a list of the characteristics that distinguish a chimpanzee from a bonobo; or even, if the doubts raised would involve a stricter standard, he could test the DNA of the supposed chimpanzee to be surer. Anyway,

---

[34] This concept is ought to Elgin (2004).

every challenge moved to the epistemic position of Thomas would set an epistemic standard that, among other things, would also define which error possibilities could be properly ignored, which propositions could be smoothly assumed as true and which methods to catch truth could be considered as reliable.

The structure of the contextualist anti-skeptical argument seems to support this reading of EC. Indeed, according to the contextualist skepticism is in a certain sense a licit practice: When the epistemic standard is allowed to raise until a skeptical level in fact, according to EC we know quite nothing. In a skeptical context the warrant required to know a proposition is in fact generally out of reach, but also, in that context we cannot rely on propositions which we would otherwise assume as true, as well as on methods to catch truth which in an ordinary context would be approved; according to the contextualist then, in his own context the skeptic efficaciously undermines our confidence in the truth of many propositions which we would have assumed as true in an ordinary context. Therefore, in the skeptical context we cannot know that certain propositions are true, and then we cannot know that propositions. But if so, then (*iii*), the truth condition, should be understood by the contextualist in a moderate way: indeed, arguing that, if we would know that *p* is true *simpliciter* then the skeptic could not undermine our confidence in *p* would mean, for the contextualist, to endorse a Moorean approach towards skepticism which seems extraneous to EC; on the other hand, also an odd proposition as "I know that *p* is true, but in this skeptical context I don't know that *p*" should be avoided by the supporter of EC. However, maintaining that the truth of "*S* know that *p*" implies that *p* is true enough according to the standard of the knowledge attributor would let the contextualist to preserve his classic approach towards skepticism.

To be clear, arguing that the truth of a knowledge-attributing sentence as "*S* knows that *p*" implies that *p* is true enough - and not true *simpliciter*, - doesn't mean that, if it would come out that *p* is false we would still have to say that "*S* knows that *p*" is true. However, if we would not know if *p* is a true proposition or not, we could still acknowledge that "*S* knows that *p*" is true according to the epistemic standard of the subject who has attributed knowledge to *S*. Let's recall our example set in the natural reserve: in his demanding context Thomas doesn't know if the proposition (*a*) is true; nevertheless, he knows - as a contextualist, - that one of the conditions for the truth of a proposition as ‹"Lucy knows that (*a*)" is true in the safari context› is that (*a*) is considered true enough in the safari context. Because Thomas knows that Lucy's trip mates have attributed knowledge to her about (*a*), and since he cannot argue that (*a*) is false, Thomas could at least acknowledge that "Lucy knows that (*a*)" is true in the safari context. Of course, this would not mean that Lucy count as knowing that (*a*) according to standard at stake in the context of Thomas. In this way the contradiction of the FP could be avoided, and it's open to the contextualist to acknowledge that some knowledge-attributions made in low standard contexts are true. The contextualist should then reject the traditional factivity principle (F) in favour of a contextualist friendly factivity principle as:

(F$_i$).   *S* knows$_X$ that $p \rightarrow p$ is reasonable true according to the standard of the subject who attributed knowledge to *S*

Therefore from:

(3).   Thomas knows$_H$ that Lucy knows$_L$ that p

We cannot infer (4) anymore; what we could obtain is:

(4$_i$).   Thomas knows$_H$ that p is reasonably true according to standard at stake in Lucy's context

5   Concluding Remarks

Admitting the factivity principle seems to be a really dangerous step for EC: The idea that from a knowledge attribution sentence about p we can deduce that p is true *simpliciter* seems to be inconsistent *per se* with the formulation of EC according to which to the contextualist should be granted the possibility to know, in his more demanding context, that some knowledge attributions made in less demanding contexts are true.

Our reading of EC, rejecting the traditional factivity principle, if by one hand let us to solve the FP, on the other describes the knowledge-attributing practice as concerned mainly with the practical aspects of knowledge and, mostly, undoubtedly saddles the theory with a certain kind of relativism: Indeed, the practical interests and purposes of the knowledge attributor assume a really heavy role in the theory. However, it's in doubt if this consequence could have ever been avoided maintaining the main characteristics of EC; after all, as Stewart Cohen, John Greco and Leonid Tarasov have observed,[35] in a way or in another, relativism seems to be a companion of EC. The relativism consequence could perhaps be due to EC's "practical" nature: The theory is in fact patterned upon the so called ordinary language's evidences, which show an understanding of knowledge that seems to be committed especially to the needs of our practical reasoning; indeed, the practical interests of the knowledge attributor play a much than a pivotal role in the dynamics of the theory per se, and perhaps we have taken this premises to their more radical consequences. If then we conclude that EC entails a certain kind of relativism, much has to be done to put this relativism "under rigorous restraint"; after all, the ordinary language speakers' use of "know" doesn't seems to be totally "disparately varied and undisciplined, individual-dependent and arbitrary"[36] as has been argued by Tarasov. An inquiry in that direction is then especially needed.

References

Ashfield, M.D.: Against the minimalistic reading of epistemic contextualism: a reply to wolfgang Freitag. Acta Anal. **28**(1), 111–125 (2013)

---

[35] Cohen (1987) p. 15, Greco (2008), Tarasov (2013).

[36] Tarasov (2013), pp. 574, 575.

Baumann, P.: Contextualism and the factivity problem. Philos. Phenomenological Res. **76**(3), 580–602 (2008)

Baumann, P.: Factivity and contextualism. Analysis **70**(1), 82–89 (2010)

Brendel, E.: Why contextualists cannot know they are right: self-refuting implications of contextualism. Acta Anal. **20**(2), 38–55 (2005)

Brendel, E.: Contextualism, relativism, and the semantics of knowledge ascriptions. Philos. Stud. **168**(1), 101–117 (2014)

Brueckner, A.: The elusive virtues of contextualism. Philos. Stud. **118**, 401–405 (2004)

Brueckner, A., Buford, C.T.: Contextualism, SSI and the factivity problem. Analysis **69**(3), 431–438 (2009)

Brueckner, A., Buford, C.T.: Reply to Baumann on factivity and contextualism. Analysis **70**(3), 486–489 (2010)

Cohen, S.: Knowledge, context, and social standards. Synthese **73**(1), 3–26 (1987)

Cohen, S.: Contextualist solutions to epistemological problems: scepticism, gettier, and the lottery. Australas. J. Philos. **76**(2), 289–306 (1998)

Cohen, S.: Contextualism, skepticism, and the structure of reasons. Philos. Perspect. **13**(s13), 57–89 (1999)

Cohen, S.: Knowledge, speaker and subject. Philos. Q. **55**(219), 199–212 (2005)

Davis, W.A.: On non indexical contextualism. Philos. Stud. **163**, 561–574 (2013)

DeRose, K.: Solving the skeptical problem. Philos. Rev. **104**(1), 1–52 (1995)

DeRose, K.: The case for contextualism. Oxford University Press, Oxford (2009)

Dinges, A.: Epistemic contextualism can be stated properly. Synthese **191**(15), 3541–3556 (2014)

Elgin, C.Z.: True enough. Philos. Issues **14**(1), 113–131 (2004)

Freitag, W.: Epistemic contextualism and the knowability problem. Acta Anal. **26**(3), 273–284 (2011)

Freitag, W.: In defence of a minimal conception of epistemic contextualism: a reply to M. D. Ashfield's response. Acta Anal. **28**(1), 127–137 (2013)

Greco, J.: What's wrong with contextualism? Philos. Q. **58**(232), 416–436 (2008)

Hannon, M.: The practical origins of epistemic contextualism. Erkenntnis **78**(4), 899–919 (2013)

Hazlett, A.: The myth of factive verbs. Philos. Phenomenological Res. **80**(3), 497–522 (2010)

Kallestrup, J.: Contextualism between scepticism and common-sense. Grazer Philosophische Studien **69**(1), 247–266 (2005)

Lewis, D.: Elusive knowledge. Australas. J. Philos. **74**, 549–567 (1996)

Luper, S.: The epistemic closure principle. In: Zalta, E.N. (ed.) The Stanford Encyclopedia of Philosophy (2006). http://plato.stanford.edu/archives/spr2006/entries/closure-epistemic/

Kompa, N.: Knowledge in context. Rivista Internazionale di filosofia e psicologia **5**(1), 58–71 (2014)

Montminy, M., Skolits, W.: Defending the coherence of contextualism. Episteme **11**(3), 319–333 (2014)

Rysiew, P.: Epistemic contextualism. In: Zalta, E.N. (ed.) The Stanford Encyclopedia of Philosophy (2011). http://plato.stanford.edu/archives/win2011/entries/contextualism-epistemology/

Steup, M.: Contextualism and conceptual disambiguation. Acta Anal. **20**(1), 3–15 (2005)

Tarasov, L.: Contextualism and weird knowledge. Philos. Q. **63**(252), 565–575 (2013)

Unger, P.: Philosophical Relativity. University of Minnesota Press, Minnesota (1984)

Williamson, T.: Comments on Michael Williams' contextualism, externalism and epistemic standards. Philos. Stud. **103**(1), 25–33 (2001)

Wright, C.: Contextualism and scepticism: even-handedness, factivity and surreptitiously raising standards. Philos. Q. **55**(219), 236–262 (2005)

Analysis of Geographical Proper Names in Terms of the Indexicality Account of Proper Names

Tomoo Ueda[1,2,3](✉)

[1] Tokyo Metropolitan University, Tokyo, Japan
ueda-tomoo@tmu.ac.jp
[2] Institut Jean-Nicod, Paris, France
[3] Japan Society for the Promotion of Science, Tokyo, Japan

Abstract. This paper focuses on geographical names, which are names of geographical regions. Geographical and personal names will be distinguished with reference to evidence from linguistics as well as cognitive science. Since the entire debate is on the relationship between geographical names and their bearers, the reference of geographical names in the context of use will be examined and two possible views on geographical name reference will be proposed. As a test case, the cases of geographical developments will be examined. The basic idea involves extending the indexicality account with generic names. If a geographical name is used in the standard context, it refers to specific geographical territory, whose development is tracked by discursive participants through (a) the causal history of communications of the name in question and (b) its current referred territory.

Keywords: Proper names · Geographical names · Indexicality account · Pragmatics · Semantics

1 Introduction: Geographical Names and Their Reference

This paper focuses on geographical names, a type of names of geographical regions including rivers, countries, cities, city districts, and so on. To characterize them, two topics are discussed in this section: First, geographical and personal names are distinguished with reference to evidence from linguistics as well as cognitive science.[1] Second, since the entire debate is on the relationship between geographical names and their bearers, the reference to geographical names in context of use will be examined and three possible views on geographical name reference will be proposed.

1.1 Indexicality vs. Homonymy

Before focusing on geographical names, let me briefly review the discussions on the nature of proper names. The discussion on the nature of proper names

---

[1] I do not address the issue of temporal names, including whether such names exist.

© Springer International Publishing Switzerland 2015
H. Christiansen et al. (Eds.): CONTEXT 2015, LNAI 9405, pp. 316–327, 2015.
DOI: 10.1007/978-3-319-25591-0_23

is discussed in the framework of whether they are indexical. In particular, the discussion has focused on the cases in which two people share the same linguistic expression (which Kaplan [4] calls "generic name"[2]), say, "Thomas", as their names. There are two positions on the issue: On the one hand, there is the indexicality account (defended, e.g., by Recanati [15, cp. 10], Stojanovic [20] and Rami [14]), according to which these people share the same proper name, and "Thomas" works like indexical expressions, such as personal pronouns. On the other hand, there is the homonymy account (defended, e.g., by Almog [1], Korta and Perry [5], as well as Fujikawa [2]), according to which the names of different Thomases are homonyms and they merely share the linguistic surface.

The core of the debate is in naming conventions that connect a certain generic name with an entity in the world. In the homonymy account, a specific name (which Kaplan [4] calls "common currency name") is the combination of a generic name and an entity. In the case of two Thomases above, there are two names, and it is the role of contexts to determine which name is relevant. In contrast, in the indexicality account, the naming convention of a name is relativized to the context of use. One single name (i. e., a generic name) is associated with a class of bearers,[3] and naming conventions associate names with a relevant object in a certain context of utterance. In other words, according to the indexicality account, there is no common specific names.

This paper does *not* aim at solving this discussion; rather, it focuses on a particular version of the indexicality to deal with specific features of geographical names. In the following discussion, the distinction between personal and geographical names will be justified. Then, two clusters of questions that arise with geographical names will be introduced.

1.2 Distinction Between Geographical and Personal Names

In literature, studies on proper names often take personal names as the standard examples and assume that the same consequences are applicable to all types of proper names. However, discussions in linguistics (e. g., [11]) and clinical studies (e. g., [7,18,22]) strongly suggest otherwise; that is, geographical names require special examinations because they behave differently. First, in English,[4] geographical names that include those for regions, famous buildings, rivers, straits, seas, oceans, and deserts (but not personal names) are weak proper names, in which the definite article marks the definiteness redundantly [11, pp. 517–518]. The redundancy of the definite article is justified by observing that "weak proper

---

[2] The notion of generic names will be explained in Sect. 4.1.

[3] One can see some disanalogies between personal pronouns and names here, because the lexical meanings of pronouns work as intensional characterization of the class in question. Contrastingly, as far as personal names are concerned, there is no properties shared among referents of a name except for being named such and such.

[4] The strong/weak distinction is "a rather arbitrary matter" [11, fn. 77 (p. 517)] from the cross-lingual perspectives. For example, Russian personal names are strong as well as geographical names.

names normally lose the definite article when they don't constitute a full NP—when they are modifying the head of an NP or are themselves modified" [11, p. 517]; for example, "two United States warships."[5] However, note that most country and city names are strong in English.

Second, while many dictionaries that exclude proper names (particularly personal), there are dictionaries that contain certain entries for particular geographical proper names, such as *DUDEN: Das große Wörterbuch der deutschen Sprache* (third edition) [22b, sect. A]. As Marconi [8] discusses, the deletion of proper names from dictionaries was a historical coincidence; however, it is the point to be observed that some editors intuitively saw the difference between geographical and proper names.

Finally, clinical studies on anomia suggest that personal and geographical names "seem to behave in the brain differently from one another" [18] (see, e.g., FH reported in [7]). Furthermore, there is a study [22] that hypothesizes that personal and geographical names are processed in different brain hemispheres.

Of course, these data are not decisive with regard to whether personal and geographical names belong to different linguistic categories; however, it is not self-evident any more whether they belong to the same linguistic category. Thus, it must be examined how geographical names must be examined. From these considerations, this paper focuses on geographical names.

When we use a geographical name, we talk about a geographical territory or a continuum, such as a metaphysical space-time worm [19], that exists through history. As far as the referent of personal names is concerned, a person is not something that extends over time and, in major metaphysical positions (see [10, particularly sec. 3] for the overview), there is something that goes beyond human organisms.

1.3 Questions About Reference of Geographical Names

In the following account, I focus on geographical names instead of discussing proper names in general. The first question is whether there is something that is persistent as referents of geographical names. For example, if we take, "Paris," a name of city, it can be used in the following manner:

(1)　a. Alice came to <u>Paris</u>.
　　　b. Bob sued <u>Paris</u>.

Since the name "Paris" is used referentially in (1), the question is what is the referent of "Paris" if each of (1) is uttered. While the former usage of "Paris" is about certain geographical territory, the latter usage is about the juridical person that governs this territory.

---

[5] Naïve Millians, such as Salmon [16], treated proper names as syntactically simple. Thus, they appear to claim that English weak proper names, like "the Pacific Ocean," are not proper names but definite descriptions. However, this distinction is not justified both linguistically and philosophically.

The second cluster of questions that is discussed is whether geographical names are nouns with a complex type, which are logical polysemies. The situation becomes complicated because you can combine both sentences in (1) and assert that

(2) Alice came to Paris and Bob sued it.

Since "it" is anaphoric, "Paris" appears to be polysemic rather than just metonymic. Similar cases can be often observed for printed matter. For example, compare (1) with logical plysemies, for example:

(3)   a. Eno the cat is sitting on <u>this book</u>.
      b. <u>This book</u> really got me upset.

"This book" in (3a) refers to a specific copy as a physical object, while "this book" in (3b) refers to the content that is printed in such copies. Although both singular terms refer to two different things, you can use anaphora, for example:

(4) This book, on which Eno is sitting, really got me upset.

The cases like "this book" will be explained because "book" is a nominal with a complex type [12, 13]. In other words, a book is a type of both information and physical objects (Pustjevsky [12, 13] calls meanings of this type of nouns "dot-objects"). This parallelism between (2) and (4) could be understood as suggesting that, like "this book," geographical names, like "Paris," are polysemic and they refer to different things depending on the context.

There are two things that must be noted: First, as far as the personal names are concerned, they are not polysemic. To clarify this, compare the following two sets of examples:

(5)   a. I've read Dickens.
      b. My great grandmother met Dickens.
      c. I've read Dickens but my great grandmother met him.

(6)   a. *I read the author.
      b. My great grandmother met the author.
      c. *I've read the author but my great grandmother met him.

While every sentence of (5) is acceptable, it is highly questionalble whether the first sentence (hence, the last, too) of (6) is acceptable, because "the author" is conventionally not a book.[6] This data suggests that this "Dickens" used in (5) is not polysemic.[7]

---

[6] Furthermore, it is highly doubtful whether the following sentence is acceptable:

(7) My great grandmother met Dickens, whome I read.

[7] Of course, there are metonymical usages of geographical names. For example, observe the following article title:

(8) Seoul Surprises with Interest Rate Cut. [24]

Second, and more importantly, (1) is only the case because "Paris" is a city. Thus, the following examples are acceptable:

(1)   a. Alice came to the city.
      b. Bob sued the city.

And it appears (2) is only possible for geographical names whose territories constitute communities. Since there is no such community that is representative of this region, for other geographical names, such as Mt. Everest, this case cannot be generalized. Thus, it appears to be safe to assume that geographical names are not polysemic in general. And for this reason, in the remaining of the paper, I ignore the cases in which "Paris" refers to the juristic person. Even if geographical names are not polysemic, it is still worth considering whether there is something that gives geographcial entities their persistence; in other words, whether the referent of a geographical name is a space-time worm.

2   Geographical Developments: A Test Case

The discussion thus far can be summarized in the following manner: A geographical name can refer either to some geographical territory or a socially constructed continuum, if it is used. Through the following discussion on geographical developments, I argue that geographical names can refer to certain geographical territory as its default referent. Let me start with some difficulties of the thesis that geographical names refer to a certain territory in their default cases. Namely, taking certain geographical territory as the default referent[8] (or the bearer) of a geographical name—for example, "Paris"—faces the following problem of growing cities.

Some regions, for example, Paris in France, grow, or change the border from time to time. Suppose I read the statistical development of Paris,[9] and assert the following today:

(11) Paris had a population density of 15,205 people per square kilometer in 1790.

"Seoul" used here is neither the geographical territory nor the city hall. For example, it is doubtful whether the following is acceptable:

(9)   ?The city surprises with interest rate cut.

Rather, "Seoul" is used to metonymically refer to the Bank of Korea that has its headquarter in Seoul.

[8] I use the term "default referent" as a neutral term between the indexicality and homonymy accounts.

[9] From "Demographia": http://www.demographia.com/dm-par90.htm (seen on Jan. 26, 2014).

This is a perfectly informative claim and the truth of (11) should be evaluated with respect to the territory of the time that is explicitly marked in (11). However, the situation is complicated if we take territory as the default referent of a geographical name. For, if "Paris" is used today, it refers to some larger territory than that in 1790. If the current territory would be the default referent of the name, (11) would express something different from my intended meaning. In particular, if "Paris" is interpreted in today's sense, the population density will be different from the population density of Paris in the late eighteenth century. This implies that the name "Paris" becomes ambiguous depending on context. Thus, in the evaluation context, we have one name—that is, "Paris"—and two different territories attached to the generic name. That is, the situation appears to be comparable to the pair of Paris, France and Paris, Texas.

There are three things to be noted: First, and most importantly, this is not only a problem for communities including nations, states, and cities. Rather, it is a general point due to geographical developments, whose cause may be artificial as well as natural. For example, an island can change its territory through its volcanic activities. Second, this is problematic only if we take some territory as the default referent of a geographical name. If a geographical would refer to an abstract entity, territory is one contingent property of the named object. The case is similar if a geographical name refers to a historical continuum; this is because, then, a specific geographical territory is, then, a temporal part or a stage of the entire continuum. Finally, the current issue must be distinguished from that related to vagueness. The referent of "Paris" is vague in neither an epistemic [21] nor a metaphysical [17] sense. In both eighteenth and twenty-first centuries, the territories are clearly defined, so that there is no borderline region and you can identify them.

3   Relevance of Geographical Developments

One might argue that geographical developments are irrelevant to the analysis of geographical names since the type of cases appear non-standard. For example, in the case of personal names, the growth of a person does not affect keeping track of the referent of her name even though a personal name refers to the specific physical object, namely, a person. In this section, I argue that socially constructed entities are not good candidates for the bearers of geographical names; therefore, the cases of geographical developments remain relevant.

3.1   The Default Referent of a Geographical Name Is Not An abstract Social Entity

If "Paris" refers to a socially constructed entity, it seems to follow that the geographical territory is a contingent property of the city, and hence, the referent of a city's name,[10,11] This is good because, then, the cases of geographical developments are not problematic at all.

However attractive the social construction thesis is, it is not tenable. For it is not exhaustive since there are many regions—including the North Sea, the Tokyo Bay, Mt. Everest—that is not represented by any juristic person nor any communities. For such names, it is mysterious how the relevant social entities are constructed. City and country names are rather exceptional as geographical names. Therefore, most geographical names neither refer to abstract social entities nor is every such name a dot-object.

3.2 Metaphysics is Not Helpful

One of the most natural solutions to the geographical development cases is to commit to the metaphysics that geographical regions are a sort of continuum with temporal parts. As I briefly stated above, the cases with geographical developments, for example, of Paris, appear to be easily explained if a city name refers to a continuum that has temporal parts; this is because the change in territory is a metaphysical development of a city and it is the city as a whole we are talking about in each utterance, like (5). However, certain criticisms are raised for such a metaphysical viewpoint. For example, the four-dimensional worms really change [3, sect. 2].

Even if we take the stage view [19] about the geographical name reference, there are further problems that are specific to the geographical territories as the referents of geographical names. Namely, exactly the same territory can be and is indeed referred to by geographical names without standing to the part-whole relations to each other. For example, the northern part of Saint Martin island belongs to the *Collectivité de Saint-Martin*, which includes some other islands. Thus, the referent of the name the "Saint Martin Island" is not a part of the referent of the *"Collectivité de Saint-Martin"* and vice versa. However, it is normally considered that there are no two objects that occupy the same space at the same time (or they cannot coincide). This coincidence of stages appears to imply that referents of geographical names are parasitic to some more fundamental entities (or ontologically dependent on them).

Furthermore, it appears that if the disputes between the worm view and the stage view is applied to geographical entities, it is necessary to solve the current problem. Sider summarizes this aspect in the following manner:

> The central claim of the stage view is supposed to be that a speaker refers to stages of worms sliced at the time of utterance [19, p. 199].

---

[10] However, it is not the aim of this paper to discuss the metaphysics of social institutions (see, e.g., [9] for some overview).

[11] This notion of social entity, whatever it is, must be clearly separated from that of stereotype. For example, a district in Paris, in which there are only Starbuck's instead of cafes, is not stereotypical to Paris such that we would like to claim: "This is not Paris."

Discussing stereotypes is not relevant for the issue because one can construct equivalent examples for all sorts of expressions, including common nouns or verbs.

This is exactly the problem that we face. Therefore, metaphysics does not solve our problem. Rather, to solve the metaphysical dispute between the worm view and the stage view, one needs to solve the question regarding the reference of geographical names.

3.3 Geographical Territory as the Geographical Name Default Reference

These considerations lead to endorse the position that a geographical name refers to some specific geographical territory in the standard context. That is, the geographical developments are a relevant puzzle in the appropriate theoretical analysis of geographical names.

This is also supported by another class of observations that similar cases can be constructed for any kind of geographical development, namely, including referents of geographical names that are not represented by any social institutions.

4   Analyzing Geographical Names with the Indexicality Account

Now, we are coming back to the point where we were at the end of Sect. 2; that is, if a geographical name, "Paris," is used, it refers to certain specific territory, as the city of Paris has different territories in eighteenth century and today.

It is now clear why this paper does not take the homonymy account, which treats the combination between a genuine name and its bearer as a part of a specific name, into consideration. This is because, according to homonymy accounts, it does not make sense to claim what is conveyed in the following sentence:

(12) Since sixteenth century, Paris has continued to expand.

On the one hand, asserting (12) assumes that there is one and the same name that is used through time. On the other hand, there are many specific names in this time period. This undermines the existence of common currency names [4].[12]

It is worth noting that, while (12) is problematic for the indexicality account of geographical names, equivalent examples containing personal names, such as (13), is not:

(13) Until 21, Tomoo has kept growing.

In this section, the geographical development cases will be analyzed in terms of the indexicality account with generic names.

---

[12] From the same reason, the Weaker view of the indexicality account (which Stojanovic [20] examined) does not come into consideration here.

4.1   The Stronger View

To keep track of the referents of names for growing regions, one needs to track two things: one is the current bearer of the name, and the other is the history of the development of usage of the name. The former amounts to what Stojanovic calls the "Stronger View":

> All there is to semantic content is the lexically encoded content. Proper names are not associated with any lexically encoded content. As a consequence, proper names make no contribution to semantic content [20, p. 156].

If this is applied to geographical names, the semantic content of a geographical name is nothing but its referent in a certain context.

Another aspect is the causal history of communication regarding the name in question (see Kripke's discussion [6]). This lets one track the usage of a certain geographical name. However, this is not a part of semantic content of a geographical name. Thus, the view in question is no form of descriptivism, because the information the speaker uses for tracking can well be wrong without making the entire report false.

These two aspects enable participants of discourse identify the referent of the name in the specific context of utterance. Suppose I say to you,

(14) Paris is sunny today.

To determine the content of this utterance, you and me need to be located at the end of the chain of communication about "Paris." Further, you know that I know the current bearer of the name; hence, that I am located in the same chain, too. Since there are (at least) two chains of communication with regard to "Paris," which leads to two distinct current geographical territories (one in France and one in Texas, United States), a generic name "Paris" in (14) has two distinct potential referents depending on the context.

The difference between the Stronger View and the account that this paper attempts to establish consists in the fact that generic names belong to the vocabulary of each language. First, in many countries, there are regulations that constrain what kind of generic names can be used for newborn children. Second, for many names, the relationship between two equivalent names, such as, "Nihon" (pronounced as /nihõɴ/) "Japan" (pronounced as /dʒəˈpæn/ in English), is not merely the transcription.

That is, if one lacks knowledge of certain important (or common) generic names, this could undermine the person's linguistic competence in a certain language. However, this is not the knowledge of naming conventions that describe the relationship between generic names and their bearers, which is its default referent. Rather, it is the knowledge of whether a specific sign can be used as a generic name in the certain language. It is important to note that there is a class of vocabulary that is not associated with lexically encoded content. Indeed, there are many conversational particles that are used in one language, but they have no descriptive content.

4.2 Evaluation of Geographical Development Cases

To distinguish between Paris in France and Paris in the United States as referents of the generic name "Paris," the participants of the discourse must be a part of the communicative chain of the names and be aware of the current bearers of the names as well as the context that provides clues; this information enables one to keep track of the reference. In this picture, it no longer makes sense to ask whether they have different names. If we now turn to the cases of geographical developments, the analysis can be easily applicable to them. Because the bearer is not any part of the meaning of the name "Paris," the knowledge of the current bearer of the name "Paris" (which is applied to the specific region in France) merely serves to identify the relevant historical chain of communication, and there is some specific geographical territory assigned for the name. Thus, for example, "Paris" used in (15) refers to the geographical territory of 1790:

(11) Paris had a population density of 15,205 people per square kilometer in 1790.

To sum up, through the examination of the geographical development cases, this paper analyzed the geographical names and their relationship to geographical territories, and maintains that the referent of a geographical name is some specific territory that varies from context to context.

The version of account, which has been defended in this paper, can deal with the following example very well:

(12) Since sixteenth century, Paris has continued to expand.

This can be described again in the following manner: for every pair time point <t1, t2>, if t1 is later than t2, the referent of "Paris" at t1 is larger than that at t2. An instance of this can be made explicit in the following manner:

(15) Paris today is larger than Paris in 1790.

5 Summary

This paper analyzed geographical names, which we saw behave differently from personal names. The basic idea involves extending the indexicality account with generic names. If a geographical name is used in the standard context, it refers to specific geographical territory, whose development the discursive participants keep track with (a) the causal history of communications of the name in question and (b) its current referred territory.

Taking geographical territories as the default referents of geographical names appears to be beset with the problem of geological development. Examining this sort of cases provides some reasons for preferring a version of the indexicality accounts to the homonymy account.[13]

References

1. Almog, J.: Semantical anthropology. Midwest Stud. Philos. **9**(1), 478–489 (1984)
2. Fujikawa, N.: Namaeni nanno imiga aru noka: Koyumeino tetsugaku [What's in a Name?: Philosophy of Proper Names]. Keiso Shobo, Tokyo (2014). (In Japanese)
3. Hawley, K.: Temporal parts. In: Zalta, E.N. (ed.) Stanford Encyclopedia of Philosophy, Winter 2010 edn. (2010). http://plato.stanford.edu/archives/win2010/entries/temporal-parts/
4. Kaplan, D.: Words. Proc. Aristotelian Soc. (Supplementary Volumes) **64**, 93–119 (1990)
5. Korta, K., Perry, J.: Critical Pragmatics. Cambridge University Press, Cambridge (2011)
6. Kripke, S.A.: Naming and Neccesity. Basil Blackwell, Oxford (1980). Originally deliverd as a series of lectures at Princeton University in 1970
7. Lyons, F., Hanley, J.R., Kay, J.: Anomia for common names and geographical names with preserved retrieval of names of people: a semantic memory disorder. Cortex J. Devoted Study Nerv. Syst. Behav. **38**(1), 23–35 (2002). http://www.ncbi.nlm.nih.gov/pubmed/11999331
8. Marconi, D.: Dictionaries and proper names. Hist. Philos. Q. **7**(1), 77–92 (1990)
9. Miller, S.: Social institutions. In: Zalta, E.N. (ed.) Stanford Encyclopedia of Philosophy, Winter 2014 edn. (2014). http://plato.stanford.edu/archives/win2014/entries/social-institutions/
10. Olson, E.T.: Personal identity. In: Zalta, E.N. (ed.) The Stanford Encyclopedia of Philosophy, Winter 2010 edn. (2010). http://plato.stanford.edu/archives/win2010/entries/identity-personal/
11. Payne, J., Huddleston, R.: Nouns and noun phrases (Chap. 5). In: Huddleston, R., Pullum, G.K. (eds.) The Cambridge Grammar of the English language, pp. 323–523. Cambridge University Press, Cambridge (2002)
12. Pustejovsky, J.: The Generative Lexicon. MIT Press, Cambridge (1995)
13. Pustejovsky, J.: A survey of dot objects. Artif. Intell. **2005**(2), 1–9 (2005)
14. Rami, D.: The use-conditional indexical conception of proper names. Philos. Stud. **168**(1), 119–150 (2013)
15. Recanati, F.: Direct Reference: From Language to Thought, Paperback edn. Blackwell, Oxford (1993)

---

[13] The earlier versions of this paper were presented at Nihon University (July, 2014), Universität Stuttgart (December, 2014), and University of Tokyo (April, 2015). I am grateful for the comments and questions raised by the participants. Luca Gasparri, Masaya Mine, and Isidora Stojanovic provided some materials in preparing this paper. I learned much from discussions with Masahide Asano, Naoya Fujikawa, Luca Gasparri, Yoichi Matsusaka, Yutaka Morinaga, Christoph Michel, Francois Recanati, Tsubasa Tsushima, Masahide Yotsu (in alphabetical order) and three anonymous reviewers. Despite all the fruitful discussions, the remaining mistakes are of course all mine. This research is funded by JSPS-Kakenhi (26-7077).

16. Salmon, N.: A Millian Heir rejects the wages of sinn (Chap. 9). In: Anderson, C.A., Owens, J. (eds.) Propositional Attitudes: The Role of Content in Logic, Language, and Mind. CSLI Lecture Notesm, vol. 20, pp. 215–247. Center for the Study of Language and Information, Stanford (1990)

17. Schiffer, S.: Two issues of vagueness. Monist **81**(2), 193–215 (1998)

18. Semenza, C.: The neuropsychology of proper names. Mind Lang. **24**(4), 347–369 (2009)

19. Sider, T.: Four-Dimensionalism. Oxford University Press, Oxford (2000)

20. Stojanovic, I.: Referring with proper names: towards a pragmatic account (Chap. 6). In: Baptista, L., Rast, E.H. (eds.) Meaning and Context, pp. 139–160. Peter Lang, Bern (2010)

21. Williamson, T.: Vagueness. Routledge, London (1994)

22. Yasuda, K., Ono, Y.: Comprehension of famous personal and geographical names in global aphasic subjects. Brain Lang. **61**(61), 274–287 (1998)

23. Wissenschaftlicher Rat der Dudenredaktion (ed.): Anordnung und Behandlung der Stichwörter. In: DUDEN - Das große Wörterbuch der deutschen Sprache. Die CD-ROM edition. Duden Verlag, Mannheim (2000)

24. Zhong, I.: Seoul surprises with interest rate cut. In: Barron's Take (2015). http://online.barrons.com/articles/SB52018153252431963983004580513272127494624. Accessed 16 Mar 2015

Logic and Meaning

Interactively Illustrating the Context-Sensitivity of Aristotelian Diagrams

Lorenz Demey[(✉)]

Center for Logic and Analytic Philosophy, KU Leuven, Leuven, Belgium
lorenz.demey@hiw.kuleuven.be

Abstract. This paper studies the logical context-sensitivity of Aristotelian diagrams. I propose a new account of measuring this type of context-sensitivity, and illustrate it by means of a small-scale example. Next, I turn toward a more large-scale case study, based on Aristotelian diagrams for the categorical statements with subject negation. On the practical side, I describe an interactive application that can help to explain and illustrate the phenomenon of context-sensitivity in this particular case study. On the theoretical side, I show that applying the proposed measure of context-sensitivity leads to a number of precise yet highly intuitive results.

Keywords: Aristotelian diagram · Context-sensitivity · Background logic · Syllogistics · Information visualization

1 Introduction

Aristotelian diagrams are compact visual representations of the elements of some logical or conceptual field, and the logical relations holding between them. Without a doubt, the oldest and most widely known example is the so-called 'square of oppositions' [32]. The history of Aristotelian diagrams is well-documented: their origins can be traced back to the logical works of Aristotle, and they have been used extensively by medieval and modern authors such as William of Sherwood [23], John Buridan [36], John N. Keynes [22], George Boole and Gottlob Frege [33]. In contemporary research, Aristotelian diagrams have been used in various subbranches of logic, such as modal logic [4], intuitionistic logic [29], epistemic logic [24], dynamic logic [9] and deontic logic [28], and also even in metalogical investigations [12]. Furthermore, because of the ubiquity of the logical relations that they visualize, these diagrams are also often used in fields outside of pure logic, such as cognitive science [2,30,34], linguistics [1,17,41,43], philosophy [27,44], law [20,31,45] and computer science [10,13,15]. In sum, then, it seems fair to conclude that Aristotelian diagrams have come to serve

Thanks to Hans Smessaert, Margaux Smets and three anonymous referees for their feedback on earlier versions of this paper. The author holds a Postdoctoral Scholarship from the Research Foundation–Flanders (FWO).

© Springer International Publishing Switzerland 2015
H. Christiansen et al. (Eds.): CONTEXT 2015, LNAI 9405, pp. 331–345, 2015.
DOI: 10.1007/978-3-319-25591-0_24

"as a kind of *lingua franca*" [19, p. 81] for a highly interdisciplinary community of researchers who are all concerned, in some way or another, with logical reasoning.

Logical geometry systematically investigates Aristotelian diagrams as objects of independent interest (regardless of their role as lingua franca), for example, in terms of their information content [42]. One of the major insights to come out of these investigations is that Aristotelian diagrams are *context-sensitive*: the exact details of an Aristotelian diagram are highly dependent on the precise logical system with respect to which this diagram is constructed.[1] Although this logical context-sensitivity has numerous and far-reaching consequences, it seems to be relatively unknown—or at least insufficiently appreciated—by contemporary researchers working on Aristotelian diagrams.

The main aim of this paper is therefore to further illustrate and study the context-sensitivity of Aristotelian diagrams. We will consider a single 8-formula fragment (consisting of the categorical statements with subject negation), and study the Aristotelian diagrams that it gives rise to in various logical systems. This context-sensitivity can be concretely illustrated by means of an online available application, which allows users to define their own logical system (by selecting the axioms they want to 'activate'), and instantaneously shows them how their choices affect the resulting Aristotelian diagram. Although this application was primarily developed for broadly pedagogical purposes, it has also played an important heuristic role in more theoretical investigations, for example, concerning the relation between logical strength and context-sensitivity.

It might be objected that the terms 'context' and 'context-sensitive' are used here in a highly abstract fashion, since the relevant contexts here are logical systems, which can be seen as mere lists of axioms. Indeed, the context-sensitivity of an Aristotelian diagram (with respect to the background logic that is being used) seems to be of a fundamentally different nature than the more canonical cases of context-sensitivity, such as the deictic words 'I', 'you', 'now', 'here', etc. (with respect to context of utterance), or the words 'to know' and 'knowledge' (with respect to epistemic standards) [38]. However, this objection fails to take into account that the acceptance or rejection of a certain axiom is often itself the manifestation of a substantial position in some philosophical or empirical debate. Consider, for example, the formulas Kp and $\neg KKp$ (where $K\varphi$ stands for 'the agent knows that φ'). The Aristotelian relation holding between these formulas depends on the background logic: they are contradictory in the system S4, but subcontrary in the system T. However, these two systems only differ from each other with respect to whether the positive introspection principle for knowledge ($K\varphi \rightarrow KK\varphi$) is accepted as an axiom, and thus reflect different positions in the epistemological debate on the nature of knowledge [47].

---

[1] Strictly speaking, the term 'context-sensitive' does not apply to the Aristotelian diagram itself, but to the fragment of formulas occurring in that diagram. Throughout this paper, however, I will be using this term both in a strict sense (as applying to fragments of formulas) and in a looser sense (as applying to Aristotelian diagrams).

The paper is organized as follows. Section 2 introduces some basic notions that will be used throughout the paper, and proposes a new account of measuring the logical context-sensitivity of Aristotelian diagrams. The next three sections deal with a single fragment of 8 formulas, and the Aristotelian diagrams it gives rise to under various logical systems. First, Sect. 3 introduces the fragment and the various logical systems, and discusses their conceptual and historical importance. Next, Sect. 4 describes the interactive application that was developed to illustrate the context-sensitivity of this 8-formula fragment. Finally, Sect. 5 shows how the context-sensitivity measure proposed in Sect. 2 can be applied to the 8-formula fragment, and analyzes the results of this application. To conclude, Sect. 6 wraps things up, and mentions some questions for further research.

2 Measuring Logical Context-Sensitivity

We begin by introducing the central notions that will be studied in this paper:

Definition 1. *Let* S *be a logical system, which is assumed to have connectives expressing classical negation ($\neg$), conjunction ($\wedge$) and implication ($\rightarrow$), and a model-theoretic semantics ($\models$). The* Aristotelian relations *for* S *are defined as follows: two formulas φ and ψ are said to be*

S-contradictory	*iff*	$S \models \neg(\varphi \wedge \psi)$	*and* $S \models \neg(\neg\varphi \wedge \neg\psi)$,
S-contrary	*iff*	$S \models \neg(\varphi \wedge \psi)$	*and* $S \not\models \neg(\neg\varphi \wedge \neg\psi)$,
S-subcontrary	*iff*	$S \not\models \neg(\varphi \wedge \psi)$	*and* $S \models \neg(\neg\varphi \wedge \neg\psi)$,
in S-subalternation	*iff*	$S \models \varphi \rightarrow \psi$	*and* $S \not\models \psi \rightarrow \varphi$.

Definition 2. *Let* S *be a logical system as specified in Definition 1 and let* $\mathcal{F}$ *be a fragment of* S*-contingent and pairwise non-*S*-equivalent formulas that is closed under negation.[2] An* Aristotelian diagram *for* $\mathcal{F}$ *in* S *is a diagram that visualizes an edge-labeled graph* $\mathcal{G}$. *The vertices of* $\mathcal{G}$ 0 *are the formulas of* $\mathcal{F}$, *and the edges of* $\mathcal{G}$ *are labeled by the Aristotelian relations holding between those formulas, i.e. if* $\varphi, \psi \in \mathcal{F}$ *stand in some Aristotelian relation in* S, *then this is visualized according to the code in Fig. 1(a).*

Definition 1 is a formalized version of the traditional perspective on the Aristotelian relations, according to which two formulas are, for example, contrary iff they cannot be true together, but can be false together. Note that the seemingly *absolute* statement "φ and ψ can be false together" corresponds to the statement "there exists an S-model that satisfies $\neg\varphi \wedge \neg\psi$" (formally: $S \not\models \neg(\neg\varphi \wedge \neg\psi)$), which refers to the logical system S, and is thus *logic-dependent*. The restrictions made in Definition 2 (S-contingent, pairwise non-equivalent, closed under negation) are motivated by historical as well as technical reasons (see [42, Subsect. 2.1] for details). Figure 1(b) shows a typical example of an Aristotelian

---

[2] So for all distinct $\varphi, \psi \in \mathcal{F}$, it holds that $S \not\models \varphi$, $S \not\models \neg\varphi$, $S \not\models \varphi \leftrightarrow \psi$, and there exists a $\varphi' \in \mathcal{F}$ such that $S \models \varphi' \leftrightarrow \neg\varphi$.

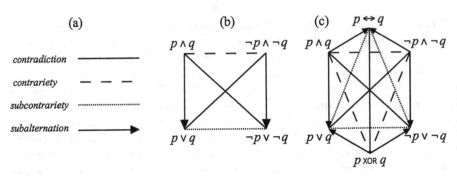

Fig. 1. (a) Code for visualizing the Aristotelian relations, (b) an Aristotelian square in CPL, and (c) its Boolean closure.

diagram, viz. a square for 4 formulas from classical propositional logic (CPL), and Fig. 1(c) shows this square's Boolean closure, i.e. the Aristotelian diagram that consists of all contingent Boolean combinations of formulas in the square.

In its theoretical study of Aristotelian diagrams, logical geometry makes extensive use of *bitstrings*. Bitstrings are representations of formulas that allow us to easily determine the Aristotelian relations holding between these formulas. A systematic technique for assigning bitstrings to any given finite fragment $\mathcal{F}$ of formulas in any logical system S is described in detail in [11]; here we will focus on those aspects that are relevant for our current purposes. We define the partition $\Pi_S(\mathcal{F}):=\{\bigwedge_{\varphi\in\mathcal{F}} \pm\varphi\} - \{\bot\}$ (where $+\varphi = \varphi$ and $-\varphi = \neg\varphi$),[3] and note that every formula $\varphi \in \mathcal{F}$ is S-equivalent to a disjunction of elements of $\Pi_S(\mathcal{F})$, viz. $\varphi \equiv_S \bigvee\{\alpha \in \Pi_S(\mathcal{F}) \mid S \models \alpha \to \varphi\}$.[4] The number $|\Pi_S(\mathcal{F})|$ is the number of bit positions, i.e. the *bitstring length*, that is required to represent the formulas of $\mathcal{F}$ by means of bitstrings. If $\mathcal{D}$ is an Aristotelian diagram for the fragment $\mathcal{F}$ in the system S, then the Boolean closure of $\mathcal{D}$ contains $2^{|\Pi_S(\mathcal{F})|} - 2$ formulas. Consider, for example, the fragment $\mathcal{F}:=\{p\wedge q, \neg p\wedge\neg q, p\vee q, \neg p\vee\neg q\}$ of CPL-formulas, and its Aristotelian diagram, which is the square in Fig. 1(b). It can be shown that $\Pi_{\mathsf{CPL}}(\mathcal{F}) = \{p\wedge q, \neg p\wedge\neg q, p \text{ XOR } q\}$, and hence, the Boolean closure of the square in Fig. 1(b) should be a diagram containing $2^{|\Pi_{\mathsf{CPL}}(\mathcal{F})|} - 2 = 2^3 - 2 = 6$ formulas, which is exactly the hexagon in Fig. 1(c).

If a fragment $\mathcal{F}$ contains only S-contingent and pairwise non-S-equivalent formulas, and is closed under negation, then the relation between the fragment's size (i.e. $|\mathcal{F}|$) and the bitstring length required to represent it (i.e. $|\Pi_S(\mathcal{F})|$) can be characterized as follows: $\lceil \log_2(|\mathcal{F}| + 2)\rceil \leq |\Pi_S(\mathcal{F})| \leq 2^{\frac{|\mathcal{F}|}{2}}$ [11, Subsect. 3.3]. Defining the *n-range* to be the set $R_n:=\{x \in \mathbb{N} \mid \lceil\log_2(n + 2)\rceil \leq x \leq 2^{\frac{n}{2}}\}$, this can trivially be reformulated as follows: if $\mathcal{F}$ contains the formulas appearing in

---

[3] The set $\Pi_S(\mathcal{F})$ is called a 'partition' because its elements are (i) jointly exhaustive ($S \models \bigvee \Pi_S(\mathcal{F})$), and (ii) mutually exclusive ($S \models \neg(\alpha\wedge\beta)$ for distinct $\alpha, \beta \in \Pi_S(\mathcal{F})$).

[4] The bitstring representation of φ is meant to keep track which formulas of $\Pi_S(\mathcal{F})$ enter into this disjunction. For example, if $\Pi_S(\mathcal{F}) = \{\alpha_1, \alpha_2, \alpha_3, \alpha_4\}$, then φ is represented by the bitstring 1011 iff $\varphi \equiv_S \alpha_1 \vee \alpha_3 \vee \alpha_4$.

some Aristotelian diagram, then $|\Pi_S(\mathcal{F})| \in R_{|\mathcal{F}|}$ (informally: $\mathcal{F}$ can be represented by bitstrings of length $\ell \in R_{|\mathcal{F}|}$). Furthermore, it can be shown that all values in the n-range R_n will be 'needed' at some point, in the sense that for every $\ell \in R_n$, there exists a fragment/logic pair whose representation requires bitstrings of length exactly ℓ. Formally:

for all $\ell \in R_n$, there exists a fragment $\mathcal{F}$ (such that $|\mathcal{F}| \leq n$) and
there exists a logical system S such that $|\Pi_S(\mathcal{F})| = \ell$.

Note that in order to reach every $\ell \in R_n$, the statement above allows us to choose specific values for both the 'fragment parameter' and the 'logical system parameter' (cf. the existential quantification over $\mathcal{F}$ as well as S). This observation leads to the following proposal to measure the context-sensitivity of a given Aristotelian diagram/fragment with respect to a set $\mathcal{S}$ of logical systems.

Proposal. The logical context-sensitivity of a given fragment $\mathcal{F}$ with respect to some set $\mathcal{S}$ of logical systems is positively correlated to the number of values in the $|\mathcal{F}|$-range that are reached if

1. the 'fragment parameter' is fixed to $\mathcal{F}$, and
2. the 'logical system parameter' varies within $\mathcal{S}$.

This proposal has two limiting cases:

- $\mathcal{F}$ is *minimally* context-sensitive with respect to $\mathcal{S}$.
  This means that for all logical systems S, T $\in \mathcal{S}$, it holds that $|\Pi_S(\mathcal{F})| = |\Pi_T(\mathcal{F})|$. This is equivalent to there being some $\ell \in R_{|\mathcal{F}|}$ such that for all logical systems S $\in \mathcal{S}$, it holds that $|\Pi_S(\mathcal{F})| = \ell$. Informally: by fixing the fragment parameter to $\mathcal{F}$, only a *single* value in the $|\mathcal{F}|$-range is reached.
- $\mathcal{F}$ is *maximally* context-sensitive with respect to $\mathcal{S}$.
  This means that for all $\ell \in R_{|\mathcal{F}|}$, there exists a logical system S $\in \mathcal{S}$ such that $|\Pi_S(\mathcal{F})| = \ell$. Informally: even though the fragment parameter is fixed to $\mathcal{F}$, varying the logical system parameter within $\mathcal{S}$ suffices to reach *all* values in the $|\mathcal{F}|$-range. In other words, all bitstring lengths that might theoretically be necessary to represent fragments of the same size as $\mathcal{F}$, are already needed to represent $\mathcal{F}$ itself, under the different logical systems in $\mathcal{S}$.[5]

To illustrate this account of context-sensitivity, we will consider the case of 4-formula fragments, i.e. the case of Aristotelian *squares*. Note that the 4-range is $R_4 = \{3, 4\}$, which means that every Aristotelian square (regardless of the formulas it contains, regardless of the logical system in which it is constructed) can be represented by bitstrings of length either 3 or 4. Now consider the specific 4-formula fragment $\mathcal{F}^\dagger := \{all(A, B), some(A, B), all(A, \neg B), some(A, \neg B)\}$. Informally, these formulas read as "all As are B", "some As are B", "all As

---

[5] Note the subtly different quantification patterns corresponding to these two limiting cases: minimal context-sensitivity corresponds to $\exists \ell \in R_{|\mathcal{F}|} : \forall S \in \mathcal{S} : |\Pi_S(\mathcal{F})| = \ell$, while maximal context-sensitivity corresponds to $\forall \ell \in R_{|\mathcal{F}|} : \exists S \in \mathcal{S} : |\Pi_S(\mathcal{F})| = \ell$.

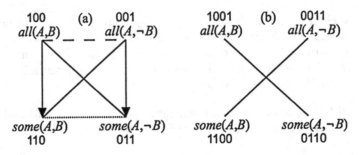

Fig. 2. (a) 'Classical' square for $\mathcal{F}^{\dagger}$ in SYL, (b) 'degenerated' square for $\mathcal{F}^{\dagger}$ in FOL. Each square is decorated with both $\mathcal{F}^{\dagger}$ and its bitstring representation.

are not B" and "some As are not B", respectively,[6] and they can be interpreted in various 'reasonable' logical systems, such as FOL (contemporary first-order logic) and SYL (i.e. FOL $+ \exists x Ax$).[7] It is shown in [11, Sect. 4] that

- $\Pi_{\mathsf{FOL}}(\mathcal{F}^{\dagger}) = \{all(A, B) \wedge some(A, B), some(A, B) \wedge some(A, \neg B), all(A, \neg B) \wedge some(A, \neg B), all(A, B) \wedge all(A, \neg B)\}$,
- $\Pi_{\mathsf{SYL}}(\mathcal{F}^{\dagger}) = \{all(A, B), some(A, B) \wedge some(A, \neg B), all(A, \neg B)\}$,

and hence $|\Pi_{\mathsf{FOL}}(\mathcal{F}^{\dagger})| = 4$ and $|\Pi_{\mathsf{SYL}}(\mathcal{F}^{\dagger})| = 3$. This shows that for all $\ell \in R_4 = \{3, 4\}$, there exists a logical system $S \in \mathcal{S}^{\dagger} := \{\mathsf{FOL}, \mathsf{SYL}\}$ such that $|\Pi_S(\mathcal{F}^{\dagger})| = \ell$, and hence, the fragment $\mathcal{F}^{\dagger}$ is maximally context-sensitive with respect to $\mathcal{S}^{\dagger}$. This context-sensitivity can also clearly be seen in the Aristotelian diagrams themselves: in SYL, the fragment $\mathcal{F}^{\dagger}$ gives rise to a 'classical' square of opposition, which is shown in Fig. 2(a) and can be represented by bitstrings of length 3, whereas in FOL, the same fragment gives rise to a 'degenerated' square or "X of opposition" [3, p. 13], which is shown in Fig. 2(b) and can be represented by bitstrings of length 4.

3   Categorical Statements and Subject-Negation

At the end of the previous section, I introduced the fragment $\mathcal{F}^{\dagger}$, and showed it to be maximally context-sensitive with respect to the reasonable logical systems in $\mathcal{S}^{\dagger}$. The next three sections of this paper will be devoted to studying and illustrating the context-sensitivity of a larger fragment $\mathcal{F}^{\ddagger}$ (which includes $\mathcal{F}^{\dagger}$ itself) with respect to a larger set of logical systems $\mathcal{S}^{\ddagger}$ (which includes $\mathcal{S}^{\dagger}$ itself). I start by introducing the fragment $\mathcal{F}^{\ddagger}$ and the logical systems in $\mathcal{S}^{\ddagger}$.

The statements in the original fragment $\mathcal{F}^{\dagger}$ are *categorical statements*, which are of the form *quantifier(subject, predicate)*. They are among the oldest sentences to be studied from a logical perspective [32], and traditionally, they

---

[6] As is well-known, in the language of first-order logic, these formulas can be formalized as $\forall x(Ax \to Bx)$, $\exists x(Ax \wedge Bx)$, $\forall x(Ax \to \neg Bx)$ and $\exists x(Ax \wedge \neg Bx)$, respectively.

[7] Later in the paper, I will have more to say about when exactly a logical system can be considered 'reasonable' for a given fragment.

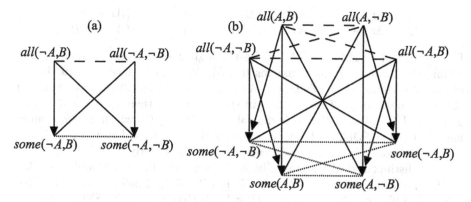

Fig. 3. (a) Aristotelian square for $\mathcal{F}^\ddagger - \mathcal{F}^\dagger$ in FOL($\{A2\}$), (b) Aristotelian octagon for $\mathcal{F}^\ddagger$ in FOL($\{A1, A2, A3, A4\}$).

are classified according to their *quantity* and *quality*. With respect to quantity, we distinguish between *universal* and *particular* statements, whose quantifiers are *all* and *some*, respectively. With respect to quality, we distinguish between *affirmative* and *negative* statements, whose predicates are of the form B and $\neg B$, respectively. The traditional classification according to quality is thus exclusively based on whether the statements' *predicates* are negated. Over the course of history, however, logicians have also become interested in the effects of *subject negation* [7,8,16,21,22,25,35,37], thereby obtaining the new statements $all(\neg A, B), some(\neg A, B), all(\neg A, \neg B), some(\neg A, \neg B)$. The 8-formula fragment $\mathcal{F}^\ddagger$ is defined to contain exactly these 4 new statements, together with the 4 original statements of $\mathcal{F}^\dagger$. It thus trivially holds that $\mathcal{F}^\dagger \subseteq \mathcal{F}^\ddagger$.

The 4 new statements can themselves be used to construct a second Aristotelian square. Figure 3(a) shows this square, as constructed in the logical system FOL($\{A2\}$), which will be described below. Note that this square is classical (i.e. not degenerated) iff the underlying logic contains $\exists x \neg Ax$ as an axiom, which is analogous to the first square (for $\mathcal{F}^\dagger$) being classical iff the underlying logic contains $\exists x Ax$ as an axiom. More interestingly, we can also consider Aristotelian diagrams for the entire 8-formula fragment $\mathcal{F}^\ddagger$. Some authors have proposed an *octagon* [16,21,22], while others have made use of a *cube* [8,25,37]. In the current paper, the fragment $\mathcal{F}^\ddagger$ will always be visualized by means of an octagon. Figure 3(b) shows this octagon, as constructed in the logical system FOL($\{A1, A2, A3, A4\}$), which will also be described below. Because of the logical context-sensitivity of Aristotelian diagrams, other logical systems will lead to other versions of this octagon. I will therefore now introduce the various logical systems in which the Aristotelian octagon for $\mathcal{F}^\ddagger$ will be constructed.

The logical systems that will be relevant for our purposes all consist of first-order logic (FOL), with additional axioms coming from the set $\mathcal{AX} := \{A1, A2, A3, A4, A5, A6\}$, which contains the following statements:

A1	$\exists x Ax$	A3	$\exists x Bx$	A5	$\exists x \neg (Ax \leftrightarrow Bx)$
A2	$\exists x \neg Ax$	A4	$\exists x \neg Bx$	A6	$\exists x \neg (Ax \leftrightarrow \neg Bx)$

For any set $\mathcal{A} \subseteq \mathcal{AX}$, let FOL($\mathcal{A}$) be the logical system that is obtained by adding the formulas in $\mathcal{A}$ as axioms to FOL. We will be interested in the set of logical systems $\mathcal{S}^{\ddagger} := \{\text{FOL}(\mathcal{A}) \mid \mathcal{A} \subseteq \mathcal{AX}\}$. Note that all the statements in $\mathcal{AX}$ are independent of each other, in the sense that there exist no $\mathcal{A} \subseteq \mathcal{AX}$ and $\alpha \in \mathcal{AX} - \mathcal{A}$ such that α is derivable in FOL($\mathcal{A}$) (in which case we would have FOL($\mathcal{A}$) = FOL($\mathcal{A} \cup \{\alpha\}$)); this means that the logical systems in $\mathcal{S}^{\ddagger}$ are all distinct from each other, and hence $\mathcal{S}^{\ddagger}$ contains exactly $|\wp(\mathcal{AX})| = 2^{|\mathcal{AX}|} = 2^6 = 64$ distinct logical systems. The weakest system in $\mathcal{S}^{\ddagger}$ is FOL($\emptyset$), i.e. FOL itself, while the strongest system in $\mathcal{S}^{\ddagger}$ is FOL($\mathcal{AX}$). Finally, note that since FOL = FOL($\emptyset$) and SYL = FOL($\{A1\}$), it holds that $\mathcal{S}^{\dagger} = \{\text{FOL}, \text{SYL}\} \subseteq \mathcal{S}^{\ddagger}$.

All the logical systems in $S^{\ddagger}$ are 'reasonable' to a certain degree, in the sense that all of their axioms have been defended by various logicians in relation to substantial philosophical and psychological debates. To begin with, note that all systems in $\mathcal{S}^{\ddagger}$ are extensions of the system FOL of first-order logic, which is itself by far the most widely used logical system today. Next, the statements A1–A4 can all be seen as (partial) interpretations of the traditional *existential import* principle. According to its most cautious interpretation [5,40], this principle states that the predicate occuring in the first argument position of a categorical statement should not have an empty extension, which is captured by A1. However, another interpretation is that *all* predicates should have non-empty extensions, regardless of whether they occur in the categorical statement's first or second argument position [39]; this means that both A1 and A3 should be accepted as axioms. Furthermore, based on psychological considerations, authors such as Seuren [39] have argued that just as a predicate's extension should not be allowed to be empty, it should not be allowed to encompass the entire universe either; this means that A2 and/or A4 should be accepted as axioms. The most liberal interpretation of the existential import principle, then, which is held by authors such as Keynes [22], Johnson [21] and Hacker [16], takes this principle to state that all of A1, A2, A3 and A4 should be accepted as axioms.

Finally, the statements A5 and A6 have been defended by Reichenbach [37]. Informally, the former states that the predicates A and B should not be perfect *synonyms*, while the latter states that A and B should not be perfect *antonyms*. More precisely, these statements impose a strict correlation between syntactic differences and semantic differences. For example, A5 states that since there is a syntactic difference between the predicates A and B (viz. they are symbolized using different letters), there should also be a semantic difference between them (viz. they should have different extensions). A similar principle is at work in Wittgenstein's version of predicate logic, in which distinct variables are taken to have distinct values [46,48].[8] From a more empirical perspective, principles such

---

[8] For example, in this system, the sentence 'there are at least two As' would not be formalized as $\exists x \exists y (Ax \wedge Ay \wedge x \neq y)$, but simply as $\exists x \exists y (Ax \wedge Ay)$: the syntactic difference between the variables x and y suffices to indicate that there is also a semantic difference between them, i.e. that they have distinct values.

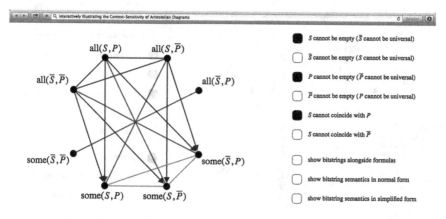

Fig. 4. Screenshot: the Aristotelian diagram for $\mathcal{F}^{\ddagger}$ in the system $\mathsf{FOL}(\{A1, A3, A5\})$.

as $A5$ and $A6$ seem to be related to linguistic work on language evolution and language acquisition, in which it is often assumed that humans have an (innate) tendency to avoid perfect synonyms as much as possible [6,18,26].

4   An Interactive Illustration

Earlier research on the logical context-sensitivity of Aristotelian diagrams has focused on (families of) logics such as epistemic logic [14] and metalogic [12]. In contrast to these more advanced case studies, the fragment $\mathcal{F}^{\ddagger}$ and the logics in $\mathcal{S}^{\ddagger}$ presented in this paper are quite elementary, which renders them particularly suitable for explaining the phenomenon of context-sensitivity to a broader audience. In order to support and facilitate this pedagogical goal, an interactive application has been created and made available online, at the following location: http://www.logicalgeometry.org/octagon_context.html.

The application was developed using the XML-based Scalable Vector Graphics format (for the graphical aspects) and JavaScript (for the interactivity). The user interface has been kept very simple: the screen is vertically divided into a left half and a right half. The left half shows the Aristotelian diagram for $\mathcal{F}^{\ddagger}$, based on the logical system that is currently 'activated'. The right half contains 6 'axiom buttons' and 3 'auxiliary buttons'. The former correspond exactly to the statements in $\mathcal{AX}$, and each of them can be activated or deactivated. In this way, the user can select any of the logical systems $\mathsf{FOL}(\mathcal{A}) \in \mathcal{S}^{\ddagger}$, by activating exactly the axiom buttons corresponding to the statements in $\mathcal{A}$. As the user activates or deactivates a particular axiom (and thus goes from one logical system to another one), she can immediately observe the effects of this change on the Aristotelian diagram for $\mathcal{F}^{\ddagger}$ on the left half of the screen. For example, the screenshot in Fig. 4 shows that the user has activated the axiom buttons corresponding to $A1$, $A3$ and $A5$, and hence the application shows the Aristotelian diagram for $\mathcal{F}^{\ddagger}$ in the system $\mathsf{FOL}(\{A1, A3, A5\})$.

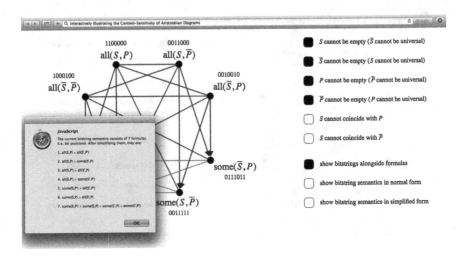

Fig. 5. Screenshot: the Aristotelian diagram for $\mathcal{F}^{\ddagger}$ in the system FOL($\{A1, A2, A3, A4\}$), its bitstring representation, and a popup window showing the formulas of $\Pi_{\mathsf{FOL}(\{A1, A2, A3, A4\})}(\mathcal{F}^{\ddagger})$ (in simplified form).

In addition to the 6 axiom buttons, the right half of the screen also contains 3 auxiliary buttons. The first button allows the user to decide whether the bitstring representations of the $\mathcal{F}^{\ddagger}$-formulas should be shown next to these formulas. Recall that the bitstring representation of a given formula depends on the logical system that is currently activated; for example, a single formula might correspond to a bitstring of length 6 in one logical system, and to a bitstring of length 9 in another logical system. Consequently, if the 'show bitstrings' button is switched on and the user activates or deactivates a particular axiom (and thus goes from one logical system to another one), this will not only affect the Aristotelian diagram for $\mathcal{F}^{\ddagger}$ itself, but also the bitstring representations of the formulas in that diagram. In this way, the user can easily experiment with the various logical systems in $\mathcal{S}^{\ddagger}$, and explore which systems give rise to which kinds of bitstrings, etc. Finally, there are two buttons that launch popup windows showing the *bitstring semantics*, i.e. the formulas in $\Pi_{\mathsf{S}}(\mathcal{F}^{\ddagger})$, with respect to which the bitstrings for $\mathcal{F}^{\ddagger}$ are defined (where $\mathsf{S} \in \mathcal{S}^{\ddagger}$ is the logical system that is currently activated). Recall from Sect. 2 that these formulas are formally defined as conjunctions of (negations of) the formulas in $\mathcal{F}^{\ddagger}$; however, in many cases, these long conjunctions can be simplified to equivalent, but much shorter formulas. One button shows the bitstring semantics in normal form (i.e. as the original conjunctions), while the other one shows it in simplified form.

For example, the screenshot in Fig. 5 shows that the user has activated the axiom buttons corresponding to $A1$, $A2$, $A3$ and $A4$, and hence the application shows the Aristotelian diagram for $\mathcal{F}^{\ddagger}$ in the system FOL($\{A1, A2, A3, A4\}$)—also see Fig. 3(b). Furthermore, the 'show bitstrings' button has been switched on, so the formulas of $\mathcal{F}^{\ddagger}$ are shown together with their bitstring representations.

Since $|\Pi_{\mathsf{FOL}(\{A1,A2,A3,A4\})}(\mathcal{F}^{\ddagger})| = 7$, these are bitstrings of length 7. Finally, clicking the 'show bitstring semantics in simplified form' button has launched a popup window showing the 7 (simplified) formulas in $\Pi_{\mathsf{FOL}(\{A1,A2,A3,A4\})}(\mathcal{F}^{\ddagger})$.

5   Theoretical Analysis

Making use of the application described in the previous section, it can easily be observed that each of the 64 logical systems in $\mathcal{S}^{\ddagger}$ gives rise to a different Aristotelian diagram for $\mathcal{F}^{\ddagger}$.[9] This can be explained by noting that the octagon for $\mathcal{F}^{\ddagger}$ contains exactly 6 squares,[10] and each square corresponds exactly to a statement in $\mathcal{AX}$: the square is classical (i.e. not degenerated) iff the corresponding statement is an axiom in the logical system with respect to which the octagon is defined. For example, the octagon for $\mathcal{F}^{\ddagger}$ with respect to $\mathsf{FOL}(\{A1,A2,A3,A4\})$ in Fig. 3(b) contains 4 classical squares (viz. those corresponding to $A1$, $A2$, $A3$ and $A4$) and 2 degenerated ones (viz. those corresponding to $A5$ and $A6$). Furthermore, note that this precise correspondence between the 6 squares inside the octagon for $\mathcal{F}^{\ddagger}$ and the 6 statements $A1$–$A6$ further corroborates the claim that $A1$–$A6$ are the most natural axioms to consider when studying $\mathcal{F}^{\ddagger}$.

The fact that each logical system in $\mathcal{S}^{\ddagger}$ gives rise to a different Aristotelian diagram for $\mathcal{F}^{\ddagger}$ is already a powerful illustration of the context-sensitivity of $\mathcal{F}^{\ddagger}$ with respect to $\mathcal{S}^{\ddagger}$. To assess this context-sensitivity in a mathematically more precise way, we will now make use of the account proposed in Sect. 2. Since $|\mathcal{F}^{\ddagger}| = 8$, we are interested in the 8-range, which is easily calculated to be $R_8 = \{4, 5, \ldots, 15, 16\}$. Recall that this intuitively means that every possible Aristotelian octagon (regardless of the formulas it contains, regardless of the logical system in which it is constructed) can be represented by bitstrings of length between 4 and 16 (inclusive). Again making use of the application described in the previous section, we obtain Table 1.[11]

For example, the cell on the uppermost row and rightmost column of Table 1 tells us that if we are working in the system $\mathsf{FOL}(\{A1, A2, A3, A4\})$, the fragment $\mathcal{F}^{\ddagger}$ can be represented by bitstrings of length 7 (also see Fig. 5).

Comparing Table 1 with the 8-range, several observations can be made. First of all, note that even though the logical systems in $\mathcal{S}^{\ddagger}$ all give rise to different Aristotelian diagrams for $\mathcal{F}^{\ddagger}$, it is *not* the case that they all give rise to different bitstring lengths (which is impossible, since $|\mathcal{S}^{\ddagger}| = 64 > 13 = |R_8|$). It turns out that only 8 bitstring lengths are required to represent $\mathcal{F}^{\ddagger}$ under the various systems in $\mathcal{S}^{\ddagger}$, viz. 16, 12, 10, 9, 8, 7, 6 and 5. Since $\frac{8}{|R_8|} = \frac{8}{13} \approx 0.62$, this

---

[9] We already encountered a similar situation in Sect. 2, where it was shown that each system in $\mathcal{F}^{\dagger}$ gives rise to a different Aristotelian square for $\mathcal{F}^{\dagger}$; see Fig. 2.

[10] An Aristotelian octagon can be seen as consisting of 4 pairs of contradictory formulas (PCDs), and a square as 2 PCDs. The number of squares inside an octagon thus equals the number of ways in which one can select 2 PCDs out of 4 (without replacement), which is $\binom{4}{2} = \frac{4!}{2!2!} = 6$.

[11] For reasons of space, a logical system such as $\mathsf{FOL}(\{A1, A2, A3\})$ is abbreviated as '123', and the bitstring length $|\Pi_{\mathsf{S}}(\mathcal{F}^{\ddagger})|$ as ℓ_{S}.

Table 1. The 64 logical systems in $\mathcal{S}^{\ddagger}$ and the corresponding bitstring lengths.

S	ℓ_S	S	ℓ_S	S	ℓ_S	S	ℓ_S	S	ℓ_S	S	ℓ_S	S	ℓ_S	S	ℓ_S
∅	16	13	10	46	10	235	9	234	8	356	8	1356	7	1234	7
1	12	14	10	35	10	246	9	146	8	456	8	1346	7	23456	6
2	12	15	10	36	10	134	8	236	8	3456	7	1345	7	13456	6
3	12	16	10	12	9	135	8	245	8	2456	7	1256	7	12456	6
4	12	23	10	34	9	123	8	156	8	2356	7	1246	7	12356	6
5	12	24	10	56	9	124	8	256	8	2346	7	1245	7	12346	6
6	12	25	10	136	9	125	8	345	8	2345	7	1236	7	12345	6
45	10	26	10	145	9	126	8	346	8	1456	7	1235	7	123456	5

means that of all the bitstring lengths that might theoretically be necessary to represent an arbitrary 8-formula fragment with respect to an arbitrary logical system, about 62 % is already necessary to represent the particular fragment $\mathcal{F}^{\ddagger}$ with respect to the particular logical systems in $\mathcal{S}^{\ddagger}$.

In terms of extreme values, we see that the highest value in the 8-range is reached, i.e. there is a system $S \in \mathcal{S}^{\ddagger}$ such that $|\Pi_S(\mathcal{F}^{\ddagger})| = 16$, viz. $S = \mathsf{FOL}(\emptyset)$. By contrast, the lowest value in the 8-range is not reached, i.e. there is no $S \in \mathcal{S}^{\ddagger}$ such that $|\Pi_S(\mathcal{F}^{\ddagger})| = 4$. In other words, even though there exist some 8-formula fragment $\mathcal{F}$ and some logical system S such that $|\Pi_S(\mathcal{F})| = 4$, we cannot take $\mathcal{F} = \mathcal{F}^{\ddagger}$ and $S \in \mathcal{S}^{\ddagger}$.[12] Note, however, that the second lowest value in the 8-range is reached by some system in $\mathcal{S}^{\ddagger}$, since $|\Pi_{\mathsf{FOL}(\mathcal{A}\mathcal{X})}(\mathcal{F}^{\ddagger})| = 5$. The weakest logical system $\mathsf{FOL}(\emptyset)$ thus yields the highest bitstring length (16), while the strongest logical system $\mathsf{FOL}(\mathcal{A}\mathcal{X})$ yields the lowest (attainable) bitstring length (5). This suggests an inverse correlation between logical strength and bitstring length: stronger logical systems yield shorter bitstrings. The intuitive explanation of this inverse correlation is based on the fact that bitstring length is itself positively correlated to the size of the Boolean closure (cf. Sect. 2): a stronger logical system can prove more formulas in the Boolean closure of $\mathcal{F}^{\ddagger}$ to be equivalent to each other, so this Boolean closure will contain fewer formulas (up to logical equivalence), which in turn means that the bitstrings will be shorter.

In order to make this inverse correlation more precise, note that for logical systems $S \in \mathcal{S}^{\ddagger}$, the logical strength of S can be taken to be simply the number of statements in $\mathcal{A}\mathcal{X}$ that are axioms of S. In other words, for $\mathcal{A}, \mathcal{B} \subseteq \mathcal{A}\mathcal{X}$, we say that $\mathsf{FOL}(\mathcal{A})$ is stronger than $\mathsf{FOL}(\mathcal{B})$ iff $|\mathcal{A}| > |\mathcal{B}|$. The inverse correlation between logical strength and bitstring length can now be expressed as follows:

for all $\mathcal{A}, \mathcal{B} \subseteq \mathcal{A}\mathcal{X}$: if $|\mathcal{A}| < |\mathcal{B}|$, then $|\Pi_{\mathsf{FOL}(\mathcal{A})}(\mathcal{F}^{\ddagger})| \geq |\Pi_{\mathsf{FOL}(\mathcal{B})}(\mathcal{F}^{\ddagger})|$ (InvCor)

---

[12] There certainly do exist systems S such that $|\Pi_S(\mathcal{F}^{\ddagger})| = 4$. This is the case, for example, for the system S^* that is obtained by adding to $\mathsf{FOL}(\mathcal{A}\mathcal{X})$ the additional axiom $all(A, B) \vee all(A, \neg B) \vee all(\neg A, B) \vee all(\neg A, \neg B)$. Note, however, that $S^* \notin \mathcal{S}^{\ddagger}$, and, more importantly, S^* is far less reasonable than any of the systems in $\mathcal{S}^{\ddagger}$.

The truth of (INVCOR) can be checked by means of Table 1. Furthermore, note that we almost have an even stricter version of this inverse correlation principle, in the sense that for almost all sets $\mathcal{A}, \mathcal{B} \subseteq \mathcal{AX}$, the comparison operator $\geq$ in the consequent of (INVCOR) can be replaced by $>$. As can be verified by means of Table 1, the only counterexamples to this stricter claim involve $\mathcal{A} \in \{\{A1, A2\}, \{A3, A4\}, \{A5, A6\}\}$ and $\mathcal{B} \in \{\{A1, A3, A6\}, \{A1, A4, A5\}, \{A2, A3, A5\}, \{A2, A4, A6\}\}$, in which case we have $|\mathcal{A}| = 2 < 3 = |\mathcal{B}|$ and yet $|\Pi_{\mathsf{FOL}(\mathcal{A})}(\mathcal{F}^{\ddagger})| = 9 = |\Pi_{\mathsf{FOL}(\mathcal{B})}(\mathcal{F}^{\ddagger})|$.

6 Conclusion

This paper has studied the logical context-sensitivity of Aristotelian diagrams, focusing on the fragment $\mathcal{F}^{\ddagger}$ of categorical statements with subject negation, and the set $\mathcal{S}^{\ddagger}$ of logical systems based on the axioms in $\mathcal{AX}$. I have described an interactive application that can help to illustrate the context-sensitivity of $\mathcal{F}^{\ddagger}$ with respect to $\mathcal{S}^{\ddagger}$. On the theoretical side, I have proposed a new account to measure the context-sensitivity of Aristotelian diagrams, and shown that it leads to precise yet highly intuitive results in the case of $\mathcal{F}^{\ddagger}$ and $\mathcal{S}^{\ddagger}$. In future work, this account will be applied to other fragments and logical systems.

References

1. van der Auwera, J.: Modality: the three-layered scalar square. J. Seman. **13**, 181–195 (1996)
2. Beller, S.: Deontic reasoning reviewed: psychological questions, empirical findings, and current theories. Cogn. Process. **11**, 123–132 (2010)
3. Béziau, J.Y., Payette, G.: Preface. In: Béziau, J.Y., Payette, G. (eds.) The Square of Opposition: A General Framework for Cognition, pp. 9–22. Peter Lang, Bern (2012)
4. Carnielli, W., Pizzi, C.: Modalities and Multimodalities. Logic, Epistemology, and the Unity of Science, vol. 12. Springer, The Netherlands (2008)
5. Chatti, S., Schang, F.: The cube, the square and the problem of existential import. Hist. Philos. Logic **32**, 101–132 (2013)
6. Clark, E.V.: On the pragmatics of contrast. J. Child Lang. **17**, 417–431 (1990)
7. De Morgan, A.: On the Syllogism, and Other Logical Writings. Routledge and Kegan Paul, London (1966)
8. Dekker, P.: Not only Barbara. J. Logic Lang. Inf. **24**, 95–129 (2015)
9. Demey, L.: Structures of oppositions for public announcement logic. In: Béziau, J.Y., Jacquette, D. (eds.) Around and Beyond the Square of Opposition. Studies in Universal Logic, pp. 313–339. Springer, Basel (2012)
10. Demey, L., Smessaert, H.: The relationship between Aristotelian and Hasse diagrams. In: Delaney, A., Dwyer, T., Purchase, H. (eds.) Diagrams 2014. LNCS, vol. 8578, pp. 213–227. Springer, Heidelberg (2014)
11. Demey, L., Smessaert, H.: Combinatorial bitstring semantics for arbitrary logical fragments. Ms. (2015)
12. Demey, L., Smessaert, H.: Metalogical decorations of logical diagrams. Ms. (2015)

13. Dubois, D., Prade, H.: From Blanché's hexagonal organization of concepts to formal concept analysis and possibility theory. Logica Univ. **6**, 149–169 (2012)
14. Frijters, S., Demey, L.: The context-sensitivity of epistemic-logical diagrams (in Dutch), internal report, KU Leuven (2015)
15. Gottfried, B.: The diamond of contraries. J. Vis. Lang. Comput. **26**, 29–41 (2015)
16. Hacker, E.A.: The octagon of opposition. Notre Dame J. Formal Logic **16**, 352–353 (1975)
17. Horn, L.R.: A Natural History of Negation. University of Chicago Press, Chicago/London (1989)
18. Hurford, J.R.: Why synonymy is rare: fitness is in the speaker. In: Ziegler, J., Dittrich, P., Kim, J.T., Christaller, T., Banzhaf, W. (eds.) ECAL 2003. LNCS (LNAI), vol. 2801, pp. 442–451. Springer, Heidelberg (2003)
19. Jacquette, D.: Thinking outside the square of opposition box. In: Béziau, J.Y., Jacquette, D. (eds.) Around and Beyond the Square of Opposition, pp. 73–92. Springer, Basel (2012)
20. Joerden, J.C.: Logik im Recht. Springer-Lehrbuch. Springer, Heidelberg (2010)
21. Johnson, W.: Logic: Part I. Cambridge University Press, Cambridge (1921)
22. Keynes, J.N.: Studies and Exercises in Formal Logic. MacMillan, London (1884)
23. Khomskii, Y.: William of Sherwood, singular propositions and the hexagon of opposition. In: Béziau, J.Y., Payette, G. (eds.) The Square of Opposition: A General Framework for Cognition, pp. 43–60. Peter Lang, Bern (2012)
24. Lenzen, W.: How to square knowledge and belief. In: Béziau, J.Y., Jacquette, D. (eds.) Around and Beyond the Square of Opposition. Studies in Universal Logic, pp. 305–311. Springer, Basel (2012)
25. Libert, T.: Hypercubes of duality. In: Béziau, J.Y., Jacquette, D. (eds.) Around and Beyond the Square of Opposition. Studies in Universal Logic, pp. 293–301. Springer, Basel (2012)
26. Manin, D.Y.: Zipf's law and avoidance of excessive synonymy. Cogn. Sci. **32**, 1075–1098 (2008)
27. Massin, O.: Pleasure and its contraries. Rev. Phil. Psych. **5**, 15–40 (2014)
28. McNamara, P.: Deontic logic. In: Zalta, E.N. (ed.) Stanford Encyclopedia of Philosophy. CSLI, Stanford (2010)
29. Mélès, B.: No group of opposition for constructive logic: the intuitionistic and linear cases. In: Béziau, J.Y., Jacquette, D. (eds.) Around and Beyond the Square of Opposition. Studies in Universal Logic, pp. 201–217. Springer, Basel (2012)
30. Mikhail, J.: Universal moral grammar: theory, evidence and the future. Trends Cogn. Sci. **11**, 143–152 (2007)
31. O'Reilly, D.: Using the square of opposition to illustrate the deontic and alethic relations constituting rights. Univ. Toronto Law J. **45**, 279–310 (1995)
32. Parsons, T.: The traditional square of opposition. In: Zalta, E.N. (ed.) Stanford Encyclopedia of Philosophy. CSLI, Stanford (2006)
33. Peckhaus, V.: Algebra of logic, quantification theory, and the square of opposition. In: Béziau, J.Y., Payette, G. (eds.) The Square of Opposition: A General Framework for Cognition, pp. 25–41. Peter Lang, Bern (2012)
34. Porcaro, C., et al.: Contradictory reasoning network: an EEG and FMRI study. PLOS One **9**(3), e92835 (2014)
35. Pratt-Hartmann, I., Moss, L.S.: Logics for the relational syllogistic. Rev. Symbolic Logic **2**, 647–683 (2009)
36. Read, S.: John Buridan's theory of consequence and his octagons of opposition. In: Béziau, J.Y., Jacquette, D. (eds.) Around and Beyond the Square of Opposition. Studies in Universal Logic, pp. 93–110. Springer, Basel (2012)

37. Reichenbach, H.: The syllogism revised. Philos. Sci. **19**, 1–16 (1952)
38. Rysiew, P.: Epistemic contextualism. In: Zalta, E.N. (ed.) Stanford Encyclopedia of Philosophy. CSLI, Stanford (2011)
39. Seuren, P.: The natural logic of language and cognition. Pragmatics **16**, 103–138 (2006)
40. Seuren, P.: The cognitive ontogenesis of predicate logic. Notre Dame J. Formal Logic **55**, 499–532 (2014)
41. Seuren, P., Jaspers, D.: Logico-cognitive structure in the lexicon. Language **90**, 607–643 (2014)
42. Smessaert, H., Demey, L.: Logical geometries and information in the square of opposition. J. Logic Lang. Inf. **23**, 527–565 (2014)
43. Smessaert, H., Demey, L.: Béziau's contributions to the logical geometry of modalities and quantifiers. In: Koslow, A., Buchsbaum, A. (eds.) The Road to Universal Logic. Studies in Universal Logic, pp. 475–494. Springer, Switzerland (2015)
44. Sosa, E.: The analysis of 'knowledge that P'. Analysis **25**, 1–8 (1964)
45. Vranes, E.: The definition of 'norm conflict' in international law and legal theory. Eur. J. Int. Law **17**, 395–418 (2006)
46. Wehmeier, K.: Wittgensteinian predicate logic. Notre Dame J. Formal Logic **45**, 1–11 (2004)
47. Williamson, T.: Knowledge and its Limits. Oxford University Press, Oxford (2000)
48. Wittgenstein, L.: Tractatus Logico-Philosophicus. Routledge/Kegan Paul, London (1922)

On the Context Dependence of *Many*

Matthias F.J. Hofer[(⊠)]

Vienna University of Technology, Vienna, Austria
hofer@logic.at

Abstract. We augment the applicability of Lappin's intensional parametrization of the determiners *many* and *few* by combinatorial means, and show how to arrive at graded interpretations of corresponding natural language statements.

Keywords: Natural language semantics · Vagueness · Quantifiers · Fuzzy logic

1 Introduction

Our aim in this paper is incorporate combinatorial techniques into natural language semantics, and, particularly, to develop a concept of natural language quantification. During the last decades, there have been several approaches to context dependent interpretation regarding determiners like *many*, and we find that they can be divided into two main streams. One is carried out by linguists, like Fernando and Kamp [6], Keenan and Stavi [9], Barwise and Cooper [1], or Westerståhl [13,14], and Lappin [10–12], and they all share the feature of two-valued interpretation, that is, the expressions, involving *many*, are either true or false. Then, there is the other community, namely the fuzzy logicians, like Zadeh [15], Glöckner [8], or Hájek [2], and Fermüller [3–5]. They, accept the need of graded interpretations, while the use of contexts, that is in particular also the acceptance of intensional aspects of the matter, is still best observed in [12], where extensionality and intensionality get fit into one unifying frame. In both, the extensional and the intensional setting, one can make use of comparison classes. Lappin calls them comparison sets in the extensional case and normative situations for the intensional case. Even though we follow his terminology, we intend to show, that the underlying combinatorial pattern is the same in both cases. In so doing, we will give four additional extensional readings for *many* in Sect. 2, and eight new intensional readings in Sect. 3, and each time emphasize the combinatorial underpinnings. Also, in Sect. 2, we characterize what can be seen as a canonical structure for contexts, which we call a distinction tree. The ideas get illustrated through an easy context dependent statement, involving *many*. Eventually, in Sect. 4, we show how one can go down the road to graded interpretations of *many*, and how *many* and *few* are functionally related. Thus, this paper aims at bringing the two streams closer together.

M.F.J. Hofer–Supported by Austrian Science Fund (FWF) I1897-N25 (MoVaQ).

© Springer International Publishing Switzerland 2015
H. Christiansen et al. (Eds.): CONTEXT 2015, LNAI 9405, pp. 346–358, 2015.
DOI: 10.1007/978-3-319-25591-0_25

2 Lappin's Approach and Beyond

Our main reference throughout this paper will be Shalom Lappin's article [12], and particularly his two-valued intensional parametric interpretation of $\| \text{ many } \|$, namely this one:

$$\|B\|^{\text{sa}} \in \| \text{ many } \|(\|A\|^{\text{sa}}) \text{ iff}$$
$$S \neq \emptyset, \text{ and for all sn} \in S, \quad \| \|A\|^{\text{sa}} \cap \|B\|^{\text{sa}} \| \geqslant \| \|A\|^{\text{sn}} \cap \|B\|^{\text{sn}} \| \quad (1)$$

The statement is formulated in the following notation (cf. [12], p. 601):

- sa is an actual situation (a situation that supports only states of affairs that we identify as actual).
- S is a set of normative situations sn.
- a situation is, effectively, a non-maximal possible world[1].
- for a predicate P, we have $\|P\|^s = \{a : s \vDash P(a)\}$, thus the set of a such that $P(a)$ holds in s^2.

Thus, assuming we have a domain of objects, that share two crisp[3] properties A and B, and distinguished other domains where all objects have the same two properties, Lappin's semantics of *many* evaluates to true iff the original domain is at least as big, in terms of cardinality, as the distinguished other domains.

Hence, it is necessary to point out that not the range of the determiner is vague, but what is meant by vague, is that there are different possibilities for interpreting it. Contrasting it with $\| \text{ every } \|$, Lappin states this:

> "By contrast $\| \text{ many } \|$ and $\| \text{ few } \|$ allow a large number of distinct interpretations whose specification involves essential reference to contextual parameters. In fact, these quantifiers seem to be vague in a way that GQ's like $\| \text{ every } \|$ are not" ([12], p. 600).

In this section, we will follow that strategy, and come back to the possibility of augmenting it in Sect. 4. The following statement of Lappin's can be seen as a motivation for his approach:

> "The determiners *many* and *few* are problematic for generalized quantifier theory because, as has frequently been noted, their interpretations are radically context-dependent and under-determined" ([12], p. 599).

He singles out the two options of extensional readings and intensional readings, but in each case (1) is meant to give an adequate semantics. Referring to the work of Keenan and Stavi (1986, [9]) and Fernando and Kamp (1996, [6]) in this area, he gives the following assessment:

---

[1] Cf. p. 601 in [12], footnote 6.
[2] Cf. p. 602 in [12], footnote 8.
[3] Here, crisp means two-valued.

"Both types of analysis are problematic in that either they do not allow for certain readings of *many (few)*, or they generate multiple ambiguity with no apparent upper bound on the set of possible interpretations for these determiners. In Sect. 2 I present a different sort of intensional account of vague quantifiers which avoids these difficulties" ([12], p. 601).

Also, he explains how extensionality and intensionality are related:

"In addition to intensional readings like those indicated for (7) and (9), it is possible to derive from (5)[4] the full range of extensional readings which have been proposed on alternative accounts of many by extensionalising S through the requirement that the elements of $S = $ sa" ([12], p. 603).

For illustration, Lappin gives nine different extensional readings of *many*, referred to as (10), (11a–d), and (12a–d) (in [12]), and embeds them into his setting. This appears to be a neat formal frame for figuring things. Along with it, we present four additional interpretations that fit into this frame, to extend his list of nine into a list of 13. We want to employ several "comparison sets" to augment applicability. Thus, for $i \in \{1, \dots, k\}$ let C_i be the i-th comparison set. Also, we define:

$p_B^{\mathrm{sa}} := |\|A\|^{\mathrm{sa}} \cap \|B\|^{\mathrm{sa}}|$, and $\quad p_i^{\mathrm{sa}} := |\|A\|^{\mathrm{sa}} \cap \|C_i\|^{\mathrm{sa}}|$.
Be $I_a := \{i : i \in \{1, \dots, k\}, p_i^{\mathrm{sa}} > p_B^{\mathrm{sa}}\}$, and $a := |I_a|$.
Be $I_b := \{i : i \in \{1, \dots, k\}, p_i^{\mathrm{sa}} = p_B^{\mathrm{sa}}\}$, and $b := |I_b|$.
Be $I_c := \{i : i \in \{1, \dots, k\}, p_i^{\mathrm{sa}} < p_B^{\mathrm{sa}}\}$, and $c := |I_c|$.
$|C^+| := \max\{|\|C_i\|^{\mathrm{sa}} \cap \|A\|^{\mathrm{sa}}| : i \in \{1, \dots, k\}\}$

Now we can state our four additional interpretations for *many*:

(A1) $S = \{\mathrm{sn} : \mathrm{sn} = \mathrm{sa} \,\&\, p_B^{\mathrm{sa}} \geqslant p_i^{\mathrm{sa}}, \text{ for all } i \in \{1, \dots, k\}\}$
(A2) $S = \{\mathrm{sn} : \mathrm{sn} = \mathrm{sa} \,\&\, p_B^{\mathrm{sa}} \geqslant p_i^{\mathrm{sa}}, \text{ for most } i \in \{1, \dots, k\}\}$
(A3) $S = \{\mathrm{sn} : \mathrm{sn} = \mathrm{sa} \,\&\, \sum_{i \in I_a} p_i^{\mathrm{sa}} < \sum_{i \in I_c} p_i^{\mathrm{sa}}\}$
(A4) $S = \{\mathrm{sn} : \mathrm{sn} = \mathrm{sa} \,\&\, |C^+| < p_B^{\mathrm{sa}}\}$

The second one is clearly just a weakening of the first. Employing a '>'-sign, instead of the '$\geqslant$'-sign (in (A1) or (A2)), would be a strengthening. We will come back to this in a bit. The third one is apparently somewhat new, even though the meaning should be clear from a philosophical viewpoint. Having several comparison sets, it says the following: If the sum of the masses of the comparison sets smaller than the one under investigation, is bigger than the corresponding sum of the masses of the comparison sets bigger than the one under investigation, then $\|B\|^{\mathrm{sa}} \in \| \text{ many } \|(\|A\|^{\mathrm{sa}})$ can be regarded adequate. The last one can be seen as a strengthening with respect to Lappin's 12a, using several, instead of one, comparison set. One can also think of very different readings of many. For example, there

---

[4] The (1) from the present paper is referred to as (5) in [12].

may be psychological reasons to consider something as *many*. In cases in which there are no comparison classes, or at least, they are not obvious, we can think of an *information overflow* interpretation. For example, if we look at a table, that hosts exactly five coins, most of us will be able to recognize the number of coins at once, without counting them. If we put 17 coins instead, most human brains will consider that as to many to handle at once, and will need to count them. It seems interesting to encounter such readings as well, but for the sake of brevity we will not so do here. One example we consider here informally, builds upon the idea of small children playing LEGO. What could bring a child to consider the number of red LEGO-bricks as *many*, is that there are more of the red sort than of all other colors. That would be Lappin's interpretation, if he had introduced more than one allowed comparison set, since we are in the extensional case. Still, it makes sense to talk about 'many red LEGO-bricks', even if there is a color of which there are more LEGO-bricks. That is what we formalize here. The intensional case would be the one of one child playing LEGO, but, e.g., at different days, and in different rooms, and using these different information clusters as comparison classes. This shall, on empirical grounds, motivate the need to extend Lappin's approach. The next example will be our running example for the remaining part of the paper. It is a context dependent statement that will be revisited in Sect. 3:

$$\text{"Many students are German"} \tag{2}$$

For a first assessment of this, we intend to use a situation sa to which we informally refer to as Vienna. Formally, we will define sa as follows:

Be E the set of all people registered in Vienna. Thus, $E = \{p_1, \dots, p_m\}$ for some $m \in \mathbb{N}$. We assume that 'being a student' is a crisp predicate, and is decidable, for all people living in Vienna. Hence, we separate the elements of E in the following way:

$$J_{\text{st}} := \{j : j \in \{1, \dots, m\} \text{ and } \text{student}(p_j) \text{ is true}\}$$

Again, we assume that, for all such students, it is decidable whether one is German or not, thus we separate I_{st} further:

$$J_{\text{st},G} := \{j : j \in J_{\text{st}} \text{ and } \text{German}(p_j) \text{ is true}\}$$

Similarly, assuming there are k other nationalities $C_1, \dots, C_k$, apart from the Austrians themselves, and that each person only possesses one citizenship, we have:

$$J_{\text{st,At}} := \{j : j \in J_{\text{st}} \text{ and } \text{Austrian}(p_j) \text{ is true}\}$$
$$J_{\text{st},C_i} := \{j : j \in J_{\text{st}} \text{ and } C_i(p_j) \text{ is true}\} \text{ for all } C_i \text{ with } i \in \{1, \dots, k\}.$$

Thus: $J_{\text{st}} = J_{\text{st},G} \dot{\cup} J_{\text{st,At}} \dot{\cup} J_{\text{st},C_1} \dot{\cup} \dots \dot{\cup} J_{\text{st},C_k}$

Before we continue presenting the example for the extensional assessment of (2), using only one situation, namely sa (= Vienna), we capture that context structure. Here, it is crucial, that the used predicates are crisp, and fulfill the law of excluded middle with respect to the root-set E:

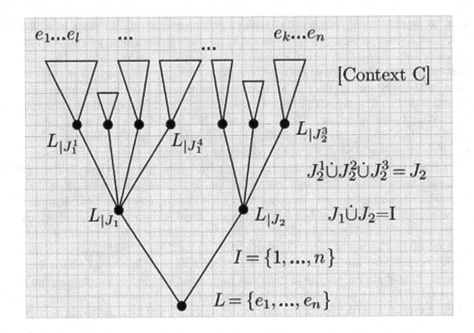

This intuitive picture captures basically the idea of a canonical structure for figuring orderings of objects in a predicative way: At each node, we add a new predicate, that all the node following objects have to fulfill forever. Thus, the leaves carry objects, that fulfill all the predicates of the preceding nodes. This tangible way of writing things down, can eventually lead to an increase of processability by machines. Lets give a definition:

Definition 1 (Distinction Tree). *Be L a finite list of objects, and T a finitely branching (finite) tree with root T_0. Be $n \in \mathbb{N}$ the number of objects in L, and $I := \{1,\ldots,n\}$. A node corresponds to a sublist of L. In particular, $L = T_0$. A node t' is a successor of node t, if $t' \subseteq t$. For this we write $t' \succ t$. If the list-inclusion is strict, we say that t' is an echt successor of t. For a list $l = (t_m, \ldots, t_0)$ of nodes with $t_m \subseteq t_{m-1} \subseteq \ldots \subseteq t_0$, we set m (the number of list-inclusions) as length of l, and denote it with $\mathrm{lh}(l)$. With $\mathrm{slh}(l)$ we denote the number of strict list-inclusions. Also, we denote with $t' \succ_m t$, that there is a list $l = (t, t_1, \ldots, t_{m-1}, t')$ with $\mathrm{lh}(l) = m$, and with $t' \succ^s_m t$ the same situation with $\mathrm{slh}(l) = m$. The list-inclusions are governed by the principle of separation, that is, if, for $r \in \mathbb{N}$, $t' \succ^s_r t$, then the objects of t' fulfill all the very same properties that the objects of t fulfill, plus exactly r more. If there is such a tree T with $L = T_0$, and the union of all leaves gives L again, we call it a distinction tree.*

Definition 2 (Context). *A context is a distinction tree with $L = E$, for some non-empty set E of finite cardinality.*

Now, if we evaluate (2), using (Aj) (with $j \in \{1, \ldots, 4\}$), and the context Vienna, where we take $C_1, \ldots, C_k$ as our comparison sets, i.e. we actually evaluate *"Many (foreign) students (in Vienna) are German."*, we have the following four results:

$j = 1$: (2) is true if and only if the number of German students in Vienna is at least as big as the number of students from C_i in Vienna, for all $i \in \{1, \ldots, k\}$.

$j = 2$: (2) is true if and only if the number of German students in Vienna is at least as big as the number of students from C_i in Vienna, for more than half of the $i \in \{1, \ldots, k\}$. (Weakening w.r.t. $j = 1$)

$j = 3$: (2) is true if and only if the sum of the fractions that are smaller than the German one, is greater than the sum of fractions that are bigger than the German one.

$j = 4$: (2) is true if and only if the fraction of German students in Vienna is greater than the one of every single comparison class. (Strengthening w.r.t. $j = 1$)

Note that, unlike later on in the intensional case, there is no need to introduce relativity counts here, since we would simply divide every term by the same normative term $\|A\|^{\mathrm{sa}}(\neq 0)^5$, and hence it cancels out again immediately.

For readability and the sake of conciseness, we introduced only the four readings (A1),....,(A4), but we will now, as a preparation for Sect. 4, show, how the structure behind unfolds into a neat combinatorial pattern. Eventually, we intend to motivate the use of this pattern for a graded interpretation of *many*.

We can have the following 13 orders of the numbers a, b, and c:

$z_1 : a > b > c$ $z_2 : b > a > c$ $z_3 : c > a > b$ $z_4 : a = b > c$

$z_5 : a > b = c$ $z_6 : b > a = c$ $z_7 : c > a = b$ $z_8 : a = c > b$

$z_9 : a > c > b$ $z_{10} : b > c > a$ $z_{11} : c > b > a$ $z_{12} : b = c > a$

$z_{13} : a = b = c$

These classes correspond to integer partitions where the order counts:

$$\mathrm{IP} := \{(a, b, c) \in \mathbb{N}^3 : a + b + c = k\}$$

We can make out three possible *majority scenarios* and one additional one:

(B1) for most $i \in \{1, \ldots, k\}$ $p_i^{\mathrm{sa}} > p_B^{\mathrm{sa}}$
(B2) for most $i \in \{1, \ldots, k\}$ $p_i^{\mathrm{sa}} = p_B^{\mathrm{sa}}$
(B3) for most $i \in \{1, \ldots, k\}$ $p_i^{\mathrm{sa}} < p_B^{\mathrm{sa}}$
(B4) neither (B1) nor (B2) nor (B3) hold.

Of course, we could replace "most" with "all" in (B1), (B2), and (B3), but this would make considerations only easier, thus we decided to go for the slightly more involved (weaker) interpretation. Also, we want to point out, that the distinction into the four above scenarios will yield a more precise consideration than the one from (A1),..,(A4). It could easily be weakened to fit the former interpretation again, though. The reason why we are not doing it, is that we

---

[5] Cf. remark on p. 9.

want to emphasize the inherent combinatorial systematics. Particularly, when we develop the graded interpretation of *many*, it will make sense to treat *few* simultaneously, and therefor we will need the full structure at our disposal. To stress this a bit further, in [12] Lappin also attempts to so do as well. This can be read off the second quote of this paper, and from the following:

> "A corresponding interpretation for $\| \text{few} \|$ is obtained by substituting $\| \text{few} \|$ for $\| \text{many} \|$ and '$<$' for '$\geqslant$' in (5)[6]" ([12], p. 601).

Singling out (B2) as a distinct set of scenarios will enable us to give an even more direct correspondence of the two determiners, in Sect. 4.

Clearly, $|\text{IP}| = \frac{(k+1)(k+2)}{2}$.

The number of possible scenarios for (B1), (B2), and (B3) is the same, namely:

If k is even: $\frac{1}{2}\left(\frac{k}{2}\right)\left(\frac{k}{2}+1\right)$. And if k is odd: $\frac{1}{2}\left(\lfloor\frac{k}{2}\rfloor+1\right)\left(\lfloor\frac{k}{2}\rfloor+2\right)$.

The remaining scenarios from (B4) will be referred to as *no-majority situations*. Of them we have:

If k is even: $\frac{1}{2}\left(\frac{k}{2}+1\right)\left(\frac{k}{2}+2\right)$. And if k is odd: $\frac{1}{2}\left(\lfloor\frac{k}{2}\rfloor\right)\left(\lfloor\frac{k}{2}\rfloor+1\right)$.

Also, we could give a detailed analysis of the distribution of the orderings of the numbers a, b, and c, and show how they can be used to determine intermediate truth values of statements involving *many*. As this would take to much space, we delegate it to future work. Here, we only emphasize the existence of this pattern. The vicinity to probabilistic approaches (like the one of Fernando and Kamp) may be apparent, even from this narrow treatment, though. Still, we give an intuition how to evaluate (2) using more than two truth values, making use only of the numbers a, b, and c:

Many$(A, B) = 0$, for scenarios from (B1).
Many$(A, B) = \frac{1}{2}$, for scenarios from (B2).
Many$(A, B) = 1$, for scenarios from (B3).
And for one of the remaining ones from (B4)
$c > a \rightarrow \text{Many}(A, B) \in \left(\frac{1}{2}, 1\right]$, and $a > c \rightarrow \text{Many}(A, B) \in \left[0, \frac{1}{2}\right)$

Later, we intend to look into that a little bit deeper and also make use of the measure of the corresponding comparison sets, as already indicated through (A3). Deeper considerations, into the direction of probability theory, can find a good starting point in [7].

3 Diversification

In this section, we will show, why Lappin's (1) is not enough to capture all readings of *many*, by giving an easy example, employing again (2). Before we

---

[6] The (1) from the present paper is referred to as (5) in [12].

will so do, lets fix some notation and have a look at the most apparent different interpretations: $(S = \{sn_i : i \in \{1, \ldots, n\}\})$

$p_{B,\text{rel}}^{\text{sa}} := \frac{|\|A\|^{\text{sa}} \cap \|B\|^{\text{sa}}|}{|\|A\|^{\text{sa}}|}$, and $p_{B,\text{rel}}^{i} := \frac{|\|A\|^{\text{sn}_i} \cap \|B\|^{\text{sn}_i}|}{|\|A\|^{\text{sn}_i}|}$, and

$p_{B,\text{abs}}^{\text{sa}} := |\|A\|^{\text{sa}} \cap \|B\|^{\text{sa}}|$, and $p_{B,\text{abs}}^{i} := |\|A\|^{\text{sn}_i} \cap \|B\|^{\text{sn}_i}|$.

Be $I_\alpha := \{i : i \in \{1, \ldots, n\}, p_{B,\text{rel}}^{i} > p_{B,\text{rel}}^{\text{sa}}\}$, and $\alpha := |I_\alpha|$.

Be $I_\beta := \{i : i \in \{1, \ldots, n\}, p_{B,\text{rel}}^{i} = p_{B,\text{rel}}^{\text{sa}}\}$, and $\beta := |I_\beta|$.

Be $I_\gamma := \{i : i \in \{1, \ldots, n\}, p_{B,\text{rel}}^{i} < p_{B,\text{rel}}^{\text{sa}}\}$, and $\gamma := |I_\gamma|$.

Be $I_{\tilde{\alpha}} := \{i : i \in \{1, \ldots, n\}, p_{B,\text{abs}}^{i} > p_{B,\text{abs}}^{\text{sa}}\}$, and $\tilde{\alpha} := |I_{\tilde{\alpha}}|$.

Be $I_{\tilde{\beta}} := \{i : i \in \{1, \ldots, n\}, p_{B,\text{abs}}^{i} = p_{B,\text{abs}}^{\text{sa}}\}$, and $\tilde{\beta} := |I_{\tilde{\beta}}|$.

Be $I_{\tilde{\gamma}} := \{i : i \in \{1, \ldots, n\}, p_{B,\text{abs}}^{i} < p_{B,\text{abs}}^{\text{sa}}\}$, and $\tilde{\gamma} := |I_{\tilde{\gamma}}|$.

$\|B\|^{\text{sa}} \in \|\text{ many }\|(\|A\|^{\text{sa}})$ iff

$$S \neq \emptyset, \text{ and for every } i \in \{1, \ldots, n\}, \text{ it holds: } p_{B,\text{rel}}^{\text{sa}} \geqslant p_{B,\text{rel}}^{i} \quad (3)$$

Other options are striking, like the following two weakenings:

$\|B\|^{\text{sa}} \in \|\text{ many }\|(\|A\|^{\text{sa}})$ iff

$$S \neq \emptyset, \text{ and for most } i \in \{1, \ldots, n\}, \text{ it holds: } p_{B,\text{abs}}^{\text{sa}} \geqslant p_{B,\text{abs}}^{i} \quad (4)$$

$\|B\|^{\text{sa}} \in \|\text{ many }\|(\|A\|^{\text{sa}})$ iff

$$S \neq \emptyset, \text{ and for most } i \in \{1, \ldots, n\}, \text{ it holds: } p_{B,\text{rel}}^{\text{sa}} \geqslant p_{B,\text{rel}}^{i} \quad (5)$$

Or similarly to what we did before, we can use these two strengthenings:

$\|B\|^{\text{sa}} \in \|\text{ many }\|(\|A\|^{\text{sa}})$ iff

$$S \neq \emptyset, \text{ and it holds: } p_{B,\text{abs}}^{\text{sa}} > \max\{p_{B,\text{abs}}^{i} : i \in \{1, \ldots, n\}\} \quad (6)$$

$\|B\|^{\text{sa}} \in \|\text{ many }\|(\|A\|^{\text{sa}})$ iff

$$S \neq \emptyset, \text{ and it holds: } p_{B,\text{rel}}^{\text{sa}} > \max\{p_{B,\text{rel}}^{i} : i \in \{1, \ldots, n\}\} \quad (7)$$

To complete our selection:

$\|B\|^{\text{sa}} \in \|\text{ many }\|(\|A\|^{\text{sa}})$ iff

$$S \neq \emptyset, \text{ and it holds: } \sum_{i \in I_{\tilde{\gamma}}} p_{B,\text{abs}}^{i} > \sum_{i \in I_{\tilde{\alpha}}} p_{B,\text{abs}}^{i} \quad (8)$$

$\|B\|^{\text{sa}} \in \|\text{ many }\|(\|A\|^{\text{sa}})$ iff

$$S \neq \emptyset, \text{ and it holds: } \sum_{i \in I_{\gamma}} p_{B,\text{rel}}^{i} > \sum_{i \in I_{\alpha}} p_{B,\text{rel}}^{i} \quad (9)$$

Remark: To not run into trouble here, we use the following common interpretation: For two sets M and N, we have: $\frac{|M \cap N|}{|N|} = 0$, if $|N| = 0$.

A natural question is now, whether we really do something new here, or whether there is a way to translate (3)–(9), by some means, into (1). This can also be so expressed:

Proposition 1. *Let* $S = \{\mathrm{sn}_1, \ldots, \mathrm{sn}_n\}$ *be a set of normative situations. Furthermore, assume we want to employ the relative reading of many captured by* (3). *It is not possible to rewrite* S *into* $S' = \{\mathrm{sn}'_1, \ldots, \mathrm{sn}'_n\}$ *such that* S' *is a set of normative situations, and we can employ* (1) *equivalently to* (3) *using this* S' *instead of* S.

Proof. Assumptions:

- $\|A\|^{\mathrm{sa}} \neq \emptyset$, and $\|A\|^{\mathrm{sn}} \neq \emptyset$ for all sn $\in S$
- $\forall_{\mathrm{sn} \in S}$: $\|\|A\|^{\mathrm{sa}} \cap \|B\|^{\mathrm{sa}}| / |\|A\|^{\mathrm{sa}}| \geqslant |\|A\|^{\mathrm{sn}} \cap \|B\|^{\mathrm{sn}}| / |\|A\|^{\mathrm{sn}}|$ (⊠)

We are looking for a S' such that:

$\forall_{\mathrm{sn}' \in S'}$: $\|\|A\|^{\mathrm{sa}} \cap \|B\|^{\mathrm{sa}}| \geqslant |\|A\|^{\mathrm{sn}'} \cap \|B\|^{\mathrm{sn}'}|$

Our setting grants us, that A and B are fixed predicates. Also, we can fix some sn from S for which we have equality in (⊠), and try to rewrite it into an sn$'$, while sa is fixed from the very beginning. Thus, generally by assumption, $q := \frac{\|A\|^{\mathrm{sa}}}{\|A\|^{\mathrm{sn}}}$ is a non-negative rational number, depending particularly on the actual situation sa. Hence, it can be, that $q = \frac{1}{3}$, as a special case, as well as it may be the case, that $|\|A\|^{\mathrm{sn}} \cap \|B\|^{\mathrm{sn}}| = 26$. Now, since $|\|A\|^{\mathrm{sn}'} \cap \|B\|^{\mathrm{sn}'}| \in \mathbb{N}$, it can not be, that:

$$|\|A\|^{\mathrm{sn}} \cap \|B\|^{\mathrm{sn}}| \frac{\|A\|^{\mathrm{sa}}}{\|A\|^{\mathrm{sn}}} = |\|A\|^{\mathrm{sn}'} \cap \|B\|^{\mathrm{sn}'}|,$$

since that would mean, that $\frac{26}{3} \in \mathbb{N}$, which is clearly wrong. This completes the argument.   □

(4) can be treated differently. Indeed, we can switch to a subset S' of S for which we have that, (4) becomes (1), if we replace S by S'. Regarding (5), (7) and (9), we can use a similar argument like for (3), hence they are not translatable into (1). (6) is just a strengthening of (1), thus, the first implies the latter, but not the other way around. Hence, if we start off with (1), we can rewrite the S from there into a S', by dropping the situations for which we have equality, and arrive at (6). Since (8) is a new plausible reading of *many*, not directly referring to $p_{B,\mathrm{abs}}^{\mathrm{sa}}$ at all, it is not translatable into (1) without losing the intended information that it should express.

This shows, that there is much more to achieve than (1) suggests. Now, again for clarity, lets look again at (2) and evaluate it with regard to the following set of normative situations $S = \{\mathrm{sn}_1, \mathrm{sn}_2, \mathrm{sn}_3\}$, with $\mathrm{sn}_1 =$ New York, $\mathrm{sn}_2 =$ London, and $\mathrm{sn}_3 =$ Paris. So, what we intend to evaluate is, actually, the following statement:

"*Many students are German (in Vienna, compared to NY, London and Paris)*".

Lets consider the following fictive numbers[7]:

City	Inhabitants	German students	Per cent	Students total	Relative count
Wien	1800000	20000	~ 0.011	60000	~ 0.33
NY	10000000	20010	~ 0.002	150000	~ 0.13
London	10000000	12000	~ 0.0012	100000	~ 0.12
Paris	2200000	10010	~ 0.0046	80000	~ 0.125

Using (1), we are bound to evaluate (2) to false, unlike with, say (3), since, even though there is a normative situation in which we have (absolutely) more German students than in the actual one, it is still true, that Vienna has (relatively) the most German students within all normative situations. This is still a plausible reading of *many*. Since (3) is just the relative version of (1), it it obvious how the interpret the two (absolute and relative) weakenings (4) and (5), as well as the two (absolute and relative) strengthenings (6) and (7). Clearly, (8) and (9) are the truly new intensional readings of *many*. Again, we point out, that, unlike before in the extensional case, here it makes a difference whether we introduce a relative version of (8), namely (9), for the normative situations are distinct. These respectively evaluate to:

W.r.t. (8): $\text{Many}(A, B) = 1$, since $10010+12000>20010$

W.r.t. (9): $\text{Many}(A, B) = 1$, since $0.12+0.125+0.13>0$

4   Fuzzy Range

In this section, we intend to arrive at first options for interpreting *many* in the semy-fuzzy case, i.e. we want our predicates to be crisp, and classically interpreted, but at the same time, we want to take up the challenge of graded interpretations of *many*. In Glöckner's [8], no intensionality is considered. Lets have a look at the following function:

$f : \mathbb{Q}_0^+ \times \mathbb{Q}_0^+ \backslash \{0, 0\} \longrightarrow [0, 1]$, with $f(x, y) = \frac{x}{x+y}$

This one fulfills the following five properties:

- $x = y$ implies, that $f(x, y) = \frac{1}{2}$
- $x > y$ implies, that $f(x, y) \in \left(\frac{1}{2}, 1\right]$
- $x < y$ implies, that $f(x, y) \in \left[0, \frac{1}{2}\right)$
- $y = 0$ implies, that $f(x, y) = 1$

---

[7] 'German students' corresponds to $|\|B\| \cap \|A\||$, 'students' corresponds to $|\|A\||$, and 'relative count' corresponds to $\frac{|\|B\| \cap \|A\||}{|\|A\||}$.

- $x = 0$ implies, that $f(x, y) = 0$

Thus, as a first option one can give the following refinement, for scenarios from (B4), of what we did at the end of Sect. 2:

$$\text{Many}(A, B) = f(c, a) = \frac{c}{c + a}$$

A deeper consideration (still with regard to the extensional case) may be this one:

$$\text{Many}(A, B) = f\left(\sum_{i \in I_c} p_i^{\text{sa}}, \sum_{i \in I_a} p_i^{\text{sa}}\right) = \frac{\sum_{i \in I_c} p_i^{\text{sa}}}{\sum_{i \in I_c} p_i^{\text{sa}} + \sum_{i \in I_a} p_i^{\text{sa}}}$$

The combinatorial structure used above allows for an analogue treatment of the intensional case:

(C1) for most $i \in \{1, \ldots, n\}$ $p_{B,\text{rel}}^i > p_{B,\text{rel}}^{\text{sa}}$
(C2) for most $i \in \{1, \ldots, n\}$ $p_{B,\text{rel}}^i = p_{B,\text{rel}}^{\text{sa}}$
(C3) for most $i \in \{1, \ldots, n\}$ $p_{B,\text{rel}}^i < p_{B,\text{rel}}^{\text{sa}}$
(C4) neither (C1) nor (C2) nor (C3) hold.

Thus (for scenarios from (C4)):

W.r.t. (5):

$$\text{Many}(A, B) = f(\gamma, \alpha) = \frac{\gamma}{\gamma + \alpha}$$

W.r.t. (9):

$$\text{Many}(A, B) = f\left(\sum_{i \in I_\gamma} p_{B,\text{rel}}^i, \sum_{i \in I_\alpha} p_{B,\text{rel}}^i\right) = \frac{\sum_{i \in I_\gamma} p_{B,\text{rel}}^i}{\sum_{i \in I_\gamma} p_{B,\text{rel}}^i + \sum_{i \in I_\alpha} p_{B,\text{rel}}^i}$$

And for the remaining ones, analogously to the extensional case:

$\text{Many}(A, B) = 0$, for scenarios from (C1).

$\text{Many}(A, B) = \frac{1}{2}$, for scenarios from (C2).

$\text{Many}(A, B) = 1$, for scenarios from (C3).

Note, that for all scenarios from (B4) and (C4) it is not possible for a and c, or α and γ respectively, to be zero at the same time. In fact, only if k, or n respectively, is even, it may happen, that one of them is zero. Otherwise both values are bound to be greater than zero.

So, through singling out the scenarios (B2) and (C2), we can now see the symmetry of the parameters a and c, or α and γ respectively. This gives us the opportunity to define *few* complementary to *many*, by just swapping the denotation of the respective values a and c, or α and γ respectively. This is also an argument why one may prefer the strict inequalities over the non-strict ones, and at the same time a singling out of the third case to which we assigned the intermediate truth value $\frac{1}{2}$. A full characterization of *few* is not part of this paper, but can be carried out straightforwardly, particularly by means of this functional equality:

$$\text{Few}(A, B) = 1 - \text{Many}(A, B)$$

5 Conclusion and Future Work

We have seen how Lappin's parametrization of *many* provides a neat frame to evaluate natural language statements involving vague quantifier expressions. On top of that, we augmented the applicability through giving additional plausible readings, for both the extensional and the intensional case. We have seen how combinatorics can facilitate considerations, but interesting parts had to be delegated to future work. The full characterization of all mentioned distributions is clearly our attempt, as well as formally making out all relevant parameters in play. One class of readings that we did not touch in this paper are the subjective ones. For example, further information about our comparison classes, such as topological distance, and even sympathy values, can enter the setting. How to rank these evaluation techniques is a very interesting question, that we will pursue in the near future. Game semantics, in particular with more than two players, will surly play an important role. This will bring us to the true fuzzy scenario, in which the predicates will not necessarily be classically interpreted anymore. Also, of course, we will treat more complex statements involving logical fuzzy connectives such as $\wedge$, $\vee$.

References

1. Barwise, J., Cooper, R.: Generalized quantifiers and natural language. Linguist. Philos. **4**(2), 159–219 (1981)
2. Cintula, P., Hájek, P., Noguera, C. (eds.): Handbook of Mathematical Fuzzy Logic. College Publications, London (2011)
3. Fermüller, C.G., Roschger, C.: Randomized game semantics for semi-fuzzy quantifiers. In: Greco, S., Bouchon-Meunier, B., Coletti, G., Fedrizzi, M., Matarazzo, B., Yager, R.R. (eds.) IPMU 2012, Part IV. CCIS, vol. 300, pp. 632–641. Springer, Heidelberg (2012)
4. Fermüller, C.G., Roschger, C.: Bridges between contextual linguistic models of vagueness and t-norm based fuzzy logic. In: Montagna, F. (ed.) Petr Hájek on Mathematical Fuzzy Logic. Outstanding Contributions to Logic, vol. 6, pp. 91–114. Springer, Switzerland (2014)
5. Fermüller, C.G., Roschger, C.: Randomized game semantics for semi-fuzzy quantifiers. Log. J. IGPL **223**(3), 413–439 (2014)
6. Fernando, T., Kamp, H.: Expecting many. In: Semantics and Linguistic Theory, pp. 53–68 (1996)
7. Flajolet, P., Sedgewick, R.: Analytic Combinatorics. Cambridge University Press, Cambridge (2009)
8. Glöckner, I.: Fuzzy Quantifiers: A Computational Theory. Studies in Fuzziness and Soft Computing, vol. 193. Springer, Heidelberg (2006)
9. Keenan, E.L., Stavi, J.: A semantic characterization of natural language determiners. Linguist. Philos. **9**(3), 253–326 (1986)
10. Lappin, S.: The semantics of many as a weak determiner. Linguistics **26**(6), 977–1020 (1988)
11. Lappin, S.: Many as a two-place determiner function. SOAS Working Papers in Linguistics and Phonetics **3**, 337–358 (1993)

12. Lappin, S.: An intensional parametric semantics for vague quantifiers. Linguist. Philos. **23**(6), 599–620 (2000)
13. Peters, S., Westerståhl, D.: Quantifiers in Language and Logic. Oxford University Press, USA (2006)
14. Westerståhl, D.: Quantifiers in formal and natural languages. In: Westerståhl, D. (ed.) Handbook of Philosophical Logic. Synthese Library, vol. 167, pp. 1–131. Springer, Netherlands (1989)
15. Zadeh, L.A.: A computational approach to fuzzy quantifiers in natural languages. Comput. Math. Appl. **9**(1), 149–184 (1983)

What the Numbers Mean? A Matter of Context!

Jean-Pierre Müller[✉]

CIRAD GREEN, Montpellier, France
jean-pierre.muller@cirad.fr

Abstract. In formal languages, it is generally assumed that the symbols having the form of numbers just denote the usual mathematical numbers. In this paper, we argue that a numeric symbol is a symbol as any other and therefore may have different meanings in different contexts. Enumerating various kind of contexts in which the numeric symbols can make sense and which one, we come up with a generalized framework in which relationships between contexts, unit conversion and geometric transformations are treated in a uniform way. It also reveals the epistemological richness of what the numbers conceptually capture. Finally, this framework raises the question of contexts as types.

1 Introduction

All formal languages are built on a set of names, also called words or symbols. Those names have a meaning, or denotation, that depends on context. This context can be implicit like the interpretation of a logical formula [13], multiple as with possible world semantics [7], explicit like in modular ontologies [2,3] or in multi-context systems [4]. The interpretation of a logical formula defines a domain of discourse and a recursive mapping of the names and constructs into this domain of discourse. Therefore, the names acquire a different meaning for each possible interpretation. The possible world semantics is built on a set of connected possible worlds, each being endowed with its own domain of discourse and mapping. This semantics allows the interpretation of constructs talking about structures of possible worlds as, for example, sequences of periods of time. The names can have different meaning in different possible worlds. For example, the name "president" may denote a different person in different periods of time. Montague semantics introduces the notion of intentional names, i.e. names having different denotations in different contexts [5]. In modular ontologies, the names are prefixed with explicit context names. Modular ontologies introduce names for contexts as well as names for things. In programming languages, the notion of package or namespace also allows prefixed names defining distinct computational objects in different contexts.

When using these formal languages, one very often needs to use arithmetic. Consequently, one introduces to the set of names the set of numeric names (integers and/or reals made of the characters '0', '1', etc.). In pure logics, these numeric names are constrained by the suitable axioms (e.g. the Peano's axioms) and can only denote objects behaving like numbers for any choice of

© Springer International Publishing Switzerland 2015
H. Christiansen et al. (Eds.): CONTEXT 2015, LNAI 9405, pp. 359–372, 2015.
DOI: 10.1007/978-3-319-25591-0_26

interpretation or possible world. It is even more constrained with the so-called rigid designators, introduced with possible-world semantics, where the numeric names denote nothing but the corresponding number (the name 1 denotes 1). The contextual ontologies have introduced the same artifact with the so-called concrete domains, where some names are constrained to denote elements of an underlying algebraic structure [1].

Back to pure logics, the numeric names have different meaning in different contexts even if the denotations behave in the same way. However, what can be a context for numeric names, and therefore what could be their possible denotations, is left unexplored. The objective of this paper is to investigate the nature of the different contexts in which numeric names can acquire different meaning and, therefore what can be these meanings. The outcome is a generic conceptual framework in which name contexts, units, reference frames for coordinates can be seen in a uniform way. Such a framework can be used for specifying a concrete domain to integrate to a concept language (see a first attempt in [10]), an API (Application Programming Interface) to manipulate quantities and coordinates, or for introducing more expressivity for natural language processing. Additionally, many mathematical operations (arithmetic, affine transforms, etc.) can be interpreted as mapping names into names, and more precisely as defining name equivalences. These operations are linked to the contexts and, as a consequence, the contexts appear as defining which operations can be applied on the number names.

We will first describe the possible meanings of the numeric names when contextualized and consequently the nature of these contexts. We will then further explore the difference between numbers for quantities and numbers for positions, introducing an extended notion of coordinates and their related contexts. We will conclude with the resulting generic picture and its consequences.

2   Notations

In the following, we will use the notation introduced for modular ontologies in [2]. We introduce the set N of *names* that denote either objects in a given domain of discourse, or sets of those objects. The names S made of alphabetic characters will be called the *symbols*. The integers $\mathbb{Z}$ and reals $\mathbb{R}$ will be called the *numeric names* to distinguish them from the mathematical *numbers* supposed to have a transcendental existence (that we personally deny, but it is not the point of this paper). Therefore, the set A of numeric names is defined as $\mathbb{Z} \cup \mathbb{R}$. The use of the mathematical numbers themselves as names has two advantages:

1. we do not have to care about the number notation that depends on the base, or whether it is in a decimal or scientific notation,
2. the arithmetic operators can be used to specify the semantics more easily.

It is at the expense of possible confusion that we will try to avoid as much as possible. One way to avoid this difficulty is to introduce notations as contexts: 101 in the binary context would be equivalent to 5 in the decimal context but we will not pursue this path in this paper. To sum up, the set of (simple) names N is $S \cup A$.

The modular ontologies introduce a set C of context names that are just symbols, possibly indexed for facility. We will have to extend this set in the following. A *contextualized name* is written $c{:}n$ where $c \in C$ is a context name and $n \in N$ is a symbol or numeric name. We will use this notation for now on and explore what it means when n is a numeric name. We note CN the set of contextualized names.

3 Meaning of Numbers

We will distinguish three uses of numeric names, namely as identifiers, ordinals and quantities. In each case, we will qualify the nature of the context and the way we can relate a context to another. We will add a comment on mathematical numbers and will open to the next section as conclusion.

3.1 Identifiers

In formal logics, we have to name all the objects we want to talk about individually. Usually, we use the symbols for that. When there are an important number of objects, it is easier to enumerate them using integers. For example, if we have a lot of tables, we can have the table 1, the table 754, etc. An everyday life example is the phone numbers. It also is the role of the identifiers in databases; the integers are just names. A mathematical example is the gödelization where a separate number is computed for each distinct logical formula, allowing numbers to denote those formulas and therefore theorems about numbers to be theorems about those logical formulas. For this reason, we propose to talk about *identifiers* when using the numeric names this way. Obviously, these numeric names can denote different objects in different contexts: tables when we want to enumerate tables, people when we want to enumerate people, etc. Moreover, the number attributed to an object is arbitrary. Therefore, the collection of the objects we want to name is unordered: the table 32 is not before the table 173 in any sense. In this case, the identifiers behave as any name. In particular, there is no arithmetic associated to it. For example, there is no sense to add neither two phone numbers nor two identifiers in a database.

In this case, a context is just a domain of discourse, i.e. a collection of objects we want to talk about using, among others, identifiers. Let us just write these contexts in the modular ontology way. For example, we will have contextual names as $c_1{:}582$ or $c_4{:}17$. If we want to express that an identifier in a given context denotes the same object as an identifier in another context, we have to explicitly specify this relationship. We introduce the notation $N_1 \equiv N_2$ where $N_i \in CN$ to express that the two contextualized names are equivalent (i.e. always denote the same object for whatever interpretation). We will use $=$ only for strict equality, i.e. with the same context name and the same symbol or numeric name. For example, we can express that $c_1{:}582 \equiv c_4{:}17$, i.e. that the object named 582 in the context c_1 is the same as the object named 17 in the context c_4. Consequently, the relations among the contexts must be specified in extension by an enumeration of such equivalences. Given a set of commutative equivalences $\{c_i{:}n_j \equiv c_k{:}n_l\}$ where $j \neq l$ and i may be distinct from k but not necessarily (equivalence intra-context), we can define two sets:

- the equivalence class of a name $c_i{:}n_j$, written $E(c_i{:}n_j)$ is the set $\{c_k{:}n_l | c_k{:}n_l = c_i{:}n_j \vee \exists q.c_i{:}n_j \equiv c_{m_1} : n_{p_1} \wedge \cdots \wedge c_{m_q} : n_{p_q} \equiv c_k{:}n_l\}$;
- the projection (or translation) of a name $c_i{:}n_j$ in a context c_k, written $T(c_i{:}n_j, c_k)$ is the set $\{c_k{:}n_l \in E(c_i{:}n_j)\}$, i.e. the restriction of $E(c_i{:}n_j)$ to the names with context c_k. This set can be empty. An obvious property is that $T(T(c_i{:}n_j, c_k), c_l) = T(c_i{:}n_j, c_l)$

These definitions are valid for any names endowed with equivalence definitions, not only numeric names. For example, if we have $\{en{:}sun \equiv fr{:}soleil, fr{:}soleil \equiv de{:}sonne, es{:}sol \equiv en{:}sun\}$, $E(en{:}sun) = \{en{:}sun, fr{:}soleil, de{:}sonne, es{:}sol\}$ and $T(en{:}sun, es) = \{es{:}sol\}$.

3.2 Ordinals

A step further is to add order. We want to name a collection of objects in a certain order and therefore, there is a first object, a second object, and so on. We define that a numeric name is an *ordinal* or an *index*, if the order on the numeric names reflects the order of the denoted objects, i.e. if $c{:}n_i$ denotes an object o_i and $c{:}n_j$ denotes another object o_j in the same context c, then if $n_i = n_j$, $o_i = o_j$ and if $n_i < n_j$, o_i is before o_j in the underlying order (e.g. the object $c{:}3$ is before the object $c{:}5$). In this case, the same number can denote a different object in different contexts, depending on the order. For example, a first object in a context can be the last object in another. As for the identifiers, there is no arithmetic but comparison makes sense relative to a context. For example, there is no sense to add the second to the fourth, but the second is before the fourth. Therefore a context is both a domain of discourse and an order on this domain, or at least an order on the part of the domain named by ordinals.

We still have to enumerate the correspondences. The subset of equivalences between two contexts is called a *permutation* (or *k-permutation* if they are defined on a subset of size k). In mathematics the permutations are endowed with a group structure by defining the composition of permutations. A permutation P is bijective, therefore $|P(c_i : n_j, c_k)| = 1$ for all i, j, k. For example, students may be ordered by their size in the context s and the student $s{:}n_i$ is bigger than the student $s{:}n_j$ if $n_i > n_j$ (here the first student is the smaller), or they may be ordered by their weights in the context w and the student $w{:}n_i$ is heavier than the student $w{:}n_j$ if $n_i > n_j$ (here the first student is the lighter). These ordinals have meaning in the same set of students and which index corresponds to which between the two contexts s and w (which student ordered by weight is which one when ordered by size) can be specified by a permutation.

3.3   Quantities

Measures to Quantities. A *measure* is a numeric name resulting from a measurement of an object or a counting operation on a set of objects. When measuring the height of an object with a measurement tool, we obtain a quantity (in this case, the height of the object) that depends on the measurement tool. If it is calibrated in inches (i.e. made such that the tool gives 1 for an inch), we will get a given number (say 12). If calibrated in meters, we will get another (namely 0.3). These two measures denote the same height. Notice that the height cannot be described in itself but only through a measurement tool[1], different measurement tools giving different measures for the same height. The difference is related to the unit, not the measurement precision or granularity.

In the same way, we can also count the number of objects in a collection (another kind of measurement). If we have a basket with three apples and five pears, counting the number of apples gives 3, while counting the number of fruits gives 8. The number we get depends on the category used for counting. Here, the measurement tool is calibrated by the mental categories (i.e. what we consider the same from a given point of view).

However, there are a variety of measurement tools (rules, laser, balance, etc.) and we need to abstract their common feature regarding the measure. This abstraction is called a unit for a measurement and a category (considered as a kind of unit) for the a count. Initially, a *unit*, also called a *standard*, was an object used as a reference to measure the corresponding feature. For example, the meter has been for a long time a bar of iridium deposited in the Pavillon of Breteuil in France (Headquarter of the "Bureau International des Poids et Mesures").

The unit can be considered as the context in which the number makes sense. 5 means nothing unless we say whether it is $5\,kg$ or 5 pounds. Therefore, 5 do not denote the same thing (here the same weight) in different contexts, i.e. with different units. In the same way, different numbers may denote the same thing (e.g. the same height) in different contexts (e.g. $12\,in.$ and $0.3\,m$ denote the same height? or length). In the following, we will use the modular ontology notation $in:12$ or $m:0.3$, using the standardized symbols for the units when they are defined. A measure together with its unit, i.e. a contextualized measure, will be called a *quantity*. A quantity may be a magnitude or a multitude. A *magnitude* is measured as the height, width, weight, monetary value, etc. and the context is a unit. The notation is of the form $u:r$ where $u \in C$ is a unit name and $r \in \mathbb{R}$ is a real numeric name. A *multitude* is counted as the number of apples or fruits and the context is a category. The notation if of the form $c:r$ where $c \in C$ is a category name and $r \in \mathbb{R}$ is a real numeric name, e.g. *apple*:3 or *fruit*:8.

The quantities can be added to one another: $m:1 + in:12$ gives $m:1.3$ or $in:52$ (the same quantity with different measures), as well as subtracted. However, it is not possible for arbitrary units. For example, $m:1 + kg:3$ does not make sense.

---

[1] It is worse for a number that cannot be "seen" but by writing it: here the height can be seen but can only be named through a measurement tool that produces numeric names that can then be read.

However, they can be freely multiplied or divided, but the result is of a different unit. For example, m:1$/s$:1 is 1 in the context m/s that appears to be a speed. The possibility to combine units to form new units has to be formalized. Given the set $\mathbb{U}^a \subset CN$ of unit symbols for the magnitudes, and $\mathbb{C} \subset CN$ of categories for the multitudes, it is possible to define a syntax of the unit names as the set $\mathbb{U}$ recursively defined as follows:

- $U \in \mathbb{U}$ if U is in the set $\mathbb{U}^a$;
- $C \in \mathbb{U}$ if U is in the set $\mathbb{C}$;
- if $U_1 \in \mathbb{U}$, $U_2 \in \mathbb{U}$ and n an integer:
  - $U_1 * U_2 \in \mathbb{U}$ for the product;
  - $U_1/U_2 \in \mathbb{U}$ for the division;
  - $U_1^n \in \mathbb{U}$ for the power.

For example, m/s^2 is a unit (namely for acceleration) if m and s are unit symbols (and they are!). Therefore, we have to extend CN to include $\mathbb{U}$

The relation among the units is called a *conversion* and is defined by multipliers. For example, any quantity (i.e. the numeric name denoting a length) in inches can be converted into a quantity in meters by the multiplier 0.025, and conversely by $1/0.025 = 40$. The set of conversions (i.e. the context relationships) endowed with multiplication forms an algebraic group.

Units to Dimensions. As already mentioned some conversions are not possible. For example, it is not possible to convert meters in kilograms (an operation between contexts) as it is not possible to add meters and kilos(an operation between quantities). Therefore, there is not a single algebraic group allowing converting anything to anything (i.e. relating the names in any context to the names in another context), but a collection of algebraic groups. For example, meter, inch, mile, etc. form an algebraic group, and kilogram, pound, ton, etc. form another algebraic group, the various ways to measure speed as well. The first one is related to distances, the second to weights and the last to speed. The nature of the feature of an object measured by an algebraic group is called a *dimension*. For example, the feature measured by a meter, inch, etc. is a length (i.e. is of dimension "length"), and the feature measured by kilograms, pounds, etc. is a weight (i.e. is of dimension "weight"). The International System of Quantities [11] defines seven *atomic physical dimensions*, from which all the other *physical dimensions* can be defined (with their respective symbols): the length (L), the duration (t), the mass $(m$, to not confuse with the symbol for meter), the electric current (I), the thermodynamic temperature (Θ), the amount of substance (J) and the luminous intensity (n). The other dimensions can be obtained by multiplication, division, and, consequently, powers of these atomic dimensions. For example, the speed is a length divided by a duration (L/t), and a force is a length multiplied by a mass and divided by the square of a duration $(L * m/t^2)$, independently of the unit used to measure it. Dimensional analysis provides a formal account of dimensions [8]. The set of atomic dimensions can be arbitrarily defined (the set of atomic physical dimensions is

historically defined and debated). For example, we could add to the seven physical dimensions the monetary value or the number of apples or fruits. It is easy
to see that those can be combined in the same way with the others, e.g. the
monetary value or the number of apples divided by a duration would be a kind
of financial or apple flow or growth speed. Given a set of *atomic dimensions*,
there is an infinity of *derived dimensions* that are independent from one another
(i.e. defining distinct algebraic groups on units). Given the set $\mathbb{D}^a$ of dimension
symbols, the set $\mathbb{D}$ of all possible dimensions is recursively defined as follows:

- $P \in \mathbb{D}$ if P is in the set $\mathbb{D}^a$;
- if $P_1 \in \mathbb{D}$, $P_2 \in \mathbb{D}$ and n an integer:
  - $P_1 * P_2 \in \mathbb{D}$ for the product;
  - $P_1/P_2 \in \mathbb{D}$ for the division;
  - $P_1^n \in \mathbb{D}$ for the power;

Finally each unit has a dimension and is, therefore, related to the possible contexts (i.e. units) of this dimension. For example, L/t^2 is the dimension of the
unit m/s^2, but also of the unit km/h^2.

Formalization. Unless identifiers and ordinals where the equivalences must
be enumerated (i.e. defined in extension), it is enough to specify relationships
between contexts to define equivalence among quantities. A unit is related to
another context by giving a single multiplier m: given two contexts u_1 and u_2
related by a multiplier m (written $u_1 \equiv m * u_2$), then any name of the form
$u_1{:}r$ is equivalent to the name $u_2{:}(m * r)$ (and, conversely, $u_2{:}r$ is equivalent
to $u_1{:}(1/m * r)$), where r is a numeric name and $m * r$ (resp. $1/m * r$) is the
numeric name of the number obtained by multiplying m (resp. $1/m$) by r. In this
case, we are using the arithmetic operators applicable on A (the set of numeric
names). For example, we can specify that $in \equiv 0.025 * m$ (one inch is 0.025 m)
and consequently $in{:}12$ is the same length than $m{:}(0.025 * 12) = m{:}0.3$. $\equiv$ being
an equivalence relation, we have:

- if $u_1 \equiv m_1 * u_2$ and $u_2 \equiv m_2 * u_3$ then $u_1 \equiv m_1 * m_2 * u_3$, or,
- if $u_1 \equiv m_1 * u_2$ and $u_3 \equiv m_2 * u_2$ then $u_1 \equiv m_1 * 1/m_2 * u_3$.

    More generally:

- $E(u{:}n)$ is the set $\{u_i{:}(m * n)|(u_i{:}n_j = u{:}n \wedge m = 1) \vee \exists m, u \equiv m * u_i\}$;
- $T(u_1{:}n, u_2)$ is the set $\{u_2{:}n_l \in E(u_1{:}n)\}$. This set can be empty in which
  case there is no equivalence with this unit, or it is of cardinality 1 because the
  correspondence is a bijection.

    Additionally, the arithmetic operations are defined on the quantities in cer-
tain conditions. For multiplication, division and power, the operations are trivial
although it relies on algebra on the context names:

- $u_1{:}n_1 * u_2{:}n_2 = u_1 * u_2{:}n_1 * n_2$
- $u_1{:}n_1/u_2{:}n_2 = u_1/u_2{:}n_1/n_2$
- $u{:}n^m = u^m{:}n^m$

Notice that on the left, we build a context name expression while on the right the arithmetic operation is actually applied giving a new numeric name as a result of the operation.

For addition and subtraction, we have to distinguish between the magnitudes and the multitudes. In the case of the magnitudes, it depends on the possibility to convert a unit into another. If it is the case, we have to choose one of the units and make the operation in this context, converting the other value. If $u_1 \equiv m*u_2$, we have:

- $u_1{:}n_1 + u_2{:}n_2 = u_1{:}(n_1 + 1/m * n_2)$ or $u_2{:}(m * n_1 + n_2)$,
- $u_1{:}n_1 - U_2{:}n_2 = u_1{:}(n_1 - 1/m * n_2)$ or $u_2{:}(m * n_1 - n_2)$,

otherwise the operation is undefined.

In the case of the multitude, the categories must be endowed with a lattice structure given two operations: the union and the intersection. A category is seen as a set of objects being the same from a given point of view (e.g. the objects being apples and the objects being red). Therefore the union and intersection applies to the corresponding sets (the objects that are either red or apples and the objects that are both red and apples). In a lattice, the supremum of two categories (noted $c_1 \wedge c_2$) is the minimal set that contains both categories. Therefore, given two multitudes $c_1{:}n_1$ and $c_2{:}n_2$, we have:

- $c_1{:}n_1 + c_2{:}n_2 = c_1 \wedge c_2{:}(n_1 + n_2)$
- $c_1{:}n_1 - c_2{:}n_2 = c_1 \wedge c_2{:}(n_1 - n_2)$

For example, the supremum of $fruit$ and $apple$ is $fruit$ (the minimal category that contains both fruits and apples), therefore $fruit{:}5 + apple{:}2 = fruit{:}7$.

3.4   Mathematical Numbers

The numbers in mathematics do not have a unit. Consequently, it is considered that they are context-independent, or that they are universal in a strict sense (i.e. the numeric names denote the same corresponding number in all possible contexts). We can obtain them from what precedes with the so-called dimensionless quantities. Among the derivable dimensions, the division of any dimension by itself (e.g. L/L) produces a dimensionless "dimension" or ratio. Although they are all dimensionless, it can still be useful to make some of them distinct like the angles (a length divided by a length) or the quantity of information, when we do not want their values to be added or subtracted freely. Nevertheless, the mathematical numbers could be considered as dimensionless ratios.

Historically, the Greeks used the integers to denote the architecture of the cosmos. Hence 1 denotes the unity of the cosmos, 2 denotes the concept of duality (male and female, good and bad) that structure our perception of the cosmos (made of distinctions), $1 + 2 = 3$ is the feeling of unity behind the apparent duality, etc. Therefore the mathematics of the Greeks has an esoteric dimension that disappeared later on, but still legitimates a discourse on numbers as idealities distinct from the measurable account of the physical world. Starting from natural numbers (0, 1, etc.), the set of numbers can be completed to ensure that the results of the various operations $(+, -, *, /,$ etc.) still are in the set of numbers. $-$ requires the relative integers ($2-7$ is not a natural number, therefore we need to introduce the negative numbers),$/$requires the rational numbers ($2/7$ is not an integer, therefore we have to introduce the rational numbers), and so on.

3.5   Temperature...

A unit of $°F$ (degree Fahrenheit) is $0.555\ldots *° C$ (degree Celsius). As all other units, there is a multiplier between the two units qualifying the same dimension. However, $1°F$ is not equal to $0.555\ldots°C$ but to $-17,222\ldots°C$ because the origin (the position of the 0) is not the same. In fact, the temperature is not really a quantity as a distance or a mass, but a position on a scale. Therefore, in addition to the conversion as any other unit, there is a displacement (this time a quantity!) between the origins of the two scales. Consequently, there is not only the unit that defines what 1 means relatively to the other units but also an origin, that defines what the 0 means, also relatively to the other origins. There are actually two contexts to define the meaning of the temperature: the unit and the origin. This case will be tackled in the next section.

4   Quantities to Coordinates

The numeric names we talked about are called quantities because they correspond to the number of units to accumulate to get the measured quantity. In this sense, measuring is also counting. $m{:}5$ is the number of meters we have to accumulate to have an equivalent distance. $s{:}56$ is the number of seconds we have to count to have an equivalent duration. However, we can choose arbitrarily a point on a line, or an instant in time and measure the distance from this point to any other or the duration from this instant to any other. In the case of the line, we obtain a *position* and in the case of time, we obtain a *date*. The chosen point or instant will be called the *origin*. As for quantities, a position or a date cannot be described in itself but can only be named when giving a context (an origin) from which the measurement is made. And the obtained measurement can only be named when given the unit. As for the temperature, there is a need for a double context.

As a first step, we will extend to multi-dimensional names or *coordinates*. Here the dimension does not refer to the nature of a measured or counted feature but to the number of features we want to represent simultaneously. To avoid confusion, we will call it the *dimensionality*. An obvious case is a multi-dimensional space (e.g. a 3D space) where a position must be represented by three quantities. The dimensionality is 3. A coordinate will be written $\langle q_1, \ldots, q_n \rangle$ for a n-dimensional coordinate where each q_i is a quantity, i.e. a contextualized numeric name. For example, a 3D coordinate in a Euclidean space would be written $\langle m{:}2.3, in{:}5, km{:}6 \rangle$. Such a coordinate may denote an object situated

2.3 m ahead on x, 5 inches on y, and 6 km on z. It remains to define what is x, y and z. As the point denoted by this name can only be identified when given the origin, we use the context names to contextualize the coordinate: $O{:}\langle q_1, \ldots, q_n \rangle$ where $O \subset CN$ is a context name naming an origin. The relation among contexts must define which coordinates in two different contexts corresponds to the same position in the space, or, equivalently, how to transform a coordinate in a given context into an equivalent name in another. When the dimensionalities of the related contexts are the same, these relationships correspond to the affine transforms. If the dimensionality is different, we have to consider the projections onto spaces of lower dimensionality or injections into space of higher dimensionality.

Given a set T of transformations t between origins, we will note $o_i \equiv t * o_j$ the transformation between the two origins. Formally, matrices usually specify them. Of course the transformations can be compounded if the dimensionalities are compatible:

- if $o_1 \equiv t_1 * o_2$ and $o_2 \equiv t_2 * o_3$ then $o_1 \equiv t_1 * t_2 * o_3$, or,
- if $o_1 \equiv t_1 * o_2$ and $o_3 \equiv t_2 * o_2$ then $o_1 \equiv t_1 * t_2^{-1} * o_3$.

The transformations can be added, multiplied (including with a scalar), divided under conditions, giving a whole range of notations for defining equivalence between coordinates.

As previously, we can define the equivalence classes and the transformations as follows:

- $E(o{:}\langle q_1, \ldots, q_n \rangle$ is the set $\{o_i{:}t * \langle q_1, \ldots, q_n \rangle | (o_i = o \wedge t = id) \vee \exists t, o \equiv t * o_i \}$ where id is the identity transformation;
- $T(o_1{:}\langle q_1, \ldots, q_n \rangle, u_2)$ is the set $\{o_2{:}\langle q_{l,1}, \ldots, q_{l,n} \rangle \in E(o_1{:}\langle q_1, \ldots, q_n \rangle)\}$, in case of injections, this set can be big because there are many ways to inject an n-dimensional space into a space with a dimensionality higher than n.

Applied to the temperature (i.e. a 1-dimensional space), $°K{:}5 \equiv °F{:}9$ while $ks{:}\langle °K{:}5 \rangle \equiv fs{:}\langle °F{:}-450, 67 \rangle$ if we define the transformation T between ks (the Kelvin scale) and fs (the Fahrenheit scale) as a translation of $255, 372$ and the conversion between Kelvin and Fahrenheit as 1.8 respectively. It naturally applies to the geometric transformations.

Another example is the notion of date that is duration since an origin that depends on the used calendar (christian, islamic (Hijra), buddhist, etc.). But a duration is very often written not as a single number (e.g. the number of seconds since the birth of JC) but as a coordinate with the name of the day (between 1 and 31), the name of the month (either symbols: January, February, etc., or numbers between 1 and 12), the name of the year (from 1), the name of the hour (between 1 and 24), etc. Therefore some calendars or choices of representation can be represented as multi-dimensional coordinates and the injection of dates as single numbers into them must be specified.

5   Discussion

From the preceding analysis, we are now in a position to sum up and comment on both the syntactical and semantical outcomes of our propositions.

Introducing quantities, i.e. a measure and a context, and coordinates, i.e. tuples of quantities, we introduced syntax on numeric names:

- numbers in $\mathbb{Z} \cup \mathbb{R}$ are *numeric name*;
- $c{:}n$ where c is a context name and n is numeric name, is a *quantity*;
- $o : \langle q_1, \ldots, q_n \rangle$ where o is a context name and q_i are quantities is called *coordinate*; in this case the possible operations extends to all sorts of geometric transformations.

When the context is a unit, there is also a syntax of context names:

- a unit symbol is a context name;
- if U_1, U_2 are context names, and n an integer, $U_1 * U_2$, U_1/U_2 and U_1^n are context names as well.

At the syntactic level of names in general and numeric names in particular, we have proposed to relate contexts by specifying equivalence relations among names. Two names are equivalent within the same context, or in different contexts, if and only if they denote the same object. When using symbols, the definition of the mapping is enough and can be composed. For ordinals, these mappings appear to be permutations. For the magnitudes, multipliers between contexts are sufficient for specifying how to map equivalent quantities depending on the unit. For the multitudes, one must use the set inclusion structures hierarchies of categories. Finally, we can specify mappings among coordinates by affine, projection and injection transformations from a context to another. Formally, equivalences can be defined:

- directly on the numeric names through statements of the form $C_i{:}n_j \equiv C_k{:}n_l$ in the case of identifiers and ordinals;
- indirectly on the contexts:
  - on magnitudes by multipliers: $C_i \equiv m * C_j$ inducing that $\forall n, C_i{:}n \equiv C_j{:} m * n \vee C_j{:}n \equiv C_i{:}1/m * n$;
  - on multitudes by inclusion: $C_i \subset C_j$ inducing $\forall n, C_i{:}n \equiv C_j{:}n$ but not the other way around;
  - on coordinates by transformations: $C_i \equiv t * C_j$ inducing that $\forall \langle q_1, \ldots, q_n \rangle, C_i : \langle q_1, \ldots, q_n \rangle \equiv C_j : t * \langle q_1, \ldots, q_n \rangle \vee C_j : \langle q_1, \ldots, q_n \rangle \equiv C_i{:}t^{-1} * \langle q_1, \ldots, q_n \rangle$

Semantically, a context is a domain of discourse and a denotation. We are now at a position to say the domain of discourse could be:

- the identifiers can denote objects in general, and not just numbers, without any implicit constraint;
- the ordinals can denote ordered objects where, in addition to a domain of discourse, the order on numeric names reflects the order on the objects;
- the quantities are talking about the dimensions of the objects, i.e. countable or measurable features of objects. Consequently, a context defines the possibility to use some arithmetic on numeric names reflecting the corresponding arithmetic operations on numbers but with some restrictions related to the algebra of dimensions and units;

- of course they can even talk about numbers, in which case there is a single, universal context providing a fixed denotation with unconstrained use of arithmetics;
- finally, the coordinates can talk about positions in abstract spaces, most of them being just euclidean spaces. The contexts behave as origins and provide the possibility to use complex transformations between these spaces and the way to name the positions.

The introduced concepts are summarized by the Fig. 1 where we can see the hierarchy of names on top, the flat hierarchy of contexts on bottom, and the notions of contextualized names as the quantity (both magnitude and multitude) and the position in between, composed of both a name and a context. As an illustration the notion of date is shown as a particular case of a mono-dimensional coordinate measured with time units.

Manipulating units and reference frames as contexts is unusual but provides a conceptual framework on which name equivalences, including for numbers and coordinates, can be represented in a uniform way. A consequence is that the conversions and transformations are now definitely linked to the contexts, as the mappings are. Another consequence is that the contexts, if we exclude the denotation, do not need any intrinsic descriptions. A unit is just a unit. A meter is what a meter measures, nothing else. An origin is just an origin. An origin is a position that is named $(0, \ldots, 0)$ in its own reference frame, nothing else. The contexts only serve as support to express relationships with other contexts. We know that an inch is 0.025 m but it says nothing about what an inch or a meter is intrinsically. The description of a context is only relative to other contexts. The other counter-intuitive result is that arithmetic and mathematical transformations appear to manipulate names and not idealities. Up to this point it is

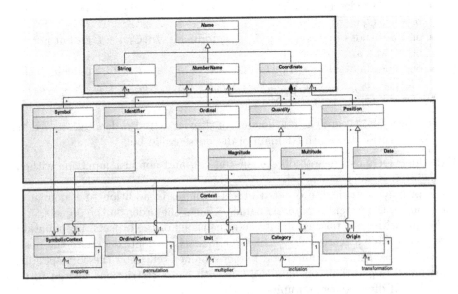

Fig. 1. The introduced concepts

not necessary to exhibit what a position, or height, or weight or even a number is to be able to manipulate it through a set of computable equivalences, hence the importance of the equations. Finally, by grouping the objects denoted by the numeric names with respect to applicable functions (comparisons, permutations, conversions, transformations), the algebraic groups of contexts seem to behave as types with respect to the denoted objects.

6   Conclusion

We explored what the numeric names are talking about, considering they are syntactic names as any other. Consequently, their meaning depends on the context whatever the nature of the context. We came up with a number of different kinds of context constraining what the numeric names can talk about: objects in general, ordered objects, measures?and even numbers! A further extension was introduced with the notion of coordinates as a accumulation of two contexts: one to qualify the magnitude and one to qualify the origin. The whole provides a uniform framework in which conversions, transformations and mappings can be freely combined up to some restrictions imposed by the contexts themselves.

Practically, an API (Application Programming Interface) has been written in Java based on the principles described in this paper (see the structure project on [9] code repository), as an alternative proposal to the JSR-275, obsolete but still is used in some advanced libraries, and JSR-363 API proposals [6]. They introduce the notions of quantity, unit and dimension but only for the physical dimensions. The BOOST unit package in C++ is more advanced, introducing the distinction between quantities and mono-dimensional coordinates [12]. None handles the notion of multitude. In these propositions, the coordinates in general are not included or part of a distinct library, and the set of possible dimensions is not easily extensible as it is in our case. For example, the currencies are handled separately when they are integrated in our case. Therefore, we provide the possibility to extend the number of dimensions and the associated units, as well as to define any kind of coordinates, reference frames and transformations between them. [10] points some hints on how to introduce quantities in description logics but did not use the notion of context. It is the next step to build and use knowledge representation languages on this framework.

Finally, we observed that the relations among contexts defining equivalences among names have a structure made of connex components that, in the case of units, corresponds to dimensions. Moreover, the different kinds of contexts also define possible operations from straight equality to geometric transformations. The next step is to investigate the nature of the context, the structure of those sets of contexts and their intuitive relationship to types. Of course, the notion of set and category in the sense of category theory may play an important role in this direction. In effect, having objects (here the contexts) only defined by the relations they have with other objects is typical of category theory. Therefore the next step is to use category theory to better specify what has been presented in this paper.

Acknowledgements. I would like to thank Pierre Livet for his comments. In particular he pointed the relationship between contexts and the possible operations, relating contexts to types as well as gödelization as a way to build numeric names to denote logical formulas. I would also like to thank Abdoulaye Diallo for his comments.

References

1. Baader, F., Hanschke, P.: A scheme for integrating concrete domains into concept languages. Technical Report RR-91-10, DFKI, April 1991
2. Caragea, D., Caragea, D., Bao, J., Bao, J., Honavar, V.G., Honavar, V.G.: Modular ontologies - a formal investigation of semantics and expressivity. In: Shi, Z.-Z., Shi, Z.-Z., Mizoguchi, R., Mizoguchi, R., Giunchiglia, F., Giunchiglia, F. (eds.) ASWC 2006. LNCS, vol. 4185, pp. 616–631. Springer, Heidelberg (2006)
3. Bouquet, P., Giunchiglia, F., Harmelen, F.: C-owl: Contextualizing ontologies. Technical report, Istituto Trentino di Cultura (2003)
4. Brewka, G., Eiter, T.: Equilibria in heterogeneous nonmonotonic multi-context systems. In: AAAI, pp. 385–390 (2007)
5. Dowty, D.R., Wall, R., Peters, S.: Introduction to Montague Semantics. Studies in Linguistics and Philosophy, vol. 11. Springer, Netherlands (2012)
6. JScience (2014). http://jscience.org
7. Kripke, S.A.: Semantical Considerations on Modal Logic. Naming and Necessity. Oxford University Press, Oxford (1971)
8. Langhaar, H.L.: Dimensional Analysis and Theory of Models. Wiley, New York (1951)
9. Müller, J.-P.: (2015). http://mimosa.sourceforge.net
10. Müller, J.-P., Rakotonirainy, H., Hervé, D.: Towards a description logic for scientific modeling. In: KEOD (2011)
11. International Bureau of Weights and Measures. The International System of Units (SI). Number, 8th edition. International Bureau of Weights and Measures (2006)
12. Schabel, M.C., Watanabe, S.: Boost unit library (2010)
13. Smullyan, R.M.: First-order Logic. Dover books on advanced mathematics, Dover (1995)

Deontic Modals with Complex Acts

Andrei Nasta[1,2]([⊠])

[1] University of Pittsburgh, Pittsburgh, PA, USA
[2] University of East Anglia, Norwich, England
and.nasta@gmail.com

Abstract. The paper identifies a key pragmatic principle that is responsible for the information-sensitivity of deontic modals. Information-sensitivity has been extensively discussed in the recent linguistic and philosophical literature, in connection with a decision problem known as the Miners' puzzle (Kolodny and MacFarlane 2010). I argue that the so-called Ellsberg paradox (Ellsberg 1961) is a more general source of information-sensitivity. Then I outline a unified pragmatic solution to both puzzles on the basis of a well-known decision procedure (MiniMax).

Keywords: Deontic modals · Information-sensitivity · Pragmatics · Acts · Outcomes · Decision-theory · Ellsberg paradox · Miners' puzzle · Minimax

What is the ideal, most desirable choice in a decision situation depends on our epistemic states. Call this the *information-sensitivity* of desirability and of similar normative concepts.

Information-sensitivity is a special form of context-sensitivity. Context-sensitivity makes the truth- or assertability-conditions of utterances dependent on features of the world. Information-sensitivity, in contrast, makes these conditions dependent on the features of the world *as known by an agent*. Often the agent does not have access to all the facts relevant for her decision, but has only limited knowledge. This knowledge can—and often does—serve the agent well in deciding what she should do. When the context makes available less than all the facts pertinent to a decision, the agent, from her limited informational standpoint, is bound to have subjective normative commitments rather than objective ones.

In this paper I am concerned with the subjective normative commitments expressed by modal vocabulary in specific contexts. I argue for an extended form of information-sensitivity of deontic modals such as *ought to* and *should* by taking a closer look at the role of certain complex actions and outcomes. I show that the information-sensitivity can be interpreted as adherence to a specific norm of decision in two well-known decision problems. While my aim is not to provide a logic of decision or a compositional semantics for deontic modals, my remarks contribute to a better understanding of the pragmatics of deontic modality, leaving the standard modal semantics untouched.

© Springer International Publishing Switzerland 2015
H. Christiansen et al. (Eds.): CONTEXT 2015, LNAI 9405, pp. 373–382, 2015.
DOI: 10.1007/978-3-319-25591-0_27

1   The Miners' Puzzle

A good starting point for my observation is the Miners' Puzzle (Kolodny and MacFarlane 2010). In the miners' setting (M) we have to decide for the best course of action regarding 10 miners who, as far as we know, can be in either one of two shafts, A or B. The two shafts are about to be flooded and we do not have enough sandbags to prevent this from happening. We have sandbags to block only one of the two shafts. If we block neither shaft, we save nine miners and lose one. If we block the right shaft, we lose no miner. But if we block the wrong shaft, we lose all the miners. In this context, the following sentences sound true.

(1) We ought to block neither shaft.
(2) If the miners are in shaft A (B), we ought to block shaft A (B).

To make sense of these sentences in one's modal semantics, it has been argued, one needs information-sensitive deontic modality and, more notably, the violation of a version of modus ponens. I will assume that a baseline semantics for deontic modals such as Kolodny and MacFarlane (2010) or Kratzer (2012) is on the right line, and my observation should be taken as a *pragmatic* complement to such a baseline semantics. However, I contend that information-sensitivity further affects the reasoning with deontic modals where *complex* actions—rather than simple actions, which have been the focus of discussion so far—are taken into consideration.

1.1   Complex Acts and Outcomes

In context (M), a complex action is a disjunction or conjunction of basic acts. Since the basic acts are block A (B_a), block B (B_b), and block neither shaft (B_n), complex acts are just compounds like $B_a \vee B_b$ and $B_a \wedge B_n$. Likewise for complex outcomes. What are then the predictions of the standard account for complex acts and outcomes?

With an enlarged set of actions, the desirability of the original acts change. The desirability of an act is measured by checking out whether the act chosen warrants that the miners' lives are saved in the greatest possible proportion. Interpreting S_0, S_1 etc. as the possible outcomes: 0 miners are saved, 1 miner is saved etc. we get the following entailments: $B_a \supset S_0 \vee S_{10}$ and $B_n \supset S_9$. The entailment of B_b is the same as that of B_a, since blocking shaft A (B) guarantees only a disjunctive outcome to the effect that we either lose all the miners or save all.[1] (The consequent-proposition—representing the outcome—serves to evaluate the deontic worth of the antecedent-proposition—representing the act.)

---

[1] In the standard modal semantics (Kratzer 2012), the consequent of such an entailment is a proposition in the ordering source, while the antecedent is the proposition in the scope of the deontic modals. Roughly, the more consequent-propositions in the ordering source follow from the antecedent-proposition, the more deontically valuable the antecedent-proposition is.

These entailments represent the original scenario, whereby B_n is preferred to either of the other two acts, because it *guarantees* S_9. The complex outcome $S_0 \vee S_{10}$, which only guarantees that 0 lives are saved, is very unappealing in comparison to the simple outcome S_9. This is indeed an intuitive result.

My question now is what happens when we consider for comparison complex acts rather than simple ones. The answer is not always straightforward. Consider, for instance, the acts $B_a \vee B_b$, $B_a \vee B_n$, and $B_b \vee B_n$. Which of these acts is best? The relevant entailments are $B_a \vee B_b \supset S_0 \vee S_{10}$ and $B_a \vee B_n \supset S_0 \vee S_9 \vee S_{10}$.[2] (The disjunction $B_b \vee B_n$ has the same entailment as $B_a \vee B_n$.) Deciding on a preference is not simple in the absence of additional theoretical assumptions, and so I will postpone discussion of this case.

However, it is much easier to decide on conjunctive acts (assuming they are available). Consider the corresponding entailment of the act of blocking both shafts, namely $B_a \wedge B_b \supset S_{10}$. Choosing to block both shafts is the best possible decision that we can imagine in setting (M), because it maximizes the lives saved (thereby minimizing the lives lost). In the miners' setting, blocking both shafts has no competitors. The other conjunctive acts, $B_n \wedge B_a$ and $B_n \wedge B_b$ are not even defined, since what would be the point of choosing e.g. to block neither shaft and to block shaft A? The act is contradictory, so useless from a deliberation standpoint.[3]

The undefinedness of certain complex acts points to a limitation of the Miners' setting. For instance, our evidence should not be restricted to cases that make most complex acts impossible. For a more comprehensive account of deontic modality we need a broader range of complex acts and outcomes. These are genuine factors in deliberation and decision making, and thus a general semantics of deontic modality should take them into account. A step towards a more general account will be taken in the following section.

2 Ambiguous Complex Acts

Daniel Ellsberg (1961) introduced a decision-theoretic setting that poses problems for some cherished principles of expected utility theory and probability theory. An Ellsberg (E) setting is the following.[4]

(E) Consider a sack containing ten coloured balls. For the present decision, the relevant colours are red, blue, and yellow. The distribution of

---

[2] In decision-theoretic terms, complex acts and outcomes are obtained by putting together the information in the estimated desirability matrix of a decision problem. More precisely, we could gather the information on a specific row of a matrix (which corresponds to an act), or by considering for deliberation a combination of rows (acts) of the matrix (cf. Jeffrey 1965/1983). For completeness, the outcomes—seen as entailments of the acts—should be weighted with the probabilities of the relevant conditions, but we shall leave this information implicit.

[3] It is natural to stipulate that acts, viewed as propositions, should be non-empty, and thus should have at least one element (i.e., world).

[4] See also Fishburn (1986) and Halpern (2005) for further discussion.

colours is as follows. Exactly three balls are red. The remainder seven balls are blue or yellow in unknown proportion. We have to decide which ball-colour to choose, knowing that we get to keep the ball if a ball of that colour is then randomly drawn from the sack. This is desirable, since the balls are all made of massive platinum on the inside.

Setting (E), unlike (M), makes probabilistic information relevant to the decision. This context, unlike the miners' context, allows for a more meaningful deliberation about complex actions. The Ellsberg setting is more general than (M), while preserving all the properties of the (M) setting. To see this, let us reformulate the (E)-assumptions to obtain the same problem noticed in the original (M) context. To this effect, it suffices to add several assumptions which are intuitively acceptable in (E).[5]

(3) We may choose either one of R, B, or Y.   (assumption)
(4) We ought to choose R (not: B or Y).   (intuitive claim)
(5) We ought to choose B or Y over R or Y.   (intuitive claim)
(6) If $B \geq Y$, we ought to choose B.   (intuitive conditional)
(7) If $Y \geq B$, we ought to choose Y.   (intuitive conditional)
(8) Either $B \geq Y$ or $Y \geq B$.   (assumption)
(9) We ought to choose either B or Y.   (from 6–8)

Assuming that if an act A_1 is not dispreferred ($\succeq$) to another act A_2, we ought to choose act A_1, the intermediary steps (6)-(7) seem right.[6] By disjunctive syllogism applied to (6)–(7), paired with (8), we get (9), which contradicts (4).This is the miners' problem resurfacing in (E). But the miners' problem is *independent* of the ones generated by the Ellsberg context, and the latter cannot be formulated in the (M) setting.

---

[5] Notation: R, B, and Y stand for red, blue, and yellow balls. $R > B$ means that there are more red balls than blue ones, or that R is more probable than B. Because the outcomes of the acts are equal, the probabilistic relation $>$ translates into a preference relation $\succ$ (but not vice-versa). So $R > B$ means that choosing red is more probable than choosing blue, but can also read as saying that the former is preferable to the latter. Finally, the disjunctions should be interpreted as inclusive, unless *either ... or*-phrases are used. E.g. opting for $R \vee B$ brings about the prize if any of the red or blue balls is then randomly drawn. (The same could be expressed in terms of conjunction, if we interpreted the letters as bets on colours, because e.g. choosing blue *or* yellow will amount to betting on blue *and* yellow.).

[6] A version of the puzzle can be stated even if we assume that the deontic necessity modal *ought to* requires a strict preference relation. To do this, we assume that one of the following should hold: $B > Y$, $Y > B$, or $Y = B$. We then formulate three conditionals having these three statements as antecedents. For instance, we'll have the new conditional: If $Y = B$, we ought to be indifferent between the three options. (The other two conditionals will be like (6)-(7), but formulated in terms of the strict relation, $>$.) It then follows by disjunctive syllogism that we ought to choose B, Y, or R, which contradicts (3).

To see one of these additional problems, it is essential to take (4) and (5) to be true.[7] This shouldn't be a problem, as most people presented with this case find them true.[8] The basic intuition is that R and $B \vee Y$ warrant clear-cut (expected) desirabilities, whereas the other acts don't. The problem appears when these preferences are coupled with the reasonable assumption that the addition of equal amounts (of probability or desirability) to each member of an inequality should not change the direction of the inequality sign. Nonetheless, if we assume *additivity*, (5) is not consistent with (4). If (4) entails that $R \succ B$, by additivity we get that $R \vee Y \succ B \vee Y$, which is exactly the opposite of (5).

Additivity is not a principle to be lightly dispensed with. Additivity is for probabilistic and decision-theoretic settings as important as modus ponens is for logical reasoning. If we think that A is more probable than B, then it is intuitive to consider $A \vee C$ more probable than $B \vee C$. Moreover, if A is more probable than B or C, it does not follow that A is more probable than both taken together. These intuitive judgements follow from additivity, and concern not only probabilities but also desirabilities.[9] Therefore, it is worth exploring the cause of non-additivity in the (E) setting, since the deontic modals in (E) are sensitive to probabilistic information. I will argue that the diagnostic of non-additivity reveals a common feature about deontic modality in both (E) and (M) settings.

The role of complex acts in our decision problem is related to their *ambiguity* status. We say, for instance, that B is ambiguous because we don't know how many blue balls there are in context (E). So we know neither the precise probability of drawing a blue ball nor its precise desirability. Y is also ambiguous. R, however, is not, because we know precisely the number of red balls, and consequently their probability and desirability. Now, by including complex acts in our decision problem, we are faced with the possibility that a non-ambiguous act can be formed from an ambiguous one, and vice-versa. This affects what we know about the probabilities involved in assessing the acts, and ultimately—as we will see—their relative desirabilities. An illustration of the shift in relative desirability is the unexpected transition from (4) to (5), or vice-versa. In possible words terminology, some worlds will be desirable/ideal when we are faced with one pool of simple acts and less appealing when we are faced with a different pool of complex acts.[10]

---

[7] Complementarity does not hold in (E) either. Complementarity requires that if $A \succ B$, then $\neg A \preceq \neg B$, where $\succ, \preceq$ etc. are preference relations. Yet $R \succ B$, but $\neg R = B \vee Y \succ \neg B = R \vee Y$, and so complementarity is violated.

[8] For experimental evidence that the pattern of reasoning is robust see references in Camerer and Weber (1992, 332ff.).

[9] In natural language semantics, additivity has been invoked as evidence for introducing a quantitative probability measure to account for (epistemic) probabilistic modals, and against the standard Kratzer-semantics of those modals (Lassiter 2010, Yalcin 2010).

[10] In Nasta (2015b) I call this property *instability* and show that it holds of preferences and deontic commitments in general.

The argument in the next section can be viewed as showing why ambiguity is problematic in (E). But the feature that generates the problem will turn out to be deeper than ambiguity, since it is present in the non-ambiguous case (M).

3 Solution: Risk Aversion

My proposal is that the information-sensitivity and the decision procedure that informs (M), also informs (E), and so that the same features produce, in both settings, problems with different principles of reasoning. In both (M) and (E) cases, we know that choosing one act ensures the overall better desirability given the outcomes in each uncertain condition. In (E) we have a fairly clear-cut preference for one of the acts (namely, choosing red). This suggests that the heart of the matter is risk aversion: certain options are preferred because their competitors involve uncertainty with respect to great losses, and produce very unappealing expected desirabilities. Certain acts are preferred even if some other acts dominate them in specific conditions,[11] essentially because dominance is offset by the undesirability of certain possible outcomes.

The short way of diagnosing our cases is to say that they implement a MiniMax strategy, a strategy that minimizes the worst losses in terms of expected value, or, in other words, maximizes the lowest expected outcome of a choice. MiniMax predicts that a unique action will be preferred in the (M) and (E) cases. It is critical in obtaining such a preference that the deliberating agent does not have the information needed to raise the desirability of the alternative actions which incur huge risks. So we can interpret the information-sensitivity of deontic modals as follows: *what we have to do depends on our knowledge, which must exclude or minimize the possibility of the most undesirable outcomes.*[12]

The problems concerning reasoning-principles appear because (M) and (E) implicitly introduce assumptions inconsistent with minimizing the worst outcomes. (M) introduces the problematic assumption through an application of modus ponens. (E) introduces the problematic assumption through an application of additivity.

Take additivity first. This principle requires that $A \succ B$ entail $A \lor C \succ B \lor C$, and vice-versa. For additivity to respect a decision procedure (e.g. MiniMax), it has to be the case that the introduction of disjunction in each member of the inequality ($\succ$), does not change their relevant decision-theoretic status needed by the decision procedure. What is this status? We may call this status the *acceptable desirability* of a proposition reflecting a decision, which, in line with MiniMax, ensures the minimal maximal loss (or minimal lowest gain) by committing to that proposition (or act). Thus, inference must preserve not simply certain probabilistic/truth values (as required by logic and probability theory), but also acceptable desirability (as required by the decision procedure). The

---

[11] E.g. in (M), blocking shaft A dominates the other two acts under the condition that the miners are in shaft A.

[12] This decision procedure is closely related to the minimax regret rule invoked in rational choice theory; see Levi (1980, 144ff.) for discussion and references.

acceptability status depends on the contextually relevant decision procedure, and in (E) that procedure is MiniMax. Thus, the introduction of disjunction should preserve this desirability status.

However, disjunction does not generally preserve *acceptable desirability*, and consequently the principle of additivity—which relies on disjunction introduction—is not valid relative to the MiniMax procedure. For instance, if I prefer R to B in (E), and thus find R acceptably desirable, it does not follow that $R \vee Y$ is acceptably desirable. It should now become clear where the inconsistency lies. The inconsistency is generated by the ambivalence with respect to the MiniMax procedure in reasoning about the (E) setting. On the one hand we assume MiniMax to conclude that R is desirable, and, on the other hand, we implicitly violate MiniMax to derive the desirability of $R \vee Y$. But since disjunction introduction—essential to additivity—does not preserve acceptable desirability, and MiniMax requires just that, either additivity or MiniMax should be given up.[13] It is easy to see that the common judgement of the Ellsberg cases inclines us towards keeping MiniMax. After all, very robust intuitions in (E) lead us to minimize the greater losses in desirability.

In terms of information-sensitivity, the trouble with the additivity-based reasoning is that acceptable desirability presupposes a certain knowledge, namely knowledge of the acceptable desirability of a certain proposition. But knowledge of the acceptable desirability of propositions is not closed under disjunction (unlike preservation of truth). That is, in establishing deontic claims, we cannot rely on 'unacceptably desirable' worlds, i.e., worlds in which the acts are not known to be acceptably desirable, or worlds in which, for all we know, the worst outcome occurs. So we cannot infer deontic modal claims from propositions that contain unacceptably desirable worlds as their denotations.

What trouble additivity makes for (E), modus ponens makes for (M). This time, the trouble is not that acceptable desirability is not preserved, but that modus ponens introduces a false assumption of acceptable desirability. In terms of information-sensitivity, the conditional (that is needed in the modus ponens inference) introduces the assumption that it is known that a certain proposition (characterizing a desirable outcome) has the acceptable desirability status, which means that it is known to the deliberating agent that proposition has a maximal minimal desirability.

The assumption that the deliberating agent has information that rules out the worst outcomes is false, and reasoning on its basis generates a contradiction. As before, the contradiction is generated by applying MiniMax to obtain a conclusion, while using in another part of the reasoning an inference rule (viz. modus ponens) which is incompatible with MiniMax. The inference rule thus yields a conclusion which is inconsistent with the MiniMax conclusion independently derived.

As suggested, the latter problem is present in both (E) and (M) settings. Note that the antecedents of the conditionals in the two puzzles, (E) and (M), update the modal background with epistemic information. Thus, we get the following readings.

---

[13] I discuss more extensively the role of disjunction in settings (M) and (E) in Nasta (2015a).

(10)   a. If the miners are in shaft A, we ought to block shaft A.
       b. If the miners are in shaft A, *and we know it*, we ought to block shaft A.

(11)   a. If $B > Y$, we ought to choose B.
       b. If $B > Y$, *and we know it*, we ought to choose B.

If the emphasised implications[14] wouldn't go through, and we would still want the conditional to be good in the context of deliberation, we need to modify the antecedents. Accordingly, even if the miners are in shaft A, we still need to block neither shaft. (And similarly for the Ellsberg case.) In other words, minimizing risk would require us, as decision makers, to take an option which is optimal with respect to our state of information, but would be sub-optimal with respect to a richer—and, by hypothesis, unavailable—body of information.

My diagnostic for the problematic reasoning is that when we are going through e.g. the miners' scenario, we have a proclivity for erroneously assuming, for the sake of local coherence, that the deliberating agent knows which shaft is open. So we simply add the assumption of relevant knowledge to our information state, though this assumption is not justified by the facts in the global context. Note that the proposed diagnostic—based on a local, pragmatically triggered error—does not amount to an error theory. First, such local pragmatic implications are useful in communication, and seldom generate problems. (No cooperative speaker would reason her way through these decision problems in the way indicated here!) More importantly, our diagnostic is consistent with the observation that competent speakers and reasoners become quickly aware of the tension between the assumptions in the local and global contexts. However, acknowledging this local coherence-based reasoning makes it less surprising that we can momentarily fall pray to such inconsistency.

To sum up, both (M) and (E) in their conditional formulations give rise to problems when modus ponens is applied without regard to the MiniMax decision procedure. In addition, the application of the additivity principle in (E) determines a further violation of the MiniMax decision procedure. The violation of MiniMax has been traced back to the introduction of disjunction (in the case of additivity), and to a coherence-based implication of the conditional (in the case of modus ponens). Since in the present context it is plausible that MiniMax is being used in deriving the preferred act (or the corresponding deontic modal proposition), violations of MiniMax in other parts of the reasoning generate inconsistency. Thus, taking MiniMax as fundamental, we have an explanation of the puzzling inconsistencies obtained in the (M) and (E) settings.

The information-sensitivity of deontic modals can be characterized by their contextual sensitivity to the decision procedure, which in turn requires knowledge. As we have seen, the truth-value of a deontic modal claim depends on the decision procedure used to establish that claim, which amounts to saying that an *ought*-claim is sensitive to what we know to be the best solution (according

---

[14] See von Fintel (2012, pp.28–29) for relevant evidence of this type. This evidence suggests that conditionals admit of implicit restrictors which are sensitive to local pragmatic implications. Such local pragmatic (coherence-based) implication exists in several other linguistic domains (cf. Simons 2014).

to the decision procedure). But what we know and what we *assume* to know in making inferences can come apart, since certain inferences may introduce epistemically and deontically unwarranted assumptions, as they do in our cases.

4 Concluding Remarks

In light of the previous discussion, I agree with Kolodny and MacFarlane (2010, pp.130,136) that what we ought to do depends on our knowledge. I also agree with them that modus ponens does not satisfy this desideratum. However, I have a more specific take on the sort of knowledge that matters and how the illicit assumption of relevant knowledge comes about in the case of conditionals. In contrast to previous accounts of the miners' puzzle—e.g. Kolodny and MacFarlane (2010), Cariani et al. (2013), Charlow (2013)[15]—my approach is more explicit on how to apply the decision-theoretic principles to contexts involving probabilistic information.

It is well beyond the scope of this paper to offer a complete recipe for picking out contexts where MiniMax is successfully applicable. It is nonetheless worth asking, How general is the strategy proposed here? My contention is that the strategy is more general than we might have guessed by looking at an isolated decision setting. At the very least, it applies to cases where some outcomes are certain (as in M) and to cases where the outcomes are not certain (as in E), irrespective of whether the actions evaluated are simple or complex, and (arguably) irrespective of their ambiguity status. However, I cannot make claims about the precise bounds of the cases for which the strategy would work. As I suggested, decision procedures other than MiniMax are relevant in other different settings. We cannot build preferences based on a unique decision procedure. Moreover, even keeping fixed the broad context of e.g. the (M) or (E) case, it's not clear that MiniMax should apply to whatever decision problem we may come up with, and indeed it might well be indeterminate whether any decision procedure applies at all.[16]

To conclude, my proposal gives pride of place to a decision norm in the pragmatics of deontic modality in a particular type of context, as provided by the (M) and (E) cases. In this type of context, I interpreted the information-sensitivity

---

[15] Though see Carr (2012) and Lassiter (2011), who directly approach the miners' puzzle, and Goble (1996) whose deontic logic account may constitute a good framework for dealing with both (M) and (E).

[16] On the one hand, it is easy to check that MiniMax gets a good prediction for the conjunctive act of blocking both shafts in (M), and for the act of choosing a red ball in (E). On the other hand, it is much more difficult to come up with a sharp comparison between the disjunctive acts $B_a \vee B_b$, $B_a \vee B_n$, and $B_b \vee B_n$. The latter two acts guarantee that either no miner will be saved or nine miners will be saved or all of them will be saved ($S_0 \vee S_9 \vee S_{10}$), whilst the former act guarantees that either zero or ten miners will be saved ($S_0 \vee S_{10}$). A first problem is that in order to estimate the desirabilities of the acts we have to come up with probabilities for the disjuncts. And even after doing that, it is not clear that there will be only one obvious way of choosing between the estimated desirabilities thus obtained.

of deontic modality as sensitivity to a MiniMax decision strategy. Deontic modal talk in such contexts is sensitive to what might be the worst outcomes. If our information state leaves the occurrence of the worst outcome open, the act leading to that outcome is dispreferred. This holds true for both simple and complex acts, and my introduction of (E) was in part motivated by the observation that (M) somewhat obscures this fact. Nonetheless, my primary motivation has been to provide a common diagnostic for the problems raised by (M) and (E). In virtue of capturing this unifying feature, my analysis offers guidelines for a more comprehensive pragmatics of deontic modals under uncertainty.

References

Camerer, C., Weber, M.: Recent developments in modeling preferences: uncertainty and ambiguity. J. Risk Uncertainty **5**, 325–370 (1992). (cit. on p. 5)

Cariani, Fabrizio, Kaufmann, M., Kaufmann, S.: Delib- erative modality under epistemic uncertainty. Linguist. Philos. **36**, 225–259 (2013). doi:10.1007/s10988-013-9134-4. (cit. on p. 9)

Carr, J.: Subjective Ought. ms. MIT (2012). (cit. on p. 9)

Charlow, Nate: What we know and what to do. Synthese **190**, 2291–2323 (2013). doi:10.1007/s11229-011-9974-9. (cit. on p. 9)

Ellsberg, D.: Risk, ambiguity, and the Savage axioms. The Quarterly Journal of Economics, pp. 643–669 (cit. on pp. 1, 3) (1961)

Fishburn, P.C.: The axioms of subjective probability. Stat. Sci. **1**(3), 335–358 (1986). (cit. on p. 3)

Goble, L.: Utilitarian deontic logic. Philos. Stud. **82**(3), 257–317 (1996). (cit. on p. 9)

Halpern, J.Y.: Reasoning About Uncertainty. MIT Press, Cambridge (2005). (cit. on p. 3)

Jeffrey, R.C.: The logic of decision, 2nd edn. University of Chicago Press, Chicago (1965/1983) (cit. on p. 3)

Kolodny, N., John, M.: Ifs and oughts. J. Philos. **107**(3), 115–143 (2010). (cit. on pp. 1, 2, 9)

Kratzer, A.: Modals and Conditionals. Oxford University Press, Oxford (2012). doi:10.1093/acprof:oso/9780199234684.001.0001. (cit. on p. 2)

Lassiter, D.: Gradable epistemic modals, probability, and scale structure. In: Li, N., Lutz, D. (eds.) Semantics and Linguistic Theory (SALT), vol. 20, pp. 197–215. CLC Publications, Ithaca (2010). (cit. on p. 5)

Lassiter, D.: Measurement and Modality: the Scalar Basis of Modal Semantics. Ph.D. thesis, New York University (2011) http://semanticsarchive.net/Archive/WMzOWU2O/ (cit. on p. 9)

Levi, I.: The Enterprise of Knowledge. MIT Press, Cambridge (1980). (cit. on p. 6)

Nasta, A.: Disjunctive deontic modals. ms. Pittsburgh/East Anglia (2015a) (cit. on p. 7)

Nasta, A.: Unstable preferences, unstable deontic modals. ms. Pittsburgh/East Anglia (2015b) (cit. on p. 5)

Simons, Mandy: Local pragmatics and structured contents. Philos. Stud. **168**, 21–33 (2014). doi:10.1007/s11098-013-0138-2. (cit. on p. 8)

von Fintel, K.: The best we can (expect to) get? Challenges to the classic semantics for deontic modals. In: 85th Annual Meeting of the American Philosophical Association, Chicago (2012). http://web.mitedu/fintel/fintel-2012-apa-ought.pdf (cit. on p. 8)

Yalcin, S.: Proability operators. Philos. Compass **5**(11), 916–937 (2010). doi:10.1111/j.1747-9991.2010.00360.x. (cit. on p. 5)

Context and Cognition

Investigating Methods and Representations for Reasoning About Social Context and Relative Social Power

Katherine Metcalf[✉] and David Leake

School of Informatics and Computing, Indiana University,
Bloomington, IN 47408, USA
{metcalka,leake}@indiana.edu

Abstract. Social context has a profound effect on how people interact with each other, and should have important ramifications for how intelligent systems interact with people. However, social context has received comparatively little attention in research on context-aware systems. This paper begins by highlighting possible dimensions for descriptions of social-interactional context, based on social science research. An important component is the interactants' place in the social hierarchy, and especially their relative social power. The remainder of the paper presents results on using machine learning methods to learn cross-domain classifiers for predicting relative social power. An experimental evaluation of cross-domain learning between three domains suggests that the important features for determining whether or not one interactional domain can be used to predict the relative social power of interactants in another are which social power dimension has the most influence in a given domain.

Keywords: Social-interactionalcontext · Relative social power · Social intelligence

1 Introduction

For a wide range of settings, contextual factors play a key role in both reasoning and understanding (see Brézillon and Gonzalez [4] for a rich collection of perspectives). Much of the work in context-aware systems has focused on leveraging contextual factors for predicting single user actions, to enable such systems to integrate more seamlessly and effectively into people's lives [9,14]. However, as Kalatzis et al. [15] observe, such treatments of context generally do not address communication and social interactions, which also play an important role in human behavior. Kalatzis et al. addresses this by focusing on the community and developing pervasive computing systems able to model the types of contextual factors normally considered at the level of individual users, but at the level of an entire community. This paper considers how context-aware systems can understand the social relationships themselves.

© Springer International Publishing Switzerland 2015
H. Christiansen et al. (Eds.): CONTEXT 2015, LNAI 9405, pp. 385–397, 2015.
DOI: 10.1007/978-3-319-25591-0_28

Social relationships such as relative social power may seldom be made explicit, but they provide a context that can have a profound effect on how two individuals interact in terms of what constitutes an appropriate action and how observed actions should be interpreted. The paper both presents a general discussion of key dimensions for social context representations, based on the social science literature on social context, and presents concrete experimental results on applying machine learning to the task of identifying relative social power.

The capability for systems to reason about and act intelligently in regard to observed social interactions would enable them to become more active participants and to interact in a natural and socially appropriate manner. This ability would enable them, for example, to participate proactively in a team with people. If a system is aware of the social context and the appropriate ways in which to interact, it would be able to recognize when and whom to interrupt when needed, and to propose ideas and solutions without "stepping on toes," optimizing chances of acceptance of its plan by the team.

The goal of this paper is three-fold: to demonstrate the need for reasoning about the social, interactional context, to show what a model of the social, interactional context must be able to represent, and to demonstrate an initial step towards developing representations and methods with which to reason about the social interactional context. The first section discusses why a system needs a representation of social interactions in order to understand what is being said and interactants' intentions, and to disambiguate between possible meanings. The second section presents aspects of social interactions that any social, interactional model would need to be able to represent and suggests for how to represent some of those aspects. The third section presents results on computational methods for predicting relative social power.

The discussion of computational methods for predicting relative social power builds on work by Metcalf and Leake [18] investigating representations and similarity metrics for comparing characteristics of speakers' performance in conversations, and the application of machine learning classifiers to the results to identify social power relationships (e.g., applying a random forest classifier can identify relative social power from a feature set representing the absolute difference in certain features produced between two speakers during a conversation). The paper presents new results on the generality/transfer of the resulting classifiers, testing their ability to predict relative social power in domains other then the ones on which they were trained. The results suggest that the performance of a classifier trained on one domain at predicting another depends on the specific dimensions along which the relative social power is based.

2   Justifying Modeling the Social-Interactional Context

2.1   Motivations and Desiderata

We use "interactional context" to describe the implicit factors underlying a given type of interaction and, therefore, determining those types of actions that are permissible, expected, cooperative, or deviant in the given situation. Awareness of the social-interactional context is necessary for an agent to demonstrate social cognition and to seamlessly integrate itself into an interaction [24]. Therefore, the ability to reason directly about the social-interactional context could provide a number of benefits. For example, it could improve the performance of dialogue systems, in-home assistive robots, or automatic profilers, and augment human-to-human interactions. This knowledge could allow a system to more accurately interpret social interactions by being able to account for intensions, expectations, and appropriate or script-like behaviors.

Defining the social context as a frame within which to analyze interactions is motivated by its close relationship to modes and expressions of the social self. The expression of one's social self is one aspect of the discourse a person expresses or enacts, with the other half being the cognitive self. The duality of a person's discourse reflects the interplay between internal cognition and the external social and environmental influences that alter and interact with it. It is this close relationship that will allow a system to use the social-interactional context to reason about an interactant's more cognitively related aspects, such as intentions and expectations.

When people participate in different domains of social life, they structure their speech, communicative acts, and actions in general according to the what is appropriate and expected within the current social domain [13]. Knowledge of the current social domain could allow a system to predict the types of language a person in likely to use. The accurate prediction of future speech turns by the human interactant would make it possible for a system to provide the interactant with what he or she needs before the interactant needs to explicitly ask for it. The capability to effectively and accurately reason about how a person is situating him or herself within a social situation, would greatly improve a system's accuracy at accounting for intentions, expectations, and appropriate or script-like behaviors.

Our goal for representing the social-interactional context is to develop a structured representation of the social-interactional domain enabling a computational system to identify discriminatively meaningful features and characteristics for a given social domain. A guiding principle is to develop a representation that is in line with the theoretical constructs and methodologies laypeople and scientists use to talk about and understand social interactions. The resulting representation should be one that is both in line with current theory and is structured such that a computational system can reason about formalized features and characteristics so as to identify what is expected and what is acceptable.

We focus especially on social context as manifested in discourse. Discourse is a means by which to do a social action, such as influencing knowledge, identity, or social relations, therefore, it is also a way to express and maintain social patterns [13] and can be used as a framework with which to analyze various forms of identity, for both groups and individuals. Using discourse to reason about social identity can allow a system to identify which communities someone identifies with. This social identity can then be used to contextualize the interaction. By contextualizing the interaction with information about the identity an interactant is enacting, it becomes easier to reason about types of intentions someone might have.

2.2   A Sample Application: Augmented Reality

To make concrete the benefits of reasoning about the social-interactional context, consider augmented reality (AR) systems. AR seeks to directly affect and alter the ways in which we perceive and interact with our environment, making it is an interesting domain in which to apply context aware technologies. AR can be used to augment or facilitate people's ability to interact with their environment and with other people. In order for AR to be able to effectively augment human interactions, it needs to be able to reason about the interactants' goals are for an interaction and why they are trying to achieve them [11,19,22], which in turn requires understanding of the interactional context.

A system able to reason about the social-interactional context could be used to guide and train people on how to behave in a given social environment. Such an application could be beneficial, for example, to those who frequently travel to other societies and cultures for which the users might not be overly familiar with the customs, and to those who struggle with interacting with other people. In both cases, an AR system could continue to assist the users by offering guidance and suggestions when needed. Facilitating the ability of people to transition into other interactional environments could reduce the potential for conflict and insult. The availability of systems able to reason about and exploit social context would transform the nature of interactions between people and technologies, enabling computers to respect social factors in the same way people do with other people. The first step towards developing such systems is to develop a model and concepts for describing social interactions.

2.3   Representing the Social-Interactional Context

Representations of the social-interactional context should enable a computational system to identify discriminatively meaningful features and characteristics for a given social domain. We propose that development of such representations be guided by the theoretical constructs and methodologies that social scientists use to talk about and understand social interactions. The resulting representation should be one that is both in line with current theory and such that a computational system can reason from it effectively.

As people are able to learn socially situated patterns of language through the observation of others and feedback from those with whom the interact, the representation should also facilitate learning on the part of the computational systems, structuring structure descriptions of a given social domain in a form that facilitates the application of learning methods. In other words, the representation scheme must be selected in concert with the learning method.

2.4   Components

Social scientists have developed a variety of methods for analyzing how humans interact [1,2]. Our approach is inspired by Hymes' SPEAKING model [12], a framework within which the social context and the interactional context can be represented. According to this framework, each interaction is situated within a speech situation, which is a series of speech events that take place in a given physical and psychological environment. Each speech situation is made up of speech events, a series of speech acts that all focus on the same topic or achieving the same goal. A speech act is a category that describes the intended meaning of an utterance or phrase; it describes what the speakers expects the hearer to do given that he or she has heard the speech act [3].

The SPEAKING framework has eight slots with which it represents the various aspects of an interaction, or a speech situation: Setting/Scene (physical and psychological environment), Participants/Personnel, Ends (goals), Art Characteristics (form and content), Key (tone with which something is said), Instrumentalities (e.g., spoken versus written), Norms of Interaction and of Interpretation (the behaviors that general accompany some speech act), and Genres (the categories of the speech acts and speech events). SPEAKING provides a convenient framework with which to represent and reason about a social interaction at a high level, and provides a potential first step for a representational framework for social-interactional context.

2.5   Challenges Implementing SPEAKING

It is generally very easy for humans to fill in the SPEAKING framework, especially for familiar interactions and situations. However, developing automated methods for a computational system to use to fill in this framework is not straightforward. For example, it is unclear how to represent utterances within the Art Characteristics slot. As many of the speech act schemes are focused on representing task and goal oriented interactions, they do not have a way to represent utterances that seem to have the goal of developing common ground, such as, "I am also in the CS department here," or "I like your shoes. I have a pair just like them at home." While these exact utterances might not be overly common, this type of utterance tends to occur frequently when two people meet for the first time.

According to the existing speech act schemes (e.g., [5–7]), such utterances would be classified as "information giving" and, potentially, also "response seeking." However, according to this classification, these two examples would have the same type of intended meaning as an utterance such as, "The train leaves a 10 AM," which does not seem to accurately get at the intention behind the utterance.

On the other hand, there are other slots for which low level representations seem relatively straightforward to define, for which immediately available computational techniques might be applicable. *Setting* could be represented in a number of ways, such as with coordinates, surrounding environmental attributes, people who are standing within a given proximity, or some location type (e.g., a classroom or a car). People already have many words for describing a person's tone of voice. Therefore, a set of words used to describe tone of voice could be identified with the same method used to identify affective categories [23]. The *participants* can be represented in terms of their cognitive and belief states, representations of which are being developed by [10,21], plus information about the social relationships that exists within a given interaction.

2.6 Identifying Social Relationships

Identifying social relations requires relationship type identification, social hierarchy identification, and the identification of the quality of the relationship. Each of these three subcomponents can be used to describe the relationship between interactants. Social hierarchy identification describes the social status one participant in an interaction has relative to the other participant(s). Identifying the authoritative individual in an interaction further defines and constrains the behaviors likely to be present in a given interaction. Having a bounded model of the expected behavior for a given interaction facilitates identifying deviant behavior.

There is a large literature that described various norms of interaction for a variety of cultures, but this has not been described in a way that facilitates translating them into a formal representation scheme. Therefore, in order to develop lower level representation for reasoning about this slot, expertise knowledge could be leverage to develop a formal knowledge representation and the features of an interactional that could be used to apply the representation correctly. Alternatively, we have begun to explore machine learning methods for learning how to predict relative social power.

3 Computational Identification of Relative Social Power

We have begun investigating computational methods for identifying one aspect of the social-interactional context, the participants' relative power relationship. Our current target is to develop methods for identifying each participant's relative social power based on the production of linguistic features. We are studying the feasibility of using machine learning to build classifiers able to accurately predict the relative power between two speakers. Our approach develops representations of linguistic features of interactions, to provide as input to machine learning methods, and then tests which candidate representation schemes and machine learning methods have the best performance at this task.

Our initial experiments [18] compared the predictive accuracy of a set of speaker representation metrics and classifier pairings. The features were based on Pennebaker's work on function words [20]. Two methods were considered for representing the features associated with each speaker pair, one a 14-element feature vector, for which each element corresponded to a different type of function word, and another representation which derived a single value from the individual feature values. Six different metrics were then applied to compare these representations. Finally, a Support Vector Machine (SVM) and a Random Forest Classifier were trained to predict relative social power. Results were compared to determine which representation and classifier pair was able to more accurately predict the relative social power between two speakers.

The resulting relationship type classifications assigned one of two directional categories (i.e., pointing to which speaker had more power) and one non-directional category (i.e., reporting that both speakers had the same relative social power). Identifying the existence of a difference in relative social power and the direction in which the power difference points are two different tasks. It was found that in both directional cases the 14 dimensional feature vector representation enabled the most accurate predictions, however, the method used to calculate specific feature values for the best results differed. When identifying that no difference in social power existed, a feature vector representing how many of the features differed provided the best results. When identifying the direction of the relative social power, a feature vector that represented both those features that varied and direction in which they varied was the most accurate.

The experiment reported in this paper expands on the above study by examining the cross-domain effectiveness of this approach: how well classifiers trained in one interactional domain are able to predict the relative power relationship in another interactional domain. The domains examined in this experiment were the ICSI meeting domain [16], the United States Supreme Court trial domain [8], and the tutoring domain [17]. Each source contains transcripts of face-to-face interaction events each–75 for the meeting domain, 204 for the Supreme Court domain, and 54 for the tutoring domain–each of which contain a number of interactions between participants. Each domain differs according to the types of power each of the interactants have over one another and which dimension of power is the most influential.

3.1   The Interactional Domains and Data Sets

The meeting domain was analyzed using the ICSI Meeting Corpus, which represents a set of task oriented meetings with a variety of [16] participants. Participants range from undergraduate students to Ph.D. students, Postdoctoral Fellows, and Professors. Therefore, the meetings contained interactions between people enacting different social roles with different amounts of social power. This is a domain where relative social power can be measure most consistently on at least two of the three dimensions of relative social power, power from expertise and respect and power from traditional, externally imposed statuses.

The Supreme Court domain was analyzed using the Cornell Supreme Court Corpus [8], which contains the transcripts from a number of trails that were presented before the United States Supreme Court. This corpus included interactions that involved a number of people with different amounts of social power, such as Chief Justice, Justice, Prosecutor, Witness, and Expert Witness. Relative social power is primarily measured along one dimension in this domain, traditional, externally imposed status. This domain's interactions are more structured and the ways in which relative power is recognized is through formal means of acknowledgment and reference.

The tutoring domain data set was the Talkbank Tutoring Corpus [17], which includes interactions between tutor and student pairs where the goal is for the tutor to teach the student to achieve some goal, whether it be to play a new game or to solve an algebraic equation. Although these interactions do not involve a variety of different social roles, they are interesting because the relative social power is based on the dimensions of need and expertise. In this domain, externally imposed social roles are not as meaningful as the relative social power in the tutoring domain does not always reflect the relative social power outside of the tutoring domain.

Thus the domains reflect a diversity of factors determining social power. Our previous experiments provided support for the hypothesis that within a given domain, features for predicting social power can be learned automatically. However, they left open the question of whether learned results could be transferred to enable predicting social power in new domains. That question is the focus of the following experiment.

3.2 The Experiment

The experiment tested cross-domain performance for configurations using three comparison metrics for the features of different speakers. Each of the metrics directly compared the number of instances each speaker produced for each feature. The metrics were absolute difference (Delta), positive-negative scaled feature-based Euclidean distance (EuclidPosNeg), and the binary (GLTL) speaker metrics. For the positive-negative scaled feature-based Euclidean distance metric, a vector for each speaker's value for a given feature(the number times a given feature was observed) was provided as input to the Euclidean distance formula. Each feature was assigned either a positive or negative value depending on whether speaker X had more instances of a feature (positive) or whether speaker Y did (negative). The binary metric contained a sequence of ones and zeros, where one indicated that speaker X had more instances of given feature that speaker Y. The assignment of each speaker to the roles of X and Y was necessary in order to be able to compute each of the similarity metrics. In the training and test sets, each speaker pair was classified twice so that both speakers played the role of speaker X and speaker Y. This was done to validate that the assignment of a speaker to X or Y did not impact the classification results.

Each of the similarity metrics was used to train and test a Random Forest Classifier (RFC) as the task of predicting relative social power. The selection of each of these metrics and the classifier was based on their performance in Metcalf and Leake [18].

Each of the corpora was represented using each of the speaker similarity metrics discussed above, which were then used to train a RFC. In total, 54 classifiers were trained. The classification task was broken into two tasks, predicting whether or not a difference in relative social power existed and, if it did, to determine who had more power. This meant using three binary classifiers, one to determine if two speakers had the same amount of power, one to determine if speaker X had more power, and one to determine if speaker Y had more power.

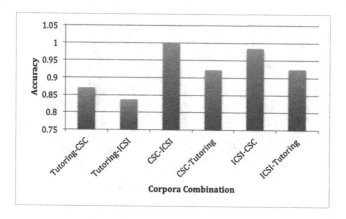

Fig. 1. The average accuracy across speaker similarity metrics with which instances where speaker X had more power than speaker Y where classified.

3.3 Results

The average accuracy across classification tasks for each training and testing corpora combination can be seen in Figs. 1, 2 and 3. Each of the metrics performed almost equally well at the task of predicting relative social power for each of the corpora combinations. Looking at the average accuracy for each of the corpora combinations for each of the tasks, it can be seen the while performance was relative similar across all of the classification tasks, there were some differences. Additionally, across all of the classification tasks it can be seen that some of the corpora combinations pairs were better able to predict one another than others. It can be seen that across the classification tasks, the Cornell Supreme Court (CSC) and the ICSI Meeting Corpora performed nearly equally well at predicting one another (accuracy = 0.94). Additionally, they both performed about equally well in predicting relative social power in the tutoring domain (CSC accuracy = 0.83, ICSI accuracy = 0.82), while the tutoring corpus nearly as well at predicting the relative social power for both the meeting and the Supreme Court domains (average accuracy = 81, CSC accuracy = 0.81, CSC accuracy = 0.80). Overall, the meeting and Supreme Court domains seem to be the most similar and the Supreme Court and tutoring domains seem to be the most dissimilar in terms of determining relative social power through function words.

3.4 Discussion

The performance of each of the domain corpora at being able to train a classifier to predict relative social power in another domain indicates similarities and differences in the types of social power that structure each of the domains. We hypothesize that the meeting and Supreme Court domains were able to accurately predict the relative social power in each other domains because their relative social power are both largely based on externally imposed statuses.

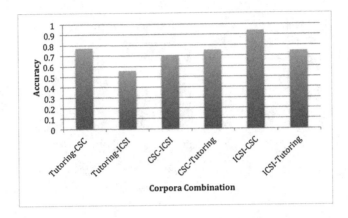

Fig. 2. The average accuracy across speaker similarity metrics with which instances where speaker Y had more power than speaker X where classified.

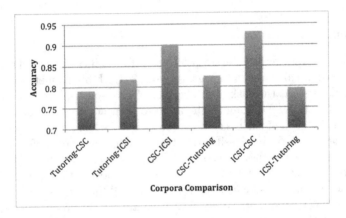

Fig. 3. The average accuracy across speaker similarity metrics with which instances where speakers X and Y had the same power where classified.

While the task-driven meeting domain, especially at an academic institution, can have a more flexible social hierarchy that the Supreme Court domain, according to expertise, and occasionally need, these shifts in relative social power are most likely temporary. The academic institution, meeting domain, while likely to be flexible within meetings, is likely to quickly revert back to its original, status-based social hierarchy immediately after the need causing a shift has been addressed. Consequently, as soon as expert offers his or her expertise, the context changes, with power shifting back to the individuals with the highest status.

The ability of the academic, institution meeting domain and the Supreme Court domain to accurately predict each other's relative social power, indicates that externally imposed statuses can play a large role that can overshadow expertise-based social power shifts.

The tutoring domain stands apart from the other two domains in terms of the accuracy with which the tutoring domain could be used to predict the relative social power of the other domains and vice versa. The differences in accuracy indicate that the task of identifying relative social power in tutoring is of a different nature than the task of identifying relative social power in the Supreme Court and academic meeting domain. A plausible explanation for this difference is that the relative social power in the tutoring domain is primarily based on the need dimension along with some influence from the expertise dimension. This suggests that there is marked difference in the nature of power that is externally imposed and, potentially, expertise-based and relative social power that is need based.

Overall, this experiment suggests that domains for which relative social power is defined along different dimensions are not the best candidates for predicting one another. In terms of developing a computational system able to reason about the social-interactional context, these results suggest that a system needs to be sensitive to nuances that differentiate different contexts. However, these nuances, or dimensions occur across interactional domains. Therefore, a system able to reason about the social-interactional context only needs as many separately trained classifiers and reasoning mechanisms as there are dimensions along which interactional contexts systematically differ. This conclusion suggestions that to identify this type of context, there does not need to be a different classifier for each interactional domain, but instead a different classifier for each category of interactional domain, where category is defined according to dimension such as need-based, expertise-based, and status-based relative social power.

4   Conclusion

Despite strong awareness of the importance of context, social context has been comparatively unexplored. This paper has argued for the (1) importance of considering the social context, both for devices meant to facilitate daily activities and for systems able to support social interactions, and (2) to enable intelligent user interfaces to perform more effectively when actively participating in an interaction. Social-interactional settings are data rich environments that can help to provide significant amounts of contextual information about the individual participants through such features as relative social power, word choice, and relationship type.

The view in this paper alligns with Kalatzis and colleagues [15] in presenting an argument for the social-interactional context as a valid and beneficial contribution to reasoning about how a device should interact with its user and its environment. However, this paper differs from their work in it proposes a social context framework that emphasizes social interactions between people and, eventually, between people and computational systems. Additionally, in our approach conclusions about non-user interactants are based on the information that is known a priori by the user or information that is directly observed by the system during an interaction. Instead of trying to select the next most likely action based on a sequence of contextualized actions, this framework aims to predict attributes about the interactants, such as their relation to one another and affect.

We have shown how the slots defined in the SPEAKING model can be used to represent features of an interaction to which machine learning methods can be applied to develop effective classifiers to predict one aspect of social context. The experiment studies the cross-task generality of the learned classifiers, providing insight into the ways in which an interactional domain is able to vary and which similarities/differences are important for predicting the relationships between interactants. The results demonstrate the specific interactional domain is not the most important factor in determining whether or not one domain can be used to accurately predict relative social power in another domain. Instead, the dimensions upon which the interactant relations vary are the most important. This suggests that it is possible to predict relative social power in a previously unobserved domain so long as the dimensions along which certain aspects of the social relations vary have been observed before.

Overall, this paper argues that reasoning about the social-interactional context is beneficial and presents support for its feasibility in intelligent systems. Next steps will be to develop methods with which to derive slot representations for features that are not straightforward and to further investigate additional types of participant relationships, beyond social power, particularly the dimensions upon which they vary and the domains with which the dimensional variations are associated.

References

1. Argyle, M., Cook, M.: Gaze and Mutual Gaze. Cambridge University Press, Cambridge (1976)
2. Atkinson, J.M., Heritage, J.: Structures of Social Action. Cambridge University Press, Cambridge (1984)
3. Austin, J.: How to Do Things with Words. Harvard University Press, Cambridge (1962)
4. Brézillon, P., Gonzalez, A.: Context in Computing: A Cross-Disciplinary Approach for Modeling the Real World. Springer, Berlin (2014)
5. Bunt, H.: The semantics of dialogue acts. In: Proceedings of the Ninth International Conference on Computational Semantics, pp. 1–13. Association for Computational Linguistics (2011)
6. Bunt, H., Alexandersson, J., Choe, J.W., Fang, A.C., Hasida, K., Petukhova, V., Popescu-Belis, A., Traum, D.R.: ISO 24617-2: a semantically-based standard for dialogue annotation. In: LREC, Citeseer, pp. 430–437 (2012)
7. Core, M.G., Allen, J.: Coding dialogs with the damsl annotation scheme. In: AAAI Fall Symposium on Communicative Action in Humans and Machines, Boston, MA, pp. 28–35 (1997)
8. Danescu-Niculescu-Mizil, C., Lee, L., Pang, B., Kleinberg, J.: Echoes of power: language effects and power differences in social interaction. In: Proceedings of the 21st International Conference on World Wide Web, pp. 699–708. ACM (2012)

9. Doolin, K., Taylor, N., Crotty, M., Roddy, M., Jennings, E., Roussaki, I., McKitterick, D.: Enhancing mobile social networks with ambient intelligence. In: Chin, A., Zhang, D. (eds.) Mobile Social Networking, pp. 139–163. Springer, New York (2014)

10. Heiphetz, L., Spelke, E.S., Harris, P.L., Banaji, M.R.: The development of reasoning about beliefs: fact, preference, and ideology. J. Exp. Soc. Psychol. 49(3), 559–565 (2013)

11. Hervas, R., Bravo, J., Fontecha, J., Villarreal, V.: Achieving adaptive augmented reality through ontological context-awareness applied to aal scenarios. J. UCS 19(9), 1334–1349 (2013)

12. Hymes, D.: The ethnography of speaking. Anthropol. Hum. Behav. 13(53), 11–74 (1962)

13. Jørgensen, M.W., Phillips, L.J.: Discourse Analysis as Theory and Method. Sage, Thousand Oaks (2002)

14. Kalatzis, N., Liampotis, N., Roussaki, I., Kosmides, P., Papaioannou, I., Xynogalas, S., Zhang, D., Anagnostou, M.: Cross-community context management in cooperating smart spaces. Pers. Ubiquit. Comput. 18(2), 427–443 (2014)

15. Kalatzis, N., Roussaki, I., Liampotis, N., Kosmides, P., Papaioannou, I., Anagnostou, M.: Context and community awareness in support of user intent prediction. In: Brézillon, P., Gonzalez, A.J. (eds.) Context in Computing, pp. 359–378. Springer, New York (2014)

16. Lamel, L., Adda, G., Bilinski, E., Gauvain, J.L.: Transcribing lectures and seminars. In: INTERSPEECH, pp. 1657–1660 (2005)

17. MacWhinney, B.: The TalkBank project. In: Beal, J.C., Corrigan, K.P., Moisl, H.L. (eds.) Creating and Digitizing Language Corpora: Synchronic Databases. Palgrave-Macmillan, Basingstoke (2007)

18. Metcalf, K., Leake, D.: Automated identification of relative social status. In: Proceedings of the Third Annual Conference on Advances in Cognitive Systems, Online Proceedings (2015). http://www.cogsys.org/conference/2015-program

19. Oh, S., Woo, W., et al.: Camar: context-aware mobile augmented reality in smart space. Proc. IWUVR 9, 48–51 (2009)

20. Pennebaker, J.W.: The secret life of pronouns. New Sci. 211(2828), 42–45 (2011)

21. Rodriguez, A., Parr, R., Koller, D.: Reinforcement learning using approximate belief states. Adv. Neural Inf. Process. Syst. 12, 1036–1042 (1999)

22. Shin, C., Kim, H., Kang, C., Jang, Y., Choi, A., Woo, W.: Unified context-aware augmented reality application framework for user-driven tour guides. In: 2010 International Symposium on Ubiquitous Virtual Reality (ISUVR), pp. 52–55. IEEE (2010)

23. Stevenson, R.A., Mikels, J.A., James, T.W.: Characterization of the affective norms for english words by discrete emotional categories. Behav. Res. Meth. 39(4), 1020–1024 (2007)

24. de Weerd, H., Verbrugge, R., Verheij, B.: Theory of mind in the mod game: an agent-based model of strategic reasoning. In: Proceedings of the European Conference on Social Intelligence (ECSI 2014), CEUR Workshop Proceedings, vol. 1283, pp. 128–136 (2014)

Neither Noise nor Signal
The Role of Context in Memory Models

Ian O'Loughlin$^{(\boxtimes)}$

Pacific University, Forest Grove, OR, USA
ian.oloughlin@pacificu.edu

Abstract. Context plays a crucial role in learning and memory, but a satisfactory characterization of this role in models of memory remains elusive. Classical and recent studies show that context cannot be meaningfully treated as either the figure or the ground, the noise or the signal, in memory models. This impasse belies certain cognitivist assumptions common to traditional cognitive science. A number of postcognitivist movements in philosophy and cognitive science have offered effective critiques of the basic framework, often borrowed in memory science, that depicts the human cognitive system as a dimensionless executive control unit receiving and transforming signals as input from the environment. These revisionary movements have also offered up alternative dynamic approaches to cognitive modeling and explanation, which can and should be deployed in memory science in order to resolve the impasse surrounding the modeling of context and memory.

1 Context and Memory

Context plays a crucial role in learning and memory. Although researchers across several domains agree that this is so, a satisfactory characterization of this role in models of memory remains elusive [1, pp. 101], [2]. In early conditioning studies, the context of a conditioning experiment was treated as a necessary evil: as a subject is conditioned to respond to a stimulus, the apparatus of the experiment also engenders responses similar to a conditional stimulus [1]. This noise in conditioning experiments eventually came to be treated as a modular stimulus in its own right, spawning a vast, albeit somewhat heterogeneous, collection of methodologies surrounding context conditioning [3,4]. In memory models, a representation of context came to play a very similar role to representations of any other target stimuli [5,6]. Recently, however, a number of studies have once more demonstrated incongruities in treating context as just another represented stimulus [7,8]. This leaves modelers at an impasse: context cannot be adequately modeled as another signal received by the learning and remembering system, but nor can it be adequately modeled as merely the noise in contrast to which such signals proceed.

This impasse, however, is predicated on an assumption about the nature of cognitive systems. This is the assumption that there is a static baseline of awareness which receives signals in a void. Everything must be either signals or noise only

H. Christiansen et al. (Eds.): CONTEXT 2015, LNAI 9405, pp. 398–409, 2015.
DOI: 10.1007/978-3-319-25591-0_29

if remembering systems are essentially zero-dimensional receivers of input, if the framework for our models is a basically computational picture of discrete phases of input, storage, and output. In cognitive science, these cognitivist assumptions are beginning to give way under their own weight. Especially in approaches within the dynamic systems family of modeling methodologies [7,9,10], as well as research trajectories informed by embodied and situated cognition [11–13], the subject as a zero-point receiver and transmitter of representations is quickly becoming obsolete [14–16]. Philosophers of mind have been debating the merits and demerits of just this framework for a century, and recent cognitive science has begun to attend to these lessons in earnest.

Although researchers in the cognitive science of human memory have not commonly been among the first to question cognitivist assumptions, whether models based on the computational architecture of encoding, storage, and retrieval can adequately serve has lately come under fire from both cognitive psychologists and philosophers [17–19]. In the emerging foundation for new models of human memory, the line between data and algorithm has been erased [20], the phases of memory processes have become continuous rather than discrete [21], and memory construction and reconsolidation have become the rule rather than the exception [22]. These developments taken together point to a plausible, albeit revisionist, resolution to the context modelers' impasse: context has resisted memory models that rely on fundamentally computationalist architectures because these frameworks insist that a stimulus is either a signal being received as "input" or the mere noise against which such signals are discerned. Relinquishing this demand may entail the dissolution of modularity and of any notion of a dimensionless "executive unit" which receives input and directs output–revisions which have been independently motivated in the philosophy of cognitive science–but the resulting frameworks will allow models to faithfully characterize the complex role that context plays in human memory.

2   Modeling Figure and Ground in the Cognitive Psychology of Memory

Although many strands still connect the two, research into *learning* and research into *memory* have been long divided into two parallel (and sometimes less parallel) trajectories in cognitive psychology. Although the primary focus of this paper is memory research, the story of the way context has come to be modeled begins with learning and conditioning experiments in the middle of the twentieth century. Early researchers in conditioning quickly discovered that the context of any given successful case of conditioning became something like its own conditioned stimulus, in addition to whatever target stimulus to which the subject had been conditioned in the context in question. Experimenters began treating the context as another CS in their models, and thus set the stage for what would become the standard treatment of context in models of human memory [1, pp. 101]. Furthermore the revisions needed to make memory models adequate for context are almost certainly analogous to needed revisions in learning models, especially since these treatments have remained more or less parallel in the divergent research traditions. Researchers who are concerned to draw conceptual connections between these, such as Mark Bouton, have made this evident:

> "In contemporary learning theory, *context* is typically defined as the background stimuli provided by the apparatus (e.g., Skinner box) in which the experiment is conducted. This definition is similar to the one used in the human memory literature, where *context* is often defined as stimuli provided by cues emanating from the room in which the experiment occurs (e.g., Smith, 1988; Smith and Vela, 2001). In either case, the context is a relatively long-duration stimulus that surrounds or embeds the target stimuli that are to be learned or remembered (whether they are conditioned or discriminative stimuli, or items in a list)." [2, pp. 233]

Memory researchers have even perhaps tended to be a bit more cautious than those modeling learning in their treatment of context. While in the conditioning literature context is often treated as a sort of necessary evil, an additional CS that always accompanies conditioning, context in the memory literature is often characterized as a *cue*, a signal that aids, perhaps crucially, in retrieval of memory. In an article from 1995–in which the cracks in the computationalist edifice are already beginning to show, and to which we shall return presently–Wesley Spencer and Raz Naftali sensibly and sensitively wrote that "The term *context* denotes all conditions and circumstances under which memory for an item was acquired." [7, pp. 528] This characterization is distinct from Bouton's in a subtle but important aspect: those researchers who treat context as "emanating cues" are essentially treating context as a signal, following the CS assignation in learning experiments, while those who follow Spencer and Naftali in treating it as the whole of the "conditions and circumstances" surrounding an encoded or retrieved memory are essentially treating context as the ground of such signals. As each of these researchers is well aware, *experimentally*, context has failed to behave well in either of these roles.

The Spencer and Naftali study demonstrated one such incongruity. It has long been known that aging has adverse effects on the memory abilities of human subjects. In their study, Spencer and Naftali showed that memory ability for context deteriorates much earlier and more severely in aging humans than does memory ability for *content*. [7] That is, aging adults show much more susceptibility to memory failure for context than would be expected were context just another stimulus that has been stored and retrieved. While semantic memory (remembering facts) and episodic memory (remembering past episodes) remain largely intact quite late in the life of the healthy remembering subject, our ability to remember the contexts associated with these retrieved memories declines dramatically long before this. It is perhaps this poor fit between context and signal that induces the authors to characterize context as "circumstances" rather than "cues", although they do not explicitly recommend any particular theoretical revisions.

Michael Fanselow has more recently demonstrated another difficulty in treating context as a memory item that is stored and retrieved. The habituation and extinction rates for remembering and forgetting in a large variety of circumstances have been thoroughly studied and documented, including basic conditioning experiments. Fanselow has found that habituation and extinction rates for *context* conditioning are not only *divergent* from what would be expected, but indeed in some cases they are *inverse* [1,8]. Fanselow's study showed that although, as is well known, the effectiveness of an instance of classical conditioning is inversely proportional to the delay between US and CS (at least above a very brief threshold, the role of which is controversial), the effectiveness of context conditioning is emphdirectly proportional, for a number of cases, to the delay between the introduction of the context "stimulus" and unconditioned stimulus. Rats introduced to a context "stimulus" *simultaneous* with, or directly preceding, unconditioned stimulus, exhibit no conditioning whatsoever to the context "stimulus", despite the fact that simultaneous and directly preceding instances of CS are standardly what produce the most reliable conditioning. Once more, treating the context as a stored, remembered item in the fashion of a conditioned stimulus or environmental cue seems directly contravened by empirical results. These are not the only difficulties faced by researchers who have tried to model context within a framework of stored and retrieved items; context is sometimes assigned multiple roles, or multiple neural bases, to try to resolve these seeming inconsistencies [5, pp. 2431].

In traditional models of human memory, it can be difficult to know how else context could be modeled. Although the details vary, the basic framework in memory research is to model human remembering on the encoding, storage, and retrieval of memory traces.[1] That is, the human rememberer is taken be fundamentally analogous to an input/output device; memories are stored data, that has been encoded, that can be transformed into various outputs under certain conditions. Stored memory traces are taken to *represent* the fact or episode that will later be recalled, and treating context as a stored item is just to treat it as a stored representation.[2] Many attempts in cognitive psychology and neuroscience to understand the role of context in memory involve an attempt to better understand the location and structure of the stored representation of the context. Fanselow as well, when describing the context conditioning in his footshock experiments, writes:

> "This exploration allows the animal to store an integrated representation of the context and once that integrated representation is formed exploration decreases. This contextual representation is also necessary

---

[1] It is true that there has been some revisionary *language* in memory science lately: 'persistence' rather than 'storage', reconsolidation instead of 'consolidation', but I have argued elsewhere that these changes–although indicative of an admirable spirit– have so far been little other than cosmetic [19].

[2] Matthew Sanders and colleagues make this explicit, writing: "In order for context conditioning to proceed, the various stimuli in the environment must be associated with one another as the context. Therefore, we have proposed that one function of the hippocampus is to assemble a contextual representation..." [6, pp. 220].

for contextual fear conditioning, so the animal must be given sufficient time to explore the chamber prior to shock delivery." [8, pp. 79]

But the notion that the human mind is an engine primarily in the business of the storage and transformation of encoded representations is a product of the 20th-century computationalism that has been fraught with basic challenges of late. In certain areas of psychology, cognitive science, and the philosophy of mind a number of developments have provided good reason to think that this picture of cognition as the computational transformations of representations is hopelessly inadequate to the data and conceptually problematic.

3 The Remembering Subject in Postcognitivist Philosophy of Mind

In recent decades, a number of researchers have successfully brought postcognitivist considerations to bear on topics in cognitive science, based on the work of several early-twentieth century philosophers. Some among these are Hubert Dreyfus, who has been applying Wittgensteinian and Heideggerian ideas to work in artificial intelligence [23], Alva Nöe, who has been fruitfully applying ideas from Merleau-Ponty and others to work on the cognitive science of perception [24], and Daniel Hutto, who has been utilizing certain insights from Wittgenstein and others in revisionist critiques of psychological explanation [25], as well Peter Hacker and Maxwell Bennett, who have devoted much, albeit controversial, work to applying Wittgenstein, critically, to cognitive neuroscience [26]. The resulting Heideggerian, Wittgensteinian, Merleau-Pontian, postcomputational, postcognitivist view of mind and cognition is far from complete or uncontroversial. These revisions have gone in different directions, among which there are inconsistencies, conceptual difficulties, and uncharted territories. Nonetheless, there are lessons to be taken from this developing characterization of mind that may help to make sense of the troublesome role that context is proving to play in models of human memory.

First, as others have noted, the postcomputational mind is not amenable to the bright line between data and algorithm that our models once drew [27, pp. 97]. This is a point that is especially emphasized by those postcognitivist researchers who owe at least moderate allegiance to connectionism and neural networks.[3] The jury is still out as to whether it makes sense, in any contexts, to talk about a neural network "representing" or "storing" anything at

---

[3] Connectionism, of the medium-strength variety, is the endorsement of neural network architectures as robustly informative for those who would study minds. That is, there are few who doubt that neural network architectures have *something* to do with cognition, a least at the "subsymbolic" level, and there are few who think that neural network architectures have already solved all of our puzzles about cognition, but many among the postcognitivist vanguard stand on the middle ground here, urging that cognitive scientists and philosophers of mind have real and revisionary lessons to learn by paying close attention to the way that neural networks actually function.

all. What is certain is that if such networks can rightly be said to store information, this must be in a very different sense (perhaps of both terms:'store' and 'information') than we are used to employing to describe traditional computational architectures and processes. Since the changing weights in a neural network act as both process and product, any "information" in the network cannot be separated from the processes acting on said information, not even in principle. The notion of representation seems even worse off in this regard. There has not yet been a convincing case presented for how any part or process of a neural network can rightly be said to represent anything at all, but even if such an account were available, it would almost certainly bear little resemblance to the inert, encoded, "stored representations of context" above. A model of human memory that respects the connectionist dissolution of the data/algorithm distinction must not treat context as an item that is stored and accessed in the processes of memory.

Another distinction that has become notoriously blurry in postcognitivist frameworks for cognition is that between subject and environment. This erasure is most well known in its guises as embodied, embedded, enactive, and extended cognition, but these branches stem, in part, from philosophical considerations about the nature of the situated subject. Classical computationalism models the human cognitive system as a dimensionless, floating, executive function that stands apart from the environment it contingently inhabits, receiving input and directing output at a distance, as it were. There have now been many arguments, from many different traditions, deployed against the possibility of such distance between subject and world, but one well known example comes from Wittgenstein. Wittgenstein seeks to convince us that once we frame the problem in terms of a gap between mind and world, *we will never be able to cross it*. Wittgenstein's grocer, upon being asked for five red apples, could not successfully recognize which apples were red on the basis of an inner mental image of redness unless he had first recognized which inner mental image was red on the basis of something else [17,28]. I take it that at least some instances of the "4E" movement in cognitive science (especially among those under the rubrics of the first three E's) are explicit attempts to resolve the difficulty framed by Wittgenstein. Notably, these are of course not attempts to resolve the difficulty by *bridging* the gap in question, but rather by obviating the apparent need for an explanation that bridges this gap. The embedded, enactive mind is already in and of the world, and so there is neither a narrow nor wide gap to be bridged in our explanations. This is especially important for models of context, since the traditional treatment of context in memory models *just is* the attempt to separate out the environment from the subject and salient stimulus in question.[4]

---

[4] Although I am here taking my philosophical cue from Wittgenstein, this point brings to mind some particularly *Heideggerian* critiques of standard cognitive science, given Heidegger's insistence that we are always already involved in the world, as Dreyfus often points out [23]. An excellent treatment of this Heideggerian/Wittgensteinian point is a recent monograph by Lee Braver [16].

Both of these developments, the erasure of the dividing line between data and algorithm, as well as the stretching, complicating, and blurring of the lines that separate subject from environment, are indicative of a more general tendency in postcognitivist philosophy of mind and cognitive science: a cautious and critical attitude toward modularity and informational encapsulation. This can be expressed as skepticism about whether cognitive phenomena can or should be explained in terms of "information processing"–once the unchallenged currency of inquiry in cognitive science. Even setting aside the notorious difficulties in getting from syntax to semantics (or, set in another key, from Shannon information to meaningful content), information processing explanations are fundamentally modular ordeals. A common theme among the approaches that seek to revise or reject the basic tenets of classic computationalism is the rejection of just this fundamental, encapsulated modularity: whether we are taking the Watt governor, the charts of expert navigators, or perceptrons as a starting point for understanding cognitive systems, our explanations and models will be relentlessly continuous, dynamic, and interdependent.[5]

This leaves those of us who would model cognitive phenomena in a new and difficult position: these models have traditionally been cast in terms of processes and (conceptually isolated) products, and they have taken the target to be a cognitive subject in an environment, receiving stimuli and performing operations on them as such. If these elements have been proscribed, one might wonder whether whatever is left can even go under the name *model*. This is a reasonable concern, but fortunately the business of dynamic, non-modular, fundamentally embedded and interdependent models has already gotten itself off to a solid start.

4   Modeling Without Modularity in Learning and Memory

The role played by context in contemporary models has been fraught with internal and external inconsistencies. Treating the "context" of a memory as a stored, discrete item that can be transformed by computation-like processes has not worked. Treating the context of a memory as though *it is not* one of the stored stimuli affecting the memory process has not worked any better. That is, although many elements of memory models demonstrate empirical tendencies that only imperfectly fit the encoding-storage-retrieval framework for these models, context, in particular, is the problem child. The resources traditionally available to cognitive modeling just seem inapt for whatever role context is playing. A mark of the recent postcognitivist movements in cognitive science and philosophy is the rejection of the computation-inspired modeling techniques that seem to frame exactly this problem. These movements have been spurred

---

[5] This is not to say that informational encapsulation has no place in models, or that a given explanans may not fruitfully be broken up into component "systems"–rather it is something like a shift in the burden of proof, a consistently suspicious attitude toward over-modularizing that which can be done in continuous and dynamic terms.

forward by similarly ill-behaved empirical cases from other research domains already, and it is reasonable to think that letting go of cognitivist assumptions about modularity, processes and products, and subjects and environments, may resolve the otherwise intractable difficulties that memory researchers continue to encounter in trying to fit the square nature of context into the round holes of their explanatory models. New models and explanatory resources have begun to resolve impasses in other areas of cognitive science; there are good reasons, as evinced above, to think that context is a *particularly* problematic candidate for classically cognitivist models. Yet it can seem that without the usual toolkit of information, modular systems, and an executive subject that performs processes on stimuli received from the environment, we are simply not left with a model–or, for that matter, perhaps not even with an explanation.

This is not a trivial or shallow concern. For many cognitive phenomena, we have a deeply felt need for the revelation of an underlying process made up of component parts and processes, if we are to have an explanation of these phenomena at all. The ability to *model* a phenomenon at all is sometimes taken *to be* the ability to underwrite a phenomenon by charting out the involved information processing in terms of modules and processes. This identification, however, should be resisted. There are certainly explananda that merit mechanical, subcomponent, information-processing explanantia, but it is carelessly myopic synecdoche to treat modeling *per se* as identical to this. Explanations come in many forms.[6] Fortunately, we need not unravel the ultimate nature of explanation before moving forward with revised frameworks for capturing context in a model of human memory. Explanations and models that fit our desiderata are abundant nearby, and these can serve as existence proofs that what we seek to do in constructing these postcognitivist models can, in fact, be done.

The models and explanations that I have in mind, of course, are predominantly those that hail from the "dynamic systems" family of methods and resources. Researchers in the dynamic systems tradition are interested in nothing if they are not interested in models–and often, in models of cognitive phenomena–yet these models persistently eschew the computational trappings that postcognitivist philosophy and science have rejected. That is, the dynamic systems model is built on continuous, interdependent mathematical tools that frame elements which co-vary, co-consist, and develop continuously in time. Patterns are studied even without being conceptually isolated, elements of the environment are modeled without separating them from a dimensionless executive control unit; any process/product separation questions are simply not well-formed queries of a dynamic model. Nor are stimuli standardly treated as "signals" that have been received or retrieved. These models' primary business is to avoid such things and yet maintain explanatory power and rigor.

John Spencer, Michael Thomas, and James McClelland compiled and edited an excellent volume of such models recently, the target domain for which was

---

[6] Indeed, adequately characterizing just what explanation consists in is a notorious enough difficulty that I take it that any identification of explanation and information processing-style models can only be a tacit, uncritical move.

human development [10]. The dynamic systems and connectionist researchers represented in this volume tackle a number of complex phenomena associated with development, and in each case succeed in building a dynamic and explanatorily powerful model without running afoul of postcognitivist insights. The special role allotted, explicitly and implicitly, to context in dynamic models of child development is nicely characterized by Paul van Geert and Kurt W. Fischer in a summary article within the Spencer et al. volume, who encourage us to follow the dynamic models and treat person and context together as a quasi-unified soft assembly:

> "If we then speak about the development of the child, we actually refer to long-term changes in the child and automatically also to the corresponding long-term changes in the child's contexts. These changes refer primarily to patterns of correlation over time (that is, in a single person's life trajectory) and not to identifiable entities. That is, development refers to a person-context assembly throughout the life span; contexts are no longer to be seen as independent variables or circumstances in which the person can be placed at pleasure, according to the whim of a researcher who treats such contexts as independent variables, the effect of which has to be estimated over many subjects" [29, pp. 327].

The dynamic development models summarized by van Geert and Fischer tend toward either the very narrow time scale or the very broad, with target phenomena usually taking after either real-time development and reaction to ongoing stimuli or taking after longitudinal, long-term studies of development in the sense of childhood development. These success stories thus flank the usual time scales of memory studies, which tend to fall somewhere in between. This is a challenge for the memory scientist who would model context using dynamic systems resources, but there is no reason to think that the challenge cannot be met. After all, the long-term development studies modeled by the researchers in the Spencer et al. volume are already operating at a remove from real-time interactions. Memory scientists will, in most cases, need to follow suit, building dynamic models based on data extracted from large numbers of remembering cases across time. In such models, *context* will be neither a stimulus nor the unassuming background against which stimuli are discerned, but will follow the soft assembly framework van Geert and Fischer describe. The line one draws between a system and its environment, in a dynamic model, is as specific or nonspecific as the line-drawer pleases. This also obviates the need for memory scientists to decide whether 'context' refers solely to external stimuli, solely to *internal* stimuli, both, or whether the term needs subscripts: memory scientists are well aware that context is presently used, sometimes imperiously, in each of these ways [30].

This also presents a challenge for dynamic systems modelers. As Karl Newell and colleagues point out, differences in time scales provide real, but not insurmountable, conceptual difficulties for dynamic system modelers [31]. The dynamic systems research program is still very much a new endeavor. Given also the fact that it is interdisciplinary in the in the truest sense–it exists *between* disciplines, in methods and approaches that do not properly belong to any of the many disciplines to which these have been applied–one can expect that applying these methods to new cases will produce some puzzlement. For example, *declarative* memory studies rely on linguistic data, which in turn require clever and still controversial techniques to model using dynamic systems resources. Human language *really is*, of course, a largely discrete, modular, information-laden endeavor, and hence experiments dependent on linguistic data can sit uneasily with the emerging dynamic systems paradigm.[7] This obstacle is easily avoided– or at least delayed–though, by choosing non-linguistic data as explananda to be modeled dynamically within memory science. These are easy enough to come by, even in studies of human memory: non-declarative memory has a rich array of experimental data, and even "declarative" varieties of memory like semantic and episodic memory are often the subject of studies that do not furnish any linguistic *data*. Indeed, dynamic and connectionist models of neuronal function have been common for some time, but even when researches treat hippocampal networks in terms of attractor dynamics, these very same researchers still tend to speak in terms of *stored representations* when they ascend to the cognitive level to discuss context memory [32,33]. One exception is the recent work of Guoqi Ling and colleagues, who have designed an attractor network model by working backward from a novel energy function *designed* for cognitive plausibility, rather than basing the model on neural dynamics and then encountering an impasse when confronting cognitive-level phenomena [34]. The modeling framework presented by Ling *et al.* has not yet been applied to context in particular, but their characterization of reconstructable (*and nonbinary*) "memory patterns" as stable equilibrium points of dynamical attractor networks already invites a context-rememberer soft assembly.

So a way forward for capturing context in our explanations of human memory has become clear and equipped with reasonably successful resources. Context does not fit the standard computational models that cognitivist frameworks have traditionally offered to memory scientists. Researchers have attempted to tweak these models, but it is clear that tweaking is insufficient. The problematic bad fit between target phenomena and modeling framework makes a significant revision seem all but inevitable, and happily the postcognitivist front in philosophy and cognitive science provides significant resources for the endeavor. There are obstacles and difficulties on all sides–not just because of the intermediate time scale involved in most memory studies, but also because memory science itself does not yet have the language to characterize its phenomena without implicitly invoking computationalist elements like encoding, storage, and retrieval. Furthermore, an interesting subsection of human memory revolves around linguistic abilities, which are resistant to dynamic modeling. Nonetheless, we have

---

[7] This is not to suggest that language-involving cognitive phenomena are necessarily beyond the ken of dynamic systems modeling techniques, only that there is good reason to expect successful models of these to be among the late-comers to the dynamic systems table.

good reason to try to construct significantly revised modeling frameworks for human memory and the clearly crucial role that context plays in these processes according to these dynamic modeling techniques. Memory models at the cognitive level need not build on computationalist foundations. This can avoid the problematic treatment of context as "content" that we find in work like that of Spencer and Naftali above. We already have many of the needed resources available, even if the initial dynamic models of human remembering phenomena are constrained to non-declarative cases, or to non-linguistic studies of semantic or episodic memory.[8] In thus relinquishing the trappings of the cognitivist framework, context–which is so clearly fundamental to the operations of human remembering–may be adequately modeled by those who seek to explain human memory.

References

1. Fanselow, M.S.: Context: whats so special about it. In: Science of Memory: Concepts, pp. 101–105. Oxford University Press, New York (2007)
2. Bouton, M.E.: The multiple forms of context in associative learning theory. In: The Mind in Context, pp. 233–258. Guilford Press, New York (2010)
3. Smith, S.M.: Context: a reference for focal experience. In: Science of Memory: Concepts, pp. 111–114. Oxford University Press, New York (2007)
4. Myers, C.E., Gluck, M.A.: Context, conditioning, and hippocampal rerepresentation in animal learning. Behav. Neurosci. **108**(5), 835 (1994)
5. Matus-Amat, P., Higgins, E.A., Barrientos, R.M., Rudy, J.W.: The role of the dorsal hippocampus in the acquisition and retrieval of context memory representations. J. Neurosci. **24**(10), 2431–2439 (2004)
6. Sanders, M.J., Wiltgen, B.J., Fanselow, M.S.: The place of the hippocampus in fear conditioning. Eur. J. Pharmacol. **463**(1), 217–223 (2003)
7. Spencer, W.D., Raz, N.: Differential effects of aging on memory for content and context: a meta-analysis. Psychol. Aging **10**(4), 527 (1995)
8. Fanselow, M.S.: Contextual fear, gestalt memories, and the hippocampus. Behav. Brain Res. **110**(1), 73–81 (2000)
9. De Bot, K., Lowie, W., Verspoor, M.: A dynamic systems theory approach to second language acquisition. Bilingualism Lang. Cogn. **10**(1), 7 (2007)
10. Spencer, J.P., Thomas, M.S.C., McClelland, J.L.: Toward a unified theory of development. In: Spencer, J.P., Thomas, M.S.C., McClelland, J.L. (Eds.) Connectionism and Dynamic Systems Theory Re-Considered, pp. 86–118. Oxford University Press, Oxford, England (2009)
11. Hutchins, E.: Cognition in the Wild. MIT press, Cambridge (1995)
12. Shapiro, L.: Embodied Cognition. Routledge, London (2010)
13. Michaelian, K., Sutton, J.: Distributed cognition and memory research: history and current directions. Rev. Philos. Psychol. **4**(1), 1–24 (2013)

---

[8] As characterized above, much of these experiments take the form of tracking the help and hindrance of "context cues" for declarative memory–simple context cue experiments such as those that study positive or negative intereference in semantic memory may be amenable, for example, in the same way that "cognitive control" was found to be in a recent study by Todd Braver [35].

14. Bringsjord, S.: Computationalism is dead; now what? J. Exp. Theor. Artif. Intell. **10**(4), 393–402 (1998)
15. Chemero, A.: Radical Embodied Cognitive Science. MIT press, Cambridge (2011)
16. Braver, L.: Groundless grounds: a study of Wittgenstein and Heidegger. MIT Press, Cambridge (2012)
17. Stern, D.G.: Models of memory: Wittgenstein and cognitive science. Philos. Psychol. **4**(2), 203–218 (1991)
18. Brockmeier, J.: After the archive: remapping memory. Cult. Psychol. **16**(1), 5–35 (2010)
19. OLoughlin, I.: The persistence of storage: language and concepts in memory research. In: European Perspectives on Cognitive Science (2011)
20. Houghton, G., Hartley, T., Glasspool, D.W.: The representation of words and non-words in short-term memory: Serial order and syllable structure. In: Models of Short-term Memory, pp. 101–127 (1996)
21. Slotnick, S.D., Dodson, C.S.: Support for a continuous (single-process) model of recognition memory and source memory. Mem. Cogn. **33**(1), 151–170 (2005)
22. Michaelian, K.: Generative memory. Philos. Psychol. **24**(3), 323–342 (2011)
23. Dreyfus, H.L.: What Computers Still Can't Do: A Critique of Artificial Reason. MIT Press, Cambridge (1992)
24. Noë, A.: Action in Perception. MIT press, Cambridge (2004)
25. Hutto, D.: Folk psychology as narrative practice. J. Conscious. Stud. **16**(6–8), 9–39 (2009)
26. Bennett, M.R., Hacker, P.M.S.: Philosophical Foundations of Neuroscience. Blackwell Publishing, Oxford (2003)
27. Clark, A.: Mindware: An Introduction to the Philosophy of Cognitive Science. Oxford University Press Inc, Oxford/New York (2013)
28. Wittgenstein, L.: Philosophical Investigations. John Wiley & Sons, New York (2010)
29. Van Geert, P.L.C., Fischer, K.W., Spencer, J.P., Thomas, M.S.C., McClelland, J.: Dynamic systems and the quest for individual-based models of change and development. In: Toward a Unified Theory of Development: Connectionism and Dynamic Systems Theory Reconsidered, pp. 313–336 (2009)
30. Eich, E.: Context: mood, memory, and the concept of context. In: Science of Memory: Concepts, pp. 107–110. Oxford University Press, New York 2007
31. Newell, K.M., Liu, Y.T., Mayer-Kress, G.: Time scales in motor learning and development. Psychol. Rev. **108**(1), 57 (2001)
32. Brunel, N.: Hebbian learning of context in recurrent neural networks. Neural comput. **8**(8), 1677–1710 (1996)
33. Wills, T.J., Lever, C., Cacucci, F., Burgess, N., O'Keefe, J.: Attractor dynamics in the hippocampal representation of the local environment. Science **308**(5723), 873–876 (2005)
34. Li, G., Ramanathan, K., Ning, N., Shi, L., Wen, C.: Memory dynamics in attractor networks. Comput. Intell. Neurosci. **2015**, 1–7 (2015)
35. Braver, T.S.: The variable nature of cognitive control: a dual mechanisms framework. Trends Cogn. Sci. **16**(2), 106–113 (2012)

Context Dependence, MOPs, WHIMs and Procedures

Recanati and Kaplan on Cognitive Aspects in Semantics

Carlo Penco[✉]

Università degli Studi di Genova, Genoa, Italy
penco@unige.it

Abstract. After presenting Kripke's criticism to Frege's ideas on context dependence of thoughts, I present two recent attempts of considering cognitive aspects of context dependent expressions inside a truth conditional pragmatics or semantics: Recanati's non-descriptive modes of presentation (MOPs) and Kaplan's ways of having in mind (WHIMs). After analysing the two attempts and verifying which answers they should give to the problem discussed by Kripke, I suggest a possible interpretation of these attempts: to insert a procedural or algorithmic level in semantic representations of indexicals. That a function may be computed by different procedures might suggest new possibilities of integrating contextual cognitive aspects in model theoretic semantics.

Keywords: Semantics · Pragmatics · Context dependence · David Kaplan · Francois Recanati · Indexicals · Procedures

1 Context of Utterance, Truth Conditions and Cognitive Significance

Indexicals are the prototypical examples of context dependent expressions.[1] Frege introduced the idea of context of utterance as a condition for interpreting what is said by a sentence:

> "[…] If someone wants to say today what he expressed yesterday using the word 'today', he will replace the word with 'yesterday'. Although the thought is the same, its verbal expression must be different, in order that the change of sense which would otherwise be effected by the different time of utterance may be canceled out" (Frege 1918, p. 64).

On this view, depending on the context of utterance, two different sentences may express the same sense. Kripke (2008) challenges this point of view; he remarks that

---

[1] Indexicals are expressions like "I", "Here", "Now", "Today"; belonging to different syntactic categories they are typically considered a semantic category characterized by context dependence and perspectival aspects (Perry 1997; Neale 2007). According to some authors they constitute the "basic set" of context dependent expressions, according to others they are just a case among a more general context dependence of lexical items (For a survey see for instance Domaneschi et al. (2010), Domaneschi and Penco (2013)).

© Springer International Publishing Switzerland 2015
H. Christiansen et al. (Eds.): CONTEXT 2015, LNAI 9405, pp. 410–422, 2015.
DOI: 10.1007/978-3-319-25591-0_30

this passage raises a problem with the compositionality principle: the expressions "Today" and "Yesterday" have different linguistic meanings, therefore, Kripke assumes, different senses. If we have two sentences "Today is F" and "Yesterday was F" uttered the following day, the two sentences should express different thoughts, given that the sense of a sentence is composed by the senses of the parts and the rules of composition; and if the senses of the parts are different, they *should* express *different* thoughts. Yet Frege – apparently against his own principles –claims that the two utterances express the *same* thought.[2]

A way out in this difficulty may come from a rational reconstruction[3] of Frege's ideas given by Wolfgang Künne (2007): we have to distinguish between "thought" (the metaphysical truth conditions of an utterance in a context) and "ways of articulating the thought" (the epistemological-cognitive aspects of an utterance). If we apply this consideration to the Fregean quotation given above, we find therefore an "easy" solution to Kripke's worry: the utterances "Today is F" and "Yesterday was F" said the subsequent day are two different ways to articulate the same truth conditions: these two utterances express the same "semantic" sense, what would be for Frege an eternal thought, true independently of time, if true at all.[4] Also in a Kripkean semantics, the two utterances should have the same truth-values in all possible worlds (in which *that* day exist). It seems therefore possible to accept Frege's claim that the two sentences express the same thought, if we regard the truth-conditional notion, inspired by Frege's ontological worries of eternal thoughts.

Still the two utterances have different cognitive significances, and the problem remains on how to connect the cognitive aspect and the truth conditional representation. Different answers have been given to the following question: how to treat the cognitive aspect of thought and of language processing inside a framework of truth-conditional semantics? How can we accommodate cognitive aspects in model theoretic semantics since these aspects are exactly those features that model theoretic semantics is designed to ignore? Context dependent expressions are a fundamental test case for this problem.

---

[2] Dummett (1989) claims that, given indexicals and other context dependent expressions, Frege's claims about the sense of a sentence should be translated in claims about the sense of an utterance in a context. We should accordingly reformulate the so-called Fregean "context principle" (the meaning of a word depends on the sense of the sentence/utterance in which it appears).

[3] Künne's reconstruction follows the acknowledgment of the presence in Frege's works of two different trends concerning the concept of sense: (i) the ontological or *semantic viewpoint*, that is centered in the definition of sense as truth conditions in *Grundgesetze* §23 and has been developed by the semantic tradition after him. (ii) the epistemological or *cognitive viewpoint* that is centred on the definition of sense as informative or cognitive content starting from the essay "Über Sinn und Bedeutung". See also Penco (2013) on the difference between Kripke and Künne.

[4] Frege's ontological worries are connected to his idea of eternal truth conditional thoughts; a sentence together with different aspects of the context of utterance (time, location, and speaker) may express different thoughts, but the content of each utterance is eternally true, if the truth conditions are satisfied. This is the basic Fregean semantic and ontological stance against the idea of a "minimal proposition" expressed by a sentence and varying in truth and falsity depending on context. See Dummett (2006), p. 12.

Traditionally direct reference theory has chosen to separate semantics and cognitive significance; recently there has been a change in this perspective, given by two of the main paradigms working inside a direct reference framework: Francois Recanati's truth conditional pragmatics and David Kaplan's semantics (revisited). Recanati (2012, 2013) and Kaplan (2012) give new suggestions on cognitive significance using, respectively, modes of presentation (MOPs) and ways if having in mind (WHIMs). In what follows I will discuss indexicals as a case study for the treatment of cognitive aspects in semantics.

2   Attempts to Find a Unified Treatment: Mops vs Whims

When we use indexicals we are dealing with singular thoughts, thoughts about an individual entity (be it a time, a place or a person). Recanati (2012) applies an idea (suggested by Evans and McDowell) according to which we may treat singular thoughts as strictly depending on the objects they are about; the sense of an indexical is not a description but a *non-descriptive* mode of presentation (MOP).[5] Recanati distinguishes linguistic and psychological MOPS, where the linguistic ones are similar to what Kaplan calls "character" and correspond to the *linguistic rule* encoded by the expression, while the psychological ones are what is activated by characters, and may be considered *cognitive constraints* on the rational subject.

Psychological MOPS have a role in thought as *Mental Files*. Mental files are mental counterpart of indexicals and other singular terms[6], and their relation to the objects they are about is not based on the information they may contain, but on the direct relation they have with the object. The structure of psychological mental files is mapped on the structure of the indexicals: at the linguistic level an expression type like "Today" encodes a linguistic rule that, in a context, connects the corresponding token to the referent (the day of the utterance); at the psychological level the mental indexical corresponding to the token of "Today" has the function of storing information derived by the context of utterance. The thought contains the mental file itself, as "vehicle", not necessarily the information that can be stored in it. If I say "Today is F" I will have a mental file where to store information of different kinds, and tomorrow I will *connect* the stored information using the mental file "Yesterday".

Recanti would consider our problem of "Today is F" and "Yesterday is F" in a way that resembles Künne's move, but with some specification and different terminology. Following our example we might say that the utterances of "Yesterday is F" and "Today is F" – entertained with two different linguistic MOPs and two different psychological MOPS – express two different thoughts, although expressing the same singular proposition (with the same truth conditions). The two different psychological MOPs or mental

---

[5] Frege himself, speaking of sense, did not give only examples with definite descriptions, but also of senses as chains of communication beginning with an initial baptism: think of the example of a mountain discovered by two travellers from two different routes and called "Afla" by one and "Ateb" by the other one.

[6] I will not treat here Recanati's view on definite descriptions, on which there are peculiar problems discussed by Vignolo (2012).

files could be coordinated when the subject realizes that the two mental indexicals are connected each other. The information stored in one file can be stored also in the other temporary mental file. In case of loss of memory of a speaker, we may have some interesting problems. Recanati (2012) (pp. 179–182) takes the example presented by Perry on Rip van Winkle, a person waking up after twenty years of sleeping. The sentence "Yesterday was F" uttered (in the same day) by a normal speaker and by Rip van Winkle would express two *different truth conditions*, because the former (uttered by a normal speaker) will be true if the day preceding the day of utterance was F, and the latter (uttered by Rip van Winkle) will be true if the day twenty years ago was F. But here, given Rip's mentally referring to a day twenty years before, the linguistic mode of presentation contrasts with the psychological mode of presentation. The objective content of the utterance of "Yesterday was F" is different from what Rip van Winkle actually refers to.

In a strong interpretation of the theory, no two speakers can share the same thought, given that no speaker can have the same psychological MOPs. A first example is given in the case of EGO-files. Partly following Perry (2000), Recanati (2013) (pp. 165–167) says that in a sentence with an indexical, like "I am F", the indexical "I" expresses the same linguistic MOP and two different psychological MOPs, one for the speaker and the other for the hearer. The hearer cannot have the same psychological MOP because she cannot entertain the thought *as* "I am F". Therefore, Recanati concludes, we have here *two* thoughts that have the *same* truth conditions for hearer and speaker, but differ for their non-descriptive psychological MOPs[7].

But what would happen with other indexicals? Given that every indexical thought is ego-centered, although a linguistic MOP-token may be shared by many speaker, its correspondent psychological MOP-token will be different for each person. This strong interpretation has the shortcoming of multiplying thoughts beyond necessity. Although the theory of mental files as indexicals is a very nice attempt to keep together truth-conditional and cognitive aspects, the risk of multiplying entities should be carefully considered. On the one hand Recanati seems to be compelled to multiply mental files in different species: every *epistemically rewarding relation* will activate different mental files of different kinds: demonstrative files, with perceptual MOPS, memory files with memory MOPS, recognitional files with recognition MOPS, and - at the two sides of his classification - proto-files and higher order files (or encyclopedia entries). On the other hand thoughts themselves would increase in number: each singular occasion or context of utterance may produce a specific thought depending on the activated mental file.

On the first multiplication of entities, the multiplication of mental files, Papineau (2013: 167 ff.) remarks that a perceptual file disappears when the epistemic rewarding perceptual contact disappears; for this reason Recanati is obliged to multiply kinds of files and epistemic rewarding relations: when a perceptual file is closed we have a memory rewarding relation that opens a memory file, and so on. Papineau suggests a

---

[7] This is also a solution of Frege's idea that a speaker gives a unique sense to the expression "I"; in a similar vein Kripke (2008: 215) suggests an utterance with the indexical "I" expresses a thought which can be *thought* or *had* only by the speaker himself but may be *understood* by a hearer who apparently «knows what type of thought is being expressed».

simpler view where the mental file, activated by a perceptual relation, outlasts the original encounter and is reactivated when remembered or re-encountered. Once opened, files become therefore permanent repository of information about the item in question.

Although this repair may help to solve some shortcomings of Recanati's view, we have to face the problem of the second kind of overabundance of entities, the multiplication of actual thoughts depending on contexts of utterance. To every use of an indexical there should correspond a psychological MOP that opens a mental file, making therefore a new thought for every occasion of utterance. Papineau's main doubt is the tendency to infer, from the use of indexicals to express a thought, that the thought itself must be similarly indexically structured. Although this criticism seems to hit the target, on the ground that "there seems no rationale for requiring that every epistemically rewarding relation generates its own file" (Papineau 2013: 171–172), yet we have to be careful, given the ambiguity of the terminology about thoughts that we have disambiguated using Künne's distinction: if we refer to *truth conditional thoughts* we have the standard case in which the same type of sentence uttered in different contexts has different truth conditions ("Today is F" is true depending on which day is uttered; "I am tired" is true depending on the time I utter the sentence, and so on). In these cases the thought does not have any indexical structure, because its truth conditions are fixed to the context of utterance and – using Frege's ideas – there is only one eternally valid thought (although Recanati would prefer to speak of the same singular proposition entertained). If we refer of *ways of articulating a thought*, the implicit suggestion is that *different* psychological MOPS, linked to different mental files, may be all connected to the same singular proposition with the same truth conditions (as in "Today is F" and "Yesterday was F"); this multiplication of mental files might be a correct rendering of the different ways of articulating a thought, where - in Recanati's view - there is always the possibility for information to "flow" among different files connected to the same source.

Recanati distinguishes *thought vehicles* and *thought contents*, and we might say that mental files as thought vehicles are ways of grasping thought contents. Pagin (2013) claims that it is not clear how the idea of thought vehicle can match the idea of mode of presentation. I think however that this criticism might be overcome remarking that we may use different vehicles of thought, different non-demonstrative MOPs to refer to the same thought content, or individual concept (maybe intended formally as a function from possible worlds to extension)[8].

Speaking of concepts as functions, we don't have yet a clear description of a possible logical form that helps formally representing mental files. On the one hand Recanati tries to give a mental counterpart of a linguistic analysis, on the other hand this mental counterpart needs to be expressed in a truth conditional semantics or pragmatics, inserting the cognitive aspects in the formal treatment of the working of language and linguistic communication. Eventually Recanati remarks that the standard Perry-Kaplan

---

[8] According to Pagin we cannot have two distinct mental files when there is no way to distinguish them but with their being in a relation of acquaintance. If we wanted to distinguish them we should use some descriptive content, contra the idea that mental files are defined non-descriptively. An answer may be that two distinct mental files for the same individual are connected with two distinct psychological non descriptive MOPs. The challange, however, remains open.

framework "is no longer influential as it used to be" (2012, p. 195), implicitly suggesting therefore that his mental file project might be developed as an alternative to the Perry Kaplan framework. Yet, at the time of writing his book on mental files, Recanati did take into consideration the new stance held by Kaplan (2012) in a paper on Keith Donnellan. It seems to me that this last paper by Kaplan is nearer to Recanati's stance that it may appear (and therefore the novelty of the approach might contrast the supposed lack of influence of the Kaplan-Perry paradigm in this new update). I will spend the rest of the paragraph to give a short summary of the new ideas presented by Kaplan.

Kaplan (2012) makes a new Fregean move in the context of direct reference theory, developing new suggestions on the background of the standard distinctions between content and character or, in Perry's terminology, *objective content* and *cognitive role* (see Perry (2000, 2013)). The distinction concerns, on the one hand, the objective semantic aspect, dealing with truth conditional content, and, on the other hand, the cognitive (epistemological) aspect, dealing with pragmatics and belief contexts. Two utterances with two different indexicals "I" and "he" may represent the same objective content, but have different characters and therefore performing different cognitive roles, as it appears in the well known examples by Perry on different behaviors depending on the use of "I" and "he" in the context of an attack by a bear or in the context of a supermarket, when seeing sugar leaking from a trolley.

Kaplan does not abandon the distinction between character and content, but thinks that it must be supplemented with the idea of different "ways of having in mind" the same objective content. Ways of having in mind are not just what is expressed by the character or linguistic meaning of an expression, but represent what he claims to be "Frege's enduring insight" that is:

"In the realm of cognitive significance, we must account not only for what is represented, but also for how it is represented" (Kaplan 2012, p.158).

What is Kaplan's new move? It is a fundamental revision of semantics, where, instead of considering only the classical truth conditional content, semantics itself should also take into account cognitive aspects. We cannot separate the theory of objective content from the theory of cognitive significance, Kaplan claims. This separation, that has extruded the problem of cognitive significance from semantics to relegate it to the domain of pragmatics or psychology, may be considered "appropriate", but it wrongly seems to imply that a systematic theory of cognitive significance "has nothing to contribute to investigations traditionally thought to be semantic." On the contrary, Kaplan claims,

"Cognitive significance is not foreign to semantics. For the maximum explanatory power, our semantic theory should countenance cognitive content, objective content, and extensions." (Kaplan 2012, p. 141).

That cognitive aspects have always been discussed by direct reference theorists is not a novelty; the novelty is to consider them as a proper part of semantics, without relegating them to pragmatic problems dealing with psychology or speakers' behavior. Beyond the difference on where to place the boundary between semantics and pragmatics, Kaplan's attitude is therefore not so distant from the proposal made by Recanati with the use of non-descriptive MOPs. Where Recanati speaks of linguistic MOPs and psychological MOPs, Kaplan speaks of character and *"ways of having in mind"* (WHIMs).

What are WHIMs? Like Recanati's MOPs they are something "non-descriptive", although they may form or may be connected to a cluster of descriptions. There may be different ways of having in mind, depending on different occasions: Donnellan taught us – with the idea of referential uses of descriptions – that we can have an individual in mind "in a way that is independent of the description that we use to refer to it" (Kaplan 2012). Descriptions used to refer are "shaped" to the occasion or to the context where we enter in cognitive touch with the referent: WHIMs might be considered "perceptual modes of presentation", and are the fundamental aspect of cognitive significance; they are therefore to be sharply distinguished from linguistic meanings (or characters). The consequences that Kaplan derives from these ideas are however slightly different from the Recanati's ones, although beginning with a striking similarity.

According to Kaplan, we may interpret Frege's sense of a singular term like "Mont Blanc" as a particular WHIM, expressing a particular cognitive perspective on a state of affair; we may then have two different *thoughts* concerning the same state of affair (for instance "Mont Blanc is higher than 4000 mt" and "*that* mountain is higher than 4000 mt"). Using different WHIMs in fact, as using different mental files, we may not be aware of referring to the same mountain. Therefore one of the main problems of "having in mind" becomes the problem of *coordination or synch of different WHIMs*, a problem that Kripke begun to discuss in a "Puzzle about Belief".[9] Indexicals are a perfect example of the problem of coordination.

In Kaplan's stance we may find an original answer to the criticism given by Kripke to the Fregean quotation discussed at the beginning of the paper. According to Kaplan "my utterance of 'Today' yesterday and my utterance of 'Yesterday' today may have the *same* cognitive significance, provided I have kept track of these days correctly" (Kaplan 2012, p. 137). This has the advantage of adhering more literally to Frege's claim of identity of thoughts in case of the two correlated utterances.

Kaplan's claim however – probably contra Frege – amounts to say than thought, intended as cognitive significance, depends on awareness: if we have a correct awareness of the flow of time, the two utterances express the same thought or cognitive significance.[10] On the influence of Evan's proposal of "dynamic thoughts"[11] the main question for Kaplan becomes a question of *awareness*: on the one hand, if I

---

[9] Kripke (2011:125-161). Kaplan (2012:156). See Perry (2013) who gives a solution of Kripke's puzzle, considered as a problem of syncing, showing that the disquotational principle is not generally valid.

[10] But in case of people with tracking or memory failures the two sentences may represent two different thoughts, as it seems to happen in general with Recanati's mental files.

[11] Kaplan seems to have accepted Evan's criticism on his earlier ideas (see Evans 1981, fn. 21). Behind Kaplan we find Evans' idea on dynamic thoughts based on the "ability to keep track" places, times and objects in time. The idea of sense of a singular term as "way of thinking" an object becomes, in the case of indexicals, way of keeping track of an object. In case of Today-Yesterday, "the thought episodes on the two days both depend upon the same exercise of a capacity to keep track of a time".

don't bother much, then there is difference in cognitive significance between the two sentences, but, on the other hand, if I bother to keep track of the passing of time, I will continue to "have in mind" the same day; therefore the two sentences will have the same cognitive significance.

This conclusion needs clarification. Kaplan shares Burge's viewpoint according to which Frege's claim of sameness of thought expressed by two utterance "Today is F" and "Yesterday is F" said the subsequent day "makes it clear that cognitive significance is not linguistic meaning" (Kaplan 2012, p. 159). From this Burge 2005 derives the idea that thoughts are abstract entities in the third realm, and difference in cognitive significance pertains to our grasping the same truth-conditional thought; for Kaplan this permits to have the *same* thought with the *same* cognitive significance, depending on our awareness. On the contrary it seems that in Recanati we will have two different mental files with different cognitive significance; what is in common with both authors is the need to discuss how two different expressions may be connected or coordinated (how two mental files may be connected to make the information content flow from one to the other).

There is however an apparent contrast between Kaplan's principles; on the one hand, *if* we take care, we may continuously be aware that the two utterances refer to the same day; in this case Kaplan speaks of the *same* thought and the *same* cognitive significance. However, following Kaplan's principle that distinct WHIMs depend of distinct occasions or contexts, when keeping track of *that* day, we are in a different context and different occasion: therefore we will change our *way* to keep it in mind, having the perception of the passing of time, probably by waking up and looking at the alarm clock. It seems therefore, *contra* Kaplan's claim on the sameness of cognitive significance given by awareness, that the difference in the occasion in which I consider the time should prompt *different* ways of having in mind, therefore different cognitive significances. In fact, if a WHIM depends on the *occasion* of utterance, the uttering of "today" and "yesterday" seems to be the stereotypical case of *different* occasions of utterance, and we should take into account the *difference* of WHIMs in order to understand how they may sync.

Besides, if we rely on awareness, how can we solve problems of syncing that arise when there is a difference of awareness between speakers? Let us go back to the case treated by Recanati of the loss of memory of Rip van Winkle. Two fully rational and coherent speakers – who don't change their mind, but may have awareness failures – may have different beliefs: where a normal speaker may believe that "Yesterday was F" is true, Rip van Winkle may believe that "Yesterday was F" is false, because they are intentionally referring to different days. To solve their disagreement they cannot rely on their awareness: both are aware of the flow of time, but one of them is wrong. They need therefore to rely on some external criteria (either an omniscient point of view, or, for the sake of simplicity, a calendar): awareness alone will not do.

The problem of failures of syncing is similar with proper names (Hesperus/Phosporus) definite descriptions (the mane drinking martini/the man greeting in the doorway) and indexicals: if somebody gives her assent to "Today is F", but refuses to give her assent to "Yesterday was F" the day after, because she does not realize that just one day passed, shall we say the she is irrational? Not really. It is simply a case of ignorance or lack of information. To say that we have the same thought only in case of "awareness" avoids the problem of explaining the differences in informational content given by two different WHIMs. "Today" and "Yesterday" are always conventionally and intentionally correlated, but two different persons may make different correlations. The *psychological origin* of the different correlation may be found in different awareness, in different days people have in mind (certainly Rip Van Vinkle is aware of what he refers to with "Yesterday"), but the *logical mistake* is due to a contrast between a *correct* and an *erroneous* use of "Yesterday"[12].

Can the lack of *correct* sync between two utterances of "Today" and "Yesterday" be explained just with lack of awareness? I have suggested that it is not the case: if two people disagree, they cannot rely on "awareness", because they both are aware to refer to a day they have in mind and both believe it is the same day they refer to as "Yesterday" during what they think it is the day after. There is something "cognitive" in the use of indexicals which is neither linguistic meaning nor awareness. What is missing in the picture is the aknowledgemt of different *ways of applying* WHIMs or MOPs to the context of utterance. Rip van Winkle's way of applying "Yesterday" is just connecting his memories of the last day he remembers; but he might also ask for information, or check on a calendar (as sometimes happens to students who have drunk a lot, and are not sure how much time passed from their binge). Ways to apply WHIMs or MOPs are not only linked to psychology, but to common social practices we learn in leaning language and social interaction.

Let me summarise where we are now. The idea of coordination or sync – that follows Perry's idea of "cognitive paths" – is an interesting new way to discuss old problems like Kripke's puzzle about belief. However making thoughts depending on awareness of people makes them very far from from Frege's idea of *cognitive* sense, that was supposed to be as objective as possible and not depending on the subjective vagaries of human psychology. With Recanati and Kaplan we seem to have a step towards a "psychologization" of Fregean thoughts. Is it the right step to take?

3   Three Levels Semantics Between Psychology and Shared Representations

Kaplan's suggestion for a three level semantics we have quotes before ("our semantic theory should countenance cognitive content, objective content and extension")

---

[12] The difference with the standard cases is that in the standard examples (Hesperus/Phosphorus, etc.) two speakers refer to the *same* object and have different beliefs about it; in the Today/ Yesterday case two speakers have the same belief about what is conventionally referred to as the same day, even if in fact - in our case with loss of memory - they *intend* to refer to *different* days. Using Kripke's terminology, in case of Rip van Winkle's mistake, we may say that the semantic reference of "Yesterday" is different from the speaker's reference. Of his two WHIMs, one is correctly expressed (when Rip was saying, e.g. "Today is F" at the time of his utterance), the other WHIM is just wrongly expressed with the term "Yesterday", because erroneously connected in Rip's mind with the "Today" said twenty years ago.

is reminiscent of the three level semantics conceived by Frege for predicates.[13] Assuming model theoretic semantics as general framework in which to take care of the cognitive dimension of semantics, we might translate Kaplan's proposal into something like the following:

	Sentence	Predicate	Singular term
Cognitive Content		*function from contexts to objective content (character plus MOPS or WHIMS)*	
Objective Content		*function from possible worlds to extensions (Intensions)*	
Extension	Truth Value	Class	Individual

In this setting, character is a function from context to content. However character alone cannot perform the entire job. Characters give general directions independently of when, where and who is speaking; however, as Kripke (2011: 268) remarks, "in any particular case, to determine the reference one needs a specification of the speaker, the time, or both". In other words, once given the general form of a semantics of indexicals, we are left with a pair (context plus character) which is supposed to give the content for semantic evaluation; however, as Predelli (2005, p.74) says, the semantic module "sits and waits" for clause-index (or caracter-context) pairs to be delivered by pragmatic processes. But how is it possible to obtain the contents from the clause-index pair? The problem is: should semantics be concerned with *how* semantic values are determined?

In his 2012 paper Kaplan tries to say something more that putting his distinction between character and content into the framework of semantics. Like the distinction between linguistic and psychological MOPS, he needs a distinction between linguistic meaning and cognitive significance: "it would be odd to end up viewing cognitive contents as nothing more than the conventional meaning of language" (2012 fn. 38). WHIMs (or psychological MOPs) are something more than linguistic meaning or character. What is exactly the difference? According to Recanati they are the mental counterpart of linguistic entities. According to Kaplan they are the way in which a referent is directly fixed by our intentions, depending on different occasions. Both are mainly dealing with psychological aspects. What is their role in semantics?

Kaplan and Recanati insist that WHIMs and psychological MOPs concern a specific cognitive access to reality, that should explain and clarify aspects of direct reference theory. They both insist on the difference between linguistic meaning and cognitive significance; however cognitive significance cannot be separated from linguistic

---

[13] See 1906 Frege's letter to Husserl (in *The Frege Reader*: 301 ff.). For a discussion see Wiggins (1984), Penco (2013a).

meaning, just because it can be defined as what is "activated" by the use of linguistic meaning in a context. The need to recognize a new level in semantics that goes beyond character or linguistic meaning is certainly a novelty in the direct reference framework, but we still have to find how to treat the problem inside a semantic theory.

The problem is which logical form – if any – to give to MOPs and WHIMs. As we have just been reminded by Kripke and Predelli, the character of an indexical is a general rule that is valid independently of any special occasion or context of utterance. What happens when the context is taken into consideration? We need to find which specific procedure may be attached to the function that given the context fixes the semantic content. In order to *understand* or *use* an indexical it is not enough to know its character, but also to master the procedures that permit its use in a context: if we know that "I" means "the speaker of the utterance" we don't know yet *how* to pick the speaker in the context; we need some specific procedures we learn when we learn language: look for where the sound comes from and pick the individual who has made that sound among others. Or, when we have to express ourselves, pronounce that sound to call attention to us, learning how to activate the right sound in the right language and at the right time.[14].

If I hear "Yesterday" I need a procedure that helps me in understanding which is the day before the day of the utterance and different procedures may attain that aim (looking at a calendar, ask a friend, keeping in mind the day looking at the sky, remembering happenings). Normally I will have the default assumption of correct memory; but if something falsifies this assumption, I may guess that something went wrong in the way of applying the indexical expression.

What should happen in the semantic model of these aspects of cognitive significance? How could we represent psychological MOPs and WHIMs? It seems that they should represent individual perspectives of individual psychologies. The move seems welcome for some authors, like Papineau, who fully supports Recanati's decision to focus on the individual rather than the community. He is highly suspect of any one notion that could do justice both to the public and individual dimensions of thought: his claim is that "there is any real work for the idea of a public concept, once we have a good account of individual mental files and the use of words to communicate them".[15]

However, when we make experiments in psychology of language we are not working on the specific ways the brain works in individuals, but on statistics on what different speakers share in the use of language, to check on the psychological plausibility of a theory

---

[14] A particularly original way to see the difference between two occurrences of "I" is given in Textor (2015). To avoid interference with human psychology, we might think of which procedures to put in an intelligent system. A robot would need a procedure that, when hearing the sound "I" makes the system individuate where the sound comes from, and brings it to the individual who has made the sound in the context. On the other hand, if the robot has to express itself, it will not look for a sound and search a person, but it will look for activating his voice with the sound "I".

[15] Although, as we have seen, he criticizes the postulation of some mental files corresponding to indexicals; he criticizes the analogy because "it encourages the view that there are token mental files corresponding to token linguistic demonstratives, when in truth there is nothing corresponding in our actual cognitive structure." (Papineau 2013: 167).

or model of language.[16] Papineau himself (2013: 166) recognizes that mental files should be conceived as a sub-personal speech production system. A study of individual competence, in this perspective, would be probably well suited for some kind of connectionist analysis of subsymbolic mechanisms and processes of the mind. As Smolensky (1988) once remarked, besides the analysis of psychological processes, we may have higher-level representations (like the ones developed in symbolic artificial intelligence) as the representations and analysis of the cognitive systems that supervene the processing of individual minds, although they must be compatible with the psychological data.

Following the three levels semantics suggested by Kaplan we might insert an algorithmic level in model theoretic semantics as a possible way to represent Recanati's MOPs and Kaplan's WHIMs, as procedures attached to functions. What is required by a semantic theory is what can be shared among speakers and how cognitive significance may affect our way of expressing and understanding thoughts and thought components. We should therefore look for different *kinds* of objective procedures that may be attached to characters or linguistic meanings (intended as functions from contexts to contents). Without a context of utterance linguistic meanings are *inert*; they need to be activated, and – without a specific procedure –characters cannot give any semantic value. Whether to treat these procedures as part of pragmatics or semantics is still an open question. Yet, speaking of procedures or algorithms we are back to the realm of objective representation of cognition; truth conditional thoughts can be articulated and grasped from different viewpoints. MOPs and WHIMs aims at showing that linguistic meanings are not enough, and we need something more for describing the working of linguistic and mental interactions. However, we should carefully distinguish between the psychological and neurophysiological search to go "inside" the working of human brain-mind, and a representation of what is open to view, a procedural representation of the different ways in which a truth conditional thought can be represented: different (kinds of) procedures in different (kinds of) contexts of utterance are kinds of things we may grasp and learn.

References

Burge, T.: Truth, Thought, Reason. Clarendon Press, Oxford (2005)
Domaneschi, F., Penco, C., Vignolo, M. (eds.): What is said. In: Proceedings of the 3rd Workshop on Context, CEUR Workshop Proceedings, vol. 594 (2010)

---

[16] I mainly refer to works in experimental pragmatics to which I have partly contributed. Today, the relation between psychology and logic is certainly very different than in Frege's times. However Frege's worries on cognitive aspects bring forward many problems in contemporary logic especially dealing with bounded rationality. In order to have a logical or algorithmic representation of cognitive significance, we probably should consider formalizations such as common sense reasoning or default reasoning strategies. On a more general perspective we may agree with Stenning-and van Lambalgen (2008, 16) that "using the formal machinery of modern logic leads to a much more insightful explanation of existing data, and a much more promising research agenda for generating further data." I hope that representing MOPs and WHIMs as algorithms attached to functions might provide fruitful developments in model theoretic semantics and offer ways to check also the psychological plausibility of different views on mental files.

Domaneschi, F., Penco, C. (eds.): What is Said and What is Not. CSLI Publications, Stanford (2013)

Dummett, M.: More about Thoughts. Notre Dame J. Formal Logic 30, 1–19 (1989). Reprinted in Dummett, M.: Frege and Other Philosophers. Clarendon Press, Oxford (1991)

Dummett, M.: Thought and Reality. Oxford University Press, Oxford (2006)

Frege, G.: Der Gedanke. Eine logische Untersuchung (The Thought. A Logical Enquiry). Beiträge zur Philosophie des Deutschen Idealismus, I: 58–77. In: Beaney, M. (ed.) The Frege Reader, pp. 325–345. Blackwell, Oxford (1997)

Kaplan, D.: An idea of Donnellan. In: Almog, J., Leonardi, P. (eds.) Having in Mind, The Philosophy of Keith Donnellan, pp. 176–184. Oxford University Press, Oxford (2012)

Kripke, S.: Frege's theory of sense and reference: some exegetical notes. Theoria 74, 181–218 (2008)

Kripke, S.: Philosophical Troubles. Oxford Unviersity Press, Oxford (2011)

Künne, W.: A dilemma in Frege's philosophy of thought and language. Riv. Estetica 34, 95–120 (2007)

Neale, S.: On location. In: O'Rourke, M., Washington, C. (eds.) Situating Semantics: Essays on the Philosophy of John Perry, pp. 251–393. MIT Press, Cambridge (2007)

Pagin, P.: The cognitive significance of mental files. Disputatio V 36, 133–145 (2013)

Papineau, D.: Comments of Francois Recanati's mental files: doubts about indexicality. Disputatio V 36, 159–175 (2013)

Penco, C.: Indexicals as demonstratives. Grazer Philosophische Studien 88, 55–73 (2013)

Penco, C.: What happened to the sense of a concept word?. Protosociology 30, 6–28 (2013a)

Perry, J.: Indexicals and demonstratives. In: Hale, B., Wright, C. (eds.) A Companion to the Philosophy of Language, pp. 586–612. Balckwell, Oxford (1997)

Perry, J.: The Problem of the Essential Indexical and Other Essays, expanded edition. CSLI, Stanford (2000)

Perry, J.: Direct discoruse, indirect discourse and belief. In: Penco, C., Domaneschi, F. (eds.) What is Said and What is Not, pp. 311–322. CSLI, Stanford (2013)

Predelli, S.: Contexts: Meaning, Truth, and the Use of Language. Oxford University Press, Oxford (2005)

Recanati, F.: Mental Files. Oxford University Press, Oxford (2012)

Recanati, F.: Reference Through Mental Files. In: Penco, C., Domaneschi, F. (eds.) What is Said and What is Not, pp. 163–178. CSLI, Stanford (2013)

Smolensky, P.: On the proper treatment of connectionism. Behav. Brain Sci. 11, 1–23 (1988)

Stenning, K., van Lambalgen, M.: Human Reasoning and Cognitive Science. MIT Press, Cambridge (2008)

Textor, M.: Frege's theory of hybrid proper names extended. Mind 124(495), 823–847 (2015)

Vignolo, M.: Referential/Attributive: the explanatory gap of the contextualist theory. Dialectica 66, 621–633 (2012)

Wiggins, D.: The sense and reference of predicates: a running repair to Frege's doctrine and a plea for the copula. Philos. Q. 34(136), 311–328 (1984)

Experimental Methods in Linguistics

Making Use of Similarity in Referential Semantics

Helmar Gust[1] and Carla Umbach[2]([⊠])

[1] IKW Universität Osnabrück, Osnabrück, Germany
helmar.gust@uos.de
[2] Zentrum für Allgemeine Sprachwissenschaft (ZAS), Berlin, Germany
umbach@zas.gwz-berlin.de

Abstract. Similarity is well-known to be a core concept of human cognition, e.g., in categorization and learning. Therefore, expressions of similarity in natural language are of special interest: How to account for their meaning including the results on similarity in Cognitive Science and Artificial Intelligence without abandoning referential semantics? In this paper we will lay out a framework connecting referential semantics to conceptual structures by generalizing the notion of measure functions known in degree semantics from the one-dimensional to the many-dimensional case mapping individuals to points in multi-dimensional attribute spaces. Similarity is then spelled out as indistinguishability with respect to a given set of attributes.

1 Introduction

Similarity is well-known to be basic in human cognition, e.g., in categorization and learning processes, and has given rise to a wide range of approaches in Cognitive Science and Artificial Intelligence in which the relation of similarity is captured in terms of, e.g., distance or feature bundles. In natural languages semantics linguistic expressions of similarity are of special interest: How to account for their meaning making use of the results in Cognitive Science and AI without abandoning the idea of referential semantics?

Following Umbach and Gust (2014) linguistic expressions of similarity include, in addition to adjectives like *similar* and verbs like *resemble*, demonstratives of manner, quality and/or degree like German *so*, Polish *tak* and English *such*. These demonstratives modify (some or all of) verbal and nominal and degree expressions, posing the problem of how to reconcile their demonstrative characteristics with their modifying capacity. Umbach and Gust argue that these demonstratives express similarity to the target of the demonstration gesture, and that the emerging similarity class constitutes an ad-hoc generated kind.

Similarity is spelled out with the help of multi-dimensional attribute spaces integrated into referential semantics by generalized measure functions mapping individuals to points in these spaces, generalizing the notion of measure functions familiar in degree semantics (cf. Kennedy 1999) from the one-dimensional to the many-dimensional case. Similarity is defined by the notion of indiscernability

H. Christiansen et al. (Eds.): CONTEXT 2015, LNAI 9405, pp. 425–439, 2015.
DOI: 10.1007/978-3-319-25591-0_31

known in rough set theory (cf. Pawlak 1998) establishing an equivalence relation. Compared to Gärdenfors' conceptual spaces (Gärdenfors 2000), this approach employs a qualitative notion of similarity (as suggested by (Tversky 1977) instead of a geometric one. More importantly, while Gärdenfors' conceptual spaces are 'stand alone' systems, the approach presented here integrates a conceptual level of representation into referential semantics, and it does that in a way that has already been paved by degree semantics.

The focus of the current paper is on the formal details of this approach, which have only briefly been touched upon in Umbach and Gust (2014). The notions of attribute spaces, measure functions, representation, indiscernability and granularity will be laid out here in detail. Similarity will be defined as a categorical predicate *sim*, with a comparative relation *more_sim* based on the categorical predicate. This notion of similarity will be compared to that in Tversky (1977) which is relational in nature. Section 2 will provide a brief summary of the linguistic data; in Sect. 3 the basic definitions will be given; in Sect. 4 similarity will be defined, and in Sect. 5 this notion of similarity will be compared to Tversky's notion of similarity.

This paper grew out of a collaboration of Artificial Intelligence and Natural Language Semantics and includes both perspectives. The AI background is in knowledge representation and ontologies, feature extraction, and reasoning with examples, cf. Randall (1993). The semantic background is in referential semantics, in particular demonstratives, degree expressions and generics (Kaplan 1989; Kennedy 1999; Carlson 1980). Rather than trying to conceal the different perspectives we will make them explicit where advisable.

2   Demonstratives of Manner, Quality and Degree

The approach in Umbach and Gust (2014) starts from German *so* ('such'/'like this'). It is one of a class of demonstratives found across languages that serve as modifiers of quality and/or manner and/or degree, including also, e.g., Polish *tak* and English *such*. Carlson (1980) proposed an analysis of English *such* as directly referring to a kind. This analysis was adopted by Anderson and Morzycki (2015) for Polish *tak*, which behaves analogous to German *so* in modifying nominal, verbal and also adjectival expressions, extending the notion of kinds to events and also degrees. Umbach and Gust (2014) argue that a directly kind-referring approach has a number of shortcomings and suggest an analysis based on similarity. While in the case of demonstratives like *this* the referent of the demonstrative phrase and the target of the demonstration gesture are identical - this is an in-build feature of the Kaplanian theory of demonstratives - in the case of manner/quality/degree demonstratives the referent and the target of the demonstration are similar (with respect to relevant features).

Consider the examples in (1). In (a), Anna's manner of dancing is characterized as being similar in certain respects to the dancing event the speaker is pointing at. In (b), Anna's cup is characterized as being similar to the cup the speaker is pointing at. Finally, in (c) Anna's height is characterized as being similar to the height of the person the speaker is pointing at. In all of these cases, by using the demonstrative *so* a similarity class is created based on the target of the demonstration. In the case of nouns and verbs the similarity class clearly exhibits kind-like properties and should be considered as an ad-hoc generated kind, e.g., in (1a) there is an ad-hoc generated sub-kind of dancing events similar to the dancing pointed at, and in (1b) there is an ad-hoc generated sub-kind of cups similar to the mug pointed at. In the case of adjectives, as in (1c), it is an open issue whether the similarity class created by the use of the demonstrative should be considered as a genuine kind (see Anderson and Morzycki and Umbach and Gust; we will not go into this issue here).

(1) a. (speaker pointing to someone dancing):
        So tanzt Anna auch.
        'Anna dances like this, too.'

     b. (speaker pointing to a cup):
        So eine Tasse hat Anna auch.
        'Anna has such a cup/a cup like this, too.'

     c. (speaker pointing to a person):
        So groß ist Anna auch.
        'Anna is this tall, too.'

The most urgent question when dealing with similarity is that of the relevant respects, or features, of similarity. Without fixing relevant respects the notion of similarity would be trivial (Goodman 1972). In the case of adjectives like *tall* there is only one dimension, which is fixed by the adjective's lexical meaning — in the case of *tall* the feature of similarity is *height*. In the case of multi-dimensional adjectives like *healthy* and *beautiful*, as well as nouns and verbs, features of similarity have to be provided by the context. There are, however, constraints on which features qualify as licit in similarity comparison. Consider the anaphoric use of *so* in (2): *so ein Auto* can easily be understood as denoting a Japanese car but not as denoting a new car. These constraints are subject of an experimental study presented in Umbach (submitted) and are traced back to the idea of principled (vs. mere statistical) connections between properties and kinds discussed in the literature on generics (see Carlson 2010).

(2) a. Anna hat ein japanisches Auto. Berta hat auch so ein Auto
        (nämlich ein japanisches Auto).

     b. Anna hat ein neues Auto. Berta hat auch so ein Auto
        (*nämlich ein neues Auto).
        'Anna has a Japanese car / a new car. Berta has such a car, too.'

One more issue when studying expressions of similarity in natural languages is the difference between demonstratives, like German *so* and English *such*, and adjectives, like German *ähnlich* and English *similar*. Although the two types of similarity expressions appear equivalent in meaning at first sight, there are fundamental differences. One of these is their behavior in additive contexts as in (3). The question-under-discussion in (3) ('Which cars do Otto and Anna drive?') has been partially answered by the preceding sentence - Otto drives a Mercedes Benz. Adding another Mercedes Benz driver should require an additive particle, which is in fact obligatory when using the demonstrative but is highly redundant when using the adjective.

(3) (Otto drives a Mercedes Benz – what about Anna?)

    a. Anna fährt auch so ein Auto.

    b. ??Anna fährt auch ein ähnliches Auto.
       'Anna drives such a car/a similar car, too.'

On the basis of this and other observations it is argued in Umbach (2014) that adjectives expressing similarity differ from demonstratives expressing similarity in carrying an in-built distinctiveness requirement on their arguments, which is the reason why the additive particle is not licensed in (3b). This entails that the similarity relation expressed by adjectives is irreflexive while that expressed by demonstratives is reflexive.

Similarity is, from the point of view of referential semantics, a simple predicate – *so ein Auto/such a car* denote an element of a set of cars similar to the target of the demonstration or antecedent (see the examples in (4,5) in Sect. 4). Since this set exhibits kind-like properties, it is justly considered as an ad-hoc created sub-kind, for example, an ad-hoc created sub-kind of cars (cf. Umbach submitted). It would be unsatisfactory, however, if similarity were just an arbitrary predicate. The challenge posed by similarity is to gain insight into the mechanism and the constraints of this relation (and thereby into the mechanism and constraints of ad-hoc kind formation). For this reason, the similarity relation is spelled out in multi-dimensional attribute spaces inspired by knowledge representation techniques familiar in Artificial Intelligence.

In the next section, multi-dimensional attribute spaces, families of contexts, predicates on attribute spaces and convex closures thereof are defined. This is the machinery used in Sect. 4 to implement a context-sensitive predicate sim and, based on that, a comparative similarity relation $more_s im$ relation. This is compared in Sect. 5 to Tversky's implementation of similarity.

3   Multi-dimensional Attribute Spaces

The basic idea of the framework presented in this paper is to have a referential semantics for natural language expressions where predicates talk about entities in the world, and a representational layer where predicates talk about abstract entities like numbers and symbols specifying attributes. Entities of the world are related to these abstract entities by processes of measurement or feature extraction or perception etc. The representational layer facilitates comparing entities in the world with respect to their attributes and, in particular, determine whether they are similar with respect to certain attributes.

Representation by attributes is familiar in AI in, e.g., knowledge representation. Multi-dimensional attribute spaces in knowledge representation make use of attributes (i.e. dimensions) with all sorts of values, for example numbers and symbols. Points in these spaces correspond to lists of attribute-value pairs. A parallel, even if simpler conception is found in referential semantics, more precisely, in degree semantics where gradable adjectives are interpreted by measure functions mapping individuals to degrees, that is, points on a dimension with metric values. Attribute spaces can be seen as a generalization of dimensions in degree semantics involving more than one dimension and allowing for values other than metrical ones. Likewise, mappings from individuals to points in attribute spaces can be seen as a generalization of measure functions from the one-dimensional to the many-dimensional case. From this point of view, making use of multi-dimensional attribute spaces in referential semantics is not a radical novelty but rather a generalization of the broadly accepted idea of degrees and measure functions.

The basic components of the framework presented in this paper are (i) *domains*, (ii) *attribute spaces* and *measure functions* and (iii) *predicate systems over attribute spaces*. They will be defined subsequently.

3.1 Domains and Representations

We define a *domain* as a subset of the universe together with a set of predicates and non-overlapping sets of positive and negative examples for each predicate. We define *families of contexts* as sets of domains with related predicates that may differ, however, in their positive and negative examples. Examples must behave consistently within a family of contexts, that is, they must not change their roles[1].

Definition 1. *Domain*
A domain is a quadruple $\mathcal{D} = \langle D, .^{+}, .^{-}, P \rangle$ with:

- *D a set representing the domain,*
- *$P = \{p_1, ..., p_n\}$ a set or family of predicates over D,*
- *$.^{+} : \{p_1, ..., p_n\} \rightarrow \mathcal{P}(D)$ a function which assigns positive examples to each predicate,*

---

[1] A side remark for semanticists: While in a classical (two-valued) truth-conditional semantics it is presumed that an individual is either p or $\neg p$, from the point of view of knowledge representation there may be individuals for which this is not (yet) known. This is the reason why the set of positive examples p^{+} need not cover the extension of p (and p^{-} need not cover $\neg p$). The idea is that with increase of information the range of indeterminateness decreases. This behavior can be accounted for in logics with a notion of *underdefinedness* (see Muskens 1995). We will, however, not go into this issue here but rather assume that the underlying logic is classical and representations are (possibly incomplete) approximations.

- $.^- : \{p_1, ..., p_n\} \to \mathcal{P}(\mathcal{D})$ *a function which assigns negative examples to each predicate,*
- $p_i^+, p_i^- \subseteq \mathcal{D}$.
- $p_i^+ \cap p_i^- = \emptyset$

Definition 2. *Family of contexts*
A family of contexts $\mathcal{C}(P) = \{\langle D_k, .^{+k}, .^{-k}, P_k\rangle | k = 1, 2, ...\}$ *evaluates P if*

- $P_k \subseteq P$
- *for any two contexts $C_i, C_j \in \mathcal{C}(P)$ and for all $p \in P_i \cap P_j$ the following conditions hold[2]*
  - *elements of p^+ and p^- cannot change roles in a different context:*
    $p^{+i} \times p^{-i} \cap p^{-j} \times p^{+j} = \emptyset$
  - *discriminative power: $p^{+i} \times p^{-i} \cap D_j \times D_j \neq \emptyset \to p^{+j} \times p^{-j} \neq \emptyset$*
    (Predicates discriminating elements in one context should be able to discriminate elements in another context.)[3]

Attribute spaces are common structures for representation. They generalize vector space approaches in allowing heterogeneous dimensions equipped with value sets of different scales (nominal, ordinal, interval, ratio), where value sets may themselves be attribute spaces (the values being points in such spaces).

An attribute space F is given by a set of attributes $A = \{a_1...a_n\}$, such that for each a_i in A there is a set of values V_{a_i}. Points in an attribute space are in $V_{a_1} \times ... \times V_{a_n}$. A *representation* includes an *attribute space F*, a (generalized) *measure function* μ mapping elements of a domain into an attribute space and a set of predicates p^* talking about points in the attribute space. These predicates serve as *approximations*[4] of the predicates P of the domain and will be detailed in Sect. 3.2.

Definition 3. *Representation*
A representation $\mathcal{F} = \langle F, \mu, .^*, \mathcal{D}\rangle$ of a domain $\mathcal{D} = \langle D, .^+, .^-, P\rangle$ is given by

- *an attribute space F,*
- *a measure function $\mu : D \to F$,*
- $.^* : P \to \Omega^F$[5]

together with the consistency conditions

- $\forall x \in p^+ : p^*(\mu(x)) = true$
- $\forall x \in p^- : p^*(\mu(x)) = false$

for all p in P.

From this we get $\mu(p_i^+) \cap \mu(p_i^-) = \emptyset$.

---

[2] This is a slightly different formalization of NR, UD, and DD in (van Rooij 2011).
[3] If a predicate *bike* discriminates between bikes and trikes in a context containing only bikes and trikes it should at least discriminate between bikes including trikes and cars in a context containing bikes, trikes and cars.
[4] More precisely: $p^* \circ \mu$ approximates p.
[5] where Ω^F is the set of characteristic functions in F.

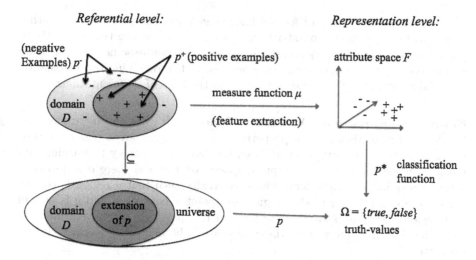

Fig. 1. Domains and representations

3.2   Predicates on Attribute Spaces

Attribute spaces are familiar methods of representation in AI and also in some branches of natural language semantics, e.g., in frame-based approaches (Barcalou 1992). What distinguishes attribute spaces and representations as proposed in this paper is the idea of predicates on attribute spaces. On the worldy side, a domain includes a set of relevant predicates $p \in P$. According to the notion of a representation in this paper these predicates have counterparts on the representational side, namely the predicates $p^* \in .^*(P) = P^*$. Counterpart predicates are required to be consistent with their originals; more precisely, they have to agree in truth value with the set of positive/negative exemplars of the original predicate (see Definition 3). Moreover, counterpart predicates will be assumed to be convex, that is, true of all points in the convex closure of (the images of) the positive exemplars.

Reasons for furnishing attribute spaces with predicates are, (i) they facilitate a straightforward definition of *indiscernability* in the sense of rough set theory (Pawlak 1998), that is, a relation establishing classes of points which count as equivalent (in a specific context, with respect to specific inferences). The indiscernability relation provides attribute spaces with a grid of granularity, allowing comparison of attribute spaces of different granularity which are identical otherwise. Indiscernability and granularity will be exploited to define a categorical and a comparative notion of similarity in the next section[6].

---

[6] It has been objected that in order to account for different granularity of representations one could use different sets of attributes and values, that is, different attribute spaces. This would require, however, a way to compare granularity directly on attribute spaces, which would be less intuitive and in the end close to what we suggest.

From the point of view of AI and knowledge representation there is another reason to have predicates on attribute spaces. If you assume that elements in a domain D are not directly accessible and instead require some sort of feature extraction or perception or measuring processes, then reasoning is possible only with the results of such processes, that is, on the level of representations.

Convex Closures on Attribute Spaces. It has been argued by Gärdenfors (2000) that natural properties correspond to convex regions in conceptual spaces, convexity being required by cognitive economy in learning and memory. Since Gärdenfors' conceptual spaces are metrical there is a notion of *betweenness* facilitating a definition of convexity such that a region is convex if for all points x and y in that region any point z between x and y is also in that region. Attribute spaces as specified in this paper do not have a metrical *between* relation. Convexity in these spaces will be defined by convex closure operators.

Definition 4. *Closure operator on attribute spaces*
An operator cl is a closure operator on an attribute space F iff

- *$cl : \mathcal{P}(F) \to \mathcal{P}(F)$*
- *For all $X \subseteq F$*
  - *$X \subseteq cl(X)$*
  - *$cl(cl(X)) = cl(X)$*
- *$cl(\emptyset) = \emptyset$*

Definition 5. *Convex closure*
A closure is convex if the anti-exchange property holds:
$$y, z \notin cl(X) \wedge y \neq z \wedge z \in cl(X \cup \{y\}) \to y \notin cl(X \cup \{z\})$$

We assume, as does Gärdenfors, that predicates in a domain are well-behaved in the sense that they cover convex regions, that is in our case, they correspond to convex subsets of an attribute space. More precisely, predicates p^* on attribute spaces are required to be true for every point in the convex closure of the predicate's positive examples when mapped to the attribute space. This is, we expect representations to be *strongly consistent*.

Definition 6. *Strong consistency*
A representation $\mathcal{F} = \langle \langle F, cl \rangle, \mu, .^*, \mathcal{D} \rangle$ of a domain $\mathcal{D} = \langle D, .^+, .^-, P \rangle$ is strongly consistent if $\forall p_i \in P : p_i^*(cl(\mu(p_i^+))) = \{true\}$[7]

We can make this even stronger by stipulating that the p_i^* itself have convex extensions, which implies strong consistency.

Above we defined a number of constraints on closure operators. That does not, however, answer the question of how to define a particular closure operator. One way of doing that is by exploiting a partial order. On a (partially) ordered set $\langle D, \leq \rangle$ we can define closure operators in a natural way:

---

[7] We make use of the convention that a function f applied to a set of arguments refers to the set of images: $f(X) = \{f(x)|x \in X\}$.

Definition 7. *Closure operations on partial orders*
For $A \subseteq D$ we define:

left closure: $cl_{\leftarrow}(A) = \{x \in D | \exists y \in A : x \leq y\}$
right closure: $cl_{\rightarrow}(A) = \{x \in D | \exists y \in A : y \leq x\}$
closure: $cl(A) = \{x \in D | \exists y, z \in A : y \leq x \leq z\}$

For all three constructions the anti-exchange property holds, so all the resulting sets are convex. Thus, if F is (partially) ordered, we can easily define a convex closure operator. In order to emphasize the aspect that an attribute space F is equipped with a closure operator cl, we add it to the representation: $\mathcal{F} = \langle\langle F, cl\rangle, \mu, .^*, \mathcal{D}\rangle$ of a domain $\mathcal{D} = \langle D, .^+, .^-, P\rangle$.

Predicate Systems. Predicates defined on attribute spaces classify points in such spaces facilitating different levels of granularity of representation. Given a set of base predicates B (e.g. *small, medium, big*) over a feature space F we define a set of predicates $\widetilde{B}$ inductively (analogous to defining a topology relative to a base):

- $B \subseteq \widetilde{B}$
- $X \in \widetilde{B} \wedge Y \in \widetilde{B} \rightarrow X \cap Y \in \widetilde{B}$
- $X \in \widetilde{B} \wedge Y \in \widetilde{B} \rightarrow cl(X \cup Y) \in \widetilde{B}$

If F is (partially) ordered we get additionally

- $X \in \widetilde{B} \rightarrow cl_{\rightarrow}(X) \in \widetilde{B}$
- $X \in \widetilde{B} \rightarrow cl_{\leftarrow}(X) \in \widetilde{B}$

We assume that the elements of B are convex ($cl(X) = X$ for $X \in B$) and we say that $\widetilde{B}$ is generated by B and cl. We can view $\widetilde{}$ as operator generating the predicate system for a give base. A good candidate for B is $\{p_1^*, ..., p_n^*\}$. For a representation $\mathcal{F} = \langle\langle F, cl\rangle, \mu, .^*, \langle D, .^+, .^-, P\rangle\rangle$, $\widetilde{P^*}$ is the predicate system generated by $P^* = \{p_1^*, ..., p_n^*\}$ and cl.

Indiscernability. Two elements in a domain D are indistinguishable on the representational level if they are mapped by μ onto the same point of the attribute space. But even if they are mapped to different points, they may nevertheless be indistinguishable in yielding the same inferences. This is the case if they agree on all predicates. We borrow the term *indiscernable* from Rough Set Theory:

Definition 8. *Indiscernable*
Given a representation $\mathcal{F} = \langle F, \mu, .^*, \langle D, .^+, .^-, P\rangle\rangle$ we define:
$x \sim_{\mathcal{F}} y \equiv \forall q \in \widetilde{P^*} : q(x) \leftrightarrow q(y)$

Granularity. For two representations $\mathcal{F}$ and $\mathcal{F}'$ sharing the same attribute space F we can ask whether one is more fine grained than the other, that is, whether two entities can be distinguished in one representation but not in the other one. Since indiscernability of entities in a representation depends on the set of predicates P provided by the domain, granularity of representations depends on the set of predicates P, too.

Definition 9. *Family of representations evaluating a system of predicates*
Given a family of contexts $\mathcal{C}(P) = \{\langle D_k, .^{+k}, .^{-k}, P_k\rangle | k = 1, 2, ...\}$ *evaluating a*
system of predicates P *we get the corresponding family of representations*
$$\mathcal{FC}(P) = \{\langle F, \mu, .^{*k}, \langle D_k, .^{+k}, .^{-k}, P_k\rangle\rangle | k = 1, 2, ...\}$$

On $\mathcal{FC}(P)$ we can define a partial order:

Definition 10. *Coarser representation*
Given two representations
$$\mathcal{F}_i = \langle F, \mu, .^{*i}, \langle D_i, .^{+i}, .^{-i}, P_i\rangle\rangle \in \mathcal{FC}(P) \ and$$
$$\mathcal{F}_j = \langle F, \mu, .^{*j}, \langle D_j, .^{+j}, .^{-j}, P_j\rangle\rangle \in \mathcal{FC}(P) \ we \ define:$$
$$\mathcal{F}_j \geq \mathcal{F}_i \ iff \ P_j \subseteq P_i \wedge \forall x, y \in F : x \sim_{\mathcal{F}_i} y \to x \sim_{\mathcal{F}_j} y$$

We say $\mathcal{F}_j$ is coarser than $\mathcal{F}_i$ ($\mathcal{F}_j > \mathcal{F}_i$) iff $\mathcal{F}_j \geq \mathcal{F}_i$ but not $\mathcal{F}_j \leq \mathcal{F}_i$.

This completes the formal machinery required for the definition of similarity in the next section.

4   Defining Similarity

4.1   Similar

We started out from the analysis of demonstratives of manner, quality and degree, in particular German *so*, in Umbach and Gust (2014). It was argued that these demonstratives express similarity to the target of the demonstration gesture, and that the emerging similarity class constitutes an ad-hoc generated kind. This suggests that the similarity relation expressed by the demonstratives is adequately captured by the notion of indiscernability as defined above.

Definition 11. *Similar*
$$\forall x, y, \in D : sim(x, y, \mathcal{F}) \equiv \mu(x) \sim_{\mathcal{F}} \mu(y)$$

Definition 11 entails that the notion of similarity expressed by the demonstratives is an equivalence relation (see also Sect. 5).

The *sim* predicate is used in Umbach and Gust (2014) for a compositional interpretation of the demonstrative *so*. When modifying nominals *so* occurs in an ad-determiner position and is thus combined with the indefinite determiner, see (4).

In the nominal case (and in the verbal case) features of comparison have to be inferred from the context of the utterance – *so eine Tasse* ('such a cup') may denote cups similar in form and/or color and/or design etc. – whereas in the adjectival case there is only one feature of comparison, which is fixed by the lexical meaning of the adjective – *so groß* ('this tall') can only mean similar in height.

We assume that features of comparison provide attributes spanning the attribute space in a representation $\mathcal{F}$. We write $\mathcal{F}(f)$ to indicate a representation with a single linguistically fixed attribute f. The variable t in (4,5) is a free variable denoting the target of the demonstration gesture.

(4) *so eine Tasse* ('such a cup')

$$[[\text{so}]] = \lambda D.\lambda P.D(\lambda x.sim(x,t,\mathcal{F}) \wedge P(x))$$
$$[[\text{so ein}]] = \lambda P.\lambda Q.\exists x.sim(x,t,\mathcal{F}) \wedge P(x) \wedge Q(x)$$
$$[[\text{so eine Tasse}]] = \lambda Q.\exists x.sim(x,t,\mathcal{F}) \wedge cup(x) \wedge Q(x)$$

(5) *so groß* ('this tall')

$$[[\text{so}]] = \lambda f.\lambda x.sim(x,t,\mathcal{F}(f))$$
$$[[\text{so groß}]] = \lambda x.sim(x,t,\mathcal{F}(height))$$

The predicate *sim*, which is a simple modifier of determiners and degree expressions in (4,5), receives its explanatory power by being defined as indiscernability in attribute spaces.

4.2   More Similar

Similarity as expressed by demonstratives is clearly not gradable - there is no way of saying that something is *more such a thing* than something else. Thus the *sim* predicate defined above is appropriate for the semantics of, e.g., English *such* and German *so*. In contrast, similarity as expressed by adjectives is gradable, cf. English *more similar* and German *ähnlicher*. We will provide a gradable notion of similarity in the way of the vague predicate analysis of Klein (1980) such that the comparative is based on the positive. We define a four-place predicate $more\_sim(x,y,z,u,\mathcal{F}C(P))$ with the help of representations of different granularity in a family of representations such that x is more similar to z than y to u, if there is a representation $\mathcal{F}'$ identifying x and z while distinguishing y and u. Any $\mathcal{F}''$ identifying y and u must not discriminate x and z.

Definition 12. *More similar*
Given a family of representations $\mathcal{F}C(P)$, x is more similar to z than y to u,
$more\_sim(x,y,z,u,\mathcal{F}C(P))$ *iff*
$$\exists \mathcal{F}' \in \mathcal{F}C(P) : sim(x,z,\mathcal{F}') \wedge \neg sim(y,u,\mathcal{F}')$$
$$\wedge \, \forall \mathcal{F}'' \in \mathcal{F}C(P) : sim(y,u,\mathcal{F}'') \rightarrow sim(x,z,\mathcal{F}'')$$

The 4-place predicate $more_sim$ provides an interpretation for natural language comparatives as in *Anna resembles her mother more than Berta resembles her father*. Definition of a 3-place version is straightforward:

$more\_sim_3(x, y, z, \mathcal{FC}(P))$ iff $more\_sim(x, y, z, z, \mathcal{FC}(P))$. Another motivation for a comparative version of similarity stems from multi-dimensional gradable adjectives like *healthy* and *beautiful* which are difficult to handle in standard degree semantics. One-dimensional adjectives like *tall* are interpreted in degree semantics as measure functions mapping individuals to degrees (Kennedy 1999). The positive form (as in *A is tall*) is assumed to include a context-dependent cut-off degree for individuals to count as tall in the context/comparison class. For many-dimensional adjectives, however, it is unclear how to define a cut-off. Sassoon (2013) suggests quantification over dimensions such that, e.g., *healthy* means *healthy in all relevant dimensions* while leaving the question open of how to define the comparative.

Umbach (2015) makes use of similarity in multi-dimensional spaces when interpreting evaluative predicates like *beautiful*. Since evaluative predicates don't have a linguistically fixed denotation there is no independently given set of positive exemplars $beautiful^+$. Whether something is beautiful or not is not a matter of fact but rather a matter of convention, that is, of negotiation of criteria. Negotiating criteria can be seen as determining a 'prototype' which is, however, not a point but rather a predicate on points in an attribute space. Suppose $proto\_beautiful$ is the class of points in F corresponding to the criteria for something to count as beautiful, that is, $proto\_beautiful$ plays the role of a cut-off. The counterpart predicate $beautiful^*$ then has to be the convex closure of $proto\_beautiful$. An individual can now be said to count as beautiful in a domain $\mathcal{D}$ and a representation $\mathcal{F}$ iff it is indiscernable from the elements of $beautiful^*$, that is, the closure of $proto\_beautiful$ (where the domain serves as comparison class).

$\forall x \in \mathcal{D}$: x is beautiful in $\mathcal{F}$ iff

$beautiful^*(\mu(x))$ in $\mathcal{F}$, that is, $cl(proto\_beautiful)(\mu(x))$ in $\mathcal{F}$. The comparative of multi-dimensional adjectives is interpreted with the help of the $more\_sim$ relation[8].

$\forall x, y \in \mathcal{D}$: x is more beautiful than y in $\mathcal{FC}(P)$ iff

$more\_sim_3(x, y, cl(proto\_beautiful), \mathcal{FC}(P))$[9]

5   Comparison to Tversky's Account of Similarity

Twersky's contrast model of similarity was developed as an alternative to the at that time dominant geometric models in which (dis)similarity of two objects is represented by distance. Tversky started from empirical findings that appeared

---

[8] This corresponds to a vague predicate interpretation of multi-dimensional adjectives. It's an open question, however, whether one-dimensional adjectives like *tall* should be interpreted in this way, too. One of our future issues is to exploit a two-way approach: While the comparative of adjectives relating to a single metrical scale makes use of that scale (in a degree semantics fashion), the comparative of multi-dimensional adjectives involves similarity to prototypes, as sketched above.

[9] We use $more\_sim_3$ also with arguments from F. This is unproblematic because μ mediates between the two levels.

incompatible with the axioms of a metric distance function. He claimed that "... the assessment of similarity between stimuli may better be described as a comparison of features rather than as the computation of metric distance between points." (p. 328), and proposed to assess similarity making use of sets of features (binary, nominal, ordinal, cardinal). Similarity of two objects a and b is measured by a matching function F accounting for communalities of and differences between their feature representations A and B:

$sim(a, b) = F(A \cap B, A - B, B - A)$.

The result of the matching function corresponds to a value on an interval scale S which makes it possible to compare the similarity of two objects to that of two other objects: $sim(a, b) \leq sim(c, d)$ iff $S(a, b) \leq S(c, d)$[10].

From the point of view of the semantics of gradable adjectives, Tversky's account of similarity is in the spirit of degree semantics where the comparative is prior to the positive. Tversky's notion of similarity maps pairs of individuals to degrees on a scale of similarity thereby facilitating comparison — *a and b are more similar to each other than c and d*. Positive judgments, as in *a and b are similar*, are not considered. In contrast to Tversky's account, the notion of similarity developed in this paper is in the spirit of the vague predicate analyses starting from the positive form and defining the comparative 'on top'.

Tversky started from empirical findings which seem incompatible with the axioms of a metric distance function, i.e. minimality, symmetry and triangle inequality. Minimality appears problematic if identification probability is interpreted as a measure of similarity because identical stimuli are not always identified by subjects as being identical. Triangle inequality is hardly compelling, following Tversky, in view of cases involving different features. For example, Jamaica is similar to Cuba (with respect to geographical proximity) and Kuba is similar to Russia (with respect to political affinity) but Jamaica and Russia are not similar at all. Finally, symmetry is apparently false considering experimental results showing that the judged similarity of North Korea to Red China exceeds the judged similarity of Red China to North Korea.

In this paper, similarity is viewed as a means of classification, that is, a relation establishing classes, or kinds, of individuals. From a classification perspective, similarity has to be reflexive, symmetric and transitive. Concerning reflexivity, Umbach (2014) shows that the interpretation of demonstratives like German *so* and English *such* requires reflexivity while the interpretation of adjectives expressing similarity, like German *ähnlich* and English *similar*, excludes reflexives pairs. Being irreflexive explains why adjectival similarity expressions lack a kind-forming capacity — kinds have to be equivalence classes.

Concerning symmetry, Gleitman et al. (1996) present a series of studies showing that Tversky's results on similarity judgments are due to figureground effects in presentation and argue that similarity is a genuinely symmetric relation. In

---

[10] where S computes the values of a model based on the weighted sums over common and distinctive features, which means, there exist f, θ, α, β with $S(a, b) = \theta f(A \cap B) - \alpha f(A - B) - \beta f(B - A)$.

fact, Tversky himself already suggested that non-directional similarity statements (*North Korea and Red China are similar*) are symmetric. Finally, transitivity is obvious in the case of demonstratives. Consider the sentence *Anna hat so ein Auto und Berta hat so ein Auto.* ('Anna has a car like this and Berta has a car like this.') where the two *so*-phrases share the same demonstration target. This sentence clearly entails that Anna's and Berta's cars belong to the same subkind of cars. For adjectives expressing similarity the situation is again less clear. It might be possible that changing the respects of similarity, as in the Jamaica/Cuba/Russia example, is licensed by adjectives (the sentence *Jamaica is similar to Cuba and Kuba is similar to Russia but Jamaica and Russia are not similar* appears acceptable in spite of different features of similarity).

Summing up, similarity as expressed by natural language demonstratives (e.g., German *so* and English *such*) is in fact an equivalence relation whereas similarity as expressed by adjectives (e.g., German *ähnlich* and English *similar*) is not. This confirms the analysis that similarity demonstratives, but not similarity adjectives, ad-hoc establish novel (sub-)kinds.

6   Summary

In this paper, a representational framework was presented featuring multidimensional attribute spaces equipped with systems of convex predicates. This framework facilitates the definition of a notion of similarity, or indiscernability, as required in classification processes, which is suited for the interpretation of demonstratives of manner, quality and degree in natural language.

The framework emerged out of a collaboration of Artificial Intelligence and Natural Language Semantics. From an AI point of view it can be seen as a means of representing (possibly incomplete) knowledge about a given domain. From a semantic point of view it can be seen as extending referential semantics in order to integrate conceptual aspects of meaning.

References

Anderson, C., Morzycki, M.: Degrees as kinds. In: Natural Language and Linguistic Theory 2015 (2015, to appear)

Barsalou, L.W.: Frames, concepts, and conceptual fields. In: Kittay, E., Lehrer, A. (eds.) Frames, Fields, and Contrasts. Lawrence Erlbaum Associates, Hillsdale (1992)

Carlson, G.N.: Reference to Kinds in English. Garland, New York & London (1980)

Carlson, G.: Generics and concepts. In: Pelletier, F.J. (ed.) Kinds, Things and Stuff, pp. 16–36. Oxford University Press, Oxford (2010)

Gärdenfors, P.: Conceptual Spaces. MIT Press, Cambridge (2000)

Gleitman, L., Gleitman, H., Miller, C., Ostrin, R.: Similar, and similar concepts. Cognition 58, 321–376 (1996)

Goodman, N.: Seven strictures on similarity. In: Goodman, N. (ed.) Problems and Projects, pp. 437–447. The Bobbs Merrill Company, Indianapolis and New York (1972)

Kaplan, D.: Demonstratives. In: Almog, J., Perry, J., Wittstein, H. (eds.) Themes from Kaplan, pp. 481–563. Oxford University Press, Oxford (1989)

Kennedy, C.: Projecting the adjective. Garland Press, New York (1999)

Klein, E.: A semantics for positive and comparative adjectives. Linguist. Philos. 4(1), 1–45 (1980)

Muskens, R.: Meaning and Partiality. CSLI, Stanford (1995)

Pawlak, Z.: Granularity of knowledge, indiscernibility and rough sets. In: Proceedings of the IEEE International Conference on Fuzzy Systems, pp. 106–110 (1998)

Davis, R., Shrobe, H., Szolovits, P.: What Is a knowledge representation? AI Mag. 14(1), 17–33 (1993)

van Rooij, R.: Measurement and interadjective comparisons. J. Semant. 28(3), 335–358 (2011)

Sassoon, G.: A typology of multidimensional adjectives. J. Semant. 30, 335–380 (2013)

Tversky, A.: Features of similarity. Psychol. Rev. 84, 327–352 (1977)

Umbach, C., Gust, H.: Similarity demonstratives. Lingua 149, 74–93 (2014)

Umbach, C.: Expressing similarity: on some differences between adjectives and demonstratives. In: Proceedings of IATL 2013, MIT Working Papers in Linguistics (2014)

Umbach, C.: Evaluative propositions and subjective judgments. In: van Wijnbergen-Huitink, J., Meier, C. (eds.) Subjective Meaning. de Guyter, Berlin (2015)

Umbach, C.: Ad-hoc kind formation by similarity (submitted)

Context-Dependent Information Processing

Towards an Expectation-Based Parsing Model of Information Structure

Edoardo Lombardi Vallauri and Viviana Masia[✉]

Università Roma Tre, Rome, Italy
viviana.masia@uniroma3.it

Abstract. The ways language encodes information depend on when and how the preceding linguistic and non-linguistic context has established it in the participants' working memory. Information-Structure categories such as Focus and Topic are used to signal that the conveyed information is, respectively, the contribution of the message to the addressee's knowledge or simply something meant to link the message to the context. Information introduced within a context ("Given") is expected to be encoded as a Topic, while "New" information is more likely to appear in Focus. The results of a dedicated EEG experiment will show that violation of such expectations causes supplementary processing costs, revealed by rhythmic changes in different frequency bands.

Keywords: Information structure · Given/new · Topic/focus · EEG · Processing costs

1 Context and Information Structure

Language develops in context, and context activates information. If we walk along a road in a rainy day and a car, passing by at high speed, splashes all the water of a puddle on me, I can say to my friend:

1. It drenched me completely!

In particular, my friend will understand that the pronoun *it* refers to the car, because the idea of the car has been activated in his short-term memory (STM) by the appearance of the car itself. If we were speaking Italian or another pro-drop language, I would say:

2. Ø mi ha infradiciato!
   me has drenched

with the car easily recognizable as the reference of the null subject. This would not be possible if the car was not previously established in our STM by the context. The same utterance, if produced in a different context would point to another referent (if activated contextually), or would simply remain impossible to understand.

A very important part of the context of any utterance is what some call its co-text, namely the rest of the text, which obviously means a fair amount of the preceding

© Springer International Publishing Switzerland 2015
H. Christiansen et al. (Eds.): CONTEXT 2015, LNAI 9405, pp. 440–453, 2015.
DOI: 10.1007/978-3-319-25591-0_32

discourse and a very small amount of the following one. For example, if we are not on the mentioned road, the pronoun in utterance (1) won't find its referent when uttered out of the blue; but a previous mention of the car would have the same effect as its physical appearance:

3. Yesterday, while I was walking under the rain, a car passed on a puddle, driving very fast. It drenched me completely!

What allows receivers to interpret an anaphoric expression (a pronoun, or a null form) is the fact that its referent is active in STM at utterance time. Both concrete events and linguistic utterances can activate contents. In Chafe's (1987, 1992, 1994) terminology, information recently introduced in discourse and active in the addressee's STM is called the Given (G), while information with no recent introduction in prior discourse or communicative situation, and therefore inactive in the addressee's STM, is the New (N). Any information encoded by an utterance is more or less Given/New at utterance time, depending on many discourse factors, among which how and how recently it has been introduced by the context.

Crucially, language encodes Given and New information differently. Anaphorical treatment of the Given but not of the New is just one example. Another very important one is known as Information Structure (IS). Different contexts require giving different emphases to utterance contents. For instance, (4) and (6) are appropriate to different contexts, according to which contents are Given and which are New, while (7) and (5), though identical, are far less acceptable because they appear in reversed contexts:

– I need a cheap laptop

4. [You can find cheap laptops]$_T$ [at MediaWorld]$_F$
5. ?[At MediaWorld]$_T$ [you can find cheap laptops]$_F$

– I will visit MediaWorld this afternoon

6. [At MediaWorld]$_T$ [you can find cheap laptops]$_F$
7. ?[You can find cheap laptops]$_T$ [at MediaWorld]$_F$

The reason is that under unmarked prosody such utterances are made of two different units (Cresti 1992, 2000, Lombardi Vallauri 2001b, 2009). The Focus (to the right) conveys information proposed by the speaker as his main contribution to the ongoing interaction; whereas the Topic (to the left) provides the semantic domain that makes the Focus understandable, and links focal information to the foregoing discourse. Correspondingly, only the Focus conveys the illocutionary force of the utterance (Austin 1962). On pragmatic grounds (cf. Grice's 1975 Maxims of Quantity), the main contribution of the message is usually expected to be information which has not been recently mentioned and is not currently active in the addressee's STM. This explains the oddity of (7), where MediaWorld is presented in Focus (F) as if it was New information, although it has been introduced by the preceding utterance. On the contrary, in the same context, (6) is perfectly acceptable, because it encodes MediaWorld as a Topic (T), thus signaling that it must be interpreted as something already active which is only proposed in order to provide an understandable semantic setting for the comprehension of the Focus unit, which encodes the real communicative contribution of the message, namely that you can find (there) cheap laptops.

A further effect of Focal vs. Topical mention is that presentation as Focus typically also contains the instruction to keep that chunk of information activated in STM (and thus available for anaphoric reference) in the subsequent discourse. In (8–10) anaphorical resuming of preceding contents made by means of the pronoun *this* has as its antecedent the content that was focused in the preceding clause (capitals in (10–11) signal prosodic prominence of marked Focus located to the left):

8.   [You read novels]$_{T(1)}$ [in translation]$_{F(2)}$, and this$_{(2)}$ is a waste of time.
9.   [In translation]$_{T(1)}$ [you read novels]$_{F(2)}$, and this$_{(2)}$ is a waste of time.
10. [You read NOVELS]$_{F(1)}$ [in translation]$_{T(2)}$, and this$_{(1)}$ is a waste of time.
11. [In TRANSLATION]$_{F(1)}$ [you read novels]$_{T(2)}$, and this$_{(1)}$is a waste of time.

The Given/New and Topic/Focus categories are related as we have briefly sketched, but by no means coincident. As Sgall et al. (1973: 17, with "comment" as a synonymous term for Focus) put it:

The distinction between topic and comment is autonomous, in the sense that it cannot be derived from the distinction between "given" (i.e. the known from the preceding context or situation, contained among the presuppositions) and "new" (not given).

Their inter-independence is shown by utterances like (12–14), which are by no means rare in ordinary communication:

12. A: What are Sue and Tim going to do tomorrow morning?
     B: [Sue]$_{T/G}$ [is going to study hard]$_{F/N}$
13. A: What are your friends going to do tomorrow morning?
     B: [Mary]$_{T/N}$ [is going to study hard]$_{F/N}$
14. A: Are Sue and Tim going to study hard tomorrow morning?
     B: Only [SUE]$_{F/G}$ [is going to study hard]$_{T/G}$

Due to the pragmatic grounds we have hinted at, (12) shows the most typical alignment - i.e. Given-Topic and New-Focus. But (13) is also possible, where new content is encoded in the sentence Topic, and (14) where given content is encoded in the Focus. Such utterances are possible, as also often shown by really-produced texts. The following passage from a public interview contains a Focus made of Given, just mentioned information, and a Topic made of New, not yet mentioned information:

15. So, I used to do freelance display jobs, 'cause a lot little stores in London – they didn't have a freelance display person, so I would do [these freelance jobs]$_{G/F}$. They were fine and there was extra cash. Then, [this guy]$_{N/T}$ came by and he said: "That's great! It's really fun! You should come work for me in L.A...".[1]

In (16) below (taken from the opening of another interview) several New Topics can be found:

---

[1] https://www.youtube.com/watch?v=GlVO87Qdm-M.

16. A: How and when did you get started as a writer?

B: [Growing up]$_{N/T}$ [I hated reading]$_{N/F}$. [My English teacher]$_{N/T}$ [announced that we'd be reading 4 books that year]$_{N/F}$. [Before I knew what was happening to me]$_{N/T}$, [I was buying books in stores and reading at least a book a month]$_{N/F}$. [When I was a freshman in high school]$_{N/T}$, [I wrote a short story about a busboy working at a party house]$_{N/F}$. [It was published in the school's annual magazine] $_{N/F}$. [My career]$_{N/T}$, you might say, [had begun]$_{N/F}$[2]

So, the occurrence of Topic and Focus can match Given and New content respectively, but it can also happen the other way round. The first case is most frequent since it is the unmarked, default option. We will call it "aligned". The second case, which we will call "misaligned", is possible but less likely to appear. For reasons we cannot expose here (cf. Lombardi Vallauri 2001a), Given Focuses are quite rare and only used for strongly contrastive effects. New Topics are more frequent, and reflect the need to avoid excessively frequent illocutionary acts of focusing contents in discourse (Lombardi Vallauri and Masia in press): if some information only has the function of providing a semantic setting for the utterance Focus (and does not need to be devoted any new mental slot), the best thing would be to package it as Topic, no matter whether it is Given or New information.

Clearly, New and Given contents have to be dealt with differently by the brain (Burkhardt and Roehm 2007, Benatar and Clifton 2014): the former require that new slots are created in memory for new entities (updating the register), while the latter basically require recognition of referents already existing in memory (linking them to the already existing register). This raises the hypothesis whether IS categories, and in particular Topic and Focus, have developed in all languages to ease the processing of upcoming information by signaling which of these two functions each chunk of information must undergo. Topical "packaging" has the main function of telling the addressee to recognize certain content just by looking for instances of the same information among the concepts that are presently active in his memory. Focal packaging, on the contrary, instructs to consider that information as a fresh contribution of that utterance, needing for a new position to be created in memory. Such different tasks are likely to require processing efforts that are different in nature, and probably also in intensity.

To sum up, Topic and Focus raise useful expectations on the information status of their content. When these expectations are not met, this cannot be without effect. Consequently, as we will argue below, the processing of misaligned Given Focus and New Topic conditions is expected to require different allocation of cognitive resources, as compared to the processing of aligned Given Topic and New Focus configurations.

2   Previous Experimental Studies

In the psycholinguistic domain, it is a common habit to trace back earlier studies on IS processing to Erickson & Mattson's Moses Illusion test (Erickson and Mattson

---

[2] http://lianametal.tripod.com/id38.html.

1981), later replicated by Bredart and Modolo (1988). The aim of the test was to show that the depth of processing of some information may change depending on the particular structure it receives in the sentence. In Erickson and Mattson's study, a number of subjects were presented with the following question: *How many animals of each kind did Moses take on the Ark?* The authors observed that almost all the subjects were likely to respond "two" without noticing that it was Noah, and not Moses, that took animals on the Ark. The effect of Information Structure in the detection of the distorted term was later assessed by Bredart and Modolo (1988) who manipulated the syntactic structure of the original sentence so as to have *Moses* once in Focus and once in Topic position (cf. *It was* [MOSES]$_F$ *who took two animals of each kind on the Ark* vs. *It was* [TWO ANIMALS]$_F$ *of each kind that Moses took on the Ark*). Not surprisingly, the subjects noticed the distortion when *Moses* was in the sentence Focus, while they tended to miss it altogether when it was conveyed as Topic, that is in the complement clause of the cleft construction. Along similar lines, later studies (Birch and Rayner 1997; Sturt et al. 2004) have further substantiated the hypothesis that Focus induced more attentive processing, thus facilitating the recognition of lexical substitutions (cf. change-detection tests) or anomalies of any kind. However, most of the paradigms adopted in these works probed into the cognitive effects of different information structures in somewhat artificial processing conditions. Precisely, testing materials in these works mostly consisted in isolated sentences with scant or no prior linguistic context grounding for the establishment of different givenness or newness degrees of contents. Indeed, only the effects of different syntactic manipulations (e.g. clefts vs. declarative sentences) were taken into account. As a result of methodological biases stemming from these paradigms, critical sentences were interpreted as conveying all new information. As Hruska and Alter (2004, p. 223) put it, "without any preceding information, the listeners [or readers] analyze each sentence as completely new and no information has to be embedded in an already given context". This also led to characterizing processing costs of contents in terms of their topical or focal nature *per se*, disregarding the impact of givenness or novelty degrees on the interpretation of IS units.

Later neurological strands on IS processing (Hruska and Alter (2004; Toepel and Alter 2004; Baumann and Schumacher 2011; Wang and Schumacher 2013) have highlighted far more intricate scenarios on the processing underpinnings of Topic-Focus and Given-New information units. Particularly, it has been noticed that the cost required to process Topic and Focus information is contingent on the activation degree of the content they carry. Generally speaking, when new information expectedly conflates with Focus and given information with Topic, processing effort appears much smaller than it is when Topic-New and Focus-Given combinations obtain. In studies utilizing Event-Related Potential (ERP) techniques, increases or decreases of processing demands in response to aligned and misaligned informational matchings have been revealed by variations in N400 signatures, with higher deflections elicited by misaligned packaging. The involvement of N400 modulations in such discourse phenomena has been delved into on both phonological and syntactic bases. An ERP study on IS processing in Japanese (Wang and Schumacher 2013) shed light on the role of syntax in graduating decoding efforts imposed by new information. Particularly, it has been observed that if in Japanese SOV declarative sentences new information is placed in the syntactic position of Focus, its processing is generally faster and less effortful than it is if it were placed in sentence-initial position, where Topics are more typically realized. This effect is particularly salient if the preceding discourse context licenses expectations on the syntactic positioning of new information. From a phonological perspective (Baumann and Schumacher 2011), the effects of aligned vs. misaligned information packaging have been measured modifying intonation contours on Given and New contents. Specifically, when given contents receive superfluous accent and are therefore interpreted as Foci, their processing elicits higher N400 peaks. Analogous neural patterns have been observed for missing accents on new contents, fostering their interpretation as Topic. Other ERP studies on IS in German (Hruska and Alter (2004; Toepel and Alter 2004) pointed towards similar findings in this respect.

Now, it stands to reason that the outcome results achieved by these latter neurological approaches are indicative of the importance of looking on sentence processing from a far wider perspective than that adopted in the psycholinguistic accounts above given. This perspective should encompass both sentence-internal and discourse-based strategies of information encoding. Apart from providing valuable insights into the role of context-driven expectations, this view also allows to tackle Information Structure and its processing in their natural, real conditions.

3   Measuring the Effects of Expectations in Context-Driven Processing of IS Units: Evidence from EEG Signals

This section describes and discusses the results of an EEG experiment aimed at measuring rhythmic changes in different frequency bands in response to topical and focal sentences in texts. The texts have been elaborated so as to have both aligned and misaligned configurations between activation statuses of contents and their topical or focal packaging in the sentence. Therefore, given contents are sometimes associated with topical, sometimes with focal packaging, and the same goes for new contents. The main purpose of the analysis is to assess the efficiency of brain processing in response to more or less expected information packagings relative to givenness and newness cues provided by the foregoing linguistic context.

Since critical regions are sentence-long, investigating EVENT-RELATED SYNCHRONIZATION (ERS) and EVENT-RELATED DESYNCHRONIZATION (ERD) in different brain rhythms proved more suitable. The elicitation of ERP components (more easily detected in response to word-length stimuli) would have been more difficult to nail down, due to the possible overlapping of components related to contiguous units within the same sentence. Before laying out the experimental procedure and the related findings, some preliminary remarks on brain oscillations (or rhythms) are in order.

Brain oscillations are generated by populations of neurons depolarizing in synchrony. Rhythms may oscillate independently of one another or may overlap, if different neural assemblies generate the same rhythm. Brain rhythms are generally distinguished for the particular shape they display, their range of oscillation and scalp distribution. Hans Berger (1873–1941), the pioneer of EEG in humans, pinpointed five different rhythms oscillating at different frequency ranges (Hz) and correlating with different cognitive operations. These are the DELTA (δ) rhythm (0.1–3 Hz), the THETA

(θ) rhythm (4–7 Hz), the A<small>LPHA</small> (α) rhythm (8–12 Hz), the B<small>ETA</small> (β) rhythm (above 12 Hz) and the G<small>AMMA</small> (γ) rhythm (40 Hz). It has been contended in a number of studies that ERS and ERD effects correlate with different mental operations or states. More particularly, cognitive overloading triggers neural synchronization in the θ and γ bands, but neural de-synchronization in the α and β bands. Put another way, while oscillatory amplitudes in θ and γ rhythms are directly related to effortful processing, those in α and β rhythms are related to it inversely (Bastiaansen et al. 2005)[3].

3.1 Predictions

The present study can be regarded as an attempt to further substantiate the role of context-driven strategies in IS processing from the perspective of rhythmic changes in different frequency ranges. Based on the considerations sketched out above, we expect misaligned combinations between activation statuses and information packaging to impose major processing demands than aligned combinations, with variations foreseen in the θ, δ, γ, β and α bands, although with different oscillatory behaviors. Notably, misaligned conditions are expected to elicit ERS effects in the θ, δ or γ bands, and ERD effects in the α or β bands, since misaligned conditions overturn packaging expectations, thus causing more difficult integration of upcoming information into the addressee's register.

3.2 Method

54 healthy subjects (20–35 years old) have participated in the experiment, after giving written informed consent. EEG signals have been acquired using a 19-channels cap (GALILEO Be Light Amplifier). For the study, only female subjects have been considered[4]. Subjects were comfortably seated in an insonorized dimly lit room. Electrodes were placed on the scalp according to the standard 10–20 montage and impedances were kept below 10 kΩ. Recordings have been referenced to the AFz position. Texts were presented auditorily and EEG recordings were time-locked to the listened utterances by synchronizing the signals marking each critical sentence on the raw traces.

3.3 Testing Material

For the experiment, four couples of specular Italian texts have been used, in which the same item of information (given and new) is Topic in one text and Focus in the other, or vice versa. Keeping the same notional content unaltered in the two specular conditions permitted to avoid cognitive biases caused by different discourse representations. Below, an illustration of the pattern described is provided.

---

[3] Ibid. (p. 530): "For both alpha and beta band activity it holds that the amplitude of these oscillations is inversely related to active processing, whereas increases stand for cortical idling and/or inhibition. The opposite holds for the two other frequency bands that have been extensively studied, that is, the theta and gamma bands. Here, it generally holds that amplitude increases are related to the active processing of information".

[4] Signals recorded from other male students were removed from the dataset, due to the high number of artifacts.

LIST A[5]:

Context:

Da adulti, siamo generalmente inclini a temere le emozioni negative. In questo senso, [*che si sviluppino dipendenze legate ai bisogni non soddisfatti*]$_{\text{NEW/TOPIC}}$ è molto frequente. Le nostre debolezze ci vengono rivelate [*dal verificarsi di questo tipo di dipendenze*]$_{\text{GIVEN/FOCUS}}$.

In questi casi, molti si rifugiano nel bere un po' di vino con un amico. [*Dopo aver sorseggiato qualche bicchiere di vino*]$_{\text{GIVEN/TOPIC}}$, [*per un po' il dolore svanisce*]$_{\text{NEW/FOCUS}}$.

LIST B:

Context:

Data la nostra inclinazione a temere le emozioni negative, spesso [*sviluppiamo dipendenze legate ai bisogni non soddisfatti*]$_{\text{NEW/FOCUS}}$ [*Quando si verifica questo tipo di dipendenze*]$_{\text{GIVEN/TOPIC}}$, scopriamo le nostre debolezze.

In questi casi, molti si rifugiano nel bere un po' di vino con un amico. E [*il dolore per un po' svanisce*]$_{\text{NEW/TOPIC}}$ [*sorseggiando qualche bicchiere di vino con qualcuno*]$_{\text{GIVEN/FOCUS}}$.

In Context A, the sentence *che si sviluppino dipendenze legate ai bisogni non soddisfatti* ("that unsatisfied needs generate dependences") carries newly activated information, which also realizes the Topic unit of the sentence. The same information item, in the same activation state, appears focalized in Context B (*sviluppiamo dipendenze legate ai bisogni non soddisfatti*, tr. "we often develop dependences related to unsatisfied needs").

Based on different alignment conditions, the following combinations obtain: Given/Topic (GT), New/Topic (NT), New/Focus (NF) and Given/Focus (GF). The number of occurrences per each condition is: 13 for G/T, 29 for N/T, 11 for G/F and 28 for N/F. All texts have been presented auditorily; one list has been listened to by a group A of 27 Italian subjects, the other list by a group B of 27 subjects (other than the former group).

---

[5] LIST A:

Context:As adults, we are generally bound to fear negative feelings. In this sense, [*that unsatisfied needs generate dependences*]$_{\text{NEW/TOPIC}}$ is very frequent. Our weaknesses are revealed [*by the manifestation of these dependences*]$_{\text{GIVEN/FOCUS}}$. In these cases, many feel comfortable drinking some wine with a friend. [*After sipping some wine*]$_{\text{GIVEN/TOPIC}}$, [*the pain disappears for a while*]$_{\text{NEW/FOCUS}}$. LIST BContext:Given our tendency to fear negative feelings, [*we often develop dependences related to unsatisfied needs*]$_{\text{NEW/FOCUS}}$. [*When these dependences come about*]$_{\text{GIVEN/TOPIC}}$ we discover our weaknesses. In these cases, many feel comfortable drinking some wine with a friend. And [*pain disappears for a while*]$_{\text{NEW/TOPIC}}$ [*sipping some wine with somebody*]$_{\text{GIVEN/FOCUS}}$.

3.4 Data Processing: Spectral Analysis

For reasons of space, only the spectral analysis of frequency bands is reported in this section. Further statistical and cross-spectrum analyses are more extensively debated in La Rocca et al. (in preparation).

In a pre-processing stage, a Common Average Referencing (CAR) has been applied to signals in order to reduce artifacts associated with inappropriate reference choices. Signals were then segmented into epochs time-locked to the onset of each critical region. Trials with non-removable artifacts for one subject have been removed from the analysis for all subjects, so the resulting epochs amounted to 40. Filtered epochs were further time-locked to the following set of contrasts: N/F vs. N/T, G/F vs. G/T, N/F vs. G/T and N/T vs. G/T. Since the purpose of the analysis was to detect differences in frequency bands' activity, a computation of the Power Spectrum Density (PSD) has been carried out. This measurement allows assessing the contribution of each EEG rhythm to the differences observed in the above-mentioned contrasts; more precisely, it indicates how the strength of a signal is distributed in the frequency domain (Stoica and Moses 2005). Another relevant measure when the study of frequency band activity is approached is the so-called Spectral Coherence (COH). In signal processing theory, Spectral Coherence indicates the cooperation of populations of neurons during various cognitive processes, whatever their nature. Neural spectra change depending on how synchronous the activity of different neural populations is, in response to given processing tasks. Technically, coherence quantifies the level of synchrony between simultaneous recorded signals at a specific frequency f.

3.5 Results

To better assess differences between the contrasts of interest, both *univariate* (single-channel) and *multivariate* (multiple-channel) analyses have been carried out, which evidenced a more prominent involvement of the α, β and θ bands, whereas in the γ band no significant variation for the aligned and misaligned conditions has been noticed, and for δ band only some spectral coherence phenomena were measured, which we cannot expose here for the sake of brevity, though they confirm the ERS and ERD results for the α, β and θ bands.

In the UNIVARIATE analysis, noticeable variations have been observed in the N/T vs. N/F contrast. Particularly, larger amplitudes in the β frequency band (mainly distributed in the right centro-parietal region) have been observed for the N/F compared to the N/T condition (Table 1). Analogous trends have been noticed in the α frequency band (central, parietal and temporal region), with larger amplitudes for the N/F condition. On a priori grounds, the processing of misaligned informational articulations in these two frequency bands seems to elicit ERD effects. Variations in θ band activity are more prominent for the G/T and G/F contrasts (Table 2). Particularly, a greater power in the left temporal region has been registered in the θ band for the G/F condition. Increasing amplitudes in this frequency band can be accounted for in terms of ERS effects. This preliminary scenario suggests (a) that the processing of aligned and misaligned information structures involves the activity of different frequency bands, and (b) that misaligned conditions entail costlier processing mechanisms, possibly due to inconsistent cues to bring about knowledge updating.

Table 1. N/F vs. N/T. PSD in the univariate analysis

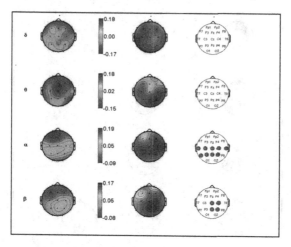

Table 2. G/F vs. G/T. PSD in the univariate anaysis

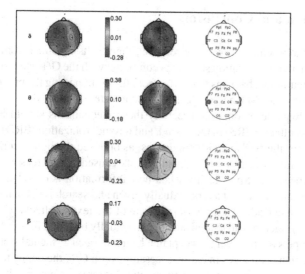

A MULTIVARIATE analysis conducted on channel pairs confirmed the single-channel analysis results, detecting larger amplitudes in the β rhythm for the N/F condition as compared to N/T, mainly distributed in the right centro-parietal and occipital region (Table 3). Similar trends have been observed in the α band for the same aligned condition, with larger amplitudes in the right centro-temporal region (Table 3). In line with the univariate analysis, the multivariate analysis confirmed higher amplitudes in the θ

rhythm (basically involving right centro-frontal regions) for the G/F condition, as compared to the G/T condition.

Table 3. Misalignment vs. Alignment. PSD in the multivariate analysis

4   Discussion and Conclusion

Although probably tentative in many respects, the results obtained from the analysis proposed reveal that information structures compliant with the G/T and N/F pairings are subserved by neural mechanisms that differ from those supporting the processing of less expected pairings (i.e. N/T and G/F). The present state of the art does not allow us to make reliable hypotheses on what precisely these mechanisms should be, but larger synchronization effects (ERS) in the θ band, and desynchronization (ERD) effects in the α and β bands for the N/T and G/F conditions, as opposed to aligned correlations (i.e. G/T and N/F), are indicative of increasing costs caused by decreasing efficiency in information processing. Specifically, when new information is encoded as Topic, the receiver is instructed to treat it as contextually given and search for an antecedent in the foregoing discourse (indeed *anaphoricity* is one of the textual effects of topicality, cf. Givón 1983). The fact that no textual anchor can actually be found for the N/T condition eventually compels the receiver to accept that he has to open a "mental slot" for the new information, which eventuates in extra processing. With G/F, the opposite happens, the addressee being instructed to create a new mental slot for some entity, which eventually reveals to already have one.

Among other things, these findings allow to account for sentence structure and information structure processing from a *Gestalt psychology* perspective (Wertheimer 1927, Koffka 1922, Köhler 1947). According to Gestalt psychology principles, similar elements are perceived as continuous and consistent with one another, which makes them processable with fewer costs. If continuity expectations are not met, more effort is involved in the processing of discontinuous elements, because they bear features which are not shared by other entities previously encountered in the discourse context. Analogous effects have been observed in the assessment of *study times* (Ackerman et al. 2014). Recent works in experimental psychology have revealed that study times are reduced when the contents to be learned can be easily related to already existing knowledge schemata, and are therefore perceived as more continuous and consistent with respect to previously or currently stored knowledge. Indeed, prediction- or expectation-based processing facilitates a person's ability to adjust his performance to a particular goal (Ackerman et al. 2014, Koriat and Bjork 2006).

Transposed to our core discussion, continuous elements are represented by contents whose activation state and packaging is consonant with what is expected depending on the linguistic and extra-linguistic context set by the communicative exchange; whereas discontinuous elements are contents whose activation state and packaging is less expected with respect to the linguistic and extra-linguistic context set. Therefore, keeping perceptual continuity amounts to reducing demands on IS processing, whereas interrupting continuity expectations increases the effort required of the receiver to fully integrate information in his mental model of discourse.

A further consideration can be made. The fact that linguistic understanding and conceptualizing strongly relies on the context in which communication takes place is probably self-evident, and can be regarded as major evidence in favor of the *Extended Mind hypothesis* (Clark and Chalmers 1998; Clark 2008). According to this hypothesis, the mind is not completely autonomous in its working, and it shares its contents with some external supports. Among these, if language as such (the language faculty, perhaps the grammar itself) can be regarded as contained in the mind, on the contrary linguistic *texts* are something whose existence (in written, persistent or in acoustic, transient form) is no doubt located outside the mind. If considered just as its products, texts do not contribute to conceive of the mind as an extended, multidimensional reality; but considered as premises steadily posing precise constraints and providing indispensable material to the working of the mind, they do.

As we have seen, during linguistic activity the mind "exchanges information" with the preceding context, especially by receiving the activation of concepts from what is said. Language is designed (probably, has evolved, cf. Lombardi Vallauri and Masia 2014, Lombardi Vallauri and Masia in press) to better deal with this function, and the categories of IS we have dealt with here are among the most important means by which this is achieved.

In particular, the mind regulates the traffic of contents from the linguistic and extra-linguistic preceding context of communication to their further encoding in new linguistic utterances, and does so by storing them temporarily in STM. In this process, there is a difference between certain information being already Given or New, and all languages use Topic and Focus as means to ease the different processing of such categories; more precisely, to anticipate and thus ease the recognition of what must be treated as Given or New. When expectations are met, things are easier; when they are not, more effort is required. The experimental findings we have exposed strongly confirm this hypothesis, and show that these conditions correspond to measurable brain events.

References

Ackerman, R., Lockl, K., Schneider, W.: The effects of goal-driven and data-driven regulation on metacognitive monitoring during learning: a developmental perspective. J. Exp. Psychol. Gen. doi:10.1037/a0031768

Austin, J.L.: How to do Things with Words. Clarendon Press, Oxford (1962)

Benatar, A., Clifton Jr, C.: Newness, givenness and discourse updating: evidence from eye movements. J. Mem. Lang. **71**, 1–16 (2014)

Bastiaansen, M.C.M., van der Linden, M., ter Keurs, M., Dijkstra, T., Hagoort, P.: Theta responses are involved in lexical-semantic retrieval during language processing. J. Cogn. Neurosci. **17**(3), 530–541 (2005)

Baumann, S., Schumacher, P.B.: (De-)accentuation and the processing of information status: evidence from event-related brain potentials. Lang. Speech **55**(3), 361–438 (2011)

Bredart, S., Modolo, K.: Moses strikes again: focalization effect on a semantic illusion. Acta Psychol. **67**, 135–144 (1988)

Burkhardt, P., Roehm, D.: Differential effects of saliency: an event-related brain potential study. Neurosci. Lett. **413**, 115–120 (2007)

Birch, S. Rayner, K.: Linguistic focus affects eye movements during reading. Mem. Cogn. **25**(5), 653–660 (1997)

Chafe, W.: Cognitive constraints on information flow. In: Russell, T. (ed.) Coherence and Grounding in Discourse, pp. 21–51. Benjamins, Amsterdam, Philadelphia (1987)

Chafe, W.: Information flow in speaking and writing. In: Downing, P., Lima, S.D., Noonan, M. (eds.) The Linguistics of Literacy, pp. 17–29. Benjamins, Amsterdam, Philadelphia (1992)

Chafe, W.: Discourse, Consciousness, and Time. The University of Chicago Press, Chicago (1994)

Clark, A.: Supersizing the Mind: Embodiment, Action, and Cognitive Extension. Oxford University Press, Oxford, New York (2008)

Clark, A., Chalmers, D.J.: The extended mind. Analysis **58**, 7–19 (1998)

Cresti, E.: Le unità d'informazione e la teoria degli atti linguistici. In: Gobber, G. (ed.) Atti del XXIV Congresso internazionale di studi della Società di Linguistica Italiana. Linguistica e pragmatica, pp. 501–529. Bulzoni, Roma (1992)

Cresti, E.: Corpus di italiano parlato. Accademia della Crusca, Firenze (2000)

Erickson, T.D., Mattson, M.E.: From words to meanings: a semantic illusion. J. Verbal Learn. Verbal Behav. **20**(5), 540–551 (1981)

Givón, T.: Topic Continuity in Discourse: A Quantitative Cross-Language Study. John Benjamins, Amsterdam, Philadelphia (1983)

Grice, H.P.: Logic and conversation. In: Cole, P., Morgan, J.L. (eds.) Syntax and Semantics. Speech Acts, vol. 3, pp. 41–58. Academic Press, New York (1975)

Hruska, C., Alter, K.: Prosody in dialogues and single sentences: how prosody can influence speech perception. In: Steube, A. (ed.) Language, Context and Cognition. Information Structure: Theoretical and Empirical Aspects, pp. 211–226. Walter de Gruyter, Berlin (2004)

Koffka, K.: Perception: an introduction to the gestalt-theory. Psychol. Bull. **19**, 531–585 (1922). Reprinted in Shipley, T. (ed.) Classics in Psychology, pp. 1128–1196. Philosophical Library, New York (1961)

Köhler, W.: Gestalt Psychology. An Introduction to New Concepts in Modern Psychology. Liveright, New York (1947). Revised version of Köhler (1929)

Koriat, A., Bjork, R.A.: Mending metacognitive illusions: a comparison of mnemonic-based and theory-based procedures. J. Exp. Psychol. Learn. Mem. Cogn. **32**(5), 1133–1145 (2006)

La Rocca, D., Masia, V., Maiorana, E., Lombardi Vallauri, E., Campisi, P.: Processing effects of aligned and misaligned Information Structures; evidence from EEG rhythmic changes (in preparation)

Lombardi Vallauri, E.: The role of discourse, syntax and the lexicon in determining focus nature and extension. Linguisticae Investigationes 23(2), 229–252 (2001)

Lombardi Vallauri, E.: La teoria come separatrice di fatti di livello diverso. L'esempio della struttura informativa dell'enunciato. In: Sornicola, R., Stenta Krosbakken, E., Stromboli, C. (eds.) Dati empirici e teorie linguistiche, Atti del XXXIII Congresso della Società di linguistica italiana, Napoli, 28–30 Ottobre 1999, Bulzoni, Roma, pp. 151–173 (2001b)

Lombardi Vallauri, E.: La struttura informativa. Forma e funzione negli enunciati linguistici. Carocci, Roma (2009)

Lombardi Vallauri, E., Masia, V.: Implicitness impact: measuring texts. J. Pragmat. 61, 161–184 (2014)

Lombardi Vallauri, E., Masia, V.: Cognitive constraints on the emergence of topic-focus structure in human communication. In: Chiera, A., Ganfi, V. (eds.) Proceedings of Codisco (2015, in press)

Sgall, P., Hajičová, E., Benešová, E.: Topic, Focus and Generative Semantics. Scriptor, Kronberg, Taunus (1973)

Stoica, P., Moses, R.: Spectral Analysis of Signals. Prentice Hall, New Jersey (2005)

Sturt, P., Sandford, A.J., Stewart, A., Dawydiak, E.: Linguistic focus and good-enough representations: an application of the change-detection paradigm. Psychon. Bull. Rev. 11, 882–888 (2004)

Toepel, U., Alter, K.: On the independence of information structural processing from prosody. In: Steube, A. (ed.) Language, Context and Cognition. Information Structure: Theoretical and Empirical Aspects, pp. 227–240. Walter de Gruyter, Berlin (2004)

Wang, L., Schumacher, P.B.: New is not always costly: evidence from online processing of topic and contrast in Japanese. Front. Psychol. 4, 1–20 (2013)

Wertheimer, M.: Gestaltpsychologische Forschung. In: Saupe, E. (ed.) Einführung in die neuere Psychologie (Handbücher der neueren Erziehungswissenschaften, 3), pp. 46–53. Zwickfeldt, Osterwieck, Harz (1927)

Cross-Linguistic Experimental Evidence Distinguishing the Role of Context in Disputes over Taste and Possibility

E. Allyn Smith[1]([✉]), Elena Castroviejo[2], and Laia Mayol[3]

[1] UQÀM, Montreal, Canada
[2] EHU/UPV, Vitoria-Gasteiz, Spain
[3] UPF, Barcelona, Spain
smith.eallyn@uqam.ca

1 Introduction

'Faultless disagreement' is exemplified in (1) and (2) for predicates of personal taste (PPT), where the intuition of philosophers and linguists has been that Sam is expressing his opinion about the cake and that Sue, in saying *No*, is either expressing that the the cake does not taste good to her or expressing that the cake should not be considered tasty more generally (following [3] and [4], i.a.).

Mary: How's the cake? Sam: It's tasty. (1)

Sue: No it isn't, it tastes terrible! (2)

Mary: How's the cake? Sam: It tastes good to me. (3)

Sue: #No it isn't/doesn't, it tastes terrible! (4)

One might think that Sam's utterance in (1) is a subjective one, essentially expressing that he personally finds the cake tasty, in which case one would not expect significant meaning differences between (1) and a similar utterance where the subjectivity is made explicit, such as (3). However, [5], arguing for a modification of [6], points out that when PPTs are explicitly relativized to the speaker, it is no longer felicitous to use *No* to directly dispute utterances in which they appear, as shown in (4). Stephenson draws a parallel between PPTs and epistemic modals (EMs) such as *might*, as in (5)–(8), on the basis of their behavior when followed by *No*, where the addition of *I don't know* to the utterance makes the uncertainty of the modal explicit.[1] The general idea is that both PPTs and EMs change behavior with speaker-restriction: in PPTs, the restriction tells us whose tastes are represented, and in EMs, it tells us whose knowledge is the basis for the statement (following a Kratzerian analysis of modals).[2]

---

[1] She actually uses *I don't know that he isn't*, which we simplified for our experiment given the known difficulty of multiple negations for naive speaker judgments as well as difficulty translating into languages where the literal form is ungrammatical.

[2] A reviewer points out that the pattern in (7) and (8) is not necessarily replicated with other ways of relativizing to the speaker, such as *As far as I know...*, an intuition we share but did not test in our experiment. This evidence supports our same general conclusion that PPTs and EMs are perhaps not as parallel as they first appear.

© Springer International Publishing Switzerland 2015
H. Christiansen et al. (Eds.): CONTEXT 2015, LNAI 9405, pp. 454–467, 2015.
DOI: 10.1007/978-3-319-25591-0_33

Mary: Where's Bill? Sam: He might be in his office   (5)

Sue: No, he can't be. He doesn't work on Fridays.   (6)

Mary: Where's Bill? Sam: I dont know whether he's in the office.   (7)

Sue: #No, he isn't/can't be, he doesn't work on Fridays.   (8)

Given the importance of these data to subsequent work in the field across a variety of disciplines and phenomena (including the contextualist/relativist debate, the treatment of polarity particles, and the use of refutation as a diagnostic elsewhere, such as for presupposition projection, just to name a few), we tested whether this pattern could be observed behaviorally in the perceptions of English speakers and whether it could be replicated cross-linguistically in French, Spanish, and Catalan (though for reasons of space, we will only discuss the English and Spanish results here). A better understanding of the psycholinguistic patterns of these "disagreements" is additionally important if they are to be used in computational or other applications given that, up to this point, they have been considered as patterns arising from linguistic structure in context and not, for example, as arising from individual or cultural context differences. The results we present here suggest that there is variability across languages but not across dialects spoken in culturally-different areas, suggesting that the patterns in (1)–(8) are indeed a product of linguistic structure and context. In order to account for the cross-linguistic differences, however, it will be important to de-couple the analysis of subjectivity in PPTs and EMs.

2   Taste and Possibility

Though authors disagree about the location and the number of sources of subjectivity in sentences with PPTs, linguists have mostly modelled these predicates as introducing judge parameters (e.g. [5–7]). For the purposes of this paper, the details of how each of these theories is formalized and the differences between contextualist and relativist approaches are less important than the fact that, in each, there is a parallel drawn with EMs (cf. [5,8]). If PPTs evaluate content/truth with respect to someone's taste, then, analogously, EMs evaluate content/truth with respect to someone's knowledge. In both cases, when that person is overtly specified grammatically (to be the speaker) rather than filled in contextually, disagreement should become infelicitous, consistent with the intuitions represented in (4) and (8) above.

[1] gives an account of PPTs and other kinds of evaluative propositions such as those with aesthetic adjectives, and in so doing, draws on [2]. We present the basics of their discourse model and return to Umbach's own analysis in the discussion section. The model contains:

- A common ground, (CG): publicly shared beliefs by all participants.
- A discourse commitment (DC) set, DC_X, for each participant X: X's public beliefs that are not (or not yet) shared by all other participants.

- The Table: similar in role to the Question Under Discussion (QUD) [11]. A stack of syntactic objects paired with their denotations that represent the discourse topic at any given time.
- A projected set of all potential CGs that could result from future additions to the current CG.
- A set of conventions, operations, and update rules for the kinds of moves permitted. For example:
  - The Table should not be empty (there should be a question)
  - The goal, however, is to empty it (we should answer questions)
  - The canonical way of removing an item from the Table is to reach a state in which the issue is decided (p or $\neg p$ follows from the CG)
  - Asserting p adds p to the speaker's DC and proposes adding it to the CG by 'pushing' it onto the Table (raising the question of 'whether p', which we will call $?p$)
  - If participants voice no objection, p moves from the DC to the CG
  - If someone does voice an objection, several things can happen, including agreeing to disagree, leaving p and $\neg p$ in their respective DC_X sets instead of the CG and popping $?p$ off the Table

This discourse model can account for why many kinds of content may not be directly refuted using *No*. [2] hypothesizes that *No* can only target propositions on the Table at the top of the stack. This accounts for why, for example, presuppositions and conventional implicatures are difficult to reject directly, because they are already in the CG and/or not the main issue at hand (cf. [9,12,13]). Here, we are concerned with other kinds of content that cannot be rejected, namely explicitly relativized PPTs (henceforth, PPTR) and explicitly relativized EMs (henceforth, EMR).

Returning to the initial data, the analysis in this kind of model would be something like the following: PPT sentences are ambiguous between a reading where someone is expressing her opinion only and not pushing for p to be added to the CG (in which case that person is just adding something to her DC without putting it on the Table) and a reading in which she is additionally suggesting that it should be added to the CG for everyone. Some have mentioned the felicity of *Poutine is tasty, but I don't like it*, suggesting that one may also put something on the Table to be accepted as a general fact while simultaneously adding its negation to her own DC. PPTR, on the other hand, have only the first of these readings, meaning that they remain exclusively discourse commitments of the speaker and are not pushed onto the Table. Because they are not on the Table, they may not be targeted by *No*.

One further related concept will be useful for our discussion. [12] define a notion of 'At-issueness' in which a proposition p is at-issue relative to a QUD if $?p$ is relevant to that QUD. In order for a question $?p$ to be relevant to a QUD, it must have an answer which contextually entails at least a partial answer to the QUD. In other words, if the question is *Who is at the library?* and there are four people there, naming all of them is a complete answer, but even giving one or two would be relevant. If someone says *Daniel is at the library*, then $?p$ is *Is*

Daniel at the library?. Either answer (Y/N) would be relevant to the QUD, thus making the proposition at-issue with respect to this QUD. On this analysis, *No* targets only at-issue meanings, giving largely the same result as with [2].

3 Experimenting with Taste and Possibility

In order to confirm the intuitions presented in (1)–(8) for English and other languages, we created an internet-based audio survey in which participants ranked the relative strangeness of various propositions in response to others across a set of constructed two-turn dialogues. While the fillers were all dialogues of agreement, the critical stimuli contained a refutation with *No*. We used English as our baseline given the preponderance of English examples in the literature, but we also tested languages from another family, including Spanish, which we focus on here. Hypothesized cross-linguistic differences included the fact that whereas English makes a difference between a polarity particle and sentence negation (*No* v. *not*), Spanish only has the lexical item *No* to carry out the two functions.

Based on the previous literature, the predictions were as follows:

- Disagreement (faultless or not) should be possible with PPTs and EMs, so their average ratings should be as high as that of assertions.
- Relativizing a PPT to a judge should limit its refutability, so the average ratings for PPTRs should be lower than for PPTs.
- EMs and PPTs should show the same pattern (so EMRs < EMs).
- These basic patterns should be consistent across languages.

Stimuli. The stimuli consisted of 88 two-turn dialogues. 44 of these dialogues were fillers, 20 were critical items and 24 were items for an experiment not reported in this paper. In the first turn, a statement is made that crucially contains a particular meaning type: PPT, PPTR, EM, and EMR or a fifth, Basic Assertion (BA), as a baseline. In the second turn, for the critical stimuli, participants heard a direct refutation. Example stimuli for the five meaning types in English and Spanish are as follows:

- BA: Daniel is at the library. No, he's at the post office.
  Dani está en la biblioteca. No, está en correos (Iberian)/el correo (Uruguayan).
- PPT: It's cold outside. No, it is rather warm.
  Hace frío afuera. No, hace más bien calor.
- PPTR: Laura's talk was boring to me. No, it was inspiring.
  A mí la charla de Laura me pareció patética. No, fue estimulante.
- EM: Matthew may join us later. No, he is going on a date.
  Alberto a lo mejor vendrá (Iberian)/Capaz que Alberto va a venir (Uruguayan) con nosotros más tarde. No, tiene une cita.
- EMR: I don't know whether the kids will take a vacation this year. No, they are going to Corsica.
  No sé si los ninos tendrán vacaciones este ano. No, se van a Córcega.

When translating the stimuli into other languages and dialects, we tried to be as faithful as possible to the original items without ending up with sentences that were vastly less natural than they were in the source language. The stimuli were recorded by two female native speakers in each language; one was always speaker A and the other, B. They read their lists of sentences separately and we spliced them together into dialogues afterwards. We chose audio rather than written presentation of stimuli to avoid the possibility that, when reading, participants might mentally assign different intonational contours and interpret *No* responses more felicitously via metalinguistic negation. We only included takes with falling intonation at the end of both turns.

Procedure. The experiment was administered via an internet survey platform called SurveyGizmo. Participants were asked to listen to short audio files containing one statement followed by either an acceptance or refutation of some part of the meaning of that statement and answer the following question: "How strange do you think it would be for someone to respond this way?" Participants saw the question and were asked to respond using a 7-point Likert scale from 'very strange' to 'not at all strange'. We chose to ask about how strange they thought it might be for someone to respond this way because it allowed us not to train our participants in the notion of entailment or give away the goal of the experiment. We were careful to focus on how strange the *response* was to cut down on judgments of the oddness of the initial sentence by itself, though we also conducted a second task (described in the *Norming* subsection below) to guard against this possibility as well. In addition to screens where they played a sound file and responded to the question of interest, after some of the items, participants had to answer a comprehension question. These questions were added to ensure that participants were paying attention, and the data of those who did not answer correctly was not analyzed. After the experiment, participants were asked to answer a feedback question about what they thought the experiment was about. The great majority of the participants did not show any awareness of the goal of the experiment.

Participants. The data presented here come from 55 American English speakers, 60 Iberian Spanish speakers, and 70 Uruguayan Spanish speakers. All were native speakers of the language in which they were tested. The places from which they were recruited were Chicago, Madrid, and Montevideo, respectively. The English participants fulfilled course requirements by participating, while the Spanish speakers in both locations, though largely also students, were entered for a chance to win one of several gift cards.

Norming task. Since our stimuli were extremely varied, a concern that one might have is that our participants were not rating how good the reply was, but how marked or frequent the first sentence was. This was especially a concern given that some types of stimuli were longer than others. In order to be able to distinguish the markedness of the first sentence from the appropriateness of the reply, we carried out a baseline task in each language with 16 to 32 participants. The procedure was the same as explained above, except that participants only heard the first sentence of any given dialogue, so they did not hear the reply. Then they were asked to rate them in a 7-point Likert scale answering the following question: "How strange do you think it would be for someone to say this?" The participants who participated in the baseline task did not participate in the other experiment. These ratings were then incorporated into our statistical model as described in the next section.

4 Results

All analyses are best-fit linear mixed effects models and were computed in R [15] using the *lmer* function in the *lme4* package [16]. The variable whose behavior we were trying to predict was the normalized Likert score that participants gave as a measure of felicity. The fixed effects included the rating of the felicity of the item without a refutation from the norming task and, when comparing meaning subtypes to one another, the meaning type. As for the random effects, the best models were always those that took item and subject differences into account (i.e. random intercepts for 'Item' and 'Participant'), and occasionally the best model was one that included the by-subject random slope for meaning subtype (which accounts for any adjustments to the effect of 'Subtype' by a difference between participants). What follows are plots of the normalized means.

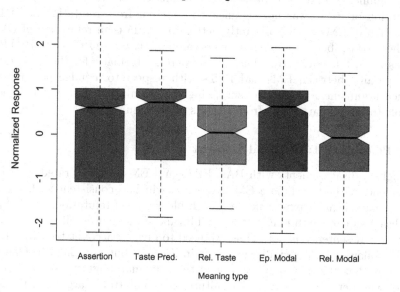

English Results. The figure above matches the statistical results: there is no significant difference between the means of assertion, PPTs and EMs. Refutation of PPTs was significantly better (p < .01) than refutation of PPTRs. Refutation of EMs was significantly better (p < .01) than refutation of EMRs. In other words, the intuitions in the literature are perfectly borne out in the English data.

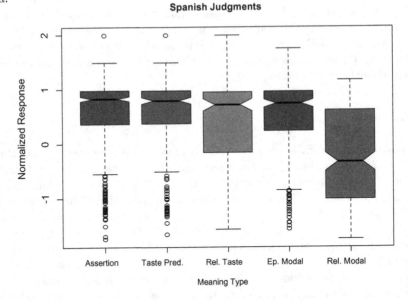

Spanish Judgments

Iberian Spanish Results. This figure again matches the statistical results: there is no significant difference between the means of assertion, PPTs and EMs. Refutation of PPTs was not significantly different from refutation of PPTRs. Refutation of EMs was significantly better (p < .01) than refutation of EMRs. In other words, Iberian Spanish speakers can directly refute PPTs and PPTRs as well as EMs, but not EMRs. These results are problematic for theories arguing for a parallel between EMs and PPTs with respect to a judge parameter or implicit argument, and they are surprising if we think that disagreement should be able to tease apart PPTs from PPTRs cross-linguistically.

4.1   Possibility 1: Cultural Context

In English, disagreements with BAs, PPTs and EMs are well rated, while disagreements with PPTRs and EMRs are not, which is consistent with the idea that an explicit judge experience argument blocks direct refutation. But in Spanish, disagreements with BAs, PPTs, PPTRs and EMs are well rated, while refuting an EMR sentence is not. Thus, we need to give an account for why PPTRs can be followed by direct disagreement in Spanish but not English. One possibility is that this is a question not of the grammar/linguistic context per se but of whether people in a given culture feel comfortable arguing with other people's judgments. After all, it is possible to say *No, the soup is not tasty to you, you're lying to spare the cook's feelings*, so participants could potentially hear the *No* in our stimuli as being anaphoric to something like that such that they understand it to be *No, the soup is not tasty to you, you're lying to spare the cook's feelings; it tastes terrible!*. Arguing with someone about his/her own tastes is certainly impolite in English and has been called a 'faulty disagreement' by [17] who consider it infelicitous (though personally, the first author of this paper feels that there is a difference between such a disagreement and ones like (4)). But it is possible that this would not be seen as a violation of politeness norms in Iberian Spanish. If we had independent evidence that speakers of a language (in this case, Spanish) are more indirect in one area as compared to another, then re-running the same experiment with the other population should change the results. Such evidence exists for Iberian v. Uruguayan Spanish, from anthropologists and linguists such as [18]. Uruguayan Spanish is claimed to be much more indirect than Iberian Spanish, especially in terms of requests, but also criticism, which would likely decrease not only the rates with PPTR, but perhaps the felicity of PPTs and other categories as well. As we see in the figure below, however, the Uruguayan Spanish speakers have an identical pattern to the Iberian speakers.

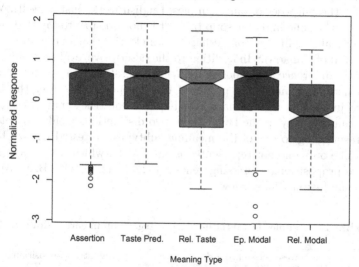

Uruguayan Spanish Judgments

In addition to verifying the same significant differences or lack thereof (between PPTs and PPTRs), we also conducted a series of t-tests comparing each category across the two dialects, resulting in no significant differences. We take this as evidence that the cross-linguistic difference in acceptability of *No* with PPTR is based on the linguistic context (how the grammar interacts with what has been said and affects what will/can be said). This is what we turn to in the next section.

4.2   Possibility 2: Linguistic Context

Recall from Sect. 2 that [1], among others, would explain the difference between PPT and PPTR as being that PPT have a reading on which p is pushed to the Table, whereas PPTR do not and stay in the DC of the speaker. *No* can only target content on the Table. In order to explain why *No* in Spanish could target both PPTs and PPRTs, one would need to either hypothesize that *No* in Spanish is different from *No* in English in that it targets propositions from any part of the discourse model, perhaps, or we would need to say that PPTRs in Spanish but not English put p on the Table. Another possibility would be to claim that in Spanish, but not English, accommodation of a proposition not on the Table could take place. We prefer, however, to rethink the definition of what each type of content contributes to the discourse model in the first place, as discussed below.

Returning to [1], Umbach makes two different kinds of distinctions, one which she considers to be semantic, and the other pragmatic. In an earlier version of the paper, the semantic distinction is between descriptive propositions like *Daniel is at the library* and evaluative propositions like *Poutine is tasty*. The general idea is that descriptive propositions are asserted as facts, while evaluative propositions represent opinions. Her diagnostic for the difference has to do with whether a given proposition can be embedded under the verb *finden* 'find' in German (though for the purposes of our examples, English works just as well). While it is fine to say that one finds the soup tasty, it is not possible to say that one finds Daniel to be at the library (or perhaps more clearly, that one finds Ottawa to be the capital of Canada).[3] In addition to the descriptive/evaluative distinction, a pragmatic difference is proposed between general and subjective judgments in [1]. General judgments are proposed by the speaker to enter the CG and subjective judgments do not have this requirement (they may enter the CG, but they can also stay in the DC). These two distinctions yield four categories, each corresponding to one of the meaning subtypes we tested (shown below in Table 1). The only one not represented is the EMRs, which we propose are also descriptive propositions expressing general judgments, as with BAs. We return to this in the justifications below.

Table 1. Examples instantiating each category of an earlier draft of [1]

	Descriptive propositions	Evaluative propositions
General Judg.	BA: Daniel is at the library	PPT: Poutine is tasty
Subjective Judg.	EM: Daniel might be at the library	PPTR: Poutine is tasty to me

In what follows, we use the same distinctions as in [1] but tweak the functions of the categories. We hypothesize that both distinctions are related to discourse moves: the type of proposition determines what is added to the speaker's DC (or, alternatively, the semantics of each of these types leads to such a distinction), and

---

[3] This gets changed in the final version because *finden* does embed some uses of non-evaluative dimensional adverbs like *tall* and does not work with some uses of evaluative propositions (the non-meta-linguistic ones, cf. [19]). Because we do not have room to go into the interesting differences among kinds of gradable predicates and are focusing on just one subtype, we ignore this distinction here (see also [20] for further work on whether gradable predicates should be considered PPTs).

the type of judgment determines what is pushed to the Table. For the purposes of this section, we need to define the following notation: p-individual (p-ind) is a proposition relativized to an individual (usually the speaker), and p-generic or p-general-knowledge (p-gen) is a proposition relativized to society more broadly. In the case of PPT/Rs, the fact that the soup tastes good to the speaker is p-ind and the fact that it is considered delicious more generally is p-gen.

We propose that descriptive propositions add only a single proposition to the discourse commitment set of the speaker (here we are not counting the join of various propositions as a single proposition, there can be additional propositions contributed by non-restrictive relative clauses, etc., that do not count toward this limit). This proposition is always pushed to the Table and it is the full literal meaning of the sentence. In the case of the BA, that is p, and in the case of the EM, that is $\Diamond p$ (because p is embedded and so not the full meaning). For EMR, it is the proposition that the speaker has no knowledge of p. This is why EMR is categorized with BA: it is a factual statement, it just doesn't assert p, $\neg p$, their disjunction, or anything other than an ignorance claim. Evaluative propositions, on the other hand, by their nature, involve an individualized and generalized version and, thus, add these two propositions to the speaker's DC by default. For PPTs and PPTRs, these are p-ind and p-gen. The reason that we say they are added 'by default' is because, whereas the descriptive proposition is an assertion, with evaluative propositions, one of the two propositions is an assertion and the other, a conversational implicature (as can be clearly tested using cancellation). Depending on whether the PPT is explicitly relativized or not, which of these (p-ind or p-gen) is an implicature changes:

$$\text{The soup is tasty} \tag{9}$$

$$\text{...but I don't like it   ...\#but I'm the only one who likes it} \tag{10}$$

$$\text{The soup tastes good to me} \tag{11}$$

$$\text{...but I'm the only one who likes it   ...\#but I don't like it} \tag{12}$$

It may seem unusual to think of conversational implicatures as default DCs and things that can be put on the Table, but they do seem to be targetable using polarity particles as in *A: Is your friend handsome? B: He has a nice personality. C: No, he's good-looking, too*, where C responds to B's implicature. We are additionally proposing that for the implicature in the pair of (evaluative) propositions, it can be either the positive or negative form (depending on what has been cancelled, general knowledge, etc.). For example, in a context where someone says *The soup tastes good to me but perhaps not to others*, the implicature is that the soup's tastiness will not generally hold up.

One other note about EMRs relates to the notion of at-issueness and relevance to a QUD. It would be easy to conflate the notion of relevance from Grice (and generally the notion that discourses require coherence) with relevance to a QUD, and they are indeed quite similar [14]. However, here we have claimed that *I don't know p* never introduces p or $\neg p$ to the DC, which means they cannot be pushed onto the Table. It also means that they are not at-issue, because the way the notion of relevance is defined, an answer to a question of whether you have knowledge of p does not eliminate an alternative in the question $?p$ (of course, it could eliminate an alternative to $?q$ if that were a question like *Do you know whether p?*). However, in the broader sense of relevance, *I don't know* is clearly a pertinent response to the question. Thus, our analysis is proposing that there are felicitous discourse moves with no at-issue content, simultaneously illustrating one difference between at-issueness and what is on the Table.

The following chart (Table 2) summarizes the hypotheses presented above in its first column, and we now turn to the contribution of the different judgment types, shown in the second column.

Table 2. Update rules for each combination of proposition/judgement type

	Added to the speaker's DC	Pushed to the table
Gen/Desc: BA	p	p
Gen/Eval: PPT	p-gen and either p-ind or $\neg p$-ind	p-gen
Subj/Desc: EM	$\diamond p$	$\diamond p$ and p
Subj/Eval: PPTR	p-ind and either p-gen or $\neg p$-gen	p-ind and either p-gen or $\neg p$-gen

The essence of our analysis of the difference between general and subjective judgments is that general judgments put just one proposition on the Table, and a general one at that, while subjective judgements push two propositions to the table, giving the more subjective proposition pride of place (the secondary nature of the general proposition ties into Umbach's original distinction). The idea draws on inspiration from [10] who propose that EMs put both p and $\diamond p$ on the Table and that which is targeted as at-issue depends upon various factors (thereby accounting for EMs differing behavior in matrix and embedded contexts, etc.). They explicitly mention data not unlike our own as a reason to posit such an analysis, expressing an intuition that both B and B' are felicitous answers to A. *A: That might have been a mistake. B: Yes, it might have been. B': Yes, it was.* We propose that this is true of all subjective judgments, which then means that this claim extends to PPTRs. This might seem strange at first given that PPTRs are thought to have a more restrictive range of meaning than PPTs but that PPTs, as general judgments, only push one proposition to the table (after all, they are supposedly explicitly relativized). However, much as the p in *might p* is considered a prejacent, the fact that something tastes good is a proper subpart of a PPTR sentence where something tastes good to the speaker. Because a cancellation of p-gen entails $\neg p$-gen and vice versa, there will always be at least one proposition in addition to p-ind that is put in the DC, pushed onto the Table and can be a target for negation. We argue that this the real difference between English and Spanish: English (often?) implicates $\neg p$-gen by default with PPTRs, whereas Spanish (often?) implicates p-gen. We usually think of implicature calculation as being universal, but there are a number of ways to compute the relevance of one sentence to another, and these can be opposed. When someone says that something tastes good to her, one could think that the extra effort involved in relativizing to herself signals that she does not think it holds more generally, or one could think that people often assume their own preferences are shared by others and that her statement is being used to establish good taste. If, for example, one of these calculations is made more often than the other by some group of speakers, it could become a generalized conversational implicature (GCI), and it is then plausible to think that two different GCIs could form in different populations.

One piece of evidence potentially supporting this idea is the following: English speakers, as per our results, cannot refute p-gen after the utterance of a PPTR. However, though it was not tested in the original experiment, it seems that $\neg p$-gen is refutable in English in the same environments were p-gen didn't work, as in *A: The soup tastes good to me. B: No, other people like it, too.* The natural question is then whether Spanish speakers, as a whole or in certain dialects, find dialogues like this one infelicitous. One of our reviewers, a native speaker of a third Spanish dialect, does find it infelicitous, though the second author of this paper does not. A few Uruguayan Spanish speakers were consulted and found the dialogue odd, but we hope to do further follow-up work on this point. In the meantime, it seems that the linguistic difference among English and Spanish with respect to PPTRs could boil down to a difference in the default polarity of an implicature.

In review, the following are our proposed update rules for each meaning type we tested in our experiment:

- BA: p is added to the speaker's DC and pushed to the Table.
- PPT: p-gen is added to the speaker's DC, as is p-ind (unless cancelled). Only p-gen is pushed to the Table.
- EM: $\Diamond p$ is added to the speaker's DC. p and $\Diamond p$ are pushed to the Table.
- PPTR: p-ind is added to the speaker's DC as well as either p-gen or $\neg p$-gen. Both p-ind and p-gen/$\neg p$-gen (whichever is in the DC) are pushed to the Table.
- EMR: neither p nor $\neg p$ are added to the speaker's DC, but the proposition that the speaker has no knowledge of p (q) enters the DC. Neither p nor $\neg p$ exists in the DC to be pushed onto the Table, but q does and is.

With this theory, we do not need to postulate that *No* works differently in the two languages, nor that *No* targets different types of content across languages. And under this analysis, it is just a fluke that PPTRs and EMRs are both unable to be directly refuted in English because there are different reasons posited for their non-refutability. The non-refutability of EMRs is absolute in the sense that they never introduce p or $\neg p$. We predict that this is true cross-linguistically, matching our results. With PPTRs, on the other hand, the lack of refutability has to do with the fact that their p-gen or $\neg p$-gen meaning blocks the possibility of denying the other. This is something we expect to vary cross-linguistically, and, as such, we argue against a parallel between PPT and EM.

5   Conclusion

In this paper, we have presented the results of an experimental study collecting naturalness judgments from English and Spanish speakers to determine whether PPTs and EMs can be refuted in dialogue. We validated the hypotheses of the philosophical/linguistic literature on faultless disagreement in English, but we also showed that PPTs and EMs do not pattern alike with respect to refutation cross-linguistically when relativized. We argued against the possibility that these effects were primarily due to differences in politeness norms across cultures by replicating our results for another dialect of Spanish, and we argued for an explanation in terms of pragmatic differences. We showed how such differences could arise on the basis of a simple polarity difference in the implicature drawn from a relativized PPT, and we specified the discourse function of each type of meaning separately, also drawing out the patterns that they share. Further work is naturally required to decide among the various theoretical consequences of these results, which we are in the process of conducting. We are particularly interested in looking for similarities and differences within the class of evaluative items, separating aesthetic and taste predicates as [21] has done. This article represents one of the first steps in the domain of psycholinguistic philosophy/pragmatics on this subject, and we believe that it shows the benefits of an experimental approach in, for example, teasing apart EMRs and PPTRs very clearly, showing that they should not be given a unified analysis.

Acknowledgments. This research has been funded in part by the Conseil de recherches en sciences humaines du Canada (430-2013-000804), EURO-XPRAG, the Juan de la Cierva Program (JDC-2009-3922), the Ramn y Cajal Program (RYC-2010-06070), and by project FFI2012-34170 (MINECO). We also thank our three anonymous reviewers whose comments were extremely useful.

References

1. Umbach, C.: Evaluative propositions and subjective judgments. In: van Wijnbergen-Huitink, J., Meier, C. (eds.) Subjective meaning. Berlin, de Guyter (2015, In press, to appear)
2. Farkas, D., Bruce, K.: On reacting to assertions and polar questions. J. Semant. **27**(1), 81–118 (2010)
3. Kölbel, M.: Faultless disagreement. In: Proceedings of the Aristotelian Society (Hardback), vol. 104(1) Blackwell, Oxford (2004)
4. Stojanovic, I.: Talking about taste: disagreement, implicit arguments, and relative truth. Linguist. Philos. **30**(6), 691–706 (2007)
5. Stephenson, T.: Judge dependence, epistemic modals, and predicates of personal taste. Linguist. Philos. **30**(4), 487–525 (2007)
6. Lasersohn, P.: Context dependence, disagreement and predicates of personal taste. Linguist. Philos. **28**(6), 643–686 (2005)
7. Pearson, H.: A judge-free semantics for predicates of personal taste. J. Semant. **30**(1), 103–154 (2013)

8. Egan, A.: Epistemic modals, relativism and assertion. Philoso. Stud. **133**(1), 1–22 (2007)
9. von Fintel, K.: Would you believe it? The King of France is back! Ms. (2004)
10. von Fintel, K., Gillies, A.: An opinionated guide to epistemic modality. In: Oxford Studies in Epistemology, vol. 2, pp. 32–62. Oxford University Press, Oxford (2007)
11. Roberts, C.: Information structure in discourse: towards an integrated formal theory of pragmatics. Semant. Pragmatics **5**(6), 1–69 (2012)
12. Simons, M., Tonhauser, J., Beaver, D., Roberts, C.: What projects and why. In: SALT Proceedings (2011)
13. Castroviejo, E., Mayol, L., Smith, E.: Conventional implicatures and direct refutation. Ms. (2014)
14. Grice, H.P., Cole, P., Morgan, J.: Logic and Conversation. In: Syntax and semantics, vol. 3, pp. 41–58. Academic Press, New York (1975)
15. R Core Team, R.: A language and environment for statistical computing. In: R Foundation for Statistical Computing, Vienna, Austria, 2012 (2014)
16. Bates, D., Maechler, M., Bolker, B.: lme4: linear mixed-effects models using S4 classes. R package version 0.999999-0. 2012 (2013)
17. Gunlogson, C., Carlson, G.: Predicates of experience. subjective meaning. In: van Wijnbergen-Huitink, J., Meier, C. (eds.) Subjective Meaning, Berlin, de Guyter (2015, In press, to appear)
18. Reiter, R.A.: Contrastive study of conventional indirectness in Spanish: evidence from Iberian and Uruguayan Spanish. Pragmatics **12**(2), 135–151 (2002)
19. Barker, C.: The dynamics of vagueness. Linguist. Philos. **25**(1), 1–36 (2002)
20. Bylinina, L.: The Grammar of Standards: judge-dependence, purpose-relativity and Comparison Classes in Degree Constructions. PhD dissertation. LOT Dissertation Series, 347 (2014)
21. McNally, L., Stojanovic, I.: Aesthetic adjectives. In: Young, J. (ed.) Semantics of Aesthetic Judgment, Oxford University Press (2015, To appear)

Short Papers

An Ontology-Based Reasoning Framework for Context-Aware Applications

Christoph Anderson[✉], Isabel Suarez, Yaqian Xu,
and Klaus David

Chair of Communication Technology, University of Kassel,
Wilhelmshöher Allee 73, 34121 Kassel, Germany
{anderson,isabel.suarez,david}@uni-kassel.de
yaqian.xu@comtec.eecs.uni-kassel.de
http://www.comtec.eecs.uni-kassel.de/

Abstract. Context-aware applications process context information to support users in their daily tasks and routines. These applications can adapt their functionalities by aggregating context information through machine-learning and data processing algorithms, supporting users with recommendations or services based on their current needs. In the last years, smartphones have been used in the field of context-awareness due to their embedded sensors and various communication interfaces such as Bluetooth, WiFi, NFC or cellular. However, building context-aware applications for smartphones can be a challenging and time-consuming task. In this paper, we describe an ontology-based reasoning framework to create context-aware applications. The framework is based on an ontology as well as micro-services to aggregate, process and represent context information.

Keywords: OWL · Android · Ontology · Context · Framework

1 Introduction

In the last few years, mobile phones have evolved from devices, used for voice communication and sending text messages only, to powerful smartphones with multiple embedded sensors and communication interfaces such as WiFi, Bluetooth, NFC or cellular. With their increasing processing capabilities and sensors such as accelerometers, gyroscopes, and magnetometers, smartphones are now being used for internet browsing, social networking, playing games, watching videos or listening to music [4]. Other fields of smartphone applications are activity recognition and context-awareness. In the area of activity recognition, for example, smartphones are used to recognize Activities of Daily Living (ADL) [13,15]. Context-aware applications can be utilized in the field of Ambient-Assisted-Living (AAL) and home automation. These types of applications monitor the environment to aggregate context information to provide recommendations or services to the user. Typically, this information is extracted

© Springer International Publishing Switzerland 2015
H. Christiansen et al. (Eds.): CONTEXT 2015, LNAI 9405, pp. 471–476, 2015.
DOI: 10.1007/978-3-319-25591-0_34

from sensor data, represented as time series, by using machine-learning and data processing algorithms. Existing frameworks in this field can be divided into two categories based on the way contexts are represented. The first category includes frameworks that describe context information without semantic e.g. as plain programming objects such as strings or class objects using object-oriented models [1,5,12,14]. However, these object-oriented models are not suitable for knowledge and data sharing in heterogeneous pervasive environments [2]. The second category consists of frameworks supporting semantic representations of context information [8,10,11]. These frameworks exploit ontologies, first-order logics or other description technologies to represent context information semantically. Building context-aware applications on smartphones that represent context information semantically as well as aggregate and process contexts through sensor information is still a challenging task.

In this paper, we present an ontology-based framework to create context-aware applications. In addition to the ontology, we extend the current state of the art frameworks by integrating micro-services to aggregate and process context information. By supporting reasoners such as Pellet [17], HermiT [6] and JFact [18], the framework can deduce complex contexts from already aggregated context information by using the reasoning paradigm of the Web-Ontology-Language (OWL). In the following sections, we present the architecture of our framework including a schematic overview over its components. A discussion about the limitations of the framework and a conclusion with a summary of future work is given at the end of this paper.

2   Framework-Architecture

The framework is based on an OWL ontology to model and represent context information. By using an ontology, complex contexts can be deduced by using the reasoning paradigm of OWL. In combination with the ontology, the framework exploits micro-services to aggregate and process context information from embedded smartphone or environmental sensors. Micro-services refer to stand-alone applications providing a background service only. These background services implement one or multiple functionalities such as context classification, prediction or sensor data aggregation, which can be used by applications or other micro-services. A schematic view of the framework architecture is given in Fig. 1.

2.1   Micro-Services

The micro-service architectural style is a mechanism, where a dedicated application provides one or multiple services, each one of them running in its process. In our framework, services communicate over an Inter-Process-Communication interface (IPC) with other services or the framework core. Figure 1 depicts three possible types of micro-services. Since the framework has to exploit sensors to aggregate and process sensor data, the first type of micro-services (*Sensing*) is

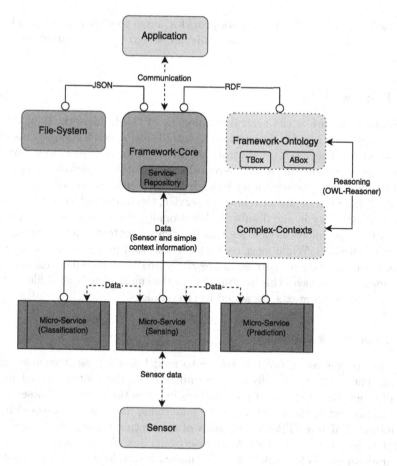

Fig. 1. Schematic view of the framework architecture

used to establish a connection to different sensors. Either embedded smartphone sensors such as the accelerometer, gyroscope or magnetometer or environmental sensors such as movement sensors or door contacts can be used. Establishing a connection to environmental sensors, for example, can be done by implementing a micro-service requesting a web-service abstracting sensor information e.g. in a smart home environment. Already aggregated sensor data can be distributed to other micro-services or the framework core by IPC. The second type of micro-services (*Classification*) is used to classify context information from sensor data. Existing approaches such as [9] can be implemented as services to classify context information such as standing, sitting, walking, and lying. The third type of micro-services (*Prediction*) is used for context prediction. Generally, existing context prediction algorithms such as [16] or [7] exploit context histories to build prediction models. These algorithms predict future contexts providing further information about the next context of the user. To support context prediction algorithms, the framework provides access to already aggregated context information in form of a context history stored on the smartphone's filesystem.

2.2 Framework Core

The framework core is the central component of the framework. It maintains the ontology and provides an interface to establish IPC connections to installed micro-services. The interface controls the life-cycle of micro-services. For this reason, the core maintains a service-repository of installed micro-services in the form of a database. The repository is needed to establish connections to installed services. If the user installs or removes a service, the framework core updates the corresponding entry in the database. Additionally, the framework core process information from micro-services such as aggregated contexts, raw sensor data or context predictions. While most context classification or prediction approaches rely on sensor or context histories to classify or predict new context information, the framework core saves this information on the smartphone's local file-system. Already classified contexts are added to the ontology.

2.3 Framework Ontology

The framework uses an OWL ontology to model and represent context information. The ontology describes the terminology of the context model in the form of conceptual classes and relationships between these classes. These classes and relations define the structure of the underlying model and are stored inside a terminological box (TBox). Instances of conceptual classes, their attributes and relationships are stored inside an assertional box (ABox). With the reasoning paradigm of OWL ontologies, the framework can deduce complex contexts from already aggregated information. In a smart environment, for example, the ontology can infer that a user is making coffee by using dense sensing [3].

3 Limitations

An experimental evaluation of the framework was carried out by implementing a context-based communication filter for Android based smartphones. The filter blocks certain communication such as phone calls, text messages or emails by aggregating and processing context information of the user in a smart home environment located in our department. Similar to the evaluation in [19], we experienced memory and performance issues on the tested smartphones (Google Nexus 4, 5 and 6) when reasoning larger ontologies[1]. Besides memory and performance issues, applications using the framework may have time constraints when deducing new context information from already aggregated contexts. These time constraints may be violated when using reasoners that are not optimized for large

---

[1] More than 6.000 Axioms.

ontologies or when using power-limited smartphones. Besides the limitations caused by different ontology sizes and reasoners being used, efforts regarding the privacy and security of context information have to be made. Although, context information is stored on the smartphone's local file-system, the exchange of data e.g. using environmental sensors, has to be secured by using current state of the art encryption technologies such as TLS (Transport Layer Security).

4   Conclusion

Context-aware applications process context information to support users in their daily tasks and routines. Building context-aware applications for smartphones that represent context information semantically as well as aggregate and process contexts through smartphone or environmental sensor is still a challenging task. In this paper, we presented an ontology-based framework to create context-aware applications. By utilizing an ontology, context information can be described semantically. In addition to the ontology, we implemented micro-services to aggregate and process context information from embedded smartphone or environmental sensors. By utilizing reasoners such as Pellet, HermiT or JFact, the framework can deduce complex contexts from already aggregated context information by using the reasoning paradigm in OWL. In the future, we plan to publish the framework, making it available to the community. Also, we plan to take into account semantic rule languages such as SWRL (Semantic-Web-Rule-Language), as well as other reasoners to extend the functionalities of the framework.

Acknowledgements. This work has been co-funded by the Social Link Project within the Loewe Program of Excellence in Research, Hessen, Germany.

References

1. Carlson, D., Schrader, A.: Dynamix: an open plug-and-play context framework for android. In: IOT, pp. 151–158. IEEE (2012)
2. Chen, H., Perich, F., Finin, T., Joshi, A.: Soupa: standard ontology for ubiquitous and pervasive applications. In: The First Annual International Conference on Mobile and Ubiquitous Systems: Networking and Services, 2004, MOBIQUITOUS 2004, pp. 258–267, August 2004
3. Chen, L., Hoey, J., Nugent, C., Cook, D., Yu, Z.: Sensor-based activity recognition. IEEE Trans. Syst. Man Cybern. Part C Appl. Rev. **42**(6), 790–808 (2012)
4. Falaki, H., Mahajan, R., Kandula, S., Lymberopoulos, D., Govindan, R., Estrin, D.: Diversity in smartphone usage. In: Proceedings of the 8th International Conference on Mobile Systems, Applications, and Services, MobiSys 2010, pp. 179–194. ACM, New York (2010). http://doi.acm.org/10.1145/1814433.1814453
5. Ferreira, D., Kostakos, V., Dey, A.K.: Aware: mobile context instrumentation-framework. Front. ICT **2**(6), 1–9 (2015)
6. Glimm, B., Horrocks, I., Motik, B., Stoilos, G., Wang, Z.: Hermit: an owl 2 reasoner. J. Autom. Reasoning **53**(3), 245–269 (2014). http://dx.doi.org/10.1007/s10817-014-9305-1

7. Gopalratnam, K., Cook, D.J.: Active lezi: an incremental parsing algorithm for sequential prediction. Int. J. Artif. Intell. Tools **13**(04), 917–929 (2004). http://www.worldscientific.com/doi/abs/10.1142/S0218213004001892

8. Gu, T., Pung, H.K., Zhang, D.: Toward an osgi-based infrastructure for context-aware applications. IEEE Pervasive Comput. **3**(4), 66–74 (2004)

9. Kwapisz, J.R., Weiss, G.M., Moore, S.A.: Activity recognition using cell phone accelerometers. SIGKDD Explor. Newsl. **12**(2), 74–82 (2011). http://doi.acm.org/10.1145/1964897.1964918

10. Meditskos, G., Dasiopoulou, S., Efstathiou, V., Kompatsiaris, I.: Sp-act: a hybrid framework for complex activity recognition combining owl and sparql rules. In: 2013 IEEE International Conference on Pervasive Computing and Communications Workshops (PERCOM Workshops), pp. 25–30, March 2013

11. Paspallis, N., Papadopoulos, G.A.: A pluggable middleware architecture for developing context-aware mobile applications. Pers. Ubiquit. Comput. **18**(5), 1099–1116 (2014). http://dx.doi.org/10.1007/s00779-013-0722-7

12. Raento, M., Oulasvirta, A., Petit, R., Toivonen, H.: Contextphone: a prototyping platform for context-aware mobile applications. Pervasive Comput. IEEE **4**(2), 51–59 (2005)

13. Roy, N., Misra, A., Cook, D.: Infrastructure-assisted smartphone-based adl recognition in multi-inhabitant smart environments. In: 2013 IEEE International Conference on Pervasive Computing and Communications (PerCom), pp. 38–46. IEEE Computer Society, March 2013

14. Sridevi, S., Bhattacharya, S., Pitchiah, R.: Context aware framework. In: Sénac, P., Sénac, P., Seneviratne, A., Seneviratne, A., Ott, M., Ott, M. (eds.) MobiQuitous 2010. LNICST, vol. 73, pp. 358–363. Springer, Heidelberg (2012). http://dx.doi.org/10.1007/978-3-642-29154-8_40

15. Shoaib, M., Scholten, H., Havinga, P.: Towards physical activity recognition using smartphone sensors. In: 2013 IEEE 10th International Conference on Ubiquitous Intelligence and Computing, Autonomic and Trusted Computing (UIC/ATC), pp. 80–87, December 2013

16. Sigg, S., Haseloff, S., David, K.: An alignment approach for context prediction tasks in ubicomp environments. Pervasive Comput. IEEE **9**(4), 90–97 (2010)

17. Sirin, E., Parsia, B., Grau, B.C., Kalyanpur, A., Katz, Y.: Pellet: a practical owl-dl reasoner. Web Semant. Sci. Serv. Agents World Wide Web **5**(2), 51–53 (2007). http://www.sciencedirect.com/science/article/pii/S1570826807000169. software Engineering and the Semantic Web

18. Tsarkov, D., Tsarkov, D., Horrocks, I., Horrocks, I.: FaCT++ description logic reasoner: system description. In: Furbach, U., Furbach, U., Shankar, N., Shankar, N. (eds.) IJCAR 2006. LNCS (LNAI), vol. 4130, pp. 292–297. Springer, Heidelberg (2006)

19. Yus, R., Bobed, C., Esteban, G., Bobillo, F., Mena, E.: Android goes semantic: Dl reasoners on smartphones. In: Bail, S., Glimm, B., Gonçalves, R.S., Jiménez-Ruiz, E., Kazakov, Y., Matentzoglu, N., Parsia, B. (eds.) ORE. CEUR Workshop Proceedings, vol. 1015, pp. 46–52 (2013). CEUR-WS.org

Ontology-Based Roles Association Networks for Visualizing Trends in Political Debate

Troels Andreasen[1], Henning Christiansen[1]([✉]),
and Mads Kæmsgaard Eberholst[2]

[1] Research Group PLIS: Programming, Logic and Intelligent Systems,
Roskilde University, Roskilde, Denmark
{troels,henning}@ruc.dk
[2] Department of Communication, Business and Information Technologies,
Research Group: Communication, Journalism and Social Change,
Roskilde University, Roskilde, Denmark
makaeb@ruc.dk

Abstract. Online resources, large data repositories and streaming social network messages embed plenitudes of interesting knowledge, often of associative nature. A specific communicative context, such as the political debate in a given country, has groupings of actors, with changing attitudes and stances towards each other and external, real or invented, threats and opportunities. A new form of associative network is introduced, that integrate flexible ontologies for complex contexts of roles and hierarchies with a labelled association structure representing observed strengths and attitudes. Twitter messages from the political landscape in Denmark up to the general election 2015 are used as a both current and relevant illustrative case.

1 Motivation and Background

@bansoe And so it begins....#fv15[1]
This tweet, like many others on May 27th 2015, initiated the Danish national elections on the social media platform Twitter. This paper outlines a graphical representational formalism that allows modeling how the election and its main actors have represented various subjects in the debate on Twitter, and with what sentiment. Twitter in Denmark is a primarily political (as opposed to a social) platform [1,2] and as both media and politicians is very aware of this, the platform can generate a large amount of data very fast when there is an election. This is primarily due to the nature of the Twitter platform making it highly suited for breaking news [3] and thus rapidly developing events and attitudes. This is especially true when an election is in process, where tweets from political actors mostly are used to campaigning, spreading information or self promotion [4]. For many end users and indeed both media professionals and politicians, making grasps of such a large amount of data can be cumbersome bordering impossible.

---

[1] #fv for Danish "Folketingsvalg", the election for the Danish national parliament.

© Springer International Publishing Switzerland 2015
H. Christiansen et al. (Eds.): CONTEXT 2015, LNAI 9405, pp. 477–482, 2015.
DOI: 10.1007/978-3-319-25591-0_35

Therefore a modeling approach to visualize and understand some of the mechanics and contents of Twitter data in a political and electoral context is very much needed. For this paper we used the open source tool Yourtwapperkeeper to harvest all tweets from the start of the election period in Denmark to the end of the campaign. This tool has proven a viable way of gathering Twitter data which due to API limitations can be hard to do [5]. We selected the appropriate electoral hashtags that appeared to be trending within the first 24 hours. These were #dkpol (Danish general politics) and a number of hashtags all referring to the election (#fv15, #valg15, #valg2015, #drdinstemme, #tv2valg, #ft15, #dkvalg, #ftvalg15 and #dkmedier). This gave a sample of 260,000 tweets on which our methodology is tested.

Co-occurrence networks and semantic networks based on co-occurrences have been used in many areas and in different shapes. We propose an integration with an ontological concept lattice to a general model for representing complex contexts including large sets of actors with many different roles and organizational groupings – as is the case in politics. The concept lattice allows to zoom in and out, from single individual to, say, the known members of a given political party, and vice-versa, and to express asymmetric and annotated relationships, anticipating different visualizations.

2   ORAN: Ontology-Based Roles Association Networks

Ontology-Based Roles Association Networks, for short ORANs, are association networks for context representation, whose nodes are concepts taken from a *concept lattice* serving as an ontology. Concepts may be extended with *roles*, so for example, concept "Smith" may refer to any message relating somehow to a single person, whereas predicating with the role "author", forming concept "author:Smith" may refer to messages written by that person. Formally, we assume finite sets of *atomic concepts* $\mathcal{A}$, and roles $\mathcal{R}$; a *concept* is of one of the forms, a or $r{:}a$ where $a \in \mathcal{A}$, $r \in \mathcal{R}$. The concept lattice is a partially ordered set of concepts $\langle \mathcal{C}, < \rangle$ where $<$ is transitive, and if $r{:}a \in \mathcal{C}$, then $a \in \mathcal{C}$ and $r_a < a$. The relation $<$ is read "more specialized than". A concept $c \in \mathcal{C}$ has an *extension*, written $[\![c]\!]$, which is a subset of some universe $\mathcal{U}$ (e.g., of messages) such that, for any c, c' with $c < c'$, it holds that $[\![c]\!] \subseteq [\![c']\!]$. The lattice is not fully generative, as not all roles can apply to all atomic concepts, e.g., refugees are often mentioned (concept about:refugees) but do not have voices as authors.

An *ORAN* is a directed or undirected, labelled graph, described as a triplet of a concepts lattice, a subset of its atomic concepts and one or more association types, $\langle \langle \mathcal{C}, < \rangle, A, T \rangle$. An *association type* is of the form $P_1 \rightarrow P_2$ in case of a directed network and $P_1 - P_2$ for an undirected one, where P_1, P_2 are *generic concepts* of form $c \in \mathcal{C}$, $*$ or $r{:}*$, where $r \in \mathcal{R}$. The association type generates the actual set of nodes N in the graph and the edges. Concept c indicates $c \in N$, $*$ that $A \subseteq N$ and $r{:}*$ that $\{r{:}a \mid a \in A\} \cap \mathcal{C} \subseteq N$. All edges whose endpoints $\in N$ match the association type are included, so e.g., author:$* \rightarrow *$ may create, among others, author: MembersOfDanishPeoplesParty → refuges. When drawing these edges, they are typically shown between the given atomic concepts.

The edges are referred to as *associations*, and each edge $c_1 \to c_2$ or $c_1 - c_2$ has one or more *labels*, one of which is its *association degree*, also called *strength*, defined for the directed, resp. undirected case, as follows.

$$d(c_1 \to c_2) = \begin{cases} \frac{|[[c_1]] \cap [[c_2]]|}{|[[c_1]]|}; & \text{if } |[[c_1]]| > 0 \\ 0 & \text{otherwise} \end{cases}$$

$$d(c_1 - c_2) = \begin{cases} \frac{|[[c_1]] \cap [[c_2]]|}{|[[c_1]] \cup [[c_2]]|} & \text{if } |[[c_1]] \cup [[c_2]]| > 0 \\ 0 & \text{otherwise} \end{cases}$$

Here $|set|$ refers to the number of elements in *set*. For our present application, the extension of a concept is given as the set of messages that refers to that concept (possible taking roles into account), and thus $d(c_1 \to c_2)$ is the fraction of the messages referring to c_1 that also refers to c_2. For example, d(author: Hansen → Jensen) is the proportion of Hansen's messages that refer to Jensen (in some way or another).

Other possible labels attached to edges $c_1 \to c_2$ can be defined from properties of $[[c_1]]$ and $[[c_2]]$. In our example, we use a *sentiment* that measures a degree of positiveness/negativeness for sets of tweets.

Extracting Concept Extensions and Rules from Twitter Data

Twitter messages include sender identification and other metainformation, and the text may includes nametags (when specific Twitter users are mentioned), hashtags as well as ordinary words. These can all be mapped to concepts, specifically sender identification into author roles, and references in the text into mentions (e.g., indicated by role 'about:'). This is not an easy task, as these tags are not a controlled vocabulary, and the same person may be referred to by a nametag, different hashtags, the name written in different ways and with spelling errors. To create concepts corresponding to members (given by nametags and user id) of specific political parties, we have used an external resource, provided by the online resource http://www.twittervalgkortet.dk. Our sentiment analysis is based on a list of common Danish words, each having a "joy index" learned by statistical means,[2] ranging from −5 for the most negative to +5 for the most positive. We added our own stemming algorithm to extend covering and take the average over words identified in each tweet, followed by averaging over a set of tweets.

Visualizing Ontology-Based Roles Association Networks

A vast collection of software of software tools are available for presenting associative information graphically, e.g., d3.js[3] which is an impressive java script library suited for application development. For our own experiments, we stored

---

[2] This list has been prepared by Finn Årup Nielsen, Technical University of Denmark, referred is referred to in a student project [6].

[3] See http://d3js.org.

ORANs extracted from Twitter messages in a relational database, from which information were exported to the NodeXL tool,[4] which is a plug-in to Excel. Typical views are complete networks, showing all possible associations between the selected nodes, and so-called ego-centric views, with a single concept c in the middle, captured by the association type $c \rightarrow *$ (or $c \rightarrow r{:}*$).

Association degrees are typically shown as thickness of the edges, and other labels can be shown, e.g., by the colour of edges or be available in pop-up menus. A zooming facility is under development which can be accessed by clicking the mouse over the nodes of the network: zooming amounts to replace one or several nodes in A by other, above or below wrt. $<$.

When data are sufficiently dense and subject of fast, dynamic changes as is the case for tweets about politics in an election, it is interesting also to take into account the time dimension: for the chosen view, images are produced for the tweets from each day, and these images are put together to an animation. This may give a very clear illustration of the development of attitudes.

3   The Danish General Election 2015

The following two views have nodes corresponding to political parties, measured for the entire period, and illustrate the overall sentiments in how the parties *Alternativet* and *Konservative* writes about the other parties. This corresponds to the ego-centric association patterns

author:Konservative $\rightarrow *$ and   author:Alternativet $\rightarrow *$.

The sentiment is shown by colour, ranging from red for the most negative, and bright green for the most positive. The strengths of the associations are given by their thickness.

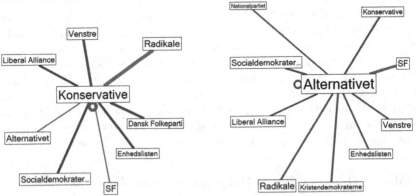

It appears that the Conservative party is very negative in its attitude towards Alternativet and SF (Socialist Peoples' Party) and basically kind for the rest. Their best friends seem to "Radikale" and then follows "Venstre".[5] We see that

---

[4] See http://research.microsoft.com/en-us/projects/nodexl/.

[5] For understanding: "Radikale" is not very radical anymore, and "Venstre" literally, "left", is a clearly right wing party. The Conservative party is, however, conservative.

Alternativet is generally kind, except towards the most leftist and the most rightist parties.

Next, we illustrate a dynamic, potentially animated view, here with snapshots for two different weeks of the distribution of the most common topics that the party Alternativet is tweeting about. I.e., the nodes are {author:Alternativet} ∪ *HotTopics*, where *HotTopics* are generated from the actual hashtag set. The association pattern is as above, but with a changed set of atomic concepts.[6] Strength is now indicated by the size of the topics boxes.

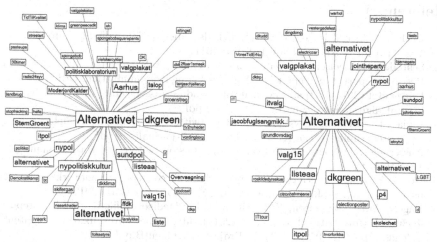

4 Conclusion and Future Work

We demonstrated a novel model for context representation, based on an ontology suited for complex systems of actors with different roles and hierarchical and overlapping groupings (multiple inheritance is inherent in the model). We showed how knowledge can be extracted from social media data streams, mapped into such a model and visualized accordingly. Our test example was Twitter messages with political contents from the announcement of and the hectic days until the Danish national election were held June 18, 2015. This study will be continued into the period after the election, more data will be acquired and analyzed in order to give a complete, zoomable and animated picture of the communication around an election. As part of this, improved sentiment analysis and topic extraction will be developed, taking into account synonymy, tweet idioms and typos, as well as semantic-pragmatic considerations.

There is a lot of recent work on summarizing and visualizing data from Twitter in sophisticated ways, e.g., [7–10], but a tight coupling to an ontology that allows easily shifting view of the data, as we have described, is not common. The work of [8] traces trends in Twitter data over time using a co-occurence of

---

[6] The selection of the associations with strengths higher than some threshold is not part of the ORAN formalism, but a facility in the prototype implementation.

482   T. Andreasen et al.

topics based model. A generic approach for monitoring message streams such as Twitter, for recognizing interesting events automatically, is described by [11], which may be interesting to combine with the present approach. Automatic of learning associative networks from Twitter is used by [12]. More detailed emotions are used in a study that correlates Twitter sentiments to socio-economic phenomena [13]. This group has also identified a high correlation of sentiment in Twitter messages with stock prices [14].

References

1. Buch, R.: Special issue on: Sociale nyhedsmedier og journalistik. Journalistica - Tidsskrift for Forskning I Journalistik (2011)
2. Schrøder, K., Nielsen, R.: Danskernes brug af digitale medier og nyheder i 2014. Research report from Center for Magt, Medier og Kommunikation, CBIT, Roskilde Universitet (2014). https://www.ruc.dk/fileadmin/assets/cbit/MMK/Danskernes_brug_af_digitale_medier_og_nyheder_i_2014.pdf
3. Elkjær, B.: Breaking news! journalistik på twitter (2009). http://medieblogger.dk/wp-content/uploads/Bagger-sagen-Om-twitter-journalistik.pdf
4. Ausserhofer, J., Maireder, A.: National politics on twitter. Inform. Commun. Soc. 16, 291–314 (2013)
5. Bruns, A.: How long is a tweet? mapping dynamic conversation networks on twitter using gawk and gephi. Inform. Commun. Soc. 15, 1323–1351 (2012)
6. Engels, M.F., Jørgensen, S.B., Holm, H.L.: Medieoverblik.dk, Student report, Technical University of Denmark (2014). http://medieoverblik.dk/wp-content/uploads/s103182_s103201_s103214_ProjectAssignmentB.pdf
7. Ozgur, A., Cetin, B.: H.B.: co-occurrence network of reuters news. Int. J. Mod. Phys. C 19, 689–702 (2011)
8. Malik, S., Smith, A., Hawes, T., Papadatos, P., Li, J., Dunne, C., Shneiderman, B.: Topicflow: visualizing topic alignment of twitter data over time. In: Rokne, J.G., Faloutsos, C. (eds.) Advances in Social Networks Analysis and Mining 2013, ASONAM 2013, Niagara, ON, Canada - August 25–29, 2013, pp. 720–726. ACM, New York (2013)
9. Ghanem, T.M., Magdy, A., Musleh, M., Ghani, S., Mokbel, M.F.: Viscat: spatio-temporal visualization and aggregation of categorical attributes in twitter data. In: Huang, Y., Schneider, M., Gertz, M., Krumm, J., Sankaranarayanan, J. (eds.) Proceedings of the 22nd ACM SIGSPATIAL International Conference on Advances in Geographic Information Systems, Dallas/Fort Worth, TX, USA, November 4–7, 2014, pp. 537–540. ACM, New York (2014)
10. Guille, A., Favre, C.: Event detection, tracking, and visualization in twitter: a mention-anomaly-based approach. Soc. Netw. Analys. Min. 5, 18:1–18:18 (2015)
11. Andreasen, T., Christiansen, H., Have, C.T.: Tracing shifts in emotions in streaming social network data. Submitted (2015)
12. Diemert, E., Vandelle, G.: Unsupervised query categorization using automatically-built concept graphs. In: Quemada, J., León, G., Maarek, Y.S., Nejdl, W. (eds.) Proceedings of the 18th International Conference on World Wide Web, WWW 2009, Madrid, Spain, April 20–24, 2009, pp. 461–470. ACM, New York (2009)
13. Bollen, J., Mao, H., Zeng, X.: Twitter mood predicts the stock market. J. Comput. Sci. 2, 1–8 (2011)
14. Bollen, J., Pepe, A., Mao, H.: Modeling public mood and emotion: twitter sentiment and socio-economic phenomena. In: Proceedings of the Fifth International AAAI Conference on Weblogs and Social Media, pp. 450–453 (2011)

Contextual Interfaces for Operator-Simulator Interaction

Alexandre Kabil[1][✉], Patrick Brézillon[2],
and Sébastien Kubicki[1]

[1] ENI Brest, Lab-STICC UMR 6285, 29200 Brest, France
{kabil,kubicki}@enib.fr
[2] UPMC, LIP6, Paris, France
Patrick.Brezillon@lip6.fr

Abstract. We present the results of a study on the role of context in the mental representation that operators have of their task realization. This work is part of the ANR TACTIC project, which aims at proposing a migration of a simulator's interface from PC ("click-simulation") to tactile devices ("finger-simulation").

Keywords: Tactile devices · Context · Task realization · Post-wimp interfaces · Mental maps · Smartphones

1 Introduction

In the current generation of computers with the triplet <screen, keyboard, mouse>, operators often confuse the simulator with its function, the simulation. Such click-based simulation gives the feeling to directly control the evolution of the simulation. The reason is that the operator's mental representation results of a mixture of interpretation of the domain intertwined with an interpretation of the interface functioning. For example, by clicking on the pause button, the operator thinks to stop the simulation, while this action on the interface sends a command to the simulator that suspends the simulation. As a consequence, operator-simulator interaction is considered secondary to actions on the interface.

New technologies like tactile devices lead to new relationships between the operator and the simulator leading to finger-based simulation. However, the migration from the click-based simulation to the finger-based one supposes a change in the design to facilitate interfaces' transfer while operator-simulator interaction stays identical. By focusing on the process that leads to an action (including the decision-making part), and not only the result of the action execution in an isolated way, it is possible to take into account the context in which the operator works effectively. By coupling context-awareness and specific interaction techniques, we can enhance the migration from an interface to another.

This paper is organized as follows. First, we will discuss the role of context in HCI (human-computer interaction), emphasizing how cognitive maps express

© Springer International Publishing Switzerland 2015
H. Christiansen et al. (Eds.): CONTEXT 2015, LNAI 9405, pp. 483–488, 2015.
DOI: 10.1007/978-3-319-25591-0_36

mental representation of operators in their task realization. The next section will introduce the modeling of operators reasoning and its interaction with a simulator. Finally, an interaction model linking gestural grammar, domain and interface actions and contextual graphs will be presented. This paper ends with a conclusion and research perspectives.

2 Context in HCI

2.1 Post-Wimp Interaction

Interaction is a *phenomenon* between a user and a computer that is controlled by the user interface running on the computer. Designing interaction rather than interfaces means that user interfaces are the means, not the end [1]. This supposes to combine and understand the context of use [2] with a special attention to the details of the interaction.

WIMP (Windows Icons Menus Pointer) for example is the most frequent interaction paradigm, used on every desktop systems. GUI (Graphical User Interfaces) are using this paradigm, which is fitted for a keyboard and a mouse usage. In order to enhance the naturalness of computer interaction, researchers proposed several approaches allowing users to go beyond WIMP interfaces. Post-WIMP is a generic term defining all interfaces that uses at least one non-WIMP control. Usually, Post-WIMP interfaces, such as tangible ones [3] takes into account context but only as a frame. The triplet <User, Platform, Environment> [4] is for instance used to constraint the whole application, by defining sets of parameters (the context of use).

This paper presents an extended context usage, by providing a model which matches specific interaction techniques to specific task-realization cases.

2.2 Contextual Graphs

Brezillon [5] introduces the Contextual-Graphs (CxG) formalism for obtaining a uniform representation of elements of knowledge, reasoning and context. Contextual graphs are acyclic because of the time-directed representation and guarantee of algorithm termination. With a series-parallel structure, each contextual graph has exactly one root and one end node because the decision-making process starts in one state of affairs and ends in another state of affairs (generally with different solutions on the different paths) and the branches express only different contextually-dependent ways to achieve this goal. A contextual graph represents the realization of a task, and each path corresponds to a practice developed by an actor in a particular context.

The challenge to address concerns what operators are doing effectively, that is, their activity (and not their task). It is the well-known problem of distinction between task and activity [6], procedures and practices [7], logic of functioning and logic of use [8], etc. Making context explicit as contextual elements allows to consider all heterogeneous elements of context, which can be used for reasoning on scenarios.

In our research, we propose to use the contextual graph, which represents operators' behavior during their interaction with the simulator, as a "task-centered" modelling approach.

3   Task Centered Approach

3.1   Mental Maps and Contextual Graphs

Expert's mental representation depends of their experience with the tasks attached to his role. This experience contains knowledge accumulated by the expert during his practical use of the domain knowledge along a number of task realizations in different contexts. The mental representation is a cognitive expression of the contextual knowledge related to the operator (the expert), the task at hand, the situation of the work, and the local environment in which resources are available. A cognitive map is a semi-structured expression of the mental representation that can be externalized, with classical knowledge-management tool.

The expert map corresponds to the selection of the part of the domain knowledge effectively used by participants during the realization of their tasks. As a result, the cognitive map gives a tree representation of the elements considered by participants. In terms of context, the expert map is a representation of the contextual knowledge, the part of the context that participants relate more or less directly to their task realization.

3.2   Domain and Interface Actions

During task realization on a system, operators interact through an interface with a simulator that implements the simulation of a real-system evolution based on a model of this real system. Thus, the system is a model-based simulator and the interface is supposed to be part of the simulator. This is in contradiction with the goal of our project that is to allow interaction through different interfaces without changing the simulator. Confusion between simulation and simulator leads operators to assimilate specific actions of the domain (e.g. performing an action on a simulator item) with specific actions of the interface commands (e.g. stopping the simulation). We call the former *Domain actions* and the latter *Interface actions*. It leads to a serious cognitive problem because a unique domain action may be associated with several interface actions (each relevant in a specific context not made explicit), and operators would assimilate these interface actions to different domain actions for having the same information.

Thus, domain actions will be more easily associated with interface actions by taking into account the context of interaction, resulting in greater flexibility of the interface, not only with respect to the actor, but also with respect of the task realizations.

However, this approach does not describes users interaction, and we had to analyse interaction techniques that could be used on a Specific Command & Control (C2) system called SWORD[1].

4   Unified Interaction Model

4.1   Gestural Grammar

In order to determine which interaction techniques should fit a C2 use, we took inspiration from similar systems, such as Disaster Management [9] and GIS [10]. We proposed an adapted ORBAT that allows a navigation through large hierarchies on tactile devices [11].

Through experimentations made with users on the SWORD C2 system, we managed to find interaction techniques that can be applied used when performing actions.

The Table 1 below sums up possible interaction techniques for C2 on a tactile mobile device:

Table 1. Proposition of a gestural grammar for a C2 system

Action	Possible gesture
Unit selection	*tap, swipe to select, tilt to select*
Contextual menu opening	*hold tap, double tap, Force Tap*
Map panning	*one-finger drag, two-finger drag, flick, Cyclopan*
Map zooming	*pinch Cyclozoom, Spiral zoom, tilt to zoom*

Merging "task-centered" and "platform-centered" approaches will enable us to facilitate users interaction with a tactile C2 system.

4.2   Interaction Model

As described in the paragraph Sect. 3.2, Domain actions represents what users want to do whereas interface actions represents how they will interact.

Identifying different actions from operator's experiments help us to determine the ways in which the operator and the simulator interact.

Figure 1 shows the two main changes in order to simplify operator's task realization with a simulator. The first one concerns a clear distinction of the interface with operator-simulator interaction. The consequence is the separation of domain actions and interface actions and a simple mechanism of translation between domain-actions and interface-actions by shifting the main problem of translation at the level of the exchange of interfaces. The second change is to

---

[1] http://www.masagroup.net/products/masa-sword/.

use, on the one hand, the expert map as a concrete expression of the mental representation of the operator, and, on the other hand, to consider the sources of information used by the simulator as the expert map of the simulator. Thus, making compatible the expert maps of the operator and the simulator could be used to tailor the information presentation to operator in a task realization oriented way.

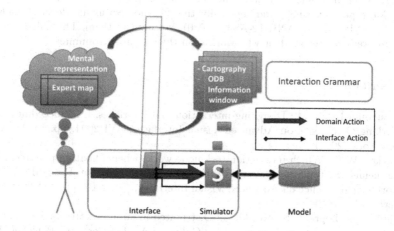

Fig. 1. A model of user-simulator interaction

We must take into account these issues in order to fit user's interaction to our model. By adapting the interaction and the visualization of the application, we could enhance its usability.

5 Conclusion

The main consequence of our work is flexibility of interfaces (say, with fingers movement in a touchscreen or mouse on PC). This approach proposes to adapt interface actions to domain actions, not to impose a translation of interface actions for each interface used. This opens the possibility to introduce easily news concepts for handling interfaces linked to new technologies. The operator-simulator interface then is physical as well as "cognitive". Thus, using "task-realization oriented" approach (or "oriented operator's behavior") for designing interfaces makes sense [5]. This translation of domain actions in interface actions is not the establishment of a simple lookup table because one must also take into account operator's preferences in this translation. Indeed, a path to explore is to identify the interface map of the operator similar to his expert map of the domain. This interface map is linked to the gestural grammar, as users adapt their interaction to the support.

This paper points out the distinction of domain actions and interface actions. A next step would be to make a model of the relationship between these two types of actions. A path to explore is to develop an interface map like an expert map can by develop in the domain. The operator makes an interface action because he wants to perform a domain action. However, the links between these two types of action are not direct, and thus not natural for the operator.

Acknowledgments. This research work was partially funded by the "Ministère de l'Enseignement Supérieur et de la Recherche", the "Conseil Général du Finistère", the "Brest Métropole Océane", and especially the "Agence Nationale de la Recherche" (TACTIC ANR Project, ANR-12-ASTR-0020). We want to thank Thibault Le Jehan, former master's degree student who worked on the Gestural Grammar.

References

1. Beaudouin-Lafon, M.: Designing interaction, not interfaces. In: Proceedings of the Working Conference on Advanced Visual Interfaces, AVI 2004, pp. 15–22. ACM, New York (2004)
2. Mackay, W.E.: Which interaction technique works when?: floating palettes, marking menus and toolglasses support different task strategies. In: Proceedings of the Working Conference on Advanced Visual Interfaces, AVI 2002, pp. 203–208. ACM, New York (2002)
3. Kubicki, S., Lepreux, S., Kolski, C.: Distributed ui on interactive tabletops: issues and context model. In: Lozano, M.D., Gallud, J.A., Tesoriero, R., Penichet, V.M. (eds.) Distributed User Interfaces: Usability and Collaboration. Human Computer Interaction Series, pp. 27–38. Springer, London (2013)
4. Calvary, G., Coutaz, J., Thevenin, D., Limbourg, Q., Bouillon, L., Vanderdonckt, J.: A unifying reference framework for multi-target user interfaces. Interact. Comput. 15(3), 289–308 (2003)
5. Brézillon, P.: Focusing on context in human-centered computing. IEEE Intell. Syst. 18(3), 62–66 (2003)
6. Clancey, W.J.: Simulating activities: relating motives, deliberation, and attentive coordination. Cogn. Syst. Res. 3(3), 471–499 (2002)
7. Brézillon, P., Brézillon, P.: Task-realization models in contextual graphs. In: Kokinov, B., Kokinov, B., Dey, A.K., Dey, A.K., Leake, D.B., Leake, D.B., Turner, R., Turner, R. (eds.) CONTEXT 2005. LNCS (LNAI), vol. 3554, pp. 55–68. Springer, Heidelberg (2005)
8. Richard, J.F.: Logique du fonctionnement et logique de l'utilisation, Rapport de recherche INRIA (1983)
9. Artinger, E., Coskun, T., Schanzenbach, M., Echtler, F., Nestler, S., Klinker, G.: Exploring multi-touch gestures for map interaction in mass casualty incidents. In: Workshop zur IT-Unterstuetzung von Rettungskraeften im Rahmen der GI-Jahrestagung Informatik (2011)
10. Ucuzal, L., Kopar, A.: Gis (geographic information systems) in ccis (command and control systems). Geogr. Inf. Convers. Manage. Syst. 30, 6 (2010)
11. Kabil, A., Kubicki, S.: RICHIE: a step-by-step navigation widget to enhance broad hierarchy exploration on handheld tactile devices. In: Kurosu, M. (ed.) Human-Computer Interaction. LNCS, vol. 9170, pp. 196–207. Springer, Heidelberg (2015)

English Vocabulary for Tourism – A Corpus-Based Approach

Jaroslav Kacetl[✉] and Blanka Klímová

Faculty of Informatics and Management, University of Hradec Kralove,
Hradec Králové, Czech Republic
{jaroslav.kacetl,blanka.klimova}@uhk.cz

Abstract. The aim of this article is to determine the most frequent and important English vocabulary used in the area of travel and tourism. This is done on the basis of an analysis of the compiled corpus consisting of the collected tokens from Deutsche Welle video clips by using the software programme called Sketch Engine. In addition, the authors of this article set several useful implications for the teaching of this specific vocabulary, particularly the head nouns.

Keywords: English for travel and tourism · Vocabulary · Corpus · Analysis

1 Introduction

This contribution is meant to determine the most important English vocabulary for students of Management of Travel and Tourism at the Faculty of Informatics and Management (FIM), University Hradec Kralove (UHK), Czech Republic. The research is based on authentic language material that is easily accessible on the Internet. The English language for students of Management of Travel and Tourism has been taught at FIM for about twenty years. I2008 FIM moved into brand new premises, where both teachers and students enjoy well-equipped classrooms and offices. Since the relocation it has been possible to use ICT in language classes.

At the beginning of 2010, about four-minute-long authentic video clips from the Internet started to be used regularly in English language classes. Since then, more than a hundred worksheets to the same number of video clips about travel and tourism have been created. First, these video clips and the corresponding worksheets were used only in classes. In 2011, however, the first textbook, called English Listening Exercises [4], was published, followed by another one, named English Listening Exercises II [5]. These textbooks may be used both in class and for self-study. They contain ninety worksheets to video clips freely and easily accessible on the Internet. Seventy of these clips feature travel and tourism, usually European cities with their tourist attractions, as their main topic. All corresponding worksheets have the same layout. They all contain the video clip title, the website where the clip can be found, an introduction, a dictionary and questions, which focus on key vocabulary and he most significant information about the topic in question.

© Springer International Publishing Switzerland 2015
H. Christiansen et al. (Eds.): CONTEXT 2015, LNAI 9405, pp. 489–494, 2015.
DOI: 10.1007/978-3-319-25591-0_37

In 2014, a group of FIM language teachers came up with an idea to study the vocabulary used in these video clips in more detail. Therefore, they started collecting transcripts of travel and tourism video clips. Since the travel and tourism video clips come from several sources – Deutsche Welle (DW), BookingHunter.com or Lonely Planet – it was necessary at this stage to determine a single source. DW was chosen because most of the video clips used in English classes were made by this German international broadcaster. As of April 2015, the authors of this article have got a file called DW Travel containing forty-one transcripts with 24,177 tokens.

2 Literature Review

2.1 Corpora

Since the 1990s one of the most progressive approaches has seemed to be a corpus-based approach [6]. This approach enables both teachers and learners to study real life language by means of naturally occurring language samples, which are stored in corpora (or corpuses). Corpora are computerized databases created for linguistic research. Flowerdew [1] provides the most recent characteristics of a corpus: authentic, naturally occurring data; assembled according to explicit design criteria; representative of a particular language or genre; and designed for a specific linguistic or socio-pragmatic purpose.

There are various types of corpora. Among the best known and influential types are the following [2]:

- General corpora, such as the British National Corpus (BNC) or the Bank of English (BoE), contain a large variety of both written and spoken language, as well as different text types, by speakers of different ages, from different regions and from different social classes of the UK.
- Synchronic corpora, such as F-LOB and Frown, record language data collected at one specific point in time, e.g. written British and American English of the early 1990s.
- Historical corpora, such as A Representative Corpus of Historical English Registers (ARCHER) and the Helsinki Corpus of English Texts, consist of corpus texts from earlier periods of time. They usually span several decades or centuries, thus providing diachronic coverage of earlier stages of language use.
- Learner corpora, such as the International Corpus of Learner English (ICLE) and the Cambridge Learner Corpus (CLC), are collections of data produced by foreign language learners and include texts types such as essays or written exams.
- Corpora for the study of varieties, such as the International Corpus of English (ICE) and the Freiburg English Dialect Corpus, represent different regional varieties of a language.
- There is also a large variety of specialized corpora, e.g. Michigan Corpus of Academic Spoken English (MICASE).

At present, there are a number of software programmes, which teachers of English as a foreign language (EFL) can use for analyses of different corpora. For example, the authors of this article use the Sketch Engine [7]. As Thomas [8] explains, the Sketch Engine is highly specialized type of search engine that can search over 60 different corpora to provide data about words, phrases and grammatical constructions. It was developed by experts at the Faculty of Informatics of Masaryk University in Brno, Czech Republic. One of the most exploited tools of the Sketch Engine is the concordancer, which gives a chance to see any word or phrase in context so that one can see what sort of company it keeps. The Sketch Engine is a programme for the analysis of texts. It reads plain text files (in different encodings) and HTML files (directly from the Internet) and it produces, among other things, word frequency lists and concordances from these files. It is very user-friendly, which means that even teachers who are relatively unskilled or inexperienced in computer use can operate it easily. Registered users can create their own corpora and analyse them afterwards, which was done in case of the DW Travel file. Users can look, for example, at the most frequent words and collocations and see them in their context [2].

3   Research

3.1   Material

As it has already been mentioned, as of April 2015, there has been a file containing forty-one transcripts with 24,177 tokens, which means there are almost 600 tokens per transcript. Even though the number of token is not by any means final, this paper works with the above-mentioned number of tokens. It is a sufficient amount of data according to Greenbaum [3]. In order to process this bulk of data, the Sketch Engine is used.

The authors strive to determine what vocabulary is commonly used in currently created (roughly 2010–2015) sources of information on travel and tourism. More specifically, they look into the vocabulary used in short DW video clips on travel and tourism in European countries. These video clips are easily accessible for language teacher and learners as well as public on the Internet.

First, a new corpus – the file with DW video clip transcripts was uploaded to the Sketch Engine. Then, a word list was checked in order to find out what words are most frequently used in the file. This research focuses on the words which have to do with travel and tourism. On the other hand, it was decided not to take into account articles (the, a, an), conjunctions (and), or prepositions (for, to, on) because they do not have any semantic value. Therefore, the following text concentrates on the most frequently used nouns. Other parts of speech significant for travel and tourism, namely the most frequently used adjectives and verbs, will be treated in another study looking into the file with DW video clip transcripts.

3.2   Results

The first list comprises of nouns (and numerals) although in some particular contexts they may represent another part of speech. This fact, however, was not taken into account as the Sketch Engine can only find words in the file based on their spelling but cannot determine their particular function in the sentence. The noun (and numeral) ranking is as follows: city (164 times, including its plural form cities, which is 6,783.3 times per one million words), people (91), year (90), place (56), building (55), century (52), town (52), time (49), world (46), art (44), tower (39), day (38), visitor (37), market (35), tourist (33), museum (30), capital (28), centre/center (27), house (25), wine (25), home (23), life (23), artist (22), London (22), today (22), water (22), street (21), architecture (20), culture (20), food (20), Europe (19), part (19), view (18), country (17), ski (17), history (16), hotel (16), Oxford (15), Palermo (15), tradition (15), beer (14), Christmas (14), million (14), thousand (14), Aarhus (13), end (13), heritage (13), hundred (13) Paris (13), San (13), Constance (12), Dijon (12), Eiffel (12), Marbella (12), number (12), Rotterdam (12), summer (11), night (10).

The list of nouns can be divided into several broad categories: places (and place names), time, architecture, people, culture, food and drink, numbers.

Words of places (city, place, town, capital, centre, and village (7)) are used 334 times. Moreover, other frequently used expressions are world, home and country (appearing 87 times together). The vocabulary of places is the most common in the file because most of the used video clips present European cities. DW focuses on the most well-known metropolises. Therefore, the word city appears much more frequently than words town or village.

The list of place names is difficult to make as there are a lot of instances of individual settlements. It also depends on video clips whose transcripts have been added to the file. Among other place names, the following appear most frequently in the file: London, Europe, Oxford, Palermo, Aarhus, Paris, San (usually only a part of place names), Constance, Dijon, Marbella, or Rotterdam. The Spanish word San, meaning Saint, appears in several different place names, like San Sebastian, or San Miguel.

Words of time (year, century, time, day, today, summer, night, winter (6), and spring (1)) can be found in 284 cases. The most frequently used expression of time is year. It is often used in the plural form after a number or one the following expressions: hundreds of, a few, each, every, this, a/per. Sometimes it is followed with the word ago. The word century is most frequency preceded by an ordinal number. Surprisingly, words representing seasons are rarely used.

Words of architecture (building, tower, market, museum, house, architecture) appear 204 times. The word building often comes after descriptive words related to architectural styles: Gothic, Renaissance, Art Nouveau, neo-classical. In other cases, it follows adjectives: old(est), new, ancient, historic, important, magnificent.

Words of people (people, visitor, tourist) feature 161 times. The word visitor collocates in the file with expressions like foreign, outside, famous, future, fashion-conscious or verbs to attract, to allow, to encourage, or to offer. The expression tourist appears after the verb to attract and before verbs to explore, to visit, to come, to flock and in phrases like (popular/major) tourist attraction (destination/information).

Words of culture (art, artist, culture, history, tradition, heritage) are mentioned 130 times. The word art is often used with adjectives like contemporary, modern, ancient, or impressive. The other terms in this group, specifically culture and heritage, refer to significant international issues like the European Capital of Culture, or UNESCO list of World Heritage Sites.

Words of food and drink (wine, water, food, beer, cuisine (6)) are present in 87 cases. Food may be great, lovely, upscale, good, and fresh, but the most frequently used word in this category is wine. Among other beverages, beer is popular, too. This is probably caused by the fact that food is referred to by means of a wide range of rather specific vocabulary as different foodstuff and meals have different names, whereas wine and beer are rather general words related to many different types of both wine and beer.

Words of numbers (million, thousand, hundred, or dozen (1)) are included 42 times. All these numerals collocate with other numbers and words like people (8 million people), tourists, visitors, residents, graduates, students, euros, of years or even post-cards.

4   Discussion of the Findings

The list of frequently used nouns has certain implications for English language teaching in classes for Management of Travel and Tourism students. DW video clips often focus on introducing places with their architectural, cultural and historical background and people who live and work there or come on holiday. All these phenomena refer to other areas. These head nouns are also connected with several dozens of useful collocations. Moreover, there can be traced certain patterns, which can be exploited in language teaching, too.

First, the high incidence of geographical names makes it advisable to teach their correct spelling and pronunciation as well as the usage of the definite article. Second, teachers should focus on the difference between pairs like: twelve hundred years – hundreds of years, 25 years ago – 16 years before. As for architecture, there is a wide range of vocabulary to be taught, including English expressions for various architectural styles and their most significant features. Concerning culture, the useful vocabulary may be categorised according to various arts (painting, sculpting, or music), then there is a group of vocabulary linked to history, traditions and (cultural) heritage, which obviously mingles with both the vocabulary of art and architecture. Food and drink is another broad area, which has recently become very popular. There are a lot of TV shows and other programmes of various types that feature cooking. Therefore, food and drink vocabulary is extremely significant. Last but not least, language learners often have problems with expressing numbers. Teaching how to say numbers is therefore another important impli-cation for language teaching.

5   Conclusion

Thus, as the findings show, the corpus-based approach can enormously contribute to determining and teaching the most common and specific vocabulary used in different areas of human activities, in this case in the teaching of head nouns and their collocations in the field of travel and tourism. Further research could concentrate on a comparison of these findings with the existing materials in the textbooks on travel and tourism, which are already available on the market.

References

1. Flowerdew, L.: Corpora and Language Education. Palgrave Macmillan, Basingstoke (2012)
2. Frydrychová-Klímová, B.: English for Academic Purposes: Developments in Theory and Pedagogy. UHK, Gaudeamus (2013)
3. Greenbaum, S.: The Development of the International Corpus of English. Longman, London (1991)
4. Kacetl, J.: English Listening Exercises. Gaudeamus, Hradec Králové (2011)
5. Kacetl, J.: English Listening Exercises II. Gaudeamus, Hradec Králové (2014)
6. McEnery, T., Xiao, R.: What corpora can offer in language teaching and learning. In: Hinkel, E. (ed.) Handbook of Research in Second Language Teaching and Learning, vol. 2, pp. 364–380. Routledge, London (2010)
7. The Sketch Engine, http://www.sketchengine.co.uk/
8. Thomas, J.: Discovering English with the Sketch Engine. Laptop Languages, USA (2014)

Classification of Interval Information with Data Drift

Piotr Kulczycki[(⊠)] and Piotr A. Kowalski

Polish Academy of Sciences, Systems Research Institute, Warsaw, Poland
{kulczycki,pakowal}@ibspan.waw.pl

Abstract. The paper deals with the classification task of interval information, when processed data is gradually displaced, i.e. they originate from a nonstationary environment. The procedure worked out is characterized by its many practical properties: ensuring the minimum expected value of misclassifications; allowing influence on the probability of errors in classification to particular classes; reducing patterns by eliminating elements with insignificant or negative influence on the results' accuracy, enabling an unlimited number of patterns and their shapes. The appropriate modifications of the classifier not only lead to an increase in the effectiveness of the procedure, but above all adapt to data drift.

Keywords: Data analysis · Classification · Interval information · Data drift · Artificial neural network · Sensitivity method · Pattern size reduction · Classifier adaptation

1 Introduction

In most of the methods used today for classification [1], one assumes invariability in time of data stream under processing. However, more and more often, in particular for those models in which new – with the most current being the most valuable – elements are continuously added to patterns, this assumption is successfully ignored [5].

The presented paper proposes the procedure for classification of information given in the form of an interval for data which may have drifted – undergoing successive changes. The idea for a solution stems from the sensitivity method used in neural networks, together with nonparametric kernel estimators. Namely, particular elements of patterns receive weights proportional to their significance for correct results. Elements of the smallest weights are eliminated. In order to account for the data drift, those elements whose weights are currently small but increase successively are kept.

This paper is a novel elaboration of research presented in the paper [6] for the interval stationary case, and in the publication [7] for the deterministic nonstationary case.

2 Preliminaries

2.1 Statistical Kernel Estimators

Consider an n-dimensional random variable, with a distribution given by the density f. Its kernel estimator $\hat{f} : \mathbb{R}^n \to [0, \infty)$ is calculated on the basis of the random sample

© Springer International Publishing Switzerland 2015
H. Christiansen et al. (Eds.): CONTEXT 2015, LNAI 9405, pp. 495–500, 2015.
DOI: 10.1007/978-3-319-25591-0_38

$$x_1, \ x_2, \ \ldots, x_m, \tag{1}$$

and defined as

$$\hat{f}(x) = \frac{1}{mh^n} \sum_{i=1}^{m} K\left(\frac{x - x_i}{h}\right), \tag{2}$$

where the positive coefficient h is known as a smoothing parameter, while the measurable function $K : \mathbb{R}^n \to [0, \infty)$ symmetrical with respect to zero, having at this point a weak global maximum and fulfilling the condition $\int_{\mathbb{R}^n} K(x)\,dx = 1$ is termed a kernel. For details see the monographs [4, 8, 10].

In this paper the one-dimensional Cauchy kernel is applied, for the multidimensional case generalized by the product kernel concept [4 – Sect. 3.1.3, 10 – Sects. 2.7 and 4.5]. For calculation of the smoothing parameter, the simplified method assuming normal distribution [4 – Sect. 3.1.5, 10 – Sect. 3.2.1] can be used, thanks to the positive influence of this parameter correction procedure applied in the following. For general improvement of the kernel estimator quality, modification of the smoothing parameter [4 – Sect. 3.1.6, 8 – Sect. 5.3.1] will be used, with the intensity $c \geq 0$. As its initial standard value $c = 0.5$ can be assumed.

2.2 Bayes Classification of Interval Information

Consider J sets

$$\{x'_1 \ x'_2, \ldots, x'_{m_1}\}, \ \{x''_1, \ x''_2, \ldots, x''_{m_2}\}, \ \ldots, \ \{x''_1 \cdots', \ x''_2 \cdots', \ldots, x''_{m_J} \cdots'\} \tag{3}$$

representing assumed classes. The sizes $m_1, m_2, \ldots, m_J$ should be proportional to the "contribution" of particular classes in the population. Because of practical aspects, one can assume that the elements from sets (3) belong to the space $\mathbb{R}^n$. Representative elements, consisting of patterns, are characterized by considerable precision and are either deterministic in nature, or of interval type with length of this interval so small that it can be identified with its midpoint without any influence on the quality of the result.

Let now $\hat{f}_1, \hat{f}_2, \ldots, \hat{f}_J$ denote kernel estimators of densities, calculated successively based on sets (3) treated as samples (1), according to the methodology from Sect. 2.1.

First consider the one-dimensional case $(n = 1)$. In accordance with the classic Bayes approach [1], ensuring a minimum of expected value of losses, the tested element $[\underline{x}, \bar{x}]$, with $\underline{x} < \bar{x}$, should be ranked to the class for which the value

$$m_1 \int_{\underline{x}}^{\bar{x}} \hat{f}_1(x)\,dx, \ m_2 \int_{\underline{x}}^{\bar{x}} \hat{f}_2(x)\,dx, \ \ldots, \ m_J \int_{\underline{x}}^{\bar{x}} \hat{f}_J(x)\,dx \tag{4}$$

is the greatest. The above can be generalized by introducing to expressions (4) the positive coefficients $z_1, z_2, \ldots, z_J$:

$$z_1 m_1 \int\limits_{\underline{x}}^{\overline{x}} \hat{f}_1(x)\, dx, \; z_2 m_2 \int\limits_{\underline{x}}^{\overline{x}} \hat{f}_2(x)\, dx, \; \ldots, z_J m_J \int\limits_{\underline{x}}^{\overline{x}} \hat{f}_J(x)\, dx. \tag{5}$$

Taking as standard values $z_1 = z_2 = \ldots = z_J = 1$, formula (5) brings us to (4). By appropriately increasing the value z_i, a decrease can be achieved in the probability of erroneously assigning elements of the i-th class to other wrong classes. Thanks to this, it is possible to favor classes which are in some way noticeable or more heavily conditioned. For the classification, these are in a natural way classes defined by non-stationary patterns, it is worth increasing coefficients relating to more varying patterns. In such case, the initial value 1.25 can be proposed for further research.

In the multidimensional case, i.e. when $n > 1$, the tested element is

$$\begin{bmatrix} [\underline{x_1}, \overline{x_1}] \\ [\underline{x_2}, \overline{x_2}] \\ \vdots \\ [\underline{x_n}, \overline{x_n}] \end{bmatrix} \tag{6}$$

with $\underline{x_k} < \overline{x_k}$ for $k = 1, 2, \ldots, n$, and criterion (5) takes the following form:

$$z_1 m_1 \int\limits_{\underline{x}}^{\overline{x}} \hat{f}_1(x)\, dx, \; z_2 m_2 \int\limits_{\underline{x}}^{\overline{x}} \hat{f}_2(x)\, dx, \; \ldots, \; z_J m_J \int\limits_{\underline{x}}^{\overline{x}} \hat{f}_J(x)\, dx, \tag{7}$$

where $E = [\underline{x_1}, \overline{x_1}] \times [\underline{x_2}, \overline{x_2}] \times \ldots \times [\underline{x_n}, \overline{x_n}]$.

For the Cauchy kernel proposed here, generalized in the multidimensional case by the product kernel concept (see Sect. 2.1), the analytical form of quantities occurring in formulas (4), (5) and (7) are possible to obtain; see the paper [6].

2.3 Sensitivity Analysis for Learning Data – Reducing Pattern Size

When modeling by artificial neural networks, particular components of an input vector most often are characterized by diverse significance of information. Using a sensitivity analysis [12], one obtains the parameters $\bar{S}_i$ describing proportionally the influence of the particular inputs ($i = 1, 2, \ldots, m$) on the output value, and then the least significant inputs can be eliminated.

To apply the above procedure, the definition of the kernel estimator will be generalized with the introduction of the nonnegative coefficients $w_1, w_2, \ldots, w_m$, normed so that $\sum_{i=1}^{m} w_i = m$, and mapped to particular elements of random sample (1). The basic form of kernel estimator (2) then takes the form

$$\hat{f}(x) = \frac{1}{mh^n} \sum_{i=1}^{m} w_i K\left(\frac{x - x_i}{h}\right).$$ (8)

The coefficient w_i value may be interpreted as indicating the significance (weight) of the i-th element of the pattern to classification correctness.

For the purpose of calculation the weights w_i values, separate neural networks are built for each investigated class. This network is submitted to a learning process using a data set comprising of the values of particular kernels for subsequent pattern elements, while the given output constitutes the value of the kernel estimator calculated for the pattern element under consideration. After this, the obtained network undergoes sensitivity analysis on learning data. The resulting coefficients $\bar{S}_i$ describing sensitivity, constitute the fundament for calculating the values

$$\tilde{w}_i = \left(1 - \frac{\bar{S}_i}{\sum\limits_{j=1}^{m} \bar{S}_j}\right) \text{ normed to } w_i = m \frac{\tilde{w}_i}{\sum\limits_{i=1}^{m} \tilde{w}_i}.$$ (9)

The shape of the formula defining the parameters $\tilde{w}_i$ results from the fact that the network created here is the most sensitive to atypical and redundant elements, which implies a necessity to map the appropriately smaller values $\tilde{w}_i$, and in consequence w_i, to them. Coefficients w_i represent the significance of particular elements of the pattern to accuracy of the classification. Because – thanks to normalization – the mean value of the coefficients w_i equals 1, the pattern set should be relieved of those elements for which $w_i < 1$.

3 Classification Procedure

This section presents the method for classification of interval information with data drift.

First one should fix the reference sizes of patterns (3), hereinafter denoted by $m_1^*, m_2^*, \ldots, m_J^*$. The patterns of these sizes will be the subject of a basic reduction procedure, described in Sect. 2.3. The sizes of patterns available at the beginning of the algorithm must not be smaller than the above referential values. These values can however be modified during the procedure's operation, with the natural condition that their potential growth does not increase the number of elements newly provided for the patterns. For preliminary research, $m_1^* = m_2^* = \ldots = m_J^* = 25 \cdot 2^n$ can be proposed.

The elements of initial patterns (3) are provided as introductory data. Based on these – according to Sect. 2.1 – the value of the parameter h is calculated (for the parameter c initially assumed to be equal 0.5). Next, corrections in the parameters c and $h_1, h_2, \ldots, h_n$ values are made by introducing $n + 1$ multiplicative correcting coefficients. Denote them as $b_0 \geq 0$, $b_1, b_2, \ldots, b_n > 0$, respectively. Their values can be calculated by a static optimization procedure in the $(n + 1)$-dimensional space, where the initial conditions at the beginning are the points of a grid, or the previous values in the following steps, while the performance index is given as the number of misclassifications. To find the minimum a modified Hook-Jeeves algorithm [2] was applied.

The next procedure is the calculation of the parameters w_i values mapped to particular patterns' elements, separately for each class, as in Sect. 2.3. Following this, within each class, the values of the parameter w_i are sorted, and then the appropriate $m_1^*, m_2^*, \ldots, m_J^*$ elements of the largest values w_i are designated to the classification phase itself. The remaining ones undergo further treatment, which will be presented later, after Bayes classification has been dealt with.

The reduced patterns separately go through a procedure newly calculating the values of parameters w_i, shown in Sect. 2.3. Next, these patterns' elements for which $w_i \geq 1$ are submitted to further stages of the classification procedure, while those with $w_i < 1$ are sent to the beginning of the algorithm for further processing in the next steps of the algorithm, after adding new elements of patterns. The final, and also the principal part of the procedure worked out here is Bayes classification, presented in Sect. 2.2. Obviously many tested elements of interval type can be subjected to classification separately. After the procedure has been finished, elements of patterns which have undergone classification are sent to the beginning of the algorithm, to further avail of the next steps, following the addition of new elements of patterns.

Now – in reference to the end of the paragraph before the last – it remains to consider those elements whose values w_i were not counted among the $m_1^*, m_2^*, \ldots, m_J^*$ largest for particular patterns. Thus, for each of them the derivative w_i' is calculated. A method based on Newton's interpolation polynomial [9] is suggested here. If the element is "too new" and does not possess enough earlier values w_i, then the gaps should be filled with zeros, which prevents premature removal. Next for each separate class, the elements w_i' are sorted. The respective

$$qm_1^*, qm_2^*, \ldots, qm_J^* \tag{10}$$

elements of each pattern with the largest derivative values, on the additional requirement that the value is positive, go back to the beginning of the algorithm for further calculations carried out after the addition of new elements. If the number of elements with positive derivative is less than $qm_1^*, qm_2^*, \ldots, qm_J^*$, then the number of elements going back may be smaller (including even zero). The remaining elements are finally eliminated from the procedure. In the above notation q is a positive constant influencing the proportion of patterns' elements with little, but successively increasing meaning. The standard value of the parameter q can be proposed as $q = 0.1$.

The above procedure is repeated following the addition of new elements. Besides these elements – as has been mentioned earlier – for particular patterns respectively $m_1^*, m_2^*, \ldots, m_J^*$ elements of the greatest values w_i are taken, as well as up to $qm_1^*, qm_2^*, \ldots, qm_J^*$ elements of the greatest derivative w_i', so successively increasing its significance, most often due to the data drift.

4 Empirical Verification and Comparison

The correct functioning of the concept under investigation have been comprehensively verified numerically, and also compared with results obtained using procedures based on the support vector machine concept. Research was carried out for data sets in various configurations and with different properties, particularly with nonseparated classes, complex patterns, multimodal and consisting of detached subsets located alternately.

Comparative analysis was submitted to detailed investigations. Due to lack of an available algorithm dedicated to interval and drifting data, comparisons were made using two concepts based on the support vector machine method with proper modifications. The first one, intended for deterministic nonstationary data, presented in the article [3], was used with respect to midpoints of classified intervals. The second, from the publication [11], for stationary interval data was applied by removing the oldest elements from patterns and replacing them with the newest. And so, in relation to the first algorithm, the number of misclassifications was up to 20 % lower, while with the second even 50 % (treating no decision – a possibility there – as an unsatisfactory result). The advantage of the procedure presented in this paper was particularly visible in the case of steady drift – taking into account the fact that its idea is based on derivatives of a predictive nature, this observation is completely understandable.

The broader description of particular aspects and the analytical form of formulas can be found in the papers [Kulczycki and Kowalski, 2011, 2015].

References

1. Duda, R.O., Hart, P.E., Storck, D.G.: Pattern Classification. Wiley, New York (2001)
2. Kelley, C.T.: Iterative Methods for Optimization. SIAM, Philadelphia (1999)
3. Krasotkina, O.V., Mottl, V.V., Turkov, P.A.: Bayesian approach to the pattern recognition problem in nonstationary environment. In: Kuznetsov, S.O., Mandal, D.P., Kundu, M.K., Pal, S.K. (eds.) PReMI 2011. LNCS, vol. 6744, pp. 24–29. Springer, Heidelberg (2011)
4. Kulczycki, P.: Estymatory jadrowe w analizie systemowej. WNT, Warsaw (2005)
5. Kulczycki, P., Hryniewicz, O., Kacprzyk, J. (eds.): Techniki informacyjne w badaniach systemowych. WNT, Warsaw (2007)
6. Kulczycki, P., Kowalski, P.A.: Bayes classification of imprecise information of interval type. Control Cybern. **40**(1), 101–123 (2011)
7. Kulczycki, P., Kowalski, P.A.: Bayes classification for nonstationary patterns. Int. J. Comput. Methods **12**(2), 19 (2015). Article ID 1550008
8. Silverman, B.W.: Density Estimation for Statistics and Data Analysis. Chapman and Hall, London (1986)
9. Venter, G.: Review of optimization techniques. In: Blockley, R., Shyy, W. (eds.) Encyclopedia of Aerospace Engineering, pp. 5229–5238. Wiley, New York (2010)
10. Wand, M.P., Jones, M.C.: Kernel Smoothing. Chapman and Hall, London (1995)
11. Zhao, Y., He, Q., Chen, Q.: An interval set classification based on support vector machines. In: 2nd International Conference on Networking and Services, Silicon Valley, USA, 25–30 September 2005, pp. 81–86 (2005)
12. Zurada, J.: Introduction to Artificial Neural Neural Systems. West Publishing, St. Paul (1992)

Intentions in Utterance Interpretation

Palle Leth[1,2(✉)]

[1] Stockholm University, Stockholm, Sweden
palle.leth@philosophy.su.se
[2] Institut Jean-Nicod, Paris, France

Abstract. Which is the role of intentions in utterance interpretation? I sketch an argument to the effect that the role of intentions is indirect; the interpreter's assignment of meaning rather depends on considerations of what meaning is most reasonably assigned and her interest. This approach often results in the assignment of intended meaning, but might also result in the assignment of non intended meaning. I consider the three basic options offered to the interpreter when, in the course of the conversation, she is confronted with further evidence about the speaker's intention.

Keywords: Intentions · Utterance interpretation · Semantics/pragmatics · Conversational interaction

1 Introduction

Which is the role of intentions in utterance interpretation? It is often assumed that once the decoding of conventional meaning has yielded its result – truth conditions or schemata for further enrichment – the goal of interpretation is the recovery of speaker intentions. Here I will present an alternative picture to the effect that the contextualist all things considered establishment of meaning, even though involving considerations about speaker intentions, is best conceived as aiming at the most reasonably assigned meaning and crucially depends on the interpreter's interest. The approach often results in the assignment of intended meaning, but may also result in the assignment of non intended meaning. The most reasonably assigned meaning may or may not correspond to intended meaning; in the latter case, the utterance may or may not, according to the interpreter's interest, be erased by her. I will consider the interpreter's options when in the course of the conversation she is confronted with further evidence about the speaker's intended meaning.

2 The Regular Course of Interpretation

Many theorists take for granted that the goal of utterance interpretation is the recovery of the speaker's intended meaning. At the same time it is pointed out that a speaker cannot mean whatever she wants by an utterance. Many theorists hold that conformity to conventions is constitutive of speaker meaning and that in cases of divergence between conventional and intended meaning, conventional meaning wins out [1–3].

© Springer International Publishing Switzerland 2015
H. Christiansen et al. (eds.): CONTEXT 2015, LNAI 9405, pp. 501–505, 2015.
DOI: 10.1007/978-3-319-25591-0_39

From a contextualist point of view, which I will here adopt, what matters to speaker meaning is not conformity to conventional meaning as such, but the speaker's making the interpreter able to figure out her intended meaning in some way or other. Conformity to conventions is one means, but there are all sorts of cues which can be exploited by the speaker [4]. This suggests that the meaning of an utterance is the meaning intended by the speaker, provided that the speaker sees to it that her intended meaning be the meaning most reasonably assigned to the utterance by the interpreter.

The interpreter approaches an utterance with the aim of making reasonable sense of it. This means that she is prepared to go beyond or against conventional meaning whenever she is so invited by the absurdity of the conventional meaning, the preceding discourse, what she knows about the speaker's attitude and aims, her background assumptions, the common ground of the conversation, the requirements of cooperation, considerations of what is interesting and relevant. All things considered, what is the most reasonable meaning to assign to this utterance in this context?

The notion of most reasonably assigned meaning is certainly quite vague. Such a notion makes a component of meaning a matter of all things considered consideration [5], discussion, argument, even dispute, in short a matter not of decoding but of decision. Which contextual cues are available? What is included in the common ground? What is reasonably taken into consideration?

The interpreter is not directly concerned with the speaker's actual intention, but with the meaning which manifests itself given all sorts of contextual factors. There is nevertheless a natural connection with the speaker's intended meaning in so far as the interpreter has every reason to assume that what she takes to be the most reasonably assigned meaning coincides with the speaker's intended meaning. The interpreter believes that the speaker in order to get her meaning across relies on the contextual factors which she takes into consideration when assigning meaning. The reason why the interpreter believes that she has arrived at the speaker's intended meaning is not some access to the speaker's mind, but that she has no reason to suppose that what is the most reasonably assigned meaning is not the intended meaning.

In the regular course of interpretation, the goal is to establish, all things considered, the most reasonably assigned meaning. The question is not what the speaker's intention is, but what is conveyed by a certain utterance in a certain context. Speakers trust interpreters will go beyond conventional meaning and interpreters trust speakers intentions are conform to most reasonably assigned meaning [6].

3 Cases of Divergence Between Intended and Assigned Meaning

It may happen of course that the speaker fails to make her intention manifest, so that there is a divergence between what is most reasonably taken to be the meaning of the utterance by the interpreter and the meaning actually intended by the speaker. Philosophers imagine such failures, but do not reflect upon the conversational interaction following upon such failures [7, 8]. Linguists are much more interested in clarifications and corrections, but do not seem to consider the variety and the reasons for the options offered to the interpreter when confronted with further evidence about the speaker's intention. There seems to be a tendency to take for granted that speaker initiated repairs are always accepted by the interpreter [9]. The consideration of the interpreter's dealing with evidence about intentions in cases of divergence will contribute to determining the role of intentions in interpretation.

Let us look at the case invented by Kaplan:

Suppose that without turning and looking I point to the place on my wall which has long been occupied by a picture of Rudolf Carnap and I say: [That] is a picture of one of the greatest philosophers of the twentieth century. But unbeknownst to me, someone has replaced my picture of Carnap with one of Spiro Agnew [10].

Philosophers discuss whether a false statement of the picture of Agnew [10, 11] or a true statement of the picture of Carnap or no statement is made in this case [12]. It seems to me that the question if or what a statement is made should be asked with respect to the interpretive strategies in the subsequent conversational interaction. The statement made depends upon kinds of action on the interpreter's part. Let us look at a possible continuation of the conversation.

S: That is the picture of the greatest philosopher of the 20th century.

I: I did not know Spiro Agnew was the greatest philosopher of the 20th century.

S: Spiro Agnew? No, I meant Rudolf Carnap. *Turning around.* Someone must have replaced the Carnap picture.

I: Yes, Carnap was a great philosopher.

Let us assume that the interpreter takes the picture of Spiro Agnew to be the value of the demonstrative, because at the time of the utterance this seems to her to be the most reasonably assigned meaning, all things considered. The speaker's reaction to her reply provides her with further evidence as to the intended meaning. The sequence above is compatible with two different interpretive strategies.

First, the interpreter may reconsider her meaning assignment. For instance, she may recall that actually a picture of Carnap used to occupy the spot of the wall where there is now a picture of Agnew. This may convince her that the meaning she assigned was not the meaning most reasonably assigned. It would have been reasonable to consider this piece of common ground when assigning a meaning to the utterance. In short, the interpreter would blame herself for having made the wrong interpretation and correct her meaning assignment.

Even though the plausibility of such a line of interpretation may not be entirely convincing in the case at hand, I believe it is not too difficult in general to imagine cases where the interpreter, in the face of further evidence of the speaker's intention, reconsiders her interpretation and agrees with the speaker that the most reasonable interpretation is the one intended by the speaker. The interpreter says to herself: "This is how, upon consideration, I should have taken the speaker's utterance."

Another option for the interpreter is to consider that at the time of utterance, the meaning assignment she made actually was the most reasonably assigned meaning. She has no reason to reconsider the interpretation she made and she proceeds at no correction. The speaker is to blame for not having made her intention apparent. Upon being informed that the speaker's intention was actually to refer to a picture of Rudolf Carnap, she is however prepared to erase the utterance made by the speaker and, as it were, replace it with an utterance to the effect that Rudolf Carnap was one the greatest philosophers of the twentieth century. The interpreter says to herself: "This is not at all – all things considered – what the speaker said, but never mind, I now see what the speaker wanted to say."

These two procedures may occur silently and may not be outwardly visible, as in the conversation above, but are nevertheless distinct in that they originate in different judgments. In reinterpretation the mistake is with the interpreter, when erasure and replacement occur the mistake is with the speaker.

In the former case, the speaker succeeds after all in communicating her intention. In the latter case, she fails. I do not think that the fact that the most reasonably assigned meaning is not intended by the speaker makes it the case that the most reasonably assigned meaning is not the meaning of the utterance (unlike [12] for example). But new evidence concerning the intended meaning may constitute a reason for the interpreter to erase the utterance and replace it with a novel one, in line with what the further evidence suggests to her. According to this conception, it makes sense to speak of the meaning of an utterance only from the viewpoint of an interpreter's assigning a meaning to it.

However, these two options are not exhaustive. The interpreter is not obliged to refuse an utterance which from the speaker's point of view is infelicitous in that the most reasonable meaning assigned to the utterance does not correspond to her intended meaning. The interpreter may deliberately preserve the utterance and hold the speaker responsible for the meaning most reasonably assigned to it even though she does not believe that the speaker intended this meaning. Why would the interpreter take an interest in assigning non intended meaning? Is not utterance interpretation always in the service of speaker intentions?

Let us go back to Kaplan's case again and modify it somewhat. Unbeknownst to the speaker, the picture of Carnap has been replaced with, not a picture of Agnew, but a nice-looking painting. Without turning and looking the speaker points to the place of the wall now occupied by the painting and says: "That is your birthday present."

Let us suppose that at the time of utterance, the most reasonable meaning assigned to the utterance is a statement to the effect that the speaker has bestowed the nice-looking painting upon the interpreter. The interpreter may later come across further evidence concerning the speaker's intentions in making that utterance which makes it clear that the speaker intended to give her a picture of Carnap. However, she is not prepared to erase the utterance and replace it with one to the effect that the speaker had given a picture of Rudolf Carnap to her, simply because the original utterance is advantageous to her. The interpreter has an interest in refusing to erase it. We may imagine further cases, involving for example implicatures, slurs and offensive talk where interpreters would not be willing to erase utterances failing to convey the intended meaning. The interpreter has an interest in holding the speaker responsible not for what she wanted to convey, but for the meaning most reasonably assigned to her utterance.

What lead up to the interpreter's meaning assignments are considerations involving not only the conventional meaning of the sentence uttered, but all sorts of contextual factors. This approach to utterance interpretation does not have as a consequence that only intended meaning is assigned. Upon being informed that the most reasonably assigned meaning does not correspond to the intended meaning, the interpreter may not withdraw her meaning assignment, if she has an interest – which may be rather private or rather general in character – in preserving it, using it for her own purposes, as it were. The justification of the interpreter's interpretation crucially depends on the legitimacy of not erasing the utterance and on the reasonableness of the meaning assigned. Both issues are matters of discussion and argumentation and are settled only within the community of interpreters.

4 Conclusion

The interpreter sets out to establish the most reasonably assigned meaning. In the regular course of interpretation she has every reason to believe that this meaning is the speaker's intended meaning. In cases of divergence between the assigned meaning and intended meaning, the interpreter may reconsider her interpretation of the utterance or erase and replace it or preserve it for her own purposes. The interpreter is generally interested in assigning intended meaning, but that does not imply that she is only interested in assigning intended meaning. An utterance means what the speaker means, provided that her intended meaning is the meaning most reasonably assigned. An utterance means what the interpreter takes it to mean, provided that her assigned meaning is the meaning most reasonably assigned.

References

1. Dummett, M.: A nice derangement of epitaphs: some comments on Davidson and Hacking. In: Le Pore, E. (ed.) Truth and Interpretation: Perspectives on the Philosophy of Donald Davidson, pp. 459–476. Blackwell, Oxford (1986)
2. Green, K.: Davidson's derangement: of the conceptual priority of language. Dialectica 55(3), 239–258 (2001)
3. Reimer, M.: What malapropisms mean: a reply to Donald Davidson. Erkenntnis 60(3), 317–334 (2004)
4. Davidson, D.: A nice derangement of epitaphs. In: Davidson, D. (ed.) Truth, Language, and History, pp. 89–107. Oxford University Press, Oxford (2005)
5. Gauker, C.: Zero tolerance for pragmatics. Synthese 165(3), 359–371 (2008)
6. Akman, V.: Rethinking context as a social construct. J. Pragmat. 32, 743–759 (2000)
7. Barwise, J.: On the circumstantial relation between meaning and content. In: The Situation in Logic, pp. 59–77. Center for the Study of Language and Information, Stanford (1989)
8. Wilson, D., Sperber, D.: Truthfulness and relevance. Mind 111(443), 583–632 (2002)
9. Ginzburg, J.: The Interactive Stance. Oxford University Press, Oxford (2012)
10. Kaplan, D.: Dthat. In: Yourgrau, P. (ed.) Demonstratives. Oxford University Press, Oxford (1990)
11. Reimer, M.: Three views of demonstrative reference. Synthese 93(3), 373–402 (1992)
12. King, J.C.: Speaker intentions in context. Noûs 48(2), 219–237 (2014)

Modelling Equivalent Definitions of Concepts

Daniele Porello[✉]

Institute of Cognitive Sciences and Technologies, CNR, Trento, Italy
daniele.porello@loa.istc.cnr.it

Abstract. We introduce the notions of syntactic synonymy and referential synonymy due to Moschovakis. Those notions are capable of accounting for fine-grained aspects of the meaning of linguistic expressions, by formalizing the Fregean distinction between *sense* and *denotation*. We integrate Moschovakis's theory with the theory of concepts developed in the foundational ontology DOLCE, in order to enable a formal treatment of equivalence between concepts.

1 Introduction

We place our analysis of concepts within the foundational ontology DOLCE [2]. Concepts are there intended to model classification of entities of a domain according to some relevant perspectives, in case the intensional aspect of the classification matters for knowledge representation. For example, in legal domains and in social ontology, concepts are fundamental for representing *roles* such as president, delegate, and student [2]. Although the extension of the role "student" provides the class of entities that can be classified as students, it is the intensional aspect of "student" that specifies the relevant information about the meaning of "student", i.e. the conditions that are necessary for classifying something as a student. Concepts are indeed intended to capture those intensional aspects. The motivation for using DOLCE is that it formalizes a rich theory of concepts that allows for relating conceptual information with other types of information concerning the entities of a given domain. In DOLCE, concepts are defined by means of *descriptions*, that are intended as semantic entities that are used to refer to concepts. Since descriptions are our way to access concepts, we are lead to focus on the meaning of the description that is associated to the concept. In Fregean terms, we are not only interested in the *denotation* of the definitions of concepts, but also in their *sense*, that is, in the way in which the denotation is given. Consider the following example. According to the Italian Constitution, the "President of the Italian republic" is also the "President of the Superior Judicial Court". Although the two descriptions refer to the same individual, we want to say that they are referring to different concepts *because* they express distinct senses.

We introduce a language for expressing definitions of concepts and we discuss a number of equivalence relations between definitions. Our approach is based on the calculus of meaning and synonymy developed by Moschovakis [1,6]. This calculus provides a formal interpretation of the crucial distinction made by Frege

© Springer International Publishing Switzerland 2015
H. Christiansen et al. (Eds.): CONTEXT 2015, LNAI 9405, pp. 506–512, 2015.
DOI: 10.1007/978-3-319-25591-0_40

between sense and denotation. The calculus is based on a typed system for expressing the meaning of natural language expressions, in the style of the Montague grammar [3,4]. We will integrate the calculus of meaning and synonymy with the treatment of concepts in DOLCE. We shall see how to deploy the calculus of meaning and synonymy to decide whether two descriptions of concepts are equivalent, i.e. they define the same concept. Since the calculus is based on type theory, it is not embeddable in first order logic, therefore, we view it as an external module that interacts with DOLCE, rather than a part of the ontology. This is motivated by the fact that a first-order theory such as DOLCE is not sufficiently expressive to talk about the meaning of first-order terms and formulas.

Possible applications of a formal theory of equivalence of concepts concern the possibility of comparing the intensional aspects of conceptualizations provided by two agents' theories. The remainder of this paper is organized as follows. The next section presents the formal grammar based on type theory and the background of the calculus of meaning and synonymy developed by Moschovakis. Section 3 discusses the application of the calculus of synonymy with DOLCE.

2 Formal Background

We present the *typed calculus of acyclic recursion* $\mathcal{L}_{ar}^{\lambda}$ defined by Moschovakis. We refer to [1,5–7] for the details. The idea of a typed system to model natural language semantics can be briefly summarized as follows. The elements of a vocabulary (e.g. words in natural language) are associated to terms of a certain type. The type encodes the properties that are required for composing complex meanings, while the term specifies the semantic contribution of the element of the vocabulary. For instance, the word "red" is associated to the type of functions from entities to truth values. The term associated to "red" is the specific function that associates "true" to red entities and false otherwise.

As in Montague grammar, types are defined recursively from the basic types e for entities and t for truth values: TYPES::=$e \mid t \mid (\tau \rightarrow \tau')$

We assume a finite set of constants K, the vocabulary of our semantic grammar. The typing relation: $c : \tau$ indicates that c has type τ. Given a choice of K, the set of terms of the language of acyclic recursion $L_{ar}^{\lambda}(K)$ is defined as follows. For each type τ, we assume two distinct infinite sets of *variables*: *pure variable*: v_0^{τ}, v_1^{τ}, ..., and *recursion variable* or *locations*: p_0^{τ}, p_1^{τ}, ... Pure variables range over the domains of their types, whereas location variables are only assigned to terms.

The main innovation in Moschovakis's calculus is the *acyclic recursion* construction. Given a sequence of location variables, $p_1, \ldots, p_n$ and a set of terms $A_1, \ldots, A_n$, a *system of assignments* is an expression $\{p_1 := A_1, \ldots, p_n := A_n\}$ that means that a term A_i is assigned to the position p_i. The acyclic recursion is a term written as follows:

$$A_0 \text{ WHERE } \{p_1 := A_1, \ldots, p_n := A_n\}$$

The meaning of this statement is that, once one sets p_1 as $A_1, \ldots, p_n$ is A_n, then one has A_0. The condition on the recursion construction is that its system of assignment must be *acyclic* [6].

Definition 1. *The set of terms of $L_{ar}^\lambda(K)$ is defined as follows:*

$$\text{TERMS} := c \mid v \mid p \mid B(C) \mid \lambda(v)(B) \mid A_0 \text{ WHERE}\{p_1 := A_1, \ldots, p_n := A_n\}$$

with the following conditions:

T1. If c is a constant of type τ, then c is a term of type τ, $c : \tau$;
T2. Variables of type τ are terms of type τ, in particular $v : \tau$ and $p : \tau$;
T3. If $C : \tau$ and $B : \tau \to \tau'$, then $B(C) : \tau'$;
T4. If $B : \tau'$ and v is a pure variable of type τ, then $\lambda(v)(B) : \tau \to \tau'$;
T5. If for $n \geq 0$, $A_i : \tau_i$ and $p_1, \ldots, p_n$ are distinct locations $p_i : \tau_i$ such that the system of assignments $\{p_1 := A_1, \ldots, p_n := A_n\}$ is acyclic, then

$$A_0 \text{ WHERE } \{p_1 := A_1, \ldots, p_n := A_n\} : \tau_0$$

For the sake of example, we treat a very simple fragment of English (Fig. 1). We fix the set of constants $K = \{$president, of, superior judicial council, italy, not, and, the$\}$[1] Next we associate semantic types to words.

Constants	Type
president	$e \to t$
of	$((e \to (e \to t)) \to (e \to t)$
superior judicial council	$e \to t$
italy	e
not	$t \to t$
and, or	$t \to (t \to t)$
the	$(e \to t) \to e$

Fig. 1. Grammar

By means of this simple grammar, we can compose terms to obtain complex terms and to check their types. For instance, we can compute that the string in natural language "the president of Italy" has type e, that is, it denotes an individual of type e. A fundamental preliminary step in order to move from natural language sentences to their correct semantic types and hence to their logical forms is to order the functional compositions in the correct way. In a number of approaches, this is done by means of a calculus that accounts for the syntactic structure of the sentence as providing the instructions to build the semantics [4]. For lack of space, we cannot enter the details here. For instance, we shall simply assume that it is possible to obtain the right order of applications

---

[1] For the sake of simplification, we treat "Superior Judicial Council" as single lexical entry.

from the order of words in natural language. Once we have the correct order of applications, we can compute the semantic of "the president of Italy" in a fully compositional way. In this case, functional applications (T3) is enough to obtain that the expression "the president of Italy" has to denote an element of e.[2] Consider now the term corresponding to "president of Italy", that is (of(italy)) (president). Its type is $e \to t$, since it denotes a class of entities. We can exemplify the acyclic constructor as follows. We allow also for possibly empty assignments, thus (of(italy)) (president) can also be written as follows:

$$(\text{of(italy)}) \ (\text{president}) \ \text{WHERE} \ \{\ \} : e \to t$$

$$(\text{of}(p_1)) \ (\text{president}) \ \text{WHERE} \ \{p_1 := \text{italy}\} : e \to t$$

$$(\text{of}(p_1)) \ (p_2) \ \text{WHERE} \ \{p_2 := \text{president}, \ p_1 := \text{italy}\} : e \to t.$$

2.1 The Calculus of Synonymy

Firstly, we introduce a syntactic notion of equivalence of terms. We say that A and B are *congruent*, $A \equiv_c B$, if and only if one can be obtained from the other by alphabetic change of bound variables (of both kinds) and re-ordering of the assignment within the acyclic recursion. For instance, (of(p_2))(p_1) WHERE $\{p_2 := \text{president}, \ p_1 := \text{italy}\}$ is congruent with (of(p_3))(p_1) WHERE $\{p_3 := \text{italy}, \ p_1 := \text{president}\}$.

Congruent terms are mere syntactic variants of one another, for that reason this is the strictest form of equivalence between terms. We are going to define the notions of *syntactical synonymity* and *referential synonymy*. The notion of syntactic synonymy is defined in [6] in terms of congruence of *canonical forms*. Canonical forms are defined by introducing a reduction calculus on terms which allows for computing effectively the canonical forms of terms [6, cf.par.3.13]. The notion of canonical form provides the following definition of syntactic synonymity. Two constants will not be syntactically synonymous since they have non-equivalent canonical forms. For instance, "Italy" and the "Supreme Judicial Court" are not syntactically synonymous.

Definition 2 (Syntactic Synonymity). *Two terms A and B are* syntactically synonymous, *$A \approx_s B$ if and only if their canonical forms are equivalent:*

$$A \approx_s B \Leftrightarrow cf(A) \equiv_c cf(B)$$

Example 1 (Canonical Forms and Syntactic Synonymity). The canonical form of the term (of(italy)) (president) is given by: (of(p_1)) (p_2) WHERE $\{p_2 := \text{president}, \ p_1 := \text{italy}\}$. The term that expresses the meaning of "president of the Supreme Judicial Council" is: (of(sjc)) (president). Its canonical form is: (of(p_1)) (p_2) WHERE $\{p_2 := \text{president}, \ p_1 := \text{sjc}\}$.

The two canonical forms are not congruent, simply because they contain distinct constant terms: italy and sjc. Hence, although the two descriptions refer to the same individuals, the two terms are *not* syntactically synonymous.

---

[2] This step is called "semantic rendering" in [6].

Syntactic synonymity actually does not account for equivalence of *meanings* of the components. We shall see how to cope with that by means of the notion of referential synonymy. Referential synonymy is based on the notion of referential intension of a term, which intuitively models the process that computes the denotation of a term in a given model. This view corresponds to the Fregean idea that the "sense" of an expression is a way to compute its denotation [8]. In the following definition, we write $den(A)(g)$ to indicate the denotation of the term A under the assignment g.

Definition 3. *A is referentially synonymous with B, $A \approx_r B$, if and only if: (1) There exist suitable terms A_0, B_0, ..., A_n, B_n such that:*

$$A \Rightarrow A_0 \text{ WHERE } \{p_1 := A_1, \ldots p_n := A_n\}$$
$$B \Rightarrow B_0 \text{ WHERE } \{p_1 := B_1, \ldots p_n := B_n\}$$

(2) $\models A_i = B_i$, that is, for all assignments g, $den(A_i)(g) = den(B_i)(g)$

Referential synonymy takes into account the meaning of the constants that occur in a term. For instance, take two individual constants a and b that have the same denotation: let g be an assignment to the variables, $den(a)(g) = den(b)(g)$ are *not* syntactically synonymous, since they have non-equivalent canonical forms, although they are referentially synonymous, since their denotations coincide.

Example 2 (Referential Synonymy). We can see that "the president of Italy" and "the president of the Supreme Judicial Court", although denotationally equivalent, are not referentially synonymous.

The expressions are rendered by (of(italy))(president) and (of(sjc))(president), whose canonical forms are:

$$(\mathsf{of}(p_1))\ (p_2) \text{ WHERE } \{p_2 := \mathsf{president},\ p_1 := \mathsf{italy}\}$$

$$(\mathsf{of}(p_1))\ (p_2) \text{ WHERE } \{p_2 := \mathsf{president},\ p_1 := \mathsf{sjc}\}.$$

The terms are not referentially synonymous because the denotation of italy and sjc are not the same. By contrast, suppose K contains unemployed and not-employed. If we treat them as two constants, their denotations coincide, thus they are indeed referentially synonymous. Actually, in order to view them as referentially synonymous, we need to classify "not employed" as a single lexical entry. This amounts to deciding, at the level of semantic rendering, for unemployed and not-employed, whether the associated functions compute their reference by a different computation or not.

3   Application to Concepts

In DOLCE, concepts are defined by *descriptions*. For instance, different agents may use different descriptions of the same concept, or a concept may be introduced by means of a description at a certain time. By relating concepts and descriptions, the view of DOLCE is that concepts manifest a certain dependence on the agents that use them to classify entities. For that reason, it is important to discuss notions of equivalence between descriptions. We only sketch the application of the notion of referential synonymy to the theory of concepts of DOLCE. Following [2], besides assuming a category for concepts C, we assume a category of descriptions DS. The categories are related by means of the relation of *definition* $DF(x, y)$: a description x defines a concept y.

We assume that every element x of the category DS is a *name* for a term t_x in a suitable fragment of the language of acyclic recursion. Moreover, we assume that (syntactically) distinct terms are associated to distinct descriptions. We introduce two binary relations between descriptions $\sim_r$ and $\sim_s$: they are relational constants in DOLCE that are intended to represent $\approx_r$ and $\approx_s$, respectively. Since $\approx_r$ and $\approx_s$ are not fist-order relations, we do not attempt to provide a definition within DOLCE. We shall view the calculus of meaning and synonymy as an external (decidable) module that allows for computing whether $\approx_r$ and $\approx_s$ hold. That is, we are putting an external constraint on the models $\mathcal{M}$ of DOLCE that forces $a \sim_r b$ and $a \sim_s b$ to hold in $\mathcal{M}$ iff $t_a \approx_r t_b$ and $t_a \approx_s t_b$ are computable in the calculus of meaning and synonymy.

We can now define when two descriptions define the same concept. Two descriptions defines the same concept if they are referentially synonymous (a1). Moreover, the same concept cannot be defined by two descriptions that are not referentially synonymous (a2).

a1. $x \sim_r y \wedge DF(x, v) \wedge DF(y, w) \rightarrow v = w$

a2. $DF(x, v) \wedge DF(y, w) \wedge v = w \rightarrow x \sim_r y$

Even if two terms are denotational equivalent, but not referentially synonymous, they may still define two different concepts. For instance, as a theorem we can infer that although (of(italy)) (president) and (of(sjc)) (president) are denotationally equivalent, they define different concepts. Suppose that d_1 is the name of (of(italy)) (president) and d_2 is the name of (of(sjc)) (president). That is, d_1 and d_2 are elements of the category DS. One can establish in the calculus of meaning and synonymy [6] that

$$(\text{of(italy)}) \ (\text{president}) \ \not\approx_r \ (\text{of(sjc)}) \ (\text{president})$$

Thus, we have that $d_1 \not\approx_r d_2$, that implies for our external constraints, that $d_1 \not\sim_r d_2$. Then, if $DF(d_1, v)$ and $DF(d_2, w)$, by axiom 2, v and w must be distinct concepts.

By contrast, a term that represent the word *student* in English and a term that represent the word student in Italian, say student and studente can be shown to be referentially synonymous, although they are not syntactically synonymous.

References

1. Kalyvianaki, E., Moschovakis, Y.N.: Two aspects of situated meaning. In: Logics for Linguistic Structures, pp. 57–86. Mouton de Gruyter (2008)

2. Masolo, C., Vieu, L., Bottazzi, E., Catenacci, C., Ferrario, R., Gangemi, A., Guarino, N.: Social roles and their descriptions. In: Proceedings of the 6th International Conference on the Principles of Knowledge Representation and Reasoning (KR-2004), pp. 267–277 (2004)
3. Montague, R.: The proper treatment of quntification in ordinary english. In: Formal Semantics: The Essential Readings, pp. 17–34, Blackwell, Oxford (1973, 2008)
4. Moot, Richard, Retoré, Christian: A logic for categorial grammars: lambek's syntactic calculus. In: Moot, Richard, Retoré, Christian (eds.) The Logic of Categorial Grammars. LNCS, vol. 6850, pp. 23–63. Springer, Heidelberg (2012)
5. Moschovakis, Y.N.: Sense and denotation as algorithm and value. In: Oikkonen, J., Väänänen, J. (eds.) Logic Colloquium '90: ASL Summer Meeting in Helsinki, vol. 2, pp. 210–249. Springer, Heidelberg (1993)
6. Moschovakis, Y.N.: A logical calculus of meaning and synonymy. Linguist. Philos. **29**, 27–89 (2006)
7. Moschovakis, Y.N.: A logic of meaning and synonymy: Course notes. ESSLLI, Copenhagen (2010)
8. Penco, C., Porello, D.: Sense as proof. In: New Essays in Logic and Philosophy. College Publications, London (2010)

Social Approach for Context Analysis: Modelling and Predicting Social Network Evolution Using Homophily

Alejandro Rivero-Rodriguez[(✉)], Paolo Pileggi, and Ossi Nykänen

Mathematics Department, Tampere University of Technology, Tampere, Finland
{Alejandro.Rivero,Paolo.Pileggi,Ossi.Nykanen}@tut.fi

Abstract. Understanding the user's context is important for mobile applications to provide personalized services. Such context is typically based on the user's own information. In this paper, we show how social network analysis and the study of the individual in a social network can provide meaningful contextual information. According to the phenomenon of homophily, similar users tend to be connected more frequently than dissimilar. We model homophily in social networks over time. Such models strengthen context inference algorithms, which helps determine future status of the user, resulting in prediction accuracy improvements of up to 118 % with respect to a naïve classifier.

Keywords: Social network analysis · Context inference · Homophily

1 Introduction

Web 2.0 technologies have been developed, enabling users to easily publish and share information on the web (e.g. *Facebook, Wikipedia*) [1]. Meanwhile, the mobile device industry has also developed tremendously. Among these developments, we highlight the inclusion of inexpensive physical sensors in mobile devices, and the opening of application programming interfaces to enable any person to develop their own application. Such developments provide a quantity of data without precedent, streaming from a number of sensors located everywhere, and from the increasing Web data. This data can be used to understand the user context and needs, providing them with the so-called context-aware services.

Although the idea of context-aware applications is brilliant, its implementation is challenging and it is often reduced in practice to services based on the users' position [2]. However, we believe that the behavior of a user within a group, e.g. online social networks, can provide meaningful user context. We study Social Network Analysis (SNA) techniques, which focus on *the discovery and evolution of relationships among entities, such as people, organizations, activities, and so on* [3]. In particular, we focus on homophily, described as the principle that a contact between similar people occurs at a higher rate that among dissimilar people [4]. Above and beyond measuring homophily for descriptive tasks, it can also be used to infer information in social networks, both by context inference and link prediction [5–7]. Most investigations assume homophily present and propose techniques to benefit inference in the social

© Springer International Publishing Switzerland 2015
H. Christiansen et al. (Eds.): CONTEXT 2015, LNAI 9405, pp. 513–519, 2015.
DOI: 10.1007/978-3-319-25591-0_41

networks. We previously proposed a homophily indicator to better represent the degree of homophily in a certain system [8], easy to understand and interpret.

Our contribution is two-fold: first, in Sect. 2, we extend our indicator of homophily [8] to measure its effect on the evolution of the network; second, in Sect. 3 we show how the proposed indicator can be used to strengthen existing inference solutions, resulting in model-driven methods to assist context inference methods. Section 4 includes an experiment using real-world data, demonstrating the performance gains achieved by using the indicator to enhance the context inference of existing solutions.

2 Homophily Indicator Over Time

We use the concept of homophily to model a system. As shown in Fig. 1, we extract system information according to modeling parameters and convert it into graphs. The graphs are used to learn about the nature of the system, which can be used to understand the nature of the system and model it better in the future.

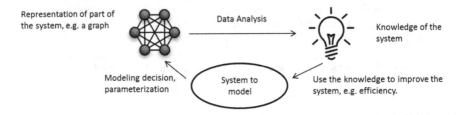

Fig. 1. Overall view of system modeling process

In concrete, we consider the network over time period D. By discretizing G into L periods, each of duration W, we obtain the sequence of successive graph states $G = (G_1, G_2, \ldots G_L)$, such that $LW = D$. We have a set of observable graph states to analyze. Representing time discretely in this way is an important parameter when modeling the system, as we shall see in the experiments in Sect. 4. We aim to study the effect of homophily in the evolution of connections in a social network over time. In other words, we aim at using the phenomenon of homophily for link prediction. For such purposes, we create the structural homophily indicator, based on our previous indicator [8], which captures the effect of homophily for the addition of new links.

We consider only relevant to measure structural homophily in graphs where at least one homogeneous and one heterogeneous edges can be potentially added in the next iteration. Otherwise, it makes no sense to measure structural homophily if all possible edges are of the same type. Consider graph G_t to be the state of the social network previously represented by graph G at some time $t \in L$. Then, $\Delta E(G_t)$ represents the set additional edges added between two consecutive graphs, i.e., $\Delta E(G_t) = E(G_t) - E(G_{t-1})$. The complement or inverse graph $\bar{G}_t$ of G_t contains all the edges of K that are absent from G_t. This set of edges at time 1 is expressed as $E(\bar{G}_t) = E(K) - E(G_t)$. $E(\bar{G}_t)$ is the set of edges that are not contained in G_t, but can be added in G_{t+1}.

As already explained, homophily suggests that some pairs of nodes are more likely to become connected in the future than others. Similarly to de definition of homophily, we consider two types of edges for structural homophily, homogeneous and heterogeneous, represented in this case as $S +$ and S-, respectively. However, determining whether an edge belongs to homogeneous set E^{S+} or to heterogeneous E^{S-} is not a trivial task. To conduct this task, we define the homophily conditions. Edges matching the homophily conditions are considered homogeneous, otherwise they are heterogeneous edges. The homophily conditions depend on the system being studied. The condition definitions apply to the added edges $\Delta E(G_t)$, as well as to the absent edges $E(\bar{G}_t)$., where $\Delta E(G_t) = E^{S+}(G_t) \cup E^{S-}(G_t)$, and $\Delta E(\bar{G}_t) = E^{S+}(\bar{G}_t) \cup E^{S-}(\bar{G}_t)$.

Following previous logic in [8], we extend the ratios r^+ and r^- to the structural homophily in the stochastic case. We define $r_G^{S+}(t), r_G^{S+}(t)$ as the ratios of added homogeneous and heterogeneous edges, respectively, present in G, with respect to the potential edges. These rations are expressed as follows.

$$r_{G_t}^{S+} = \frac{|\Delta E^{S+}(G_t)|}{|E^{S-}(\bar{G}_{t-1})|}, r_{G_t}^{S-} = \frac{|\Delta E^{S+}(G_t)|}{|E^{S-}(\bar{G}_{t-1})|}$$

Note that the denominators are never 0, since there must be at least one edge of each type to be added. We can now express the single-step structural homophily indicator $Hom_s(G_t)$ at time period t as

$$Hom_s(G_t) = \frac{r_{G_t}^{S+} - r_{G_t}^{S-}}{r_{G_t}^{S+} + r_{G_t}^{S-}}$$

Extending the single-step indicator to consider homophily from the start of the network's evolution, we finally define what we call the global structural homophily indicator Hom_s. As a function of the graph G, we have

$$Hom_s(G) = \left(\frac{1}{\sum_{t=2}^{L} |\Delta E(G_t)|} \right) \sum_{t=2}^{L} |\Delta E(G_t)| \, Hom_s(G_t)$$

The interpretation of structural homophily is analogous to the interpretation of homophily [8], where ε_s is the structural homophily threshold:

$$Hom_s \begin{cases} > \varepsilon_s, & \text{structural homophily} \\ -\varepsilon_s \leq Hom_s \leq \varepsilon_s, & \text{no structural homophily} \\ < \varepsilon_s, & \text{structural heterophily} \end{cases}$$

3 Methods for Inference

We consider Hom_s to infer successive graph in the graph sequence G, i.e., we infer graph G_{t+1} based on the information available of G_t. In order to analyze the impact of Hom_S we propose two methods based on the structural homophily indicator.

First, we present the method to be used as the baseline method in the control of our experiment afterwards. This method is called the Random Method (RM), which does not consider homophily at all. We chose this method at our discretion to make a comparison by calculating improvements our structural homophily methods give in context inference.

In all methods, we define $N \in \mathbb{Z}^+$ as the number of edges we would like to infer for the following time period, i.e., over the period t to $t + 1$. For each of the methods presented, the probability that an absent edge may be introduced into the successive graph is calculated differently. Our objective is to show that our homophily indicator can be integrated in existing methods, resulting in better inference predictions.

- The **Random Method (RM)** does not consider the effect homophily for link prediction: it is a naïve Bayesian classifier that uses no a priori information with probability $P(e, G_{t+1}) = N \frac{1}{|E(\bar{G}_t)|}, e \in E(\bar{G}_t)$
- The **Structural Homophily Randomized Method (SHRM)** considers homophily in the network; therefore, it considers two types of edges, heterogeneous and homogeneous edges. This methods simply assumes structural homophily to be constant over time, $Hom_s(G_{t+1}) = Hom_s(G_1 \ldots G_t)$., resulting

$$Hom_s(G_t) = \frac{P(e^{S+}, G_{t+1}) - P(e^{S-}, G_{t+1})}{P(e^{S+}, G_{t+1}) + P(e^{S-}, G_{t+1})}$$

The following equivalence is obvious from a simple summation of the probabilities of inferable edges of each type and the selection of N, such that

$$\sum_{e^{S+} \in E^{S+}(\bar{G}_t)} P(e^{S+}, G_{t+1}) + \sum_{e^{S-} \in E^{S-}(\bar{G}_t)} P(e^{S-}, G_{t+1}) = N$$

Solving then this system of equations, we assign the probability of being introduced into the graph, for each type of edge, according to

$$P(e^{S+}, G_{t+1}) = \frac{N}{|E^{S+}(\bar{G}_t)| + |E^{S-}(\bar{G}_t)|^{\frac{1-Hom_s(G_t)}{2}}};$$
$$P(e^{S-}, G_{t+1}) = \frac{N}{|E^{S+}(\bar{G}_t)|^{\frac{2}{1-Hom_s(G_t)}} + |E^{S-}(\bar{G}_t)|}$$

- The Deterministic Homophily Method (DHM) assumes that connections in the network appear exclusively according to homophily, i.e. different nodes will never connect. It is proposed as a simplified version of SHRM (with $Hom_s = 1$). $P(e^{C1}, G_{t+1}) = N \frac{1}{|E^{C1}(\bar{G}_t)|}; P(e^{C2}, G_{t+1}) = 0$

4 Real-World Experiment

We apply the aforementioned methods for context prediction to the *Nodobo* dataset.

Nodobo is an open and publicly-available dataset that contains social data of twenty-seven senior students in a Scottish high school [9]. The data consist of cellular tower transitions, Bluetooth proximity logs and communication events, including calls and text messages.

From *Nodobo* dataset, we construct a series of graph: We split the data into L different periods of size W. For each period i, we construct the graph G_i, obtaining the whole graph $G = (G_1, G_2, \ldots G_L)$.

For constructing each graph G_i, the users are the nodes of the graph. We include an undirected edge between nodes if they have been in proximity for an average of 60 min a day. In our case, an edge (v_i, v_j) meets the homophily condition if nodes v_i and v_j have at least f common friends. In Tab. 1, we report $Hom_S(G)$ for different values of W and f.

After measuring values of Hom_S, we consider its usage for inferring future graph status in the graph sequence. Given G_t, applying a method Θ results in the inferred graph G_{t+1}^{Θ}. Applying the respective formulas, we obtain the inferred graphs G_{t+1}^{RM}, $G_{t+1}^{SHRM} G_{t+1}^{DHM}$ for *RM*, *SHRM* and *DHR*. The accuracy of the method Θ at a period t is given as $acc_{t+1}^{\Theta} = \frac{|G_{t+1}^{\Theta} \cap \Delta E(G_{t+1})|}{|\Delta E(G_{t+1})|}$.

It is an expression of the ratio of correct predictions with respect to the added edges in the real graph. However, this is the single-step accuracy measure. For each step, we repeat the inference step R times. This makes it possible to tune the accuracy of the method. To gauge the overall accuracy of the method, we calculate the arithmetic mean of each single-step accuracy value, $acc^{\Theta} = \frac{1}{L-1} \sum_{t=1}^{L-1} acc_{t+1}^{\Theta}$.

We calculate Δacc^{SHRM} and Δacc^{DHM}, i.e., the accuracy improvements of methods *SHRM* and *DHR* with respect to *RM*, $\Delta acc^{\Theta} = \frac{acc^{\Theta} - acc^{RM}}{acc^{RM}}$.

We select three configurations with which to experiment, taken from Table 1:

Table 1. Hom_S for different values of W(days) and f (friends)

W(days) \ f (friends)	2	3	4
15	0,48	0,49	0,38
21	0,39	0,48	0,52
35	0,67	0,63	0,60

Table 2. Δacc^{SHRM} and Δacc^{DHM} for Experiments A, B and C reporting inference improvements of SHRM and DHR

N	R	100			200			500		
	Method	A	B	C	A	B	C	A	B	C
10	SHRM	0.24	0.48	0.54	0.27	0.39	0.22	0.28	0.43	0.29
	DHM	1.15	1.02	0.76	1.14	1.03	0.51	1.18	1.05	0.51
15	SHRM	0.20	0.36	0.44	0.30	0.39	0.32	0.29	0.39	0.31
	DHM	1.06	1.00	0.58	1.18	0.99	0.50	1,17	1.06	0.47
20	SHRM	0.28	0.35	0.28	0.24	0.35	0.28	0.27	0.40	0.33
	DHM	1.15	1.05	0.47	1.10	1.00	0.56	1.14	1.03	0.54

- **Experiment A:** $W = 15$, $f = 4$, Hom = 0.38 (low)
- **Experiment B:** $W = 15$, $f = 2$, Hom = 0.48 (medium)
- **Experiment C:** $W = 35$, $f = 2$, Hom = 0.67 (high)

5 Results & Conclusions

The results for Experiments A, B and C are reported in Table 2. The increases in accuracy, Δacc^{SHRM} and Δacc^{DHM} are reported for different numbers of executions of R and N parameters. The homophily-based methods improves the accuracy from 20 % to 118 % over RM (with an arithmetic overall mean improvement of 62 %), which does not engage in homophily in context inference. In this case study, DHM performs significantly better than SHRM most of the time.

Therefore, there is a clear benefit from exploiting the phenomenon of homophily. The selection of the modeling parameters is rather relevant: homophily can be modeled better when having insights of the system's behavior. Future work includes further understanding the relationship between modelling parameter and the homophily methods we presented and the definition of additional useful homophily-related metrics that can be effective for prediction tools.

6 Acknowledgements

This work was financially supported by EU FP7 Marie Curie Initial Training Network MULTI-POS (Multi-technology Positioning Professionals) under grant nr. 31652.

References

1. O'Reilly, T.: What is web 2.0: design patterns and business models for the next generation of software. Social Science Research Network, Rochester, NY (2007)
2. Rao, B., Minakakis, L.: Evolution of mobile location-based services. Commun. ACM **46**, 61–65 (2003)
3. Belov, N., Patti, J., Pawlowski, A.: GeoFuse: context-aware spatiotemporal social network visualization. In: Proceedings of the 13th International Conference on Human Computer Interaction (2011)
4. McPherson, M., Smith-Lovin, L., Cook, J.M.: Birds of a feather: homophily in social networks. Annual Rev. Soc. **27**, 415–444 (2001)
5. Mislove, A., Viswanath, B., Gummadi, K.P., Druschel, P.: You are who you know: inferring user profiles in online social networks. In: Proceedings of the Third ACM International Conference on Web Search and Data Mining, pp. 251–260, USA (2010)
6. Tang, J., Gao, H., Hu, X., Liu, H.: Exploiting homophily effect for trust prediction. In: Proceedings of the Sixth ACM International Conference on Web Search and Data Mining, pp. 53–62. ACM, New York, NY, USA (2013)

7. Scripps, J., Tan, P.-N., Esfahanian, A.-H.: Measuring the effects of preprocessing decisions and network forces in dynamic network analysis. In: Proceedings of the 15th ACM SIGKDD International Conference on Knowledge Discovery and Data Mining, pp. 747–756. ACM, New York, NY, USA (2009)
8. Rivero-Rodriguez, A., Pileggi, P., Nykänen, O.: An initial homophily indicator to reinforce context-aware semantic computing. International Conference on Computational Intelligence, Communications and Networks (CICSyN) (2015, to be published)
9. Bell, S., McDiarmid, A., Irvine, J.: Nodobo: mobile phone as a software sensor for social network research. In: 2011 IEEE 73rd Vehicular Technology Conference (VTC Spring), pp. 1–5 (2011)

Towards a Formal Model of the Deictically Constructed Context of Narratives

Richard B. Scherl[✉]

Department of Computer Science and Software Engineering, Monmouth University,
West Long Branch, NJ 07764, USA
rscherl@monmouth.edu

Abstract. This paper proposes an approach to representing the context created by the use of deictic expressions in narrative discourse. It is based on the integration of approaches to formalizing context as first-class objects, the situation calculus for representing actions, text world theory for providing a cognitive model of discourse functioning and the classical linguistic understanding of the functioning of referential indices in language. The result is a representation of the context created by the indexical expressions in the narrative. This is the context needed to interpret each utterance as well as the whole discourse.

1   Introduction

This paper proposes an approach to representing the context created by the use of deictic expressions in narrative discourse. It is based on the integration of approaches to formalizing context as first-class objects [10], the situation calculus [8,9,13] for representing actions, text world theory for providing a cognitive model of discourse functioning [19,20] and the classical linguistic understanding of the functioning of referential indices [5,15–18] in language, as well as the literature on pragmatics [7]. The result is a representation of the context created by the indexical expressions in a narrative. This is the context needed to interpret each utterance as well as the whole discourse.

The work presented here considers only the target representation of the context created by the discourse. It does not address the issues of automating the construction of the context, and reasoning with the representation. But the particular representation is chosen with the goal of making use of previous work on reasoning with aspects of the representation.

Section 2 describes the representation language and how it combines tools from the theory of context and the situation calculus from within work in artificial intelligence, and the idea of worlds and world creation from within text world theory. Most importantly, new contexts are created within contexts. The representation language is related to the linguistic analysis of deictic expressions in Sect. 3. It is through the interpretation of deictics that the context is created. An example of the use of the representation is given in Sect. 4. Finally, Sect. 5 summarizes the work and situates it within current and future work.

© Springer International Publishing Switzerland 2015
H. Christiansen et al. (Eds.): CONTEXT 2015, LNAI 9405, pp. 520–525, 2015.
DOI: 10.1007/978-3-319-25591-0_42

2 The Representation

Following McCarthy and Buvač [10], contexts are *first class objects*. They are terms. To state that a proposition p is true in context c, we write $\text{IST}(c, p)$ meaning that p is true in the context c[1].

The capability to represent (and ultimately reason about) the characteristics of contexts is crucial. Speaking situations will generally have a speaker, a hearer, a time, a place and other features. For example, $\text{SPEAKER}(c) = \text{P}_1$, $\text{TIME}(c) =$ "$4:00PM$", $\text{HEARER}(c) = \text{P}_2$, $\text{PLACE}(c) = $ "$NewJersey$". We can allow multiple values by using predicate notation such as $\text{HEARERS}(c, \text{P}_3)$, $\text{HEARERS}(c, \text{P}_4)$.

The situation calculus (following the presentation in [12]) is a first-order language for representing dynamically changing worlds in which all of the changes are the result of named *actions* performed by some agent. Here, we merge contexts and states. There is no difference.

If α is an action and s a situation or context, the result of performing α in s is represented by $\text{DO}(\alpha, s)$. The constant C_0 is used to denote the initial situation or context. Relations whose truth values vary from situation to situation, called *fluents*, are represented by a predicate symbol taking a situation term as the last argument. For example, $\text{IST}(c, \text{BROKEN}(x))$ means that object x is broken in situation c. Functions whose denotations vary from situation to situation are called *functional fluents*. They are denoted by a function symbol with an extra argument taking a situation term, as in $\text{PHONE-NUMBER}(\text{BILL}, c)$. Use of the situation calculus allows one to represent the effect of the different actions on the relevant fluents [8,9,13].

The special predicate $\text{CONTEXTCREATION}(t, c_1, c_2)$ captures the world creation notion of text world theory [19,20]. So, context c_2 is created within context c_1. The t represents the type of creation. This can be "narrative" or "cognitive" or "epistemic", or "intentional, or "hypothetical". Worlds in the sense of text world theory are represented by contexts. There is no difference between world, situation, and context.

The special predicate $\text{REFERS}(s, o)$ is used to indicate that the stretch of speech s is used by some person p (the speaker) to refer to object o. This is an initial approximation of reference having occurred and a more fine grained analysis is planned for the future given the complexity of the notion [1,15,17,18].

3 Deictic Expressions

Deictics are in the terminology of Jakobson [5], analyzed as *shifters*. These are elements of the linguistic code (C), the general meaning of which cannot be defined without reference to the message (M), hence, C/M. The message is being spoken by a particular person, at a particular time, at a particular place and in

---

[1]   The preliminary work presented here is somewhat agnostic as to the exact nature of the operator IST. It may be thought of as a modality, but reasoning along with the embedded contexts may be simpler by treating the arguments to the operator as reified formulas. See [2,4,14] for a discussion of the options available here.

the context of previous and following speech and actions. All of this is located in the context (world/situation) of the representation developed here.

Jakobson distinguishes between the narrated event (symbolized as E^n), the speech event E^s, a participant of the narrated event P^n, and a participant of the speech event P^n. In the representation developed here, E^s is the context created by the action of speaking, while E^n is the context (text world) created by the speech. It is in this context that the actions being talked about actually occur. The participants of the speech event P^s are people who exist in the context in which speaking takes place, while the participants of the narrated event E^n are people who exist in the narrated context, created as a new context within the context in which speech takes place.

Person deixis P^n/P^s relates the participants of the narrated event to those of the speech event. The use of the first-person (I in English) signals that the participant in the narrated event is identical to the speaker of the speech event. Therefore the first argument to the action SPEAK is identical to the person denoted by I in the context related by the world creation predicate. The second-person (you in English) signals the identity of a participant in the speech event with the hearer in the speaking context.

Tense, symbolized as E^n/E^s relates the time of occurrence of the narrated event to that of the speech event. The present tense may indicate that the speaking occurs at the same time, while the future tense may indicate that the narrated event occurs later than the speech event. The use of tense (along with aspect) in English and in the languages of the world is much more complex [3,7,20]. Handling the complexity is beyond the scope of this paper, but part of the larger project.

Mood, symbolized as $P^n E^n/P^n$ "characterizes the relation between the narrated event and its participants with reference to the participants of the speech event [5]." It reflects the speaker's view of the action in the narrated event. This is captured in the representation developed here by the different first arguments to the predicate CONTEXTCREATION(t, c_1, c_2).

There is also *place deixis* [3,6,7] that situates an entity in the event of narration spatially with respect to the event of speaking. Examples from English are *here*, or *there*. Background knowledge is needed to calibrate the nature of the space. For example, *there* can refer to the table in view or to some place thousands of miles away. Through the interpretation of deictic expressions, the context is constructed.

4   Example

Here is an example based on one used by Werth [20], which in turn was based on a story reported in *The Guardian* on May 14, 1992.

I read in today's Guardian, over there on the table, an interesting story. A Naples man who kept cocaine in his mother's tomb was arrested yesterday by drug agents posing as cemetery workers, police said.

The known dealer was caught red-handed as he lifted the marble slab and reached inside for two envelopes containing cocaine.

Let s_1 represent the first sentence from the above account, s_2 the second sentence, and s_3 the third sentence.

The term c_0 is used to denote the initial context. This is the context of what Werth [19,20] calls the *discourse world*. The speaker says the above paragraph. Assume that the speaker is P_1, then in the discourse world the result of the speaking of the first sentence is the context $DO(SPEAK(P_1, S_1), C_0)$, the context resulting from the second sentence is $DO(SPEAK(P_1, S_2), DO(SPEAK(P_1, S_1), C_0))$, and $DO(SPEAK(P_1, S_3), DO(SPEAK(P_1, S_2), DO(SPEAK(P_1, s_1), C_0)))$ is the result of speaking the final sentence.

A number of things are asserted within context C_0.

$IST(C_0, EXISTS(OBJ_1) \wedge NEWSPAPER(OBJ_1))$
$IST(C_0, EXISTS(OBJ_2) \wedge TABLE(OBJ_2))$
$IST(C_0, ON(OBJ_1, OBJ_2))$ $IST(C_0, EXISTS(P_1) \wedge PERSON(P_1))$
$IST(C_0, EXISTS(P_0) \wedge PERSON(P_0))$
$SPEAKER(C_0) = P_1$ $HEARER(C_0, P_0)$

There is an initial reference to the newspaper and the table. So, we have

$IST(DO(SPEAK(P_1, S_1), C_0),$
$REFERS(S_1, OBJ_1) \wedge DISTAL(P_1, OBJ_1) \wedge$
$REFERS(S_1, OBJ_2) \wedge DISTAL(OBJ_2)).$

Here *distal* is used as a rough approximation of the effect of the use of *there* rather than *here*. The speaking of the sentence has created what Werth [19,20] calls a *text world* where the reading action takes place. This is a new context. So, we have

$CONTEXTCREATION(\text{"}narrative\text{"}, DO(SPEAK(P_1, S_1), C_0), C_1)$

indicating that there is a new context created in $DO(SPEAK(P_1, S_1), C_0)$ through the process of narration and that context is denoted by C_1. Additionally, because past tense was used, we indicate that $TIME(C_1) < TIME(DO(SPEAK(P_1, S_1), C_0))$. Since, the act of reading occurred within this text world, there is a new context $DO(READ(P_1, OBJ_1), C_1)$.

Within the text world describing the act of saying another text world is established. This is the text world where the police announced the crime and the arrest. We have

$CONTEXTCREATION(\text{"}narrative\text{"}, DO(READ(P_1, OBJ_1), C_1), C_2)$

indicating that there is a new context created in $DO(READ(P_1, OBJ_1), C_1)$ and that context is denoted by C_2. Additionally, because past tense was used, the relation $TIME(C_2) < TIME(C_1)$ is added. It is necessary to specify in C_2:

$IST(C_2, EXISTS(P_5) \wedge POLICE(P_5))$

Since, the act of saying by the police, occurred within this text world, we have the new context $DO(SAY(P_5, C_3), C_2)$. As indicated above, the saying describes yet another text world. This is the text world where the criminal carried out his activities and was then arrested. So, we have

$$CONTEXTCREATION(\textit{"narrative"}, DO(SAY(P_3, C_3), C_2), C_3)$$

indicating that there is a new context created in $DO(SAY(P_3, C_3), C_2)$ through narration and that context is denoted by C_3. Additionally, because past tense was used, $TIME(C_3) < TIME(C_2)$ is added.

A number of things are asserted within context C_3.

$IST(C_3, EXISTS(P_3) \wedge MAN(P_3) \wedge DRUGDEALER(P_3) \wedge FROMNAPLES(P_3))$
$IST(C_3, EXISTS(OBJ_3) \wedge TOMB(OBJ_3))$
$IST(C_3, EXISTS(P_4) \wedge TOMBOF(OBJ_3, P_4) \wedge MOTHEROF(P_4, P_2))$
$IST(C_3, EXISTS(P_6) \wedge DRUGAGENTS(P_6) \wedge POSINGASCEMETERYWORKERS(P_6))$
$IST(C_3, EXISTS(OBJ_6) \wedge COCAINEPACKETS(OBJ_6))$ $PLACE(C_3) = \textit{"Naples"}$

A number of actions take place in context C_3. Here, using an abbreviation for a sequence of actions, we have

$$DO([HIDESIN(P_3, OBJ_3, OBJ_5), LIFTS(P_3, OBJ_3),$$
$$REACHESINSIDE(P_3, OBJ_4), ARREST(P_6, P_3)], C_3)$$

To complete this example, situation calculus axioms need to be added to represent the effects of all of the actions.

5   Conclusion and Future Work

This paper has proposed an approach to representing the context created by the use of deictic expressions in narrative discourse. The work discussed here is just the beginning of a larger project. The representation of context needs to be expanded to include a wider variety of discourses and world building constructs as discussed in the text world literature [19,20]. Methods for automatically parsing natural language texts and creating the contexts described here remain to be developed. Certainly the construction of the contexts from natural language texts will need to incorporate reasoning about the context that has been constructed so far. As noted in Werth [20] background information about the activities being described needs to be accessed. The representation needs to be extended to include contexts that are jointly constructed by multiple speakers. This will involve analysis of conversational acts [11]. Additionally, automated reasoning methods for inferring which propositions hold at each context are to be developed from those available for the situation calculus.

Acknowledgments. The author thanks the anonymous referees for their helpful suggestions.

References

1. Agha, A.: Language and Social Relations. Cambridge University Press, Cambridge (2007)
2. Buvac, S., Buvac, V., Mason, I.A.: Metamathematics of contexts. Fundam. Inform. **23**(2/3/4), 263–301 (1995). http://dx.doi.org/10.3233/FI-1995-232345
3. Fillmore, C.: Lectures on Deixis. CSLI Publications, Stanford (1999)
4. Galton, A.: Operators vs. arguments: The ins and outs of reification. Synthese **150**(3), 415–441 (2006). http://dx.doi.org/10.1007/s11229-005-5516-7
5. Jakobson, R.: On Language. Harvard University Press, Cambridge (1990)
6. Levinson, S.: Deixis. In: Horn, L., Ward, G. (eds.) The Handbook of Pragmatics, pp. 97–121. Blackwell, Malden (2004)
7. Levinson, S.C.: Pragmatics. Cambridge University Press, Cambridge (1983)
8. McCarthy, J.: Programs with common sense. In: Minsky, M. (ed.) Semantic Information Processing, chap. 7, pp. 403–418. The MIT Press (1968)
9. McCarthy, J., Hayes, P.: Some philosophical problems from the standpoint of artificial intelligence. In: Meltzer, B., Michie, D. (eds.) Machine Intelligence 4, pp. 463–502. Edinburgh University Press, Edinburgh (1969)
10. McCarthy, J., Buvač, S.: Formalizing contexts (expanded notes). In: Computing Natural Language, pp. 13–50. Center for the Study of Language and Information, Stanford, California (1998)
11. Poesio, M., Traum, D.R.: Conversational actions and discourse situations. Comput. Intell. **13**(3), 309–347 (1997). http://dx.doi.org/10.1111/0824-7935.00042
12. Reiter, R.: The frame problem in the situation calculus: A simple solution (sometimes) and a completeness result for goal regression. In: Lifschitz, V. (ed.) Artificial Intelligence and Mathematical Theory of Computation: Papers in Honor of John McCarthy, pp. 359–380. Academic Press, San Diego (1991)
13. Reiter, R.: Knowledge in Action: Logical Foundations for Specifying and Implementing Dynamical Systems. The MIT Press, Cambridge (2001)
14. Shoham, Y.: Varieties of context. In: Lifschitz, V. (ed.) Artificial Intelligence and Mathematical Theory of Computation, pp. 393–407. Academic Press Professional Inc., San Diego (1991). http://dl.acm.org/citation.cfm?id=132218.132241
15. Silverstein, M.: Shifters, linguistic categories, and cultural description. In: Meaning in Anthropology, pp. 11–55. University of New Mexico Press, Albuquerque, New Mexico (1976)
16. Silverstein, M.: The three faces of "function": preliminaries to a psychology of language. In: Hickmann, M. (ed.) Socal and Functional Approaches to Language and Thought, chap. 2, pp. 17–38. Academic Press, Orlando (1987)
17. Silverstein, M.: The indeterminacy of contextualization: When is enough enough? In: Auer, P., DiLuzio, A. (eds.) The Contextualization of Language. John Benjamins, Amsterdam (1992)
18. Silverstein, M.: Indexical order and the dialectics of social life. Lang. Commun. **23**, 193–229 (2003)
19. Werth, P.: How to build a world. In: New Essays in Deixis. Rodolphi, Amsterdam-Atlanta (1995)
20. Werth, P.: TEXT WORLDS: Representing Conceptual Space in Discourse. Longman, Harlow, Essex (1999)

The Influence of Context in Meaning: The Panorama of Complement Coercion

Alexandra Anna Spalek[✉]

Universitetet i Oslo, Niels Henrik Abels vei 36, 0313 Oslo, Norway
a.a.spalek@ilos.uio.no

1 Introduction

This work addresses the phenomenon of aspectual verbs cross-linguistically and the possibility for event-describing readings to arise from combinations with object-denoting complements. Starting out with a short review of the combinatory capacity of distinct aspectual verbs, I focus on a cross-linguistic comparison of aspectual verbs that describe culmination events (English: *finish*, Spanish: *acabar*, German: *abschliessen/beenden*, Norwegian: *avslutte*) and analyze in how far they can combine with object-denoting arguments.

2 Background

It is a commonplace that establishing the meaning of a phrase represents substantively more than just putting together the content of the terms involved in the predication and that even the narrowest context can modulate meaning substantively. One of the examples where the narrow context leads to strong meaning modulation and represents a challenge for the standard compositional system is represented by aspectual verbs like *begin* or *finish*. These verbs are considered to select an event predicate which is structurally realized as an infinitive complement or a gerund, or else as an event nominal, as illustrated in examples (1).

(1) a. Johan began to sing/singing.
     b. Johan began the speech.

The influence of the narrow context becomes evident when these kinds of verbs combine with entity-denoting nouns. In the following example, our understanding is facilitated by a default interpretation of events associated with the objects *book* or *cigarette*, rather than a straightforward composition of the verb with its object.

(2) a. Johan began a book. (= began reading a book)
     b. Johan began a cigarette. (= began smoking a cigarette)

But the contextualization of these verbs can even be more complex, as pointed out by [1, p. 228]. Thus, in order to derive the sense in context of *begin the book* in the examples in (3), the external argument needs to be taken into account.

© Springer International Publishing Switzerland 2015
H. Christiansen et al. (Eds.): CONTEXT 2015, LNAI 9405, pp. 526–531, 2015.
DOI: 10.1007/978-3-319-25591-0_43

While example (a) clearly suggests a writing event, (b) is rather interpreted as a reading event and, finally, (c) definitely represents neither a writing nor a reading event, but is more plausibly interpreted as an eating event.

(3)   a.   The writer began the book.
       b.   The student began the book.
       c.   The goat began the book.

As readers we use elements in the sentence and discourse to determine the most appropriate interpretation of the NP into a different sense by understanding an entity in terms of an event it participates in. [3] termed this type of enriched composition complement coercion and identified it with a process by which a context-sensitive interpretation for a complement expression emerges from the interaction of semantic properties in the sentence. In fact, the contextual influence on the event interpretation can actually proceed from a much larger chunk of discourse than a single sentence. The following extract from [4] illustrates how 'finished the shirt', given the previous context, can only be interpreted as an event of ironing.

> Mom looked down and began ironing the shirt. I watched, as I had watched her iron many shirts. Always in the same order, collar, back yoke, sleeves. Button placket, left front, back, right front, In and out between the buttons, each movement neat and orderly. 'Sometimes I think about moving', she admitted. 'Your grandparents are getting up there in age. I probably should live nearer to them. Except we'd probably drive each other nuts.' She **finished the shirt**.

Theoretically, this phenomenon has been explained as a type mismatch between the selection restrictions of the predicate and its argument, and has been claimed to require an enriched form of composition. More concretely, the theory of [1] postulates more articulated types for entities like *book* (dot-types), which possesses two aspects: the physical aspect and the informational aspect. These types undergo dot-exploitation during composition, which is an operation that captures the intuition that certain nominal expression are logically polysemous and cannot straightforwardly be composed with certain verbs. Thus the formalism in [1], on the one hand, adds material to the verb's argument and, on the one hand, it revises the typing context.

Furthermore, findings from self-paced readings [5] and eye tracking experiments both involving aspectual verbs [6] provide behavioral evidence for the psycholinguistic reality of an enriched form of composition, illustrating that entity noun phrases take longer to process when they follow verbs that require event arguments than when they follow verbs that do not.

More recently, however, [2] have challenged the theoretical assumption that longer processing times and enriched composition need to be understood in terms of coercion. Taking cases as (4) and (5) into account these authors propose an analysis that broadens the selection restrictions of aspectual verbs to any kind of structured object, independently of whether the structure is physical, temporal or spacial, thus arguing against the need of coercing one entity to another.

(4)   This is the famous perch that officially begins the Appalachian Trail.

(5)   A little porcelain pot finished the row.

In these cases the objects that the aspectual verbs combine with do not have several aspects that can be referred to by distinct events and thus lay outside the theoretical explanation power of coercion for there seems no need for shifting an entity to another. [2] argue that the cost involved in processing aspectual verbs with object-denoting complements does not stem from coercion but from the need to identify the dimension along which the complement denotation is affected by the aspectual verb.

3   The Challenge Posed by Coercion

Whichever formal account is chosen to explain the composition of aspectual verbs with entity-denoting arguments the literature leaves the impression that aspectual verbs just easily select for entity-denoting objects and that coercion into an event or the selection of an appropriate dimension is just a pragmatic process that applies in these cases. A closer look at data, however, reveals that this is not the case. In fact, many questions arise concerning which specific compositional patterns are possible and why certain reinterpretations are out. For one thing, event readings are not available to aspectual verbs in composition with entities throughout. Thus verbs like *finish*, *stop* and *end*, which have clearly related meanings, display a quite different behavior with respect to event readings.

(6)   a.   Elena finished the apple.
        b.   Elena stopped the apple.
        c. ??Elena ended the apple.

Whereas both (6-a) and (6-b) are easily interpretable, (6-c) hardly has an interpretation. Moreover, *finish* and *stop* differ in the readings they give access to. Thus while (6-a) can be interpreted in terms of a creation or consumption event, (6-b) cannot refer to neither a creation nor a consumption event, but rather describes a some kind of dynamic event that involves the apple, such as the rolling. The fact that aspectual verbs do not license the same event readings becomes even clearer in the following examples, where the same entity-denoting argument is not even possible for *finish* and *stop* alike:

(7)   a.   Elena finished the kitchen.
        b. #Elena stopped the kitchen.

While (7-a) is perfectly natural and means that Elena has finished doing something to the kitchen, let it be cleaning or painting, it is absolutely unclear what (7-b) should mean at all.

    Cross-linguistic data also provide evidence that it will not do to generalize over aspectual verbs as having the capacity to combine with entity-denoting arguments as long as they are coerced into events, as suggested by [1], or as long as there is an affected complex structure in the argument, as argued by [2].

Aspectual verbs that describe culminations of events differ across languages: while English *finish* and Spanish *acabar* freely combine with object-denoting complements and describe culmination of events, as examples (8) and (9) illustrate, in German and Norwegian the closest relative to these verbs, namely *abschließen, beenden*[1] or *avslutte*, induce event interpretations to a very limited extent, as illustrated in (10) through (12).

(8)   Elena finished the book/ the sandwich/ the shirt.

(9)   Elena acabó   el   libro/ el   bocadillo/ la   camiseta.
       Elena finished the book/ the sandwich/ the shirt.
       'Elena finished the book/ the sandwich/ the shirt.'

(10)   a.   Elena hat das Buch abgeschlossen.
             Elena has the book finished
             'Elena has finished the book.'
        b. ??Elena hat das Brötchen abgeschlossen.
             Elena has the sandwich finished
             'Elena has finished the sandwich.'
        c. ??Elena hat das Hemd abgeschlossen.
             Elena has the shirt   finished
             'Elena has finished the shirt.'

(11)   a.   Elena has das Buch beendet.
             Elena has the book ended
             'Elena has ended the book.'
        b. ??Elena has das Brötchen beendet.
             Elena has the sandwich ended
             'Elena has ended the sandwich.'
        c. ??Elena hat das Hemd beendet.
             Elena has the shirt   ended
             'Elena has ended the shirt.'

(12)   a.   Elena har avslutta   boka.
             Elena has up finished the   book
             'Elena has finished the book.'
        b. ??Elena har avslutta   brødskive.
             Elena has up finished the   sandwitch
             'Elena has finished the sandwich.'
        c. ??Elena har avslutta   skjorta.
             Elena has up finished the   shirt
             'Elena has finished the shirt.'

What these examples illustrate is that while English *finish* can easily be composed with all kinds of entity-denoting arguments to describe culminations of events, even though the exact interpretation has to be filled in by context, the

---

[1] Though etymologically related the *benden* examples are not equivalent to English *end* for *end* is mainly unaccusative and otherwise requires a modifier, while German *beenden* is throughout transitive.

effect of German *abschließen* or *beenden* or Norwegian *avslutte* with entity-denoting nouns is much less easily interpretable. In fact, German and Norwegian are more prompted to express fully specific event structures rather than using an aspectual verb in combination with an entity-denoting noun. This is mostly done by means of a prepositional particle *fertig* attaching to a transitive verb which together form a telic predicate with the nominal complement, as in (13). Yet another option are prefixed verbs that refer explicitly to culminating events, such as in example (14).

(13)   Elena hat das Brötchen fertiggegessen.
        Elena has the sandwich ready eaten
        'Elena has eaten the sandwich.'

(14)   Elena hat das Brötchen aufgegessen.
        Elena has the sandwich up eaten
        'Elena has eaten up the sandwich.'

4   Generalization

First, given that all culminations of events are clearly goal-oriented they are clear accomplishment event types. Thus, if explicit reference to the event is left out, describing the event with an aspectual verb in composition with an entity, instead, the interpretable event still needs to be an accomplishment. This limits the domain of possible readings so that clear achievements are not part of the possible interpretations, such that (a) and (b) in the following examples are not equivalent.

(15)   a.   Johan finished the mountain.
        b.   Johan finished reaching the mountain top.

Second, for culmination events consisting of aspectual verbs in combination with object-denoting nouns the object expresses an incremental theme to the event referred to. Third, the analyzed data suggest that combining aspectual verbs with object-denoting nouns is clearly not just a pragmatic possibility throughout languages, but rather obeys more fine-grained restriction on the arguments.

5   Summary

Traditionally aspectual verbs have been regarded as content-weak units whose meaning skeleton was considered to comprises not more than the temporal, aspectual components. These verbs have been described to refer to the ingressive or the culminating phase of an event and therefore to select for event-denoting arguments. The data observed so far show that cross-linguistically aspectual verbs in composition with entity-denoting nouns differ in the following way: culminations of events expressed with aspectual verbs in English or Spanish are more underspecified than culmination events in German or Norwegian. English *finish* and Spanish *acabar* easily combine with all kinds of entity-denoting arguments and allow for interpretations of very distinct culminating events, although the exact event needs to be established by context. The kinds of events can be either some kinds of creation, such as *building*, consumption *eating*, or any other activity, such as *cleaning*, that can be delimited by the extension of the theme to form an accomplishment. The panorama looks very different for German and Norwegian: both languages require specification of culmination events. The composition possibilities of *abschliessen*, *beenden* or *avslutte* with entities are much more limited. These are mostly information-content entities. All other cases are most naturally expressed with activity verbs and some verb particle, typically *fertig*, which expresses the completion of an accomplishment. This means that main-land Germanic languages express culminations without leaving so much semantic content to the context. This raises the question whether in English and Spanish the contextual variation observed with *finish* and *acabar*, as well as other aspectual verbs, is rooted in an underspecified verb stem which acts like a free variable and, when supplied with a value, forms a telic predicate with the complement object as an incremental theme.

Acknowledgments. This research has been supported by the SynSem research program at the Faculty of Humanities of the University of Oslo.

References

1. Asher, N.: Lexical Meaning in Context. Cambridge University Press, Cambridge (2011)
2. Piñango, M.M., Deo, A.: Reanalyzing the complement coercion effect through a generalized lexical semantics for aspectual verbs. J. Semant. 1–50. Oxford University Press (in press)
3. Pustejovsky, J.: The Generative Lexicon. MIT Press, Cambridge (1995)
4. Steele Agosta, C.: Every Little Step She Takes. Carolyn Steele Agosta, North Carolina (2010)
5. Traxler, M.J., Pickering, M.J., McElree, B.: Coercion in sentence processing: evidence from eye-movement and self-paced reading. J. Memory Lang. **47**, 530–547 (2002). Elsevier Science
6. Traxler, M.J., McElree, B., Williams, R.S., Pickering, M.J.: Context effects in coercion: evidence from eye movement. J. Memory Lang. **53**, 1–25 (2005). Elsevier Science

A System for Automatic Classification of Twitter Messages into Categories

Alexandros Theodotou and Athena Stassopoulou[✉]

Department of Computer Science, University of Nicosia, 1700 Nicosia, Cyprus
alexandert90@gmail.com, stassopoulou.a@unic.ac.cy

Abstract. Twitter is a widely used online social networking site where users post short messages limited to 140 characters. The small length of these messages is a challenge when it comes to classifying them into categories. In this paper we propose a system that automatically classifies Twitter messages into a set of predefined categories. The system takes into account not only the tweet text, but also external features such as words from linked URLs, mentioned user profiles, and Wikipedia articles. The system is evaluated using various combinations of feature sets. According to our results, the combination of feature sets that achieves the highest accuracy of 90.8 % is when the original tweet terms are combined with user profile terms along with terms extracted from linked URLs. Including terms from Wikipedia pages, found specifically for each tweet, is shown to decrease accuracy for the original test set, however accuracy was shown to increase using a fraction of the original test set containing only tweets without URLs.

1   Introduction

Twitter is a widely used online micro-blogging service that allows users to post and read short messages limited to 140 characters. Since its creation in March 2006, it has gained enormous popularity with more than 100 million users posting 340 million tweets per day in 2012 [1]. Current research on Twitter classification is focusing on areas such as: sentiment analysis [2–4], spam detection [5–7], event identification [8,9], and classification of tweets into categories [10–12], each of which can be useful in a wide variety of applications.

Classifying Twitter messages into topics, which is the focus of our work, can be implemented as a Twitter application that allows a user to log in to their account and view a categorized version of their timeline. Depending on the number of people that the user follows, this can save a lot of time in looking for news and posts about a particular category, such as Sports, by just opting to view the Sports timeline.

This paper proposes a system for the automatic classification of Twitter messages into a pre-defined set of categories: Sports, Movies, Music, Fashion, Food, Technology, and Politics. The small length of Twitter messages is a challenge when it comes to classifying them into topics. The system is based on the idea of using additional data sources such as URL pages, Wikipedia articles, and user

© Springer International Publishing Switzerland 2015
H. Christiansen et al. (Eds.): CONTEXT 2015, LNAI 9405, pp. 532–537, 2015.
DOI: 10.1007/978-3-319-25591-0_44

profiles in the classification process, in order to supplement the small amount of information that can be extracted from each tweet, helping the classifier make a more "informed" decision. Our work differs from similar research in that it uses a unique combination of features in the classification process. The notion of expanding URLs and extracting additional features has already been proposed in another paper [12] and shows that it decreases accuracy, however our results show that including URL features increases accuracy. The notion of using Wikipedia pages has already been proposed in another paper [11], however no terms were extracted from the articles or used in the classifier, as is performed in our work. The effectiveness of using various combinations of tweet, URL, user profile, and Wikipedia terms in the classification process for Twitter messages is a new contribution in this field.

The remaining of this paper is structured as follows: In Sect. 2 we present an overview of our system. Experimental results are presented and analyzed in Sect. 3. We conclude in Sect. 4.

2   System Overview

The first step in our classification system is to extract and label tweets to be used in the training stage of the classification process, which we discuss in Sect. 3.1.

Once classification starts, the pre-labeled training tweets (the tweets used to build the classifier) are assembled as bags of features. The words "terms", "attributes", and "features" will be used interchangeably throughout this paper and refer to the features used in the classifier. These groups of features are used to build the classifier, based on the Naive Bayes algorithm. Features are also created for the tweets to be classified, and are compared against the category profiles created by the classifier. The classifier then assigns probabilities that each tweet belongs to each category, and the category with the highest probability is assigned to the tweet. The evaluation process is repeated 3 times with a different training/test set each time. More details on this process are given in Sect. 3.

2.1   Feature Selection

Each tweet's features can be thought of as a bag of words, extracted from various data sources. For each tweet, the features from the following data sources are used:

- words in the tweet
- words from posted URLs
- words from mentioned user profiles
- words from closest Wikipedia article.

If a tweet does not contain URLs and does not mention any users, but a Wikipedia page is found for it, only tweet and Wikipedia terms will be used. All terms taken are unigrams (single words), whereas duplicates and stop-words are removed. We also use entropy [2] to exclude terms with poor discriminatory power when training the classifier. The tokens are ordered by the result of: *word size * occurrences*, and only the top x terms are kept.

URL Terms. Terms extracted from the text in URLs mentioned in a tweet are also added to the bag of attributes. In [12], it is argued that including these terms degrades accuracy due to spam and unrelated links, however in this case spam is not included in the dataset used in the experiments. Terms are extracted from both title and content of the HTML page[1].

User Profile Terms. In cases where users are mentioned in a tweet (in the form of @userA), text from each mentioned user's profile name and biography are used to generate additional features. Once the usernames are extracted, the C# library Tweetinvi[2], which is based on the Twitter API is used to query for each user's profile name and biography strings.

Wikipedia Terms. Wikipedia terms are terms extracted from the closest Wikipedia article related to the tweet. The idea is that since many tweets contain names of people, places, sports teams, companies, and other entities, these entities can be looked up on Wikipedia in order to extract terms that will be useful in the classification.

The process of finding a closest Wikipedia page is shown in Fig. 1. First, search terms are created based on the tweet content. Each search term is searched for in a database of Wikipedia page titles[3] and a score is calculated for each matching page found using the term frequency−inverse document frequency measure (tf-idf) [13]. Finally, the page with the highest score is selected as the closest Wikipedia page for the tweet.

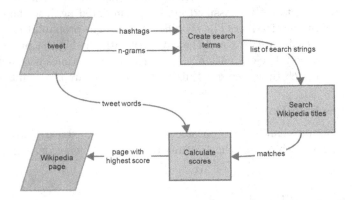

Fig. 1. The general process of finding the closest Wikipedia page for a tweet

---

[1] The NReadability library is used to take only the article text from a specific HTML file ignoring irrelevant text such as sidebar text containing other stories. https://github.com/marek-stoj/NReadability [Online; accessed 6-June-2015].

[2] https://tweetinvi.codeplex.com [Online; accessed 6-June-2015].

[3] The database contains every Wikipedia page title from the English Wikipedia, originally taken from the April 2015 Wikipedia dump. https://dumps.wikimedia.org/enwiki/20150304/ [Online; accessed 6-June-2015].

3   Experimental Results

In our system we use a Naive Bayes classifier which is a popular model for document classification due to its simplicity, robustness and its good performance even on small datasets [13]. It is a probabilistic model which assigns the class (a category in our case) that gives the maximum posterior probability given its features.

In this section we present the results of our classifier and compare its performance using various combinations of the proposed feature sets. We also conduct a further experiment to show how Wikipedia terms affect the classification of tweets that do not contain URLs.

3.1   Corpus Collection and Labelling

Tweets were fetched automatically from Twitter's public timeline using Tweet-invi[4], a C# implementation of the Twitter API. For each category, a hashtag including the category name (i.e. #sports, #movies, etc.) was added as a filter to English tweets received, followed by a check if the author has more than 50 friends and the tweet is original (not a retweet). This is to ensure that tweets saved are less likely to contain spam as well as to avoid duplicate tweets, to make it easier for manual labelling. The process was repeated periodically for various time spans over the course of three weeks in May 2015. The tweet counts in each category are shown in Table 1.

Table 1. Initial dataset size for tweets retrieved periodically over the course of three weeks in May 2015

Category	#sports	#movies	#music	#fashion	#food	#tech	#politics
Count	40,554	20,430	151,904	106,726	57,790	29,145	19,239

After the initial dataset was collected, 150 tweets from each category were manually selected for experimentation purposes, for a total of 1050 tweets. As a general rule, tweets containing news (such as sports news, politics news, etc.) were preferred over subjective tweets, when applicable.

3.2   Results

In order to test how the different feature groups affect the classification accuracy, experiments were performed on three different combinations of feature groups:

- Feature Group A (FGA): Using tweet terms only
- Feature Group B (FGB): Using tweet + user profile + URL terms
- Feature Group C (FGC): Using tweet + user profile + URL + Wikipedia terms.

---

[4] https://tweetinvi.codeplex.com [Online; accessed 6-June-2015].

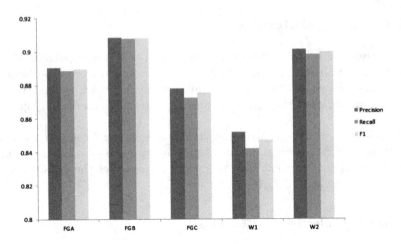

Fig. 2. Recall, precision, and F1 results for each Feature Group (FG) tested

In addition, since including Wikipedia terms showed to decrease accuracy, a separate experiment was done on tweets not containing URLs using the following feature groups:

- W1: Using tweet + user profile terms
- W2: Using tweet + user profile + Wikipedia terms.

The experiments were done based on 3-fold cross-validation [13]. Our classifier is evaluated using the standard metrics of recall, precision, and F1 score [13]. These three metrics are derived for each class, and then the overall precision and recall is calculated by macro-averaging, which takes the averages of the precision and recall on the different classes [14].

The results of recall, precision and F1 score for each Feature Group experiment are shown in Fig. 2.

As it can be seen from this figure, including URL and user profile terms in the feature set, increases accuracy from an F1 score of 88.9 % (using FGA, tweet terms alone) to 90.8 % (FGB). This increase in accuracy can be credited to the additional information available to the classifier to make a decision. When we also include Wikipedia terms (i.e. FGC), accuracy decreases with F1 score falling to 87.5 %. This can be attributed to the fact that the correct Wikipedia pages are not always found (sometimes irrelevant pages are selected which leads to misleading terms). The last two sets of bars in Fig. 2 show the impact of Wikipedia terms on tweets without URLs (i.e. Feature Groups W1 and W2 as above). The general observation is that using Wikipedia terms for tweets not containing URLs increases accuracy, with F1 score increasing from 84.7 % to 90 %.

4 Conclusion and Future Work

We presented a system to classify Twitter messages into a set of categories: Sports, Movies, Music, Fashion, Food, Technology, and Politics. The Naive Bayes classifier implemented used features from four different sources: original tweet text, URL text, user profile text, and Wikipedia article text. The effectiveness of using various combinations of these feature sets was examined, with promising results. As future direction, we plan to investigate how changing the values of parameters used throughout the system affect the classification results. On a similar note, the effect of changing number of URL, user profile, and Wikipedia terms extracted on the classification accuracy can be useful in further improving accuracy.

References

1. Twitter: Twitter turns six (2012). https://blog.twitter.com/2012/twitter-turns-six. Accessed 19 May 2015
2. Pak, A., Paroubek, P.: Twitter as a corpus for sentiment analysis and opinion mining. In: LREC, vol. 10, pp. 1320–1326 (2010)
3. Go, A., Bhayani, R., Huang, L.: Twitter sentiment classification using distant supervision. CS224N Project report, Stanford, pp. 1–12 (2009)
4. Davidov, D., Tsur, O., Rappoport, A.: Enhanced sentiment learning using twitter hashtags and smileys. In: Proceedings of the 23rd International Conference on Computational Linguistics: Posters, Association for Computational Linguistics, pp. 241–249 (2010)
5. Benevenuto, F., Magno, G., Rodrigues, T., Almeida, V.: Detecting spammers on twitter. In: Collaboration, Electronic Messaging, Anti-abuse and Spam Conference (CEAS), vol. 6, p. 12 (2010)
6. McCord, M., Chuah, M.: Spam detection on twitter using traditional classifiers. In: Calero, J.M.A., Yang, L.T., Mármol, F.G., García Villalba, L.J., Li, A.X., Wang, Y. (eds.) ATC 2011. LNCS, vol. 6906, pp. 175–186. Springer, Heidelberg (2011)
7. Song, J., Lee, S., Kim, J.: Spam filtering in twitter using sender-receiver relationship. In: Sommer, R., Balzarotti, D., Maier, G. (eds.) RAID 2011. LNCS, vol. 6961, pp. 301–317. Springer, Heidelberg (2011)
8. Becker, H., Naaman, M., Gravano, L.: Beyond trending topics: real-world event identification on twitter. In: ICWSM, vol. 11, pp. 438–441 (2011)
9. Sakaki, T., Okazaki, M., Matsuo, Y.: Earthquake shakes twitter users: real-time event detection by social sensors. In: Proceedings of the 19th International Conference on World Wide Web, pp. 851–860. ACM (2010)
10. Sriram, B., Fuhry, D., Demir, E., Ferhatosmanoglu, H., Demirbas, M.: Short text classification in twitter to improve information filtering. In: Proceedings of the 33rd International ACM SIGIR Conference on Research and Development in Information Retrieval, pp. 841–842. ACM (2010)
11. Genc, Y., Sakamoto, Y., Nickerson, J.V.: Discovering context: classifying tweets through a semantic transform based on wikipedia. In: Schmorrow, D.D., Fidopiastis, C.M. (eds.) FAC 2011. LNCS, vol. 6780, pp. 484–492. Springer, Heidelberg (2011)
12. Rosa, K.D., Shah, R., Lin, B., Gershman, A., Frederking, R.: Topical clustering of tweets. In: Proceedings of the ACM SIGIR: SWSM (2011)
13. Witten, I.H., Frank, E.: Data Mining: Practical Machine Learning Tools and Techniques. Morgan Kaufmann, San Francisco (2005)
14. Sokolova, M., Lapalme, G.: A systematic analysis of performance measures for classification tasks. Inf. Process. Manage. 45(4), 427–437 (2009)

Modeling Erroneous Human Behavior: A Context-Driven Approach

Chris Wilson and Roy M. Turner[(✉)]

School of Computing and Information Science, University of Maine,
Orono, ME 04469, USA
chris.wilson@maine.edu, rturner@maine.edu

Abstract. Artificial agents that model aspects of human behavior often model behaviors that an observer would regard as normal. In recent years, agents that exhibit observable erroneous behaviors have become common in a variety of applications, including simulated impaired agents to test assistive technologies and realistic agents in video games. In this paper, we present a context-driven approach to modeling plausible human behavior and a framework for modeling erroneous behavior which focuses on impairing an agent's ability to recognize and deal effectively with anticipated contextual changes.

Keywords: Simulation · Cognitive impairment · Context-mediated behavior · Assistive technologies

1 Introduction

Artificial agents that model aspects of human behavior can be found in a variety of applications ranging from scientific and military simulation to commercially available video games. In many cases, these agents are designed to model aspects of normative human behavior [4]. In recent years however, agents that model erroneous human behavior have become prevalent in both scientific simulation and commercially available video games.

One important use for simulated humans is in developing and testing cognitive orthotics, also known as assistive technologies for cognition (ATCs), meant to support patients with cognitive deficiencies (e.g., from dementia) [6]. Agents that can simulate erroneous human behavior save time and money and avoid the possible ethical issues with using actual patients. Tests of the Autominder system [7], for example, were done using an agent that would forget to perform activities on its daily agenda [8]. Similarly, Serna et al. [9] developed an agent that would perform a task's steps out of order, and in some cases, incorrectly.

It is also useful for non-player characters (NPCs) in some video games (e.g., first-person shooters, FPS) to exhibit erroneous behaviors to give the human player(s) a competitive advantage in order to avoid frustrating game play. For example, an NPC may often stand in the open for a period of time, exposing their position, before returning fire [5].

© Springer International Publishing Switzerland 2015
H. Christiansen et al. (Eds.): CONTEXT 2015, LNAI 9405, pp. 538–543, 2015.
DOI: 10.1007/978-3-319-25591-0_45

Most current simulations of erroneous behavior focus on forgetting or poor task performance. However, errors can also occur when a task is remembered and performed correctly, but out of context. For example, dementia patients often exhibit contextually-inappropriate behaviors such as wandering from the house in the middle of the night or leaving food unattended on the stove to engage in another activity.

We are developing an approach to modeling erroneous behavior that is based on errors in contextual reasoning. The approach starts with an agent that is able to behave appropriately for its current context and that is able to understand how that context will evolve as a result of pursuing a goal, then taking action to modify any problematic features of its context to be appropriate for the accomplishment of the goal. Unlike our previous work on context-mediated behavior (CMB) [11], which used knowledge of known contexts to decide how to behave while in them, the current approach uses contextual knowledge to drive behavior, including how to change the context to allow goal accomplishment. In addition to being a promising reasoning approach in its own right, this provides an elegant way to model compromised behavior by impairing the agent's ability to recognize and deal effectively with anticipated contextual changes.

In the remainder of the paper, we briefly discuss related work. We then discuss the problem and our overall approach. Next, we look at the kinds of contextual knowledge needed. We then turn to a discussion of how in our approach an agent formulates a contextually-appropriate plan. Finally, we look at how such an agent can be compromised to produce realistic, contextually-inappropriate plans.

2   Related Work

All agents use some form of contextual knowledge, implicit or explicit. In rule-based agents, this is encoded in rule antecedents describing when a rule is applicable. In planners, context is usually contained in preconditions or filter conditions that are part of the operator schema description.

A problem with contextual knowledge being local to rules or operators, however, is that unless care is taken, it can cause an agent to exhibit the same behavior regardless of the current context. For example, in the Fallout 3 video game, a village populated with friendly townsfolk is attacked by mutants, whom the player must repel. After the attack, surviving villagers still greet the player in the same friendly way, even though they are surrounded by the corpses of their neighbors [10]. While this could be remedied by adding additional constraints to when particular greetings are appropriate, this would tend to cause an explosion of such qualifiers. Worse, from the perspective of our current problem, the erroneous behavior does not necessarily reflect the kind that an impaired person would exhibit.

Some researchers have proposed the use of smart objects and situations to make an agent's behavior more context-appropriate [10]. For example, a smart object might, depending on the current context, inform an agent how to hold or gaze at it. These approaches do not, however, aid an agent in planning to achieve a specific goal. (In fact, since the agent can't know a priori what the object will tell it, this may actually hinder an agent's ability to plan.) For our current focus, one could imagine altering objects' actions to cause contextual errors; however, this would move the focus from the agent being modeled to the environment and could result in all agents present behaving inappropriately for the context.

In earlier work, we argued for the benefits of explicitly representing contexts and contextual knowledge with respect to acquiring, learning, reasoning about, and using such knowledge [11]. In our context-mediated behavior (CMB) approach, known contexts are represented as contextual schemas (c-schemas), which both describe the contexts as well as prescribe how to behave while in them. This can address some of the limitations of related approaches, and it provides a way to inject errors into an agent's behavior by impairing its ability to reason about its context.

In the current project, we develop this idea to create a believable impaired agent. However, we are not only interested in using contextual knowledge to mediate aspects of an agent's current behavior, but also aspects of its future behavior. To this end, we treat an agent's context as a dynamic object that continuously evolves as an agent works to achieve its goals. Similar to other approaches, we view an agent's context as a collection of smaller contextual objects that evolve at different rates.

A related approach is taken by Brézillon and colleagues [3]. They propose three types of contextual knowledge, namely *contextualized, contextual* and *external* knowledge, that can be applied to a problem-solving step. Contextualized knowledge describes any knowledge that is used by an agent during a problem-solving step, whereas contextual knowledge is any knowledge that is not explicitly used during a step, but that constrains it. External knowledge is all other knowledge that has nothing to do with the problem-solving step.

This approach allows an agent's contextual knowledge to evolve during problem solving. For example, while pursuing a goal, a piece of contextualized knowledge might become either contextual or external knowledge. In our approach, the use of c-schemas allows us to explicitly represent both contextualized and contextual knowledge and also provides, in the current work, a way to reason about future contexts in order to help create plans.

3   Overview of Our Approach

Similar to the approaches of Pollack [8] and Serna [9], we will model erroneous human behavior by implementing an artificial agent that exhibits plausible aspects of normative human behavior as it works to achieve its goals, then compromise aspects of this agent's cognitive function in order to induce the effects of a cognitive impairment.

Where our approach differs from others is in how we view, and therefore model, normative and erroneous behavior. In our approach, we regard normative agent behavior as exhibiting contextually-appropriate behavior while working to achieve its goals. We will model this type of behavior with a context-aware agent. Erroneous behavior will be achieved by compromising our agent's ability to perform basic contextual reasoning when formulating a plan for achieving a goal.

4   Required Contextual Knowledge

Our model of normative reasoning assumes that an agent uses contextual knowledge to mediate its planning process to ensure that it commits to contextually-appropriate goals. It is critical to represent context in the normative model so that it also supports impairment to give rise to erroneous behavior.

We use the term *context* to mean any identifiable configuration of environmental, goal-related, and agent-related features that has predictive power for an agent's behavior. Some features of an agent's `current context` exist as a result of the goal(s) currently being pursued. These features, which we refer to as the current *problem-solving context* (PSC), are ephemeral and are usually removed from the current context by the actions which achieved the goal (e.g., the grill cheese sandwich being prepared is not yet cooked) or by "cleanup" actions that either part of the plan or specified by the context representation (e.g., remove pan from stove, turn off stove).

Others features, which we refer to as the agent's *persistent context*, are longer-lived and persist across successive context changes. For example, once a person is dressed, being dressed will persist across successive context changes (driving to work, being at work, etc.). In our approach, we consider the agent's PSC and PC as sub-contexts of the agent's current context.

In addition to its current context, we assume that an agent has general knowledge about the context that results from pursuing a goal. For example, when considering going out to eat, most people know they will be out in public, there will be other patrons and wait staff present, and they will be expected to be appropriately dressed. This general contextual knowledge, which we refer to as an *implied context*, influences how the person plans to achieve his or her goals by ensuring that plans result in context-appropriate behaviors. In our approach, implied contexts are used during the agent's planning process.

Features of the agent's persistent context can be used to impose constraints on its future contexts, that is, on the future contexts it can be in. For example, when preparing a hot meal, the agent should focus its attention on the food in the pan to prevent it from burning or causing a kitchen fire. As long as the stove remains on and the pan remains on the stove, the agent should not enter any context in which its position is not the same as the current context (i.e., in the kitchen). Contextual constraints help the agent determine if committing to a goal will cause contextually-inappropriate behavior. In the cooking example, the agent knows that it should not commit to any goal (e.g., checking the mail) that causes it to leave the kitchen. We use contextual constraints to help the agent formulate contextually-appropriate plans.

5   Modeling Normative and Erroneous Behavior

In our approach, an agent avoids contextually-inappropriate goals and actions by using contextual knowledge to formulate what we will refer to as a *contextually appropriate plan* (CAP). A CAP is any plan that ensures the context induced as a result of pursuing the goal in consideration does not violate any contextual constraints that may be imposed by persistent features of the agent's current context.

A CAP is constructed by first considering the evolution of the agent's current context as a result of executing default procedural knowledge pertaining to a goal. This process allows the agent to identify persistent contextual features which are then merged into the implied context associated with the goal in consideration. The resulting context is a representation of the agent's future context surrounding the pursuit of the goal. Features of this context are then compared against any contextual constraints that may be imposed. This not only prevents the agent from committing to a goal that violates these constraints, but allows it to identify features of the current context that affect the appropriateness of the implied context. The latter information can be used to modify the plan for achieving a goal in a way that remedies any problematic features.

Our method of impairment induces contextually-inappropriate behavior by impairing the agent's ability to recall contextual knowledge when formulating a plan for achieving a goal. Some of the agent's contextual knowledge at any given time is in its working memory (i.e., the current context), while the rest is in its long-term memory (i.e., its c-schemas). To this end, our model of cognitive impairment borrows ideas from ACT theory [1] about how memories are stored and recalled.

Each feature in the agent's current context in working memory is assigned a strength S which is represented using the ACT equation for "memory trace" strength [2]. By increasing the decay rate of S for a contextual feature or by preventing it being committed to memory we can cause an agent to formulate a plan based on inaccurate contextual information, thus increasing the likelihood of committing to a goal in a contextually-inappropriate manner.

We can compare c-schemas in long-term memory to *chunks* in ACT-R's declarative module [1]. Borrowing from ACT-R, a c-schema c is given an *activation weight* A_c, which is represented using the ACT equation for chunk activation. We can impair the ability of an agent to retrieve appropriate c-schemas by manipulating the parameters of this equation to reduce A_c. Doing so will result in incomplete or wrong contextual knowledge being returned from long-term memory, which increases the likelihood of committing to achieve a goal in a contextually-inappropriate manner.

6   Conclusions and Future Work

Agents that simulate human behavior play an ever increasing role in a variety of scientific and commercial applications. For some applications, modeling plausible impaired human behavior is as important as modeling normative human behavior. In this paper, we presented a context-mediated approach to modeling plausible human behavior and a framework for impairing the cognitive function of a context-aware agent in order to simulate plausible erroneous behavior. Currently, our work is in the very early stages. At the time of this writing, we have implemented and conducted basic preliminary tests of our context-aware agent and we are in the process of implementing our impairment framework. In the near future, we will analyze the plausibility of the resulting erroneous behavior. We will also examine how our approach can be used to model other types of erroneous behaviors beyond those associated with a cognitive impairment like dementia.

References

1. Anderson, J.R.: The Architecture of Cognition. Harvard University Press, Cambridge, MA, USA (1983)
2. Anderson, J.R.: A spreading activation theory of memory. J. Verbal Learn. Verbal Behav. **22**(3), 261–295 (1983)
3. Brézillon, P., Pomerol, J., Saker, I.: Contextual and contextualized knowledge: an application in subway control. Int. J. Hum.-Comput. Stud. **48**(3), 357–373 (1998)
4. Kormányos, B., Pataki, B.: Multilevel simulation of daily activities: why and how? In: IEEE International Conference on Computational Intelligence and Virtual Environments for Measurement Systems and Applications (CIVEMSA), pp. 1–6. IEEE (2013)
5. Lidén, L.: Artificial stupidity: the art of intentional mistakes. AI Game Program. Wisdom **2**, 41–48 (2003)
6. O'Brien, A., Ruairi, R.M.: Survey of assistive technology devices and applications for aging in place. In: Proceedings of the Second International Conference on Advances in Human-Oriented and Personalized Mechanisms, Technologies, and Services, pp. 7–12 (2009)
7. Pollack, M.E., Brown, L., Colbry, D., McCarthy, C.E., Orosz, C., Peintner, B., Ramakrishnan, S., Tsamardinos, I.: Autominder: an intelligent cognitive orthotic system for people with memory impairment. Robot. Auton. Systems **44**(3), 273–282 (2003)
8. Rudary, M., Singh, S., Pollack, M.E.: Adaptive cognitive orthotics: combining reinforcement learning and constraint-based temporal reasoning. In: Proceedings of the Twenty-First International Conference on Machine Learning, p. 91. ACM (2004)
9. Serna, A., Pigot, H., Rialle, V.: Modeling the progression of alzheimers disease for cognitive assistance in smart homes. User Model. User-Adap. Inter. **17**(4), 415–438 (2007)
10. Sloan, C., Kelleher, J.D., Namee, B.M.: Feeling the ambiance: using smart ambiance to increase contextual awareness in game agents. In: Proceedings of the 6th International Conference on Foundations of Digital Games, pp. 298–300. ACM (2011)
11. Turner, R.M.: Adaptive Reasoning For Real-World Problems: A Schema-Based Approach. Psychology Press, New York (1994)

WHERE: An Autonomous Localization System with Optimized Size of the Fingerprint Database

Yaqian Xu[✉] and Klaus David

Chair for Communication Technology (ComTec), University of Kassel,
Wilhelmshöher Allee 73, 34121 Kassel, Germany
yaqian.xu@comtec.eecs.uni-kassel.de, david@uni-kassel.de

Abstract. Wi-Fi fingerprinting without site surveys is one interesting approach for indoor localization. Current approaches in this field either achieve high accuracy with a large fingerprint database, or yield lower accuracy when the database size is small. In this paper, we propose a novel RSS (Received Signal Strength)-range based approach for fingerprint building, which optimizes the size of the fingerprint database while maintaining the accuracy at the same level. In this approach, a fingerprint is a low-dimensional vector of RSS-ranges, which are extracted from a high-dimensional vector of Wi-Fi scans in the process of fingerprint building. The proposed approach is used and evaluated in the autonomous localization system, which we call WHERE. The evaluation results show the system can optimize the size of the fingerprint database while maintaining an accuracy of room-level.

1 Introduction

For many context-aware applications, location information is one really important context. For outdoor localization Global Navigation Satellite Systems (GNSS), such as GPS (Global Positioning System), are widely used. Unfortunately, these systems do not work or only quite badly in indoor environments. For indoor localization there are various approaches, as Wi-Fi networks are already installed in many buildings, it would be attractive if in addition to networking they could also be exploited for localization. Therefore, considerable research has and is being done on Wi-Fi based localization systems.

One promising Wi-Fi based technique for indoor localization is "*fingerprinting*". The fingerprinting technique consists of a learning phase and a positioning phase. In the learning phase, the Wi-Fi scans from surrounding access points (APs) at each visited location are sensed. Afterwards a database of fingerprints of visited locations, which is called a fingerprint database, is built. In the positioning phase, the current Wi-Fi scan is compared with the fingerprints in the fingerprint database, and the likeliest fingerprint is returned, which indicates the current location.

Wi-Fi fingerprinting-based systems with good accuracy need a large fingerprint database and therefore quite some processing time to position a user, whereas systems with a small fingerprint database reach only poor location accuracy. In order to optimize the size of the fingerprint database while keeping the room-level accuracy, this

© Springer International Publishing Switzerland 2015
H. Christiansen et al. (Eds.): CONTEXT 2015, LNAI 9405, pp. 544–550, 2015.
DOI: 10.1007/978-3-319-25591-0_46

paper proposes a novel RSS-ranged based approach. In the approach, a fingerprint is a low-dimensional vector of RSS-ranges, which are extracted from a high-dimensional vector of Wi-Fi scans in the process of fingerprint building. Thereby, the size of a fingerprint, as well as the size of the fingerprint database is significantly reduced. An RSS-range of a fingerprint is not a fixed pre-defined range. Rather, for each AP, its RSS-range is discovered by a density-based clustering algorithm [1] from a set of RSS samples available.

The RSS-range based approach is used in a localization system, which we call WHERE. WHERE is an autonomous, Wi-Fi fingerprinting-based localization system without site surveys, which provides mechanisms of automatic location learning, positioning and update of fingerprints.

2   The RSS-Range Based Approach and the System WHERE

The reason for producing a large fingerprint database is the RSS fluctuations due to small-scale fading [2] in the indoor environment. Small-scale fading [3] describes the rapid signal amplitude variation of a signal over a short period in time or over a small change in space. Small-scale fading results in numerous Wi-Fi scans which, after some time, accumulate to a large fingerprint database in these traditional Wi-Fi fingerprinting based approaches.

To address the challenges of the database size and the RSS fluctuation, we propose to use an RSS-range of an AP instead of a set of scalar RSS samples to build fingerprints in the learning phase. To discover an RSS-range of an AP, we propose to use a density-based clustering algorithm introduced in [1]. The original idea of density-based clustering is to detect clusters with points which are closer together than points outside of clusters. A cluster is a collection of points where the density of the clustered points is high. In the proposed approach, an RSS samples sensed from an AP is regarded as a point. Thus, a cluster is a set of RSS samples which are close together. The clustering process has been introduced in the publication [4–7].

In order to discover clusters, the density-based clustering algorithm in [1] utilizes concepts of *neighborhood* and *neighborhood-density*. Two parameters, a distance threshold (*Eps*) and a density threshold (*MinPts*), are specified.

Definition 1 – neighborhood and neighborhood-density: a set of RSS samples in a database D, whose Euclidean distance (e.g., the absolute RSS value of the difference of two RSS samples) to an RSS sample (RSS_k^i) is smaller than a specific distance threshold *Eps*, is the *neighborhood* of RSS_k^i, indicated by $N(RSS_k^i)$. The number of RSS samples inside the neighborhood of RSS_k^i is the *neighborhood –density*, indicated by $\rho(RSS_k^i)$.

$$N(RSS_k^i) = \left\{ RSS_t^i \in D | d(RSS_t^i, RSS_k^i) \leq Eps \right\} | (i, t, k \in \mathbb{N}_+)$$

Definition 2 – cluster: a *cluster* C_k^j is a set of RSS samples (RSS_k^i) from an AP with MAC_k, and each RSS_k^i fulfills the density criterion that $\rho(RSS_k^i)$ is equal to or higher than a density threshold (*MinPts*). j indicates the cluster C_i^j is the *j-th* cluster, because more than one cluster can be extracted. However, clusters do not overlap with one another. The high-density criterion is fulfilled when the neighborhood-density of an RSS sample (RSS_k^i) is equal to or higher than a density threshold $(MinPts)$.

$$C_k^j = \left\{ RSS_k^i \in D | \rho(RSS_k^i) \geq MinPts \right\} | (j, i, k \in \mathbb{N}_+)$$

The cluster discovering process works as follows:

- For an RSS sample (RSS_k^i), if its neighborhood-density is equal to or larger than $MinPts$, the point and his neighborhood create a cluster.
- If two clusters contain the same point, the two clusters are merged to one cluster
- For an RSS sample (RSS_k^i), which does not belong to any cluster, is regarded as noise.

After the high-density clusters are discovered, the next step is to extract an RSS-range for each cluster.

Definition 3 – RSS-range: an RSS-range R_{kj} is an interval of RSS values between a lower and upper bound $R_{kj} = [RSS_{kl}, RSS_{ku}]$, extracted from a cluster C_k^j. RSS_{kl} and RSS_{ku} are RSS values of the lower and upper bound, respectively.

$$R_{kj} = [RSS_{kl}, RSS_{ku}] | (k, j, l, u \in \mathbb{N}_+)$$

Definition 4 – fingerprint: in our RSS-range based approach, a fingerprint is a vector of n MAC and RSS-range pairs, extracted from Wi-Fi scans sensed at a location l. n is the number of extracted MAC_k: $[RSS_{kl}, RSS_{ku}]$ pairs in the fingerprint FP_l^r.

$$FP_l^r = \left\{ MAC_1: [RSS_{1l}, RSS_{1u}], \ldots, MAC_n: [RSS_{nl}, RSS_{nu}] \right\} | (r, l, u, n \in \mathbb{N}_+)$$

Figure 1 shows an example of a fingerprint in the RSS-range based approach. As observed, a fingerprint is a low-dimensional vector of RSS-ranges. The low-dimensional vector is extracted from a high-dimensional vector of Wi-Fi scans. Compared to these conventional approaches storing a high-dimension vector of Wi-Fi scans, the RSS-range based approach has an advantage of having a significantly optimized size of the fingerprint database.

```
00:24:fe:04:b2:d3  :  [-58,-66]
00:24:fe:ac:72:02  :  [-71,-79]
00:a0:57:18:c1:2f  :  [-56,-64]
00:a0:f9:37:6a:dd  :  [-84,-91]
1c:c6:3c:7d:4c:2a  :  [-72,-79]
20:4e:7f:8e:24:81  :  [-55,-63]
74:44:01:86:15:0d  :  [-81,-87]
```

Fig. 1. An example of a fingerprint in the RSS-range based approach.

The proposed RSS-range based approach is used in an autonomous localization system WHERE to build fingerprints in the learning phase. WHERE is a Wi-Fi finger-printing-based localization system with the mechanisms of automatically triggering the learning, positioning and update of fingerprints of *significant locations*. A *significant location* is a location where a user stays for a minimum period of time of e.g., at least 10 min. In this paper, WHERE uses the embedded accelerometer of a smartphone to detect significant location by detecting a smartphone's motion states (i.e., phone is in motion or in rests) [8]. When the phone is at rest for at least 10 min, the current location is regarded as a significant location.

WHERE performs the following operations.

Collection: WHERE uses the embedded accelerometer of a smartphone to detect a smartphone's motion states (i.e., phone is in motion or in rests) [8]. Once the phone is at rest, the system activates the Wi-Fi sensing function, to collect Wi-Fi scans. The system deactivates the Wi-Fi sensing function when the phone is in motion or when the collection time reaches 30 min.

Learning: If the phone is at rest for at least 10 min, the system regards the location as a significant location. The Wi-Fi scans at the significant locations are automatically learned to build fingerprints by using the RSS-range based approach. The fingerprints are stored in a database.

Positioning: When a Wi-Fi scan with n APs is sensed, the system compares it with all learned fingerprints in the fingerprint database, and calculates the likeliest fingerprint. If an RSS sample of an AP belongs to an RSS-range of a fingerprint, it is considered as a matching AP. The *matching degree* is the percentage of m matching APs among all sensed n APs of the measurement: m/n. A *likeliest fingerprint* is a fingerprint, which produces the highest matching degree, and for which the matching degree is higher than a threshold (we set 50 % in this paper). The user is positioned at the location, where the likeliest fingerprint indicates.

Update: The system either returns a response (i.e., the likeliest fingerprint) in the positioning phase, or does not respond. If the response probability at a location is lower than a threshold (e.g., 50 %), the fingerprints at that location are *invalid*. WHERE detects such situation and automatically updates the fingerprints of the location. The Wi-Fi scans sensed at that location are re-learned to generate new fingerprints, and added to the database. The system removes an invalid fingerprint when it is not recognized over a certain period (e.g., we set it to 3 months in our system).

3   Evaluation in a Real-Life Case Study

As an autonomous localization system, WHERE provides mechanisms of automatic location learning, positioning and update of fingerprints. To evaluate the system in the real-life, we carried out a case study in a shopping mall to investigate the performances of automatic learning, positioning, update of fingerprints, as well as the size of the fingerprint database.

Figure 2 shows the layout of the shopping mall. It is a one-floor building, which consists of 83 shops. A user carries a smartphone (a Nexus 5) in his pocket when he visits the shopping mall and walks around in some shops. The star-marked locations in Fig. 2 are visited several times by the user, and he stays for at least ten minutes. The visited shops and times of visit are manually recorded as ground truth for the evaluation intention. The complete case study is carried out over six weeks. Most of shops were visited at least three times.

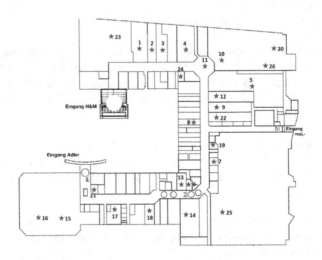

Fig. 2. The layout of a shopping mall for a case study and the measurement locations.

Learning: WHERE learns fingerprints of a location automatically when a user visits it for the first time and stays there at least 10 min.

The learning result shows that approximately 92.31 % of the locations can be learned when a user visits them for the first time. Of the 26 visited locations, 2 locations are not learned on the first visit. The two locations, which are not learned, are Location 11 and Location 22 in Fig. 2. Location 11 is a water fountain basin, where the crowd around and water of the fountain severely affect the signal stability, and in Location 22 the system only senses very weak signals because of its special magnetic security control.

Positioning: The system begins to position a user whenever he returns to a location, where its fingerprints have been learned. The system compares each Wi-Fi scan with fingerprints. The system either returns a response (i.e., the likeliest fingerprint) after the comparison, or does not respond. On average, in 73.98 % cases, WHERE provides a response immediately at a learned location. And 83.84 % of these responses indicate the correct locations by being compared with the ground truth.

Update: We summarize the update times for each location during six weeks in Table 1. The result shows that more than 60 % of the locations need to be updated only once, to keep the response rate higher than the threshold (we use 50 % in this paper). The fingerprints of these locations are robust to be used for positioning without frequent update.

Table 1. The update times for each location during the six weeks of a case study.

Update times	0	1	2	3	4 or more
Location	1, 3, 5, 9, 12, 15, 21, 23, 24	2, 4, 8, 13, 14, 16, 22	7, 18	10, 17, 19, 20, 26	6, 11, 25
Number of locations (Percentage)	9 (34.62 %)	7 (26.92 %)	2 (7.69 %)	5 (19.23 %)	3 (11.54 %)

Database Size: In the case study, 81 fingerprints of 26 locations are learned. In total, 937 RSS-ranges are stored in the fingerprint database. It means, for the positioning, each (MAC, RSS value) pair in the current scan needs to be compared with up to 937 RSS-ranges. After six weeks, the size of the fingerprint database is only 22.3 KB (22.863 Bytes).

4   Conclusion

In this paper, we propose an RSS-range based approach for fingerprinting localization. RSS-ranges, related to detectable APs sensed at a location, are extracted as components of a fingerprint. The proposed approach significantly reduces the size of the fingerprint database, while maintaining a comparable accuracy in the simulated-based study. The proposed approach is used in the autonomous localization system WHERE. WHERE provides mechanisms of automatic learning, positioning and update, without expensive site surveys. Over a six-week study in a shopping mall, approximately 92.31 % of the locations were learned when a user visited them for the first time, and the learned locations were correctly positioned with a probability of 83.84 %. The fingerprints also show strong robustness in the study. Fingerprints of more than 60 % of the locations were updated only once during the six-week time, showing very strong robustness. After the six-week study, the size of the fingerprint database was only 22.3 KB, which stores 81 fingerprints of 26 locations.

Acknowledgments. This work has been co-funded by the Social Link Project within the Loewe Program of Excellence in Research, Hessen, Germany.

References

1. Ester, E., Kriegel, H., Sander, J., Xu, X.: A density-based algorithm for discovering clusters in large spatial databases with noise. In: 2nd International Conference on Knowledge Discovery and Data Mining, Portland, OR, USA (1996)
2. Kaemarungsi, K.: Distribution of WLAN received signal strength indication for indoor location determination. In: 1st International Symposium on Wireless Pervasive Computing (ISWPC), Phuket, Thailand (2006)
3. Sklar, B.: Rayleigh fading channels in mobile digital communication systems. Part I: characterization. IEEE Commun. Mag. **35**(7), 90–100 (1997)

4. Xu, Y., Lau, S.L., Kusber, R., David, K.: DCCLA: autonomous indoor localization using unsupervised Wi-Fi fingerprinting. In: CONTEXT 2013, Annecy, France (2013)
5. Xu, Y., Kusber, R., David, K.: An enhanced density-based clustering algorithm for the autonomous indoor localization. In: MOBILe Wireless MiddleWARE, Operating Systems and Applications (Mobilware), Bologna, Italy (2013)
6. Xu, Y., Lau, S.L., Kusber, R., David, K.: An experimental investigation of indoor localization by unsupervised Wi-Fi signal clustering. In: Future Network and Mobile Summit, Berlin, Germany (2012)
7. Lau, S.L., Xu, Y., David, K.: Novel indoor localisation using an unsupervised Wi-Fi signal clustering method. In: Future Network and Mobile Summit, Warsaw, Poland (2011)
8. Lau, S.L., David, K.: Movement recognition using the accelerometer in smartphones. In: Future Network and Mobile. Florence, Italy (2010)

Author Index

Printed in the United States
By Bookmasters